Periodic Chart of the Elements

I A	II A	III B	IV B	V B	VI B	VII B	VIII B			I B	II B	III A	IV A	V A	VI A	VII A	VIII A
1 **H** 1.0080																	2 **He** 4.003
3 **Li** 6.940	4 **Be** 9.013											5 **B** 10.82	6 **C** 12.011	7 **N** 14.008	8 **O** 16.000	9 **F** 19.00	10 **Ne** 20.183
11 **Na** 22.991	12 **Mg** 24.32											13 **Al** 26.98	14 **Si** 28.09	15 **P** 30.975	16 **S** 32.066	17 **Cl** 35.457	18 **Ar** 39.944
19 **K** 39.100	20 **Ca** 40.08	21 **Sc** 44.96	22 **Ti** 47.90	23 **V** 50.95	24 **Cr** 52.01	25 **Mn** 54.94	26 **Fe** 55.85	27 **Co** 58.94	28 **Ni** 58.71	29 **Cu** 63.54	30 **Zn** 65.38	31 **Ga** 69.72	32 **Ge** 72.60	33 **As** 74.91	34 **Se** 78.96	35 **Br** 79.916	36 **Kr** 83.80
37 **Rb** 85.48	38 **Sr** 87.63	39 **Y** 88.92	40 **Zr** 91.22	41 **Nb** 92.91	42 **Mo** 95.95	43 **Tc** (99)	44 **Ru** 101.1	45 **Rh** 102.91	46 **Pd** 106.4	47 **Ag** 107.880	48 **Cd** 112.41	49 **In** 114.82	50 **Sn** 118.70	51 **Sb** 121.87	52 **Te** 127.61	53 **I** 126.[illegible]	
55 **Cs** 132.91	56 **Ba** 137.36	57 ***La** 138.92	72 **Hf** 178.50	73 **Ta** 180.95	74 **W** 183.86	75 **Re** 186.22	76 **Os** 190.2	77 **Ir** 192.2	78 **Pt** 195.09	79 **Au** 197.0	80 **Hg** 200.61	81 **Tl** 204.39	82 **Pb** 207.21	83 **Bi** 209.00	84 **Po** (210)	85 **At** (21[illegible]	
87 **Fr** (223)	88 **Ra** (226)	89 **†Ac** (227)	104 **Ku** (261)	105 **Ha** (260)													

* Lanthanides	58 **Ce** 140.13	59 **Pr** 140.92	60 **Nd** 144.27	61 **Pm** (147)	62 **Sm** 150.35	63 **Eu** 152.0	64 **Gd** 157.26	65 **Tb** 158.93	66 **Dy** 162.51	67 **Ho** 164.94	68 **Er** 167.27	69 **Tm** 168.94	7[illegible] **Y**[illegible] 173[illegible]
† Actinides	90 **Th** (232)	91 **Pa** (231)	92 **U** 238.07	93 **Np** (237)	94 **Pu** (242)	95 **Am** (243)	96 **Cm** (247)	97 **Bk** (249)	98 **Cf** (251)	99 **Es** (254)	100 **Fm** (253)	101 **Md** (256)	1[illegible] [illegible] (2[illegible]

Techniques and Experiments for Organic Chemistry

Techniques and Experiments for Organic Chemistry

Ralph J. Fessenden ›››› ‹‹‹‹ **Joan S. Fessenden**

University of Montana

Willard Grant Press
Boston

PWS PUBLISHERS

Prindle, Weber & Schmidt • Willard Grant Press • Duxbury Press •
Statler Office Building • 20 Providence Street • Boston, Massachusetts 02116

Library of Congress Cataloging in Publication Data

Fessenden, Ralph J.
Techniques and experiments for organic chemistry.

Includes index.
1. Chemistry, Organic—Laboratory manuals.
I. Fessenden, Joan S. II. Title.
QD261.F47 1983 547'.0076 82-16570

ISBN 0-87150-755-2

Printed in the United States of America

83 84 85 86 87 — 10 9 8 7 6 5 4 3 2 1

Cover photo by Bruce Iverson. We gratefully acknowledge the assistance of Prof. Philip LeQuesne, Mr. Bernard LeMire, and Mr. Jan Koso of Northeastern University.

Production editor: David M. Chelton
Text designer: Megan Brook
Text compositor: J.W. Arrowsmith, Ltd.
Art studio: Anco

Cover printer: Lehigh Press Lithographers
Text printer and binder:
Halliday Lithograph

PREFACE

Two general approaches often used in teaching the one-year introductory organic laboratory are the *investigative approach*, a tutorial method that prepares the student for research, and what we call the "*techniques approach*," a more structured method in which the student is generally first taught manipulative techniques, then synthesis, and finally organic qualitative analysis. We have chosen the techniques approach because we feel it is more efficient of time and energy (both the student's and the instructor's) in our own chemistry sequence, in which students take lecture and laboratory courses concurrently.

Techniques and Experiments for Organic Chemistry begins with the common laboratory techniques. Actual syntheses are postponed until the student has gained sufficient knowledge of organic chemistry from the lecture portion of the course to understand the experiments being performed. Synthetic work follows the technique chapters of the text and begins with alkyl halide chemistry (Williamson ether synthesis and a Grignard reaction), followed by alcohol chemistry. Spectral techniques (infrared and nmr) are placed after alcohol chemistry, approximately halfway through the text. The remaining chapters are devoted to additional synthetic procedures. The text ends with a chapter on the literature of organic chemistry, an introduction to organic qualitative analysis, and a chapter of supplemental techniques. Four appendices of assorted tables and laboratory calculations conclude the text.

Each chapter begins with a theoretical discussion of the chemistry or techniques contained in the chapter. In addition, each experiment in the chapter is preceded by a discussion of the practical aspects of the experiment, along with explanations of why each reaction or manipulation is performed as it is. Emphasis is placed on understanding *and* doing instead of just doing.

The amount of procedural detail in the experiments is reduced as the text progresses. At the start of the book, the steps to be followed are spelled out in detail. The detail provided is gradually diminished as the student proceeds through the course and gains laboratory experience. Toward the end of the text, the experimental detail is similar to that found in research journals. Therefore, the experiments in the chapters toward the end of the text call for more student ingenuity: for example, Experiment 16.4 (Synthesis of an Acetate Ester by a General Reaction Procedure) and Experiment 21.1 (Qualitative Organic Analysis: Selected Procedures).

Because the student gradually acquires more independence as the experiments progress, this text could also be used for the first part of a course in which the final few weeks consist of independent laboratory work or of a full qualitative organic analysis scheme.

It has been the authors' experience that not all experiments found in standard laboratory texts work for many students. The reasons for these failures are varied but can often be traced to the student's inexperience. Therefore, each experiment in this text has been author-tested, student-tested, and/or class-tested to ensure a high probability of student success and learning.

The cost of chemicals has become quite high in recent years. For this reason, we have tried to include only experiments that require relatively inexpensive, readily available compounds. In addition, we have tried to "recycle" as many of the student products as feasible. In some cases, products prepared or purified by students can be used in later experiments. In other cases, the student products can be used in next year's laboratory course.

Flexibility. For the most part, each technique is presented in more than one experiment or section of an experiment, with one section generally involving an unknown compound. It is not expected that the student will carry out every experiment. In many cases, the instructor may, at his or her discretion, assign the unknown compound experiment in addition to, or in place of, the principal experiment. (See the discussion of *supplemental experiments* that follows.)

To avoid delays because of equipment shortages (such as of melting-point apparatus, gas chromatographs, etc.), the instructor may want to assign different experiments to different groups of students (for example, while one group performs a melting-point experiment, another group performs the crystallization experiment), or to rearrange the order of the experiments for a portion of the class. The instructor's guide outlines various options in more detail.

Supplemental experiments. Almost all chapters in this text contain a main experiment and one or more supplemental experiments. The supplemental experiments are of several types. Some may be used *instead of* the main experiment. A few of the experiments involving unknowns are of this type. Some of the supplemental experiments are *extensions* of the principal experiment. These are included strictly for interest value and are *not* alternative to the principal experiment. We strongly urge that at least a few of these be assigned. For example, the liquid-crystal experiment (Experiment 2.3) is not time-consuming, yet it is quite fascinating. Finally, some of the supplemental experiments are included to offer *variety* in the selection of laboratory work. The instructor's guide discusses these supplemental experiments in more detail.

Chapter 22 contains three supplemental techniques (sublimation, vacuum distillation, and elution chromatography) for those instructors interested and having the proper equipment available. An instructor may wish to assign one or more of these techniques at an appropriate time during the laboratory course.

Safety considerations. Experiments in organic chemistry can be dangerous. We believe one of the purposes of the introductory organic laboratory course is to teach students how to handle hazardous chemicals in a safe and prudent manner. With this thought in mind, we have done our utmost to point out the hazards in each experiment. We have included *Safety Notes* before each experiment and precautionary phrases within the experimental instructions. Beginning with Chapter 13, we have cut back the number and detail of these precautions. Therefore, in the latter half of the book, routine hazards that the student has already encountered many times are accompanied by only brief reminders. Of course, safety notes are always included when unfamiliar or especially hazardous operations are to be performed.

Although organic experiments are inherently dangerous, we have attempted to minimize the hazards both by the choice of experiments and by the reagents that have been selected. We have attempted to avoid the use of known or suspected carcinogens, such as benzene. In some experiments, we have pointed out where alternative reagents or solvents can be used. For example, 95% ethanol can often replace methanol as a solvent. In Experiment 10.4, we give two procedures for oxidizing an alcohol: the classic procedure using a dichromate as the oxidizing agent and an alternative procedure using the less toxic hypochlorous acid.

Study problems. The study problems in this text are not as extensive as they would be in a lecture text. The problems included are intended to (1) check the student on his or her theoretical knowledge of the experiments to be performed, and (2) help the student appreciate the practical aspects of laboratory work. The answers are in the instructor's guide that accompanies this text.

Acknowledgments

We are very grateful to our colleagues who reviewed the manuscript and thank them for their many helpful suggestions: Ann M. Armour (Univ. of Alberta); Fred M. Dewey (Metropolitan State College); Ronald Kluger (Univ. of Toronto); Peggy Magde; Chester Muth (Western Virginia Univ.); Graye Shaw (North Carolina State Univ., Raleigh); William R. Stine (Wilkes College); John S. Swenton (Ohio State Univ.); Jack W. Timberlake (Univ. of New Orleans); and Leroy G. Wade (Colorado State Univ.). We must also give credit to the students who helped find the trouble spots and to our typist Laurie Palmer.

Above all, we must thank the editorial and production staffs at Willard Grant Press for their continuing guidance and support: in particular, David Chelton, Edward Murphy, and Bruce Thrasher.

Ralph J. Fessenden
Joan S. Fessenden

University of Montana
Missoula, Montana

⟫⟫ CONTENTS ⟪⟪

CHAPTER 1

The Organic Laboratory

1.1 Safety in the Laboratory

The organic chemistry laboratory can be a dangerous place. Many organic compounds are volatile and flammable. Many are toxic. Some organic chemicals can cause lung damage, some can give chemical burns, some can lead to cirrhosis of the liver, and some are carcinogenic (cancer causing). Yet, organic chemists generally live as long as the rest of the population because they have learned to be careful. When you work in an organic laboratory, you must always think in terms of safety.

A. Using Common Sense

Most laboratory safety precautions are nothing more than common sense. The laboratory is not the place for horseplay. Do not work alone in the laboratory. Do not perform unauthorized experiments. Do not sniff, inhale, or taste organic compounds, and do not pipet them by mouth. Wipe up any spilled chemicals, using copious amounts of water to wash up spilled acids and bases. (Neutralize any residual spilled acid or base with sodium bicarbonate or dilute acetic acid, respectively.) Do not put dangerous chemicals in the waste crock—the janitor may become injured. Instead, use the jugs provided for chemical disposal.

When working in the laboratory, wear suitable clothing. Jeans and a shirt with rolled-up sleeves, plus a rubber lab apron or cotton lab coat, are ideal. Do not wear your best clothing— laboratory attire usually acquires many small holes from acid splatters, and may also develop a distinctive aroma. Loose sleeves can sweep flasks from the laboratory bench, and such sleeves present the added hazard of easily catching fire. Long hair should be tied back. Broken glass sometimes litters the floor of a laboratory; therefore, shoes should always be worn. Sandals are inadequate because they do not protect the feet from spills. Wash your hands frequently, and always wash them before leaving the laboratory, even to go to the rest room.

Because of the danger of fires, smoking is prohibited in laboratories. Because of the danger of chemical contamination, food and drink also have no place in the laboratory. On the first day of class, familiarize yourself with the locations of the fire extinguishers, fire blanket, eye-wash fountain, and shower.

B. Safety Glasses

Chemicals splashed in the eyes can lead to blindness; therefore, it is imperative that **safety glasses** be worn. They must be worn *at all times*, even if you are merely adding notes to your laboratory notebook or washing dishes. You could be an innocent victim of your lab partner's mistake, who might inadvertently splash a corrosive chemical in your direction. In the case of particularly hazardous manipulations, a **full-face shield** (similar to a welder's face shield) should be worn.

Contact lenses should not be worn, even under safety glasses. The reason for this rule is that contact lenses cannot always be removed quickly if a chemical gets into the eye. A person administering first aid by washing your eye might not even realize that you are wearing contact lenses. In addition, "soft" contact lenses can absorb harmful vapors. If contact lenses are absolutely necessary, then properly fitted **goggles** should also be worn. Also, inform your laboratory instructor and neighbors that you are wearing contact lenses.

C. Chemicals in the Eye

If a chemical does get into the eye, the eye should be flushed with gently flowing water for 15 minutes. Do not try to neutralize an acid or base in the eye. Because of the natural tendency for the eyelids to shut when something is in the eye, *they must be held open during the washing*. If there is no eye-wash fountain in the laboratory, a piece of rubber tubing attached to a faucet is a good substitute. Do not take time to put together a fountain if you have something in your eye, however! Either splash your eye (held open) with water from the faucet immediately or lie down on the floor and have someone else pour a gentle stream of water into your eye. *Time* is important. The sooner you can wash a chemical out of your eye, the less the damage.

After the eye has been flushed, medical treatment is strongly advised. For any corrosive chemical, such as sodium hydroxide, prompt medical attention is imperative!

D. Acids and Bases

To prevent acid splatters, *always add concentrated acids to water* (*never add water to acids*). Concentrated sulfuric acid (H_2SO_4) should be added to ice water or crushed ice because of the heat generated by the mixing. Do not pour acids down the drain without first diluting them (by adding them to large amounts of water) and/or neutralizing them. Strong bases should also be diluted before discarding. If you splash an acid or strong base on your skin, wash with copious amounts of water as described in Section 1.1E. Concentrated hydrochloric acid (HCl) and glacial acetic acid (CH_3CO_2H) present the added hazard of extremely irritating vapors. These two acids should be used only in the fume hood.

Sodium hydroxide ("lye," NaOH) is caustic. In the solid form (usually as pellets), it is deliquescent; a pellet that is dropped and ignored will form a dangerous pool of concentrated NaOH. For this reason, solid NaOH should be handled with care. Spilled pellets should be picked up (with plastic gloves or a piece of paper) and flushed down the drain with a large amount of water.

Aqueous ammonia ("ammonium hydroxide") emits ammonia (NH_3) vapors and thus should be used in the fume hood.

E. Chemical Burns

Any chemical (whether water-soluble or not) spilled onto the skin should be washed off immediately with soap and water. The detergent action of the soap and the mechanical action of washing remove most substances, even insoluble ones. If the chemical is a strong acid or base, rinse the splashed area of the skin with *copious amounts of cool water.* Strong acids on the skin usually cause a painful stinging. Strong bases usually do not cause pain, but they are extremely harmful to tissue. Always wash carefully after using a strong base.

If chemicals are spilled on a large area of the body, they should be washed off in the safety shower. If the chemicals are corrosive or can be absorbed through the skin, contaminated clothing should be removed so that the skin can be flushed thoroughly.

F. Heat Burns

Minor burns from hot flasks, glass tubing, and the like are not uncommon occurrences in the laboratory. The only treatment needed for a very minor burn is holding it under cold water for 5–10 minutes. A pain-killing lotion may then be applied. To prevent minor burns, keep a pair of inexpensive, loose-fitting cotton gloves in your laboratory locker to use when you must handle hot beakers, tubing, or flasks.

A person with a serious burn, as from burned clothing, is likely to go into shock. He or she should be made to lie down on the floor and kept warm with the fire blanket or with a coat. Then, an ambulance should be called. Except to extinguish flames or to remove harmful chemicals, do not wash a serious burn and do not apply any ointments. However, cold compresses on a burned area will help dissipate heat.

G. Cuts

Minor cuts from broken glassware are another common occurrence in the laboratory. These cuts should be flushed thoroughly with cold water to remove any chemicals or slivers of glass. A pressure bandage can be used to stop any bleeding.

Major cuts and heavy bleeding are a more serious matter. The injured person should lie down and be kept warm in case of shock. A pressure bandage (such as a folded clean dish towel) should be applied over the wound and the injured area elevated slightly, if possible. An ambulance should be called immediately.

The use of a tourniquet is no longer advised. Experience has shown that cutting off all circulation to a limb may result in gangrene.

H. Inhalation of Toxic Substances

A person who has inhaled vapors of an irritating or toxic substance should be removed immediately to fresh air. If breathing stops, artificial respiration should be administered and an emergency medical vehicle called.

I. Avoiding Fires

Most fires in the laboratory can be prevented by the use of common sense. Before a match or burner is lit, the area should be checked for flammable solvents. Solvent fumes are heavier than air and can travel along a benchtop or a drainage trough in the bench. Hot matches (even if they have been extinguished) or any other hot substance should not be thrown into wastebaskets because some solvents have very low flash points. Conversely, do not put solvents (or filter paper soaked with solvents) in the wastebasket. The heavy solvent fumes can remain there for days.

Whenever a flammable solvent is used, all flames in the vicinity should be extinguished beforehand. Solvent bottles should always be capped when not actually in use. Flammable solvents should not be boiled away from a mixture except in the fume hood. Solvent-soaked filter paper should be placed in the fume hood to dry before it is discarded in a waste container. Spilled solvent should not be allowed simply to evaporate—if a solvent is spilled, all flames should be extinguished and the solvent cleaned up immediately with paper towels, which should be placed in the hood to dry.

Solvents should never be poured into a drainage trough (which is for water only). Highly flammable solvents should not be poured into the sink, nor should large amounts of any solvent. Small amounts of relatively harmless solvents that are miscible with water (for example, ethanol) may be flushed down the sink with water. Other solvents should be disposed of in jugs provided for solvent-disposal.

J. Extinguishing Fires

In case of even a small fire, tell your neighbors to leave the area and notify the instructor. A fire confined to a flask or beaker can be smothered with a watch glass or large beaker placed over the flaming vessel. (Try not to drop a flaming flask—this will splatter burning liquid and glass over the area.) All burners in the vicinity of a fire should be extinguished, and all containers of flammable materials should be removed to a safe place in case the fire spreads.

For all but the smallest fire, the laboratory should be cleared of people. It is better to say loudly, "Clear the room," than to scream "Fire!" in a panicky voice. If you *hear* such a shout, do not stand around to see what is happening, but stop whatever you are doing and walk immediately and purposefully toward the nearest clear exit.

Many organic solvents float on water; therefore, water may serve only to spread a chemical fire. Some substances, like sodium metal, explode on contact with water. For these reasons, water should not be used to extinguish a laboratory fire; instead, a *carbon dioxide* or *powder fire extinguisher* should be used.

If a fire extinguisher is needed, it is best to clear the laboratory and allow the instructor to handle the extinguisher. Even so, you should acquaint yourself with the location, classification, and operation of the fire extinguishers on the first day of class. Inspect the fire extinguishers. Find the sealing wire (indicating that the extinguisher is fully charged) and the pin that is used to break this sealing wire when the extinguisher is needed.

Fire extinguishers usually spray their contents with great force. To avoid blowing flaming liquid and broken glass around the room, aim toward the base and to the side of any burning equipment, not directly toward the fire. Once a fire extinguisher has been used, it will need recharging before it is again operable. Therefore, any use of a fire extinguisher must be reported to the instructor.

K. Extinguishing Burning Clothing

If your clothing catches fire, walk (do not run) to the shower if it is close by. If the shower is not near, lie down, roll to extinguish the flames, and call for help.

A clothing fire may be extinguished by having the person roll in a fire blanket. The rolling motion is important because a fire can still burn under the blanket. Wet towels can also be used to extinguish burning clothing. A burned person should be treated for shock (kept quiet and warm). Medical attention should be sought.

L. Handling Solvents

Organic solvents present the double hazard of flammability and toxicity (both short-term and cumulative). (Table III.1, page 422, in Appendix III lists the toxic levels and allowable limits of some common solvents.) *Diethyl ether* ($CH_3CH_2OCH_2CH_3$) and *petroleum ether* (a mixture of alkanes) are both very volatile (have low boiling points) and extremely flammable. These two solvents should never be used in the vicinity of a flame, and should be boiled only in the hood. *Carbon disulfide* (CS_2), which is now rarely used in the organic laboratory, is uniquely hazardous. Its ignition temperature is under 100°, the boiling point of water; therefore, fires can result even from its contact with a steam pipe. *Benzene* (C_6H_6) is flammable and also extremely toxic. It can be absorbed through the skin, and long-term exposure is thought to cause cancer. Benzene should be used as a solvent only when absolutely necessary (and then handled with great care to avoid inhalation, splashes on the skin, or fire).

Most halogenated hydrocarbons, such as *carbon tetrachloride* (CCl_4) and *chloroform* ($CHCl_3$), are toxic, and some are carcinogenic. Halogenated hydrocarbons tend to accumulate in the fatty tissues of living systems instead of being detoxified and excreted as most poisons are. In repeated small doses, they are associated with chronic poisoning and damage to the liver and kidneys. If either carbon tetrachloride or chloroform must be used, it should be handled in the fume hood. The solvent *dichloromethane* (methylene chloride, CH_2Cl_2), although narcotic, is not as toxic as carbon tetrachloride or chloroform.

Because of the dangers inherent with all organic solvents, they should always be handled with respect. Solvent vapors should not be inhaled, and solvents should not be tasted or poured on the skin. Wash any splashes on your skin immediately with soap and water. Keep solvent bottles tightly capped. Precautions to avoid fires should always be heeded.

M. Summary of Safety Rules

It may happen that you are confronted with a laboratory accident and cannot remember exactly what to do. In such a situation, just remember the following:

In case of a spill: WASH!
In case of a fire: GET OUT!

In either case, your instructor or someone in a calmer frame of mind can then decide how to handle the situation.

1.2 The Laboratory Notebook

A laboratory notebook serves several purposes. The first is for your own reference. You may think that you will remember everything you do and see in the laboratory; however, many details, such as melting points, boiling points, and weights, are easily forgotten. It is far better to record these details in a well-organized notebook than to try to memorize them. An approved notebook should be used, not little slips of paper, which seem to flutter away when your back is turned.

Another purpose of a laboratory notebook is so that someone else can review your work or repeat it *exactly*. This facet of experimental work is necessary in research, and it is also necessary in a student laboratory. If a particular experiment does not work well for you, the instructor (and the authors) would want to know why. Your detailed written procedure and observations can give clues to experimental failures.

In a research laboratory, a laboratory notebook is also important to help establish the validity of patent claims. For this reason, each page or experiment in a laboratory notebook is numbered and dated. In promising projects, some researchers will have their notebook pages signed by witnesses. This procedure is usually unnecessary in a student laboratory.

A. The Correct Notebook

The correct notebook for the laboratory is a hard-cover, bound book containing lined pages. These are available at bookstores and stationery supply stores. A loose-leaf or spiral notebook is not satisfactory because pages are easily lost. A separate notebook should be used for each laboratory course.

B. Keeping the Notebook

If the pages in the notebook are not numbered, number them before using the book. Write your name and laboratory section number on the cover of your notebook. It is also advisable to put your address or telephone number on the book in case it is lost.

Leave two blank pages at the front of the book for a table of contents. Then, enter experiments consecutively in ink; use permanent ink because your book will become splashed and stained with use.

Your laboratory notebook is the record of everything you do and observe in the laboratory. A notebook can become cluttered and illegible if an organized format is not used. For this reason, certain conventions for writing experiments have been developed.

Use only right-hand pages for writing up experiments. At the top of the page, write the date on which the experiment is performed. As you go along, leave plenty of space for notes that you might want to insert later. The empty left-hand pages may be used for calculations and jottings. If you will be running a distillation or determining more data for a particular experiment, be sure to leave blank pages as necessary before writing the procedure for the next experiment.

Write clearly so that your instructor will be able to read and grade your notebook. If you make errors, do not rip out the page. Instead, line out errors (or draw an "X" over the entire page) and go on.

C. Entering Experiments

Each experiment in the notebook should contain the following information, along with any additional material required by your instructor:

1) Title

2) Balanced chemical equation, including other information described in the following paragraphs

3) Procedure outline

4) Observations

5) Conclusions

Items (1)–(3) should be completed before you come to the laboratory. They should also be studied so that you understand not only *what* you will be doing, but *why*.

1) The **title** of the experiment should briefly describe the experiment and should also contain the experiment number and page number from the text (or other reference where appropriate). If the experimental objective is not clear from the title, a concise statement describing the experiment should be included.

2) The **balanced chemical equation** should show the formulas and names of the reactants and products. The molecular weight, the actual weight used, and the number of moles should be written under the name of each reactant. For your own convenience, pertinent physical properties, such as boiling points, may also be listed with the equation. Figure 1.1 shows a typical notebook page for the start of an experiment and also demonstrates the calculations you may have to perform.

Experiment 10.1

Synthesis of 1-Bromobutane from 1-Butanol

$CH_3CH_2CH_2CH_2OH + NaBr + H_2SO_4 \longrightarrow CH_3CH_2CH_2CH_2Br + NaHSO_4 + H_2O$

	1-butanol	sodium bromide	sulfuric acid	1-bromobutane
MW:	74.12	102.90	98.08	137.03
weight:	18.5 g	30.0 g	25.0 mL (46.0 g)	34.2 g (theory)
moles:	0.250	0.292	0.469	0.250 (theory)

calculation of numbers of moles of reactants:

for 1-butanol: $\dfrac{18.5\ g}{74.12\ g/mol} = 0.250\ mol$

for NaBr: $\dfrac{30.0\ g}{102.90\ g/mol} = 0.292\ mol$

for H_2SO_4: $\dfrac{46\ g}{98.08\ g/mol} = 0.469\ mol$

calculation of theoretical yield of product:

for 1-bromobutane: $0.250\ mol \times 137.03\ g/mol = 34.2\ g$

Figure 1.1 A typical laboratory notebook page, including a balanced equation with the pertinent data, and calculations of moles of reactants and theoretical yield of product.

The molecular weight and the **theoretical yield** should be placed under the name of each organic product. The theoretical yield of a product may be calculated from its molecular weight and the number of moles of the **limiting reagent** (the reactant present in shortest supply). In the example in Figure 1.1, NaBr and H_2SO_4 are present in excess. Therefore, 1-butanol is the limiting reagent. In the example, the maximum number of moles of product that could be obtained from 0.250 mol of 1-butanol is 0.250 mol, or 34.2 g, as shown in the last calculation in Figure 1.1. Appendix II describes yield calculations in more detail.

If the experiment is not a reaction, but is an isolation or a purification experiment, then the formulas, names, etc. of the compounds in question should be written out.

3) The **actual procedure** that will be used should be outlined *in detail.* The purpose of the outline is to provide an overview of what you will be doing in the

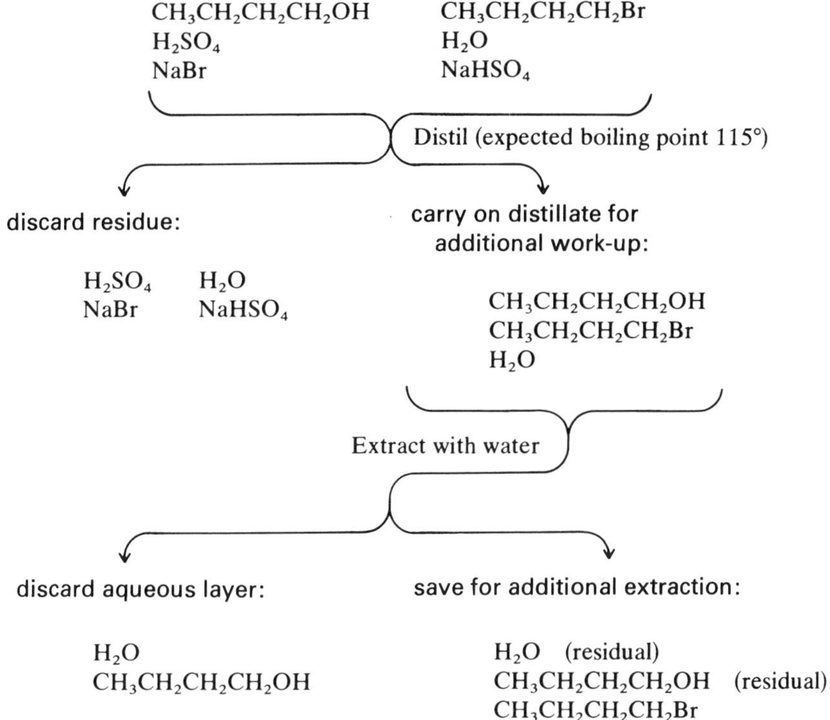

Figure 1.2 A partial flow diagram for the isolation of 1-bromobutane (Experiment 10.1).

laboratory, thus allowing you to organize your time efficiently. (Do not waste your time by copying the procedure word-for-word from the text. You can always refer back to the original procedure if necessary.) Boiling points of solvents and any special hazards, such as flammability, should be noted in your outline so they are fresh in your mind when you start the experiment.

A **flow diagram**, showing how the product will be isolated from by-products or unreacted starting material, may be included. Figure 1.2 shows a portion of such a diagram.

4) **Observations** should be recorded in your notebook *as you do the experiment.* Examples of observations might be "The ethanol solution was yellow," or "Approx. 5 mL of the reaction mixture was spilled and lost while being transferred to a separatory funnel." It is better to include too many observations than too few; however, use common sense. Entering the fact that you went to the storeroom to get a beaker wastes both your time and the time of anyone who reads your notebook. On the other hand, failure to record a weight or a physical constant may also waste your time—you may have to repeat the experiment.

5) **Conclusions** of your experiment will usually include the weight and physical properties (such as melting point or boiling point) of the isolated and purified product plus the **per cent yield**. This per cent yield is the actual per cent of the theoretical yield that you obtained.

$$\text{per cent yield} = \frac{\text{actual yield in g}}{\text{theoretical yield in g}} \times 100$$

Thus, you might write in your notebook;

> *yield*: 5.3 g (61%) of benzoic acid, mp 117°–119°;
> mp of an authentic sample of benzoic acid, 119.5°–120°;
> mixed mp of product with authentic sample, 118.5°–119°.

In most experiments, you will turn in a sample bottle of your product to the instructor. The data from the conclusions in your laboratory notebook should be used to fill out the labels placed on these sample bottles. Each instructor has a preferred label format, but a typical label might look like the following:

> COH (O above, benzene ring)
>
> Benzoic acid
> mp 117° - 119°
> 5.3g (61% yield)
> Expt. 16-1, Lab section 28
> Jane Doe

1.3 Laboratory Equipment

A. Glassware

Figure 1.3 shows the glassware that might be found in a typical student locker. Your own laboratory may not supply every item shown.

Note the ground-glass joints on the round-bottom flask, condenser, etc. These joints are ground to a **standard taper** (ST). The size of a standard-taper joint is identified by a number, such as ST 19/22, ST 14/20, or ST 24/40, which refers to the size of the joint in mm. Any 19/22 inner joint will fit any 19/22 outer joint; therefore, glassware of the same standard taper is interchangeable. The advantages of ground-glass joints are that they give a good seal between two pieces of equipment and that a laboratory set-up can be assembled quickly.

Equipment with ground-glass joints is expensive. If you break a piece of glassware containing a ground-glass joint, do not discard the joint unless you are told to do so by your instructor. A competent glass blower can recycle the joints by sealing them on ordinary, less-expensive flasks and condensers.

If your laboratory does not have ground-glass equipment, you must use corks and rubber stoppers. Your instructor will provide instructions for boring holes in corks and stoppers.

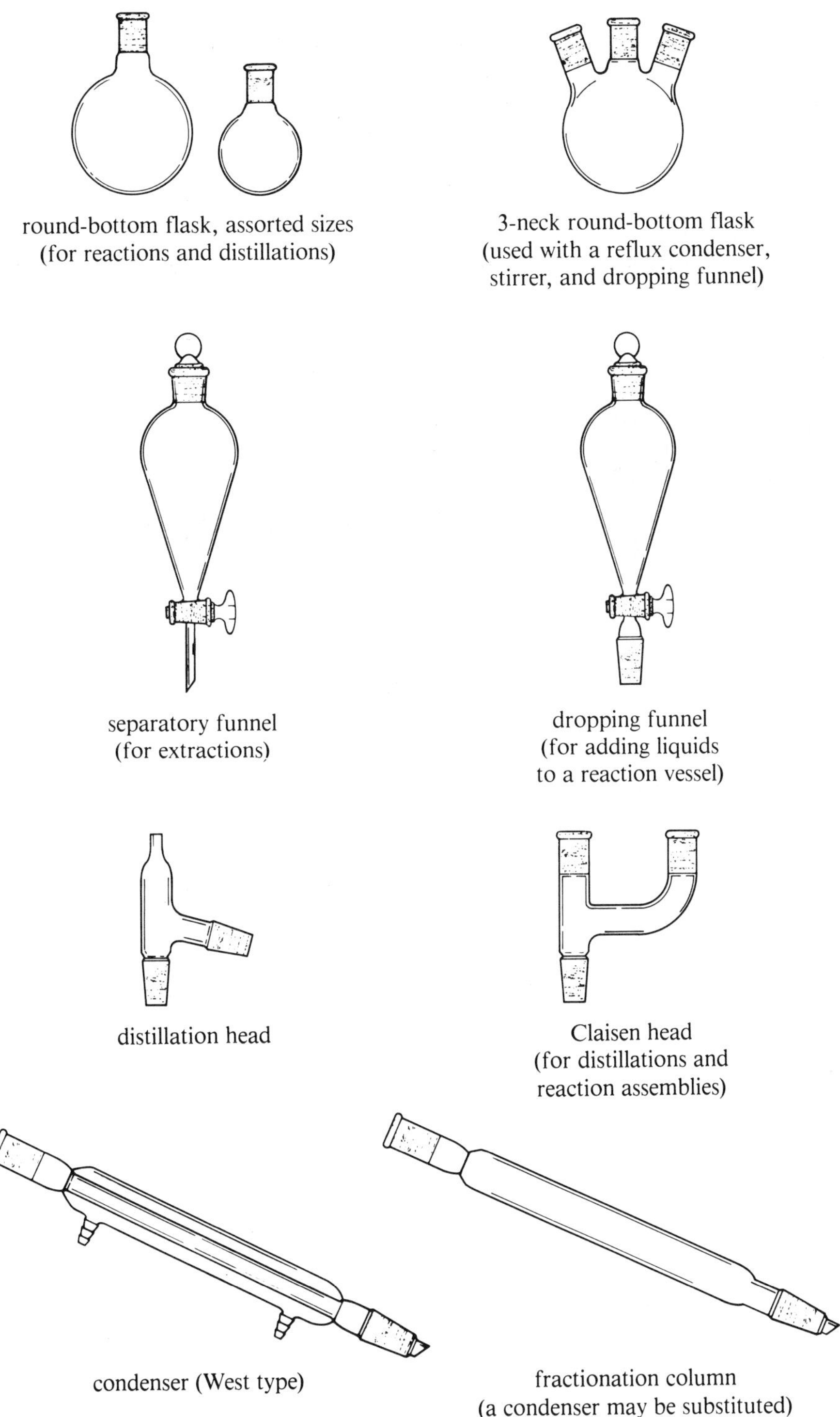

Figure 1.3 *Typical glassware stored in a student locker.*

(continued)

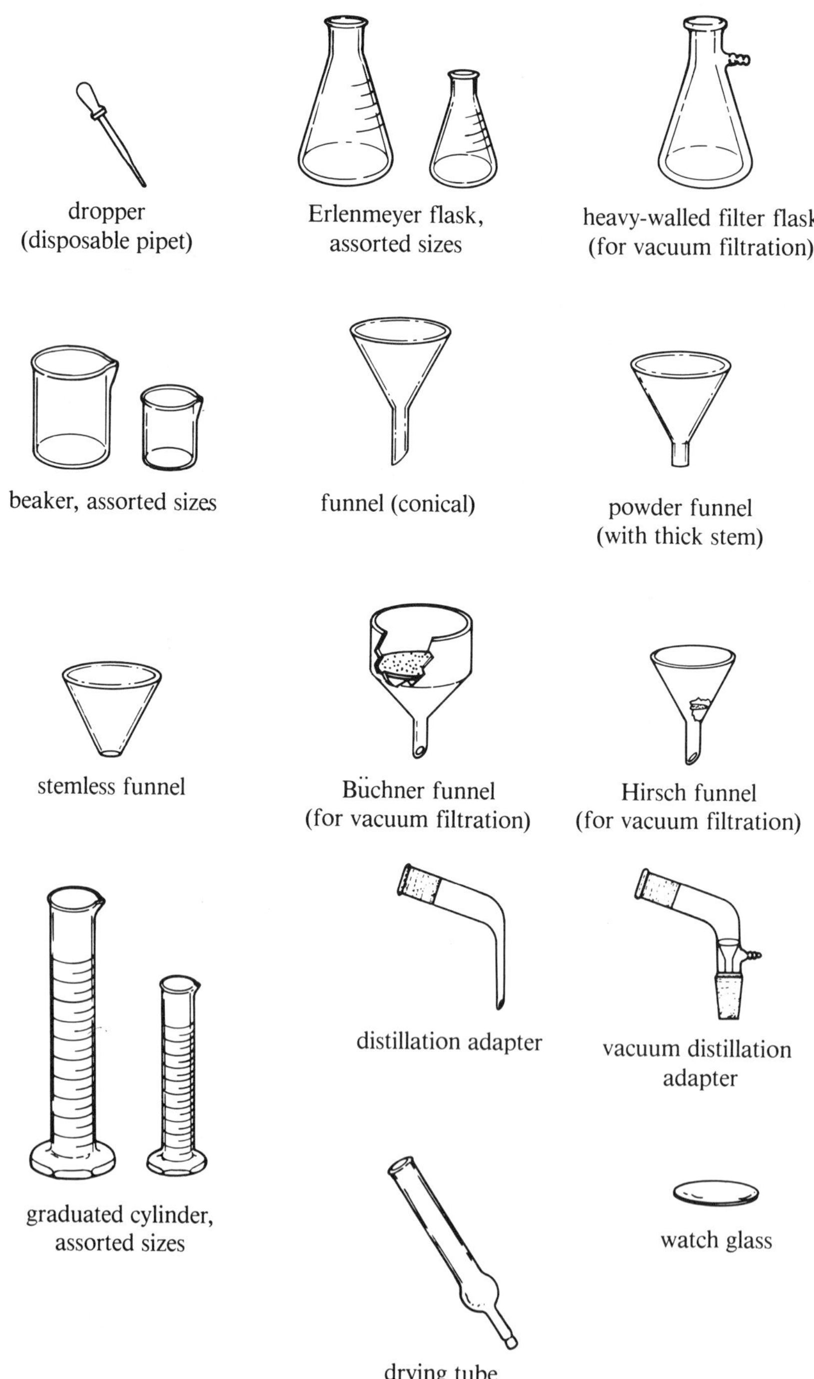

Figure 1.3 (*continued*) *Typical glassware stored in a student locker.*

When you check the equipment in your locker, check each piece of glassware carefully for cracks and "stars" (small star-shaped cracks). Each time you use a piece of glassware, recheck it. Any cracked or starred glassware should be returned to the storeroom and replaced by undamaged ware. In some cases, a glass blower will be able to repair the cracks.

B. Nonglass Equipment

Other useful items often found in a student locker are pictured in Figure 1.4. The Filtervac and neoprene adapters are used to attach a Büchner or Hirsch funnel to a filter flask. However, a one-holed rubber stopper that fits both funnel and flask gives a better seal and is easier to use.

Polyethylene wash bottles are useful for holding distilled water and most solvents. If yours is the type with a small hole in the neck, place your finger over the hole and squeeze the bottle to force liquid through the tip of the tube. Release your finger from the hole to release the pressure and stop the flow of liquid.

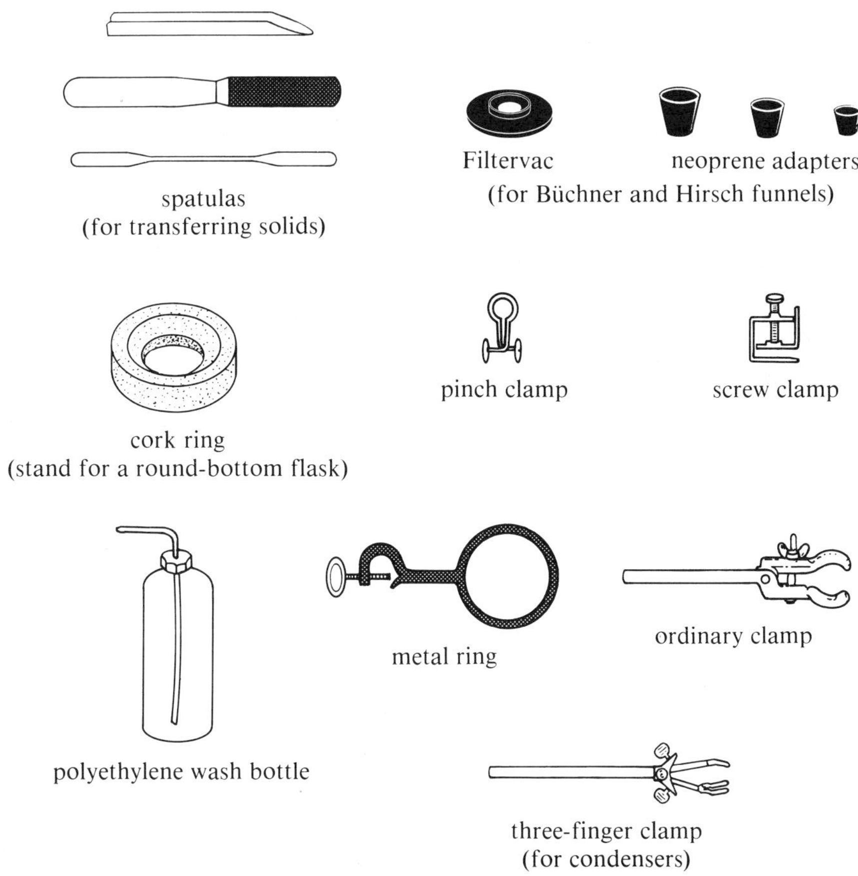

Figure 1.4 Hardware and other nonglass items typically found in a student locker.

C. Heating Equipment

Several devices for heating liquids in flasks are shown in Figure 1.5. The **steam bath** heats liquids to a maximum of about 90° and is useful for heating low-boiling solvents, especially flammable ones. To clear water from the steam line, turn the steam on forcefully. Then turn the steam to low before placing a flask on it. (Too much steam will allow water to contaminate the contents of the flask, and its noise is irritating to others in the laboratory.) A round-bottom flask can be set into the appropriately sized ring of the steam bath, while an Erlenmeyer flask can be set on top of a ring smaller than itself. If all the rings are left on top of the steam bath, two or three small Erlenmeyer flasks can be warmed simultaneously. Many modern **hot plates** are also safe for heating flammable solvents.

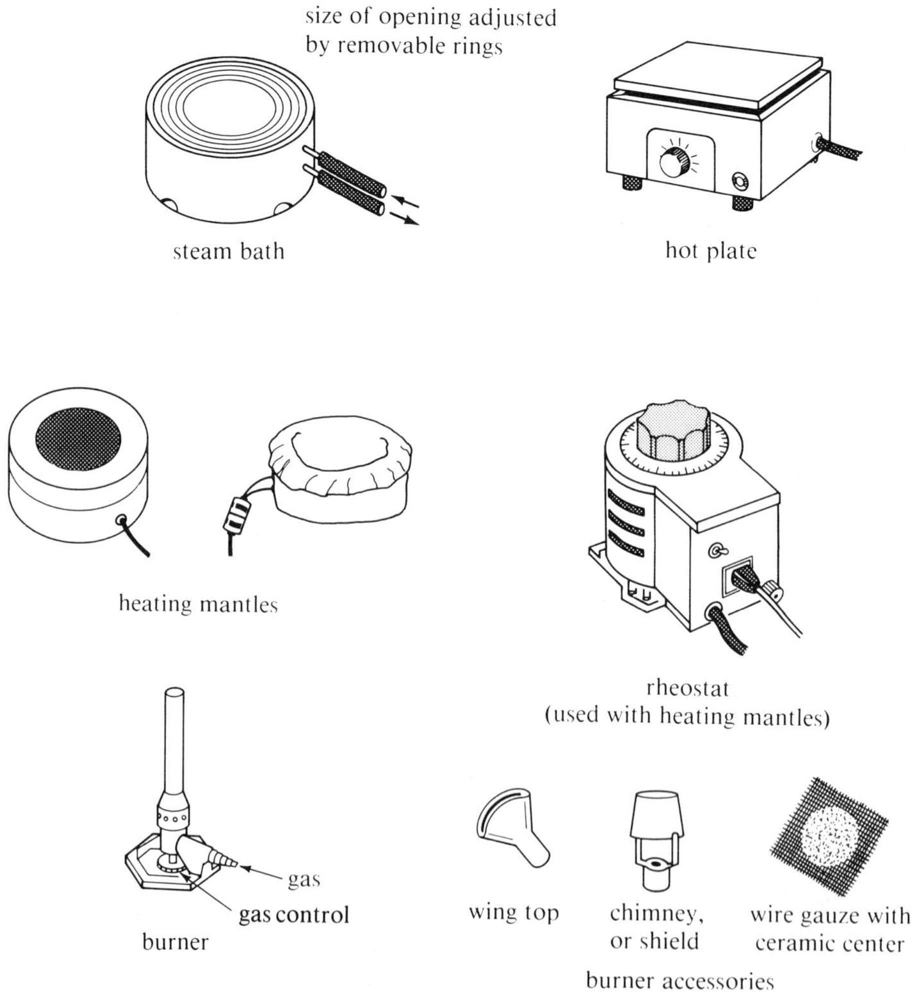

Figure 1.5 *Some typical laboratory heating devices.*

>>>> ***SAFETY NOTE*** Some hot plates develop very hot surfaces, have exposed electrical coils, or spark when the thermostat clicks on and off. These hot plates should *not* be used with flammable solvents because of the danger of fire or explosion.

A **heating mantle** is used for heating the contents of a round-bottom flask. The mantles are available in assorted sizes for different-sized flasks. A mantle should fit a flask snugly. If the heating mantle in your locker appears worn or if the heating element is exposed, return it to the storeroom. A worn heating mantle can cause a fire! The electrical plug of a typical heating mantle will not fit into the standard wall socket because heating mantles are not designed to operate at 110 volts. Instead, the mantle is plugged into a **variable transformer**, or **rheostat** (Variac, Powerstat, Powermite), which is used to adjust the voltage and thus the temperature. Figure 1.5 shows a typical transformer of this type.

Bunsen burners and microburners (small Bunsen burners) are used for bending glass tubing and for other glass-working. They may also be used to heat aqueous solutions and high-boiling liquids. The use of a wire gauze with a ceramic center is recommended to distribute the heat from a burner to a flask and to prevent localized overheating.

>>>> ***SAFETY NOTE*** Burners should be used with the utmost caution in the organic laboratory because of the ever-present vapors of flammable solvents. NEVER light a burner (or even a match) if someone is using a flammable solvent nearby. NEVER use a burner to heat a flammable solvent. NEVER open a solvent bottle without first checking the vicinity for flames.

D. Community Equipment

You will share a **balance** with your classmates. Your instructor will demonstrate the operation of the balance or balances in your laboratory. A balance is a delicate instrument: treat it with care, and always wipe up spills (liquid or solid) on or near the balance. Do *not* blow spilled powders off the balance pan—many chemicals that are ordinarily safe to handle are dangerous when inhaled. When you finish with a balance, return it to a weight of zero if this is not accomplished automatically. This will help prevent uneven wearing of the weighing mechanism.

You will probably be using analytical electronic equipment such as gas chromatographs and spectrometers in this course. These are also delicate instruments. Do not be a knob twister. Understand and follow your instructor's directions.

The **fume hood** should be operating whenever someone is working in the laboratory. It is recommended that highly flammable or toxic materials *always* be handled in the hood. Unfortunately, space limitations in a student laboratory do not always permit this. Because of the nature of the chemicals used in the hood, it is imperative that the bench in the hood be kept clean. If you find some spilled chemical of unknown origin on the hood bench, clean it up as if it were toxic, corrosive, and flammable.

The sharing of other community equipment is a matter of safety and courtesy. Keep reagent bottles tightly capped and return them to their proper spots. Do not contaminate chemicals in reagent bottles. If you see that a bottle is almost empty, report the fact to your instructor.

1.4 Cleaning Glassware

Glassware for organic chemical reactions should be both *clean* and *dry*. The presence of water in a flask can ruin many experiments (except, of course, those performed in aqueous solution). For this reason, glassware should be washed as soon as you finish using it (certainly by the end of each laboratory period), then allowed to drain and dry in your locker until the next laboratory period.

Another reason to wash your glassware promptly is that freshly dirtied glassware is easier to clean than glassware containing dried-out tars and gums. Furthermore, some compounds, like sodium hydroxide and potassium hydroxide, can etch glass and ruin ground-glass joints if left standing in flasks.

A. General Cleaning

Most glassware can be cleaned readily with strong industrial detergent or scouring powder and a bottle brush. The brush should be bent, if necessary, to reach the entire inside of a flask. The detergent should be rinsed out thoroughly, then the piece of glassware rinsed with a small amount of distilled water. A flask can be stored upside down on a crumpled towel in a breaker to drain and dry. Other pieces of equipment, like condensers, can be laid on their sides in the locker.

If it is absolutely necessary, glassware can be dried quickly by rinsing it with *acetone*, $(CH_3)_2C{=}O$. (Because of the expense, some laboratories do not allow acetone to be used for rinsing flasks.) Acetone is miscible with water and thus flushes out the water clinging to the inside of the glassware. Acetone is low-boiling (bp 56.1°) and flammable, so it should not be used near flames.

To dry a flask with acetone, first drain as much water as possible from the flask. Squirt some acetone from a wash bottle or simply pour in 5–10 mL, swirl, then drain the acetone into the "waste-acetone" jug. Never use expensive reagent-grade acetone for rinsing flasks! Use "wash acetone" instead. If it is used to rinse only water from a flask, wash acetone may be re-used several times.

A flask rinsed in acetone will dry fairly quickly in air. The drying process can be speeded up by placing a clean glass dropper, connected to a vacuum line by heavy-walled rubber tubing, in the drained flask. The vacuum will suck fresh air through the flask, sweeping acetone and water vapors into the vacuum line. Using a stream of air to dry an acetone-rinsed flask is not recommended. Compressed air is likely to contain droplets of water and oil that will contaminate the flask. Also, a noisy blast of air shot into a flask can startle other people working in the laboratory.

Never use a flame to dry a flask. If the flask contains water droplets, the flask may become unevenly heated and crack. If the flask contains solvent, the vapors may catch fire. A drying oven may be used to dry water-rinsed glassware, but not glassware rinsed with a solvent.

Ground-glass equipment should be dismantled and cleaned before it is placed in your locker. The joints should be kept free of chemicals and grit; otherwise, they may become stuck together or "frozen." A frozen joint can sometimes be unfrozen by rinsing the outer portion in hot water to cause it to expand. A *gentle* rap on the table top might also loosen it. There are some more-sophisticated techniques for unfreezing joints; your instructor can advise you.

Glass stopcocks in separatory funnels are treated like other ground-glass joints and should be cleaned of hydrocarbon greases with a tissue dampened with acetone and stored separately. (Silicone greases should be carefully removed with dichloromethane.) If you would rather keep your stopcock in the separatory funnel, regrease it before replacing it. Teflon stopcocks, which do not need greasing, should be cleaned and then replaced loosely in the joint until they are used again.

B. Hard-to-Clean Flasks

Tars and gums, which are large organic molecules called **polymers**, are formed when a large number of organic molecules react to yield very long chains or three-dimensional molecular networks. These substances are not soluble in water. Therefore, they should be scraped out of a flask with a metal spatula and discarded in a waste crock, not in the sink. Acetone is a good solvent for most organic compounds and is often useful for dissolving remnants of tars and other organic residues from dirty flasks. For these cleaning purposes, *waste acetone* should be used. In some cases, swirling 5–10 mL of acetone in a flask will dissolve the organic residues. In other cases, several hours of soaking will be necessary. If you encounter such a stubborn tar, cork the dirty flask containing a small amount of acetone and store it upright in your locker until the next laboratory period.

Suggested Readings*

Steere, N. V. *Handbook of Laboratory Safety*. 2nd ed. Cleveland, Ohio: CRC Press, 1971.

Manufacturing Chemists' Association (now the Chemical Manufacturing Association). *Guide for Safety in the Chemical Laboratory*. 2nd ed. New York: Van Nostrand Reinhold Co., 1972.

Muir, G. D., ed. *Hazards in the Chemical Laboratory*. 2nd ed. London: The Chemical Society, 1977.

Green, M. E., and Turk, A. *Safety in Working with Chemicals*. New York: Macmillan Publishing Co., Inc., 1978.

Safety in Academic Chemistry Laboratories. 3rd ed. Washington, D.C.: American Chemical Society, 1979.

Reese, K. M. *Health and Safety Guidelines for Chemistry Teachers*. Washington, D.C.: American Chemical Society, 1979.

National Research Council. *Prudent Practices for Handling Hazardous Chemicals*, Washington, D.C.: National Academy Press, 1981.

* See also the toxicology references, page 423.

Problems

1.1 Give reasons for the following safety rules:

(a) Contact lenses should not be worn in the laboratory.

(b) A chemical spill on the skin should be washed off with water, not with solvent.

(c) Solvents are not to be poured into the drainage trough.

(d) Water should not be used to extinguish laboratory fires.

(e) To dilute concentrated sulfuric acid, we pour it onto ice instead of simply mixing it with water.

1.2 What should you do in each of the following circumstances?

(a) You splash a chemical into your eye.

(b) Your neighbor splashes a chemical into his or her eye.

(c) A strong acid spills onto your hands.

(d) Your neighbor's clothing catches fire.

(e) Your neighbor's flask catches fire.

(f) You spill solvent on a hot plate.

1.3 What are the principal hazards of each of the following solvents?

(a) carbon disulfide (b) diethyl ether

(c) benzene (d) carbon tetrachloride

1.4 Make the following conversions:

(a) 5.0 g $CH_3CH_2CH_2CH_2Br$ to moles

(b) 10.0 mL conc. H_2SO_4 (96%, density 1.84) to moles

(c) 0.100 mol CH_3OH to grams

(d) 2.50 mol NaBr to grams

(e) 0.30 mol H_2SO_4 to mL of 6*N* H_2SO_4

1.5 Calculate the per cent yield when (a) the theoretical yield of a product is 15.3 g and a student obtains 6.9 g; (b) the theoretical yield is 3.1 g and a student obtains 2.7 g.

1.6 For each of the following reactions, (1) identify the limiting reagent, and (2) calculate the theoretical yield of the organic product. (*Note*: The equations as shown are not necessarily balanced.)

(a) $$\underset{25.0\text{ g}}{CH_3CO_2H} + \underset{10.0\text{ g}}{NaOH} \rightarrow CH_3CO_2Na + H_2O$$

(b) $$\underset{5.0\text{ g}}{H_2NCH_2CH_2CH_2NH_2} + \underset{10.0\text{ mL}}{HCl(12N)} \rightarrow Cl^-\ H_3\overset{+}{N}CH_2CH_2CH_2\overset{+}{N}H_3\ Cl^-$$

(c) $$\underset{8.2\text{ g}}{CH_3CH_2CH{=}CH_2} + \underset{2.0\text{ g}}{H_2O} \xrightarrow{H^+} CH_3CH_2\overset{\overset{OH}{|}}{C}HCH_3$$

1.7 Write a flow diagram for the following partial procedure for isolating Product A. "Separate the two layers. Return the lower layer (compound A, density 1.54) to the separatory funnel and extract it with 25 mL of conc. sulfuric acid (density 1.84). Discard the sulfuric acid."

1.8 What would be the heat source (or heat sources) of choice for boiling each of the following solvents?

(a) diethyl ether, $(CH_3CH_2)_2O$, bp 35°

(b) water, bp 100°

(c) ethanol, CH_3CH_2OH, bp 78°

(d) acetone, $(CH_3)_2C{=}O$, bp 56°

Melting Points

The **melting point** of a crystalline solid is the temperature at which the solid changes to a liquid at 1.0 atmosphere of pressure. The melting point is the same as the freezing point, the temperature at which the liquid becomes solid. Because liquids have a tendency to become supercooled (remain liquid below their freezing points), freezing-point determinations are only rarely performed in organic chemistry.

2.1 Characteristics of Melting Points

In the laboratory, the melting point of a solid is reported as a **melting range**. The barometric pressure, which has a negligible effect on melting points at the usual atmospheric pressures, is ignored.

The melting point is determined by heating a small sample of the solid material slowly (at the rate of about 1° per minute). The temperature at which the first droplet of liquid is observed in the solid sample is the lower temperature of the melting range. The temperature at which the sample finally becomes a clear liquid throughout is the upper temperature of the melting range. Thus, a melting point might be reported, for example, as: mp 103.5–105°.

A pure organic compound usually has a "sharp" melting point, which means that it melts within a range of 1.0° or less. A less pure compound exhibits a broader range, maybe 3° or even 10°. For this reason, a melting point can often be used as a criterion of purity. A melting range of 2° or less indicates a compound pure enough for most purposes. However, a compound to be submitted to an analytical laboratory for elemental analysis (determination of the relative weight percentages of the elements) should have a very sharp melting point.

An impure organic compound exhibits not only a broad melting range, but also a *depressed* (lower) melting point. For example, a fairly pure sample of benzoic acid might melt at 121–122°, but an impure sample might show a melting range of 115–119°.

The melting point of a compound can also be used to prove the identity of a pair of compounds. Assume that you have a compound of unknown structure that melts at 120–121°. Is the compound benzoic acid? To find the answer, you would mix the unknown with an authentic sample of benzoic acid (mp 120–121°), then take a melting point of the mixture. This melting point is called a **mixed melting point**. If the unknown is benzoic acid, the mixed melting point would remain 120–121° because the two mixed samples are the same compound. However, if the unknown is *not* benzoic acid, the mixed melting point would be depressed and show a wider range. For absolute identification purposes, additional data besides the mixed melting point are desirable.

You might wonder why a mixture of compounds melts at a lower temperature than either of the pure components. Recall from your general chemistry course that an aqueous solution of sodium chloride (or sucrose or any other compound) has a lower freezing point than pure water. The same principle is involved in melting-point depression. When a crystalline compound contains impurities in its crystal structure or is intimately mixed and heated with another compound, the components can form a solid solution. With rare exceptions, this solution has a lower melting point than its components. Furthermore, as this solid solution melts, it dissolves other solid material as well, thus spreading the effect of the melting-point depression throughout the sample.

2.2 Melting-Point Diagrams

The melting-point diagram in Figure 2.1 shows the typical melting behavior of a series of mixtures of two organic compounds, *o*-dinitrobenzene and *m*-dinitrobenzene.

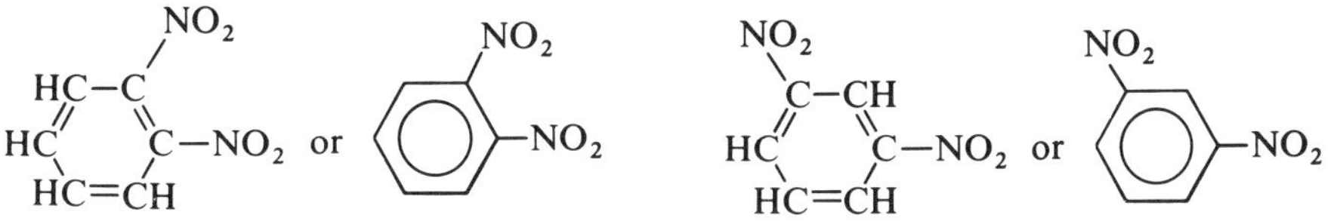

NO_2 / HC–C, HC=C–NO_2, HC=CH or benzene ring with NO_2, NO_2

o-dinitrobenzene
(1,2- or *ortho*-dinitrobenzene)

NO_2 \ C–CH, HC=C–NO_2, HC=CH or benzene ring with NO_2, NO_2

m-dinitrobenzene
(1,3- or *meta*-dinitrobenzene)

A similar graph for mixtures of two other compounds would probably appear quite similar, although different types of melting behavior are occasionally observed. The graph shows only the upper limit of the melting range, the temperature at which the mixture becomes completely liquid. From the graph, it may be seen that pure *o*-dinitrobenzene melts at 118.5°, pure *m*-dinitrobenzene melts at 90.0°, and a 50:50 mixture of the two compounds melts at about 80°.

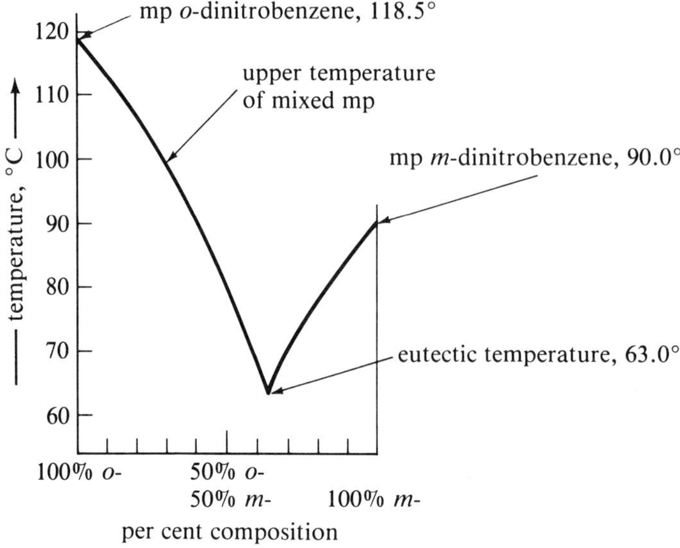

Figure 2.1 Melting-point diagram for mixtures of o-dinitrobenzene and m-dinitrobenzene.

The low point on the graph (63°) is called the **eutectic point**, or **eutectic temperature**, and represents the minimum melting point for a mixture of these two compounds. The per cent composition of the mixture that melts at the eutectic point is called the **eutectic mixture**. The values for the eutectic point and the per cent composition of a eutectic mixture depend on which compounds are being studied. For the diagram in Figure 2.1, the eutectic mixture consists of 63% *m*-dinitrobenzene and 37% *o*-dinitrobenzene. The eutectic point for any binary mixture is the temperature at which both components melt simultaneously, and is consequently a sharp melting point, rather than the broad melting range usually observed for mixtures.

A eutectic point is exhibited only by an intimate and uniform mixture of the correct composition. For all practical purposes in the organic laboratory, virtually all mixtures of two different compounds exhibit a broad melting range.

2.3 Melting-Point Apparatus

There are many types of electrically heated melting-point devices on the market today. In most of these devices, a glass capillary tube containing the sample is inserted into a heating block. With some instruments, your own thermometer is also inserted. Most instruments will accept several capillary tubes simultaneously; therefore, the melting points of a number of samples can be determined under identical conditions. Simultaneous melting-point determinations are especially useful for mixed melting points because the melting behavior of the pure compound and one or more mixtures can be compared directly.

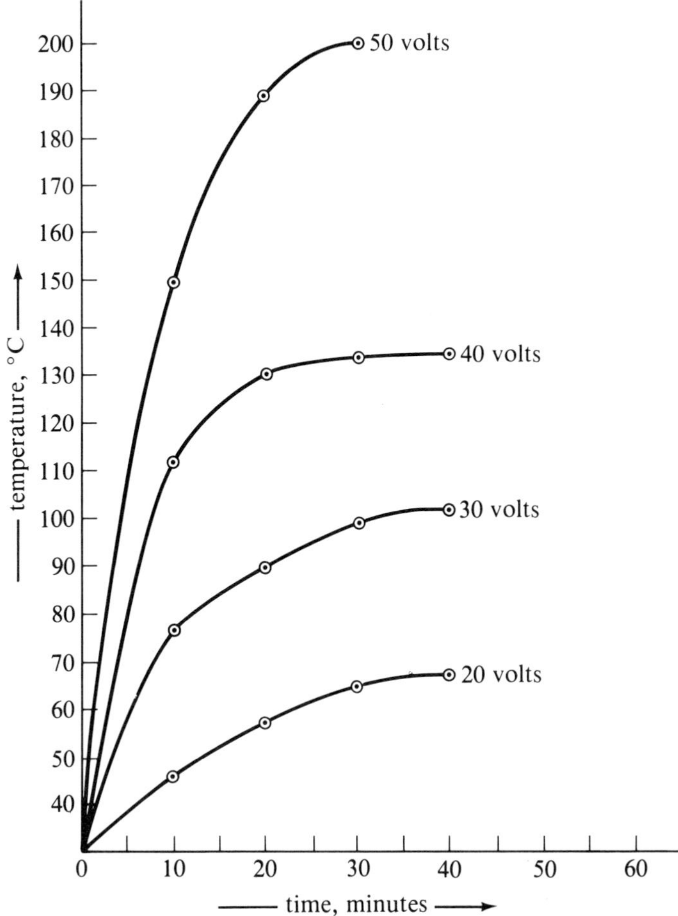

Figure 2.2 *Heating curves for a Mel-Temp® melting-point apparatus at different voltage settings.*

The rate of heating of an electrical melting-point apparatus is controlled by a rheostat. The higher the setting on the rheostat dial, the faster the block will heat. The rate of heating at different dial settings can be measured and graphed. Figure 2.2 shows typical heating curves for one type of commercial instrument. Note that the rate of heating at any one voltage setting is not linear, but drops off with increasing temperature. This variable rate is advantageous because, at one voltage setting, the sample can be heated rapidly at first and then more slowly as the melting point is approached.

If an electrical melting-point apparatus is not available, a **melting-point bath** can be easily constructed. A long-necked, heat-resistant glass flask, *securely clamped* to a rack or ring stand and containing a dry, high-boiling silicone oil or other liquid, can be used. (Silicone oil can be used for melting points up to about 250°.) When the flask is heated slowly, the liquid bath is maintained at a uniform temperature throughout by convection currents. The **Thiele tube** contains a side

top view of rubber stopper and thermometer, showing wedge-shaped opening to permit release of air when oil expands

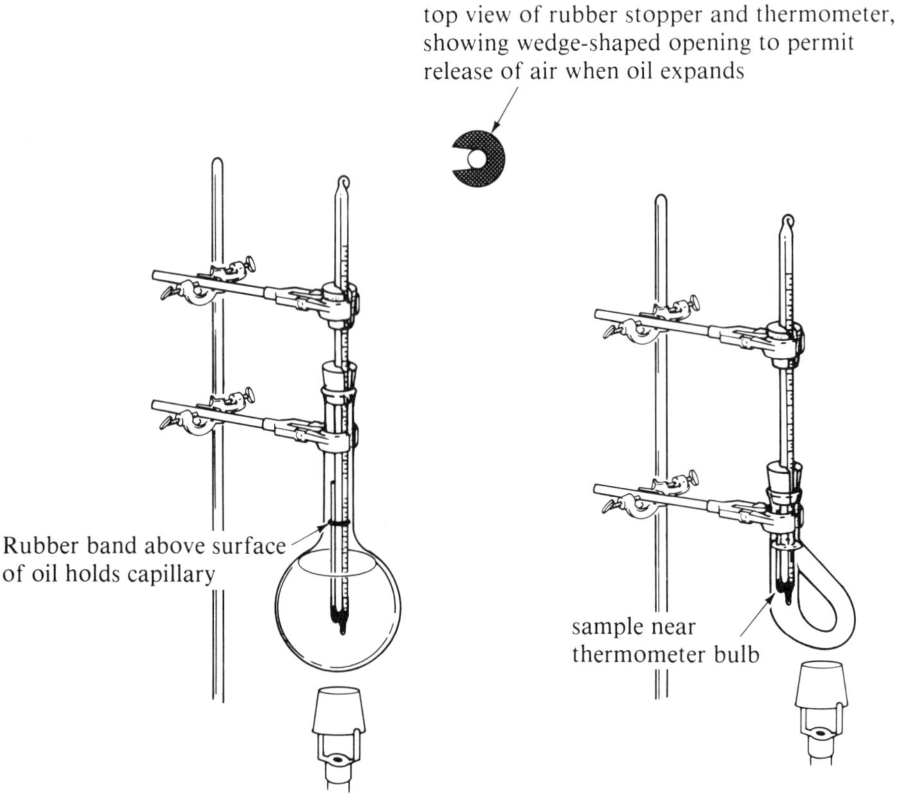

Rubber band above surface of oil holds capillary

sample near thermometer bulb

Figure 2.3 *Two types of baths for taking melting points.*

arm to increase the convection currents. Figure 2.3 illustrates these two devices. Your instructor will give you specific directions concerning the use of an oil-filled melting-point device.

2.4 Thermometer Calibration

No thermometer is accurate at every temperature reading; therefore, a thermometer to be used for melting-point determinations in a research laboratory (and sometimes in the student laboratory) should be calibrated. Calibration is accomplished by recording the melting points of five or six very pure compounds, chosen to melt at a variety of temperatures. Table 2.1 lists some suggested compounds. From these melting points, a graph similar to the one in Figure 2.4 is constructed. This graph shows the *correction factor* versus the *observed temperature* (using the upper value of the melting range). For example, pure benzoic acid melts at 121.7°. If your thermometer records 120.2° as the melting point, your correction factor at approximately 120° would be +1.5°. Any time you record a melting point near 120°, you would add 1.5° to the observed temperature. (Corrected melting point = Observed melting point + 1.5°.) A corrected melting point is reported as follows: mp 121.2–121.5° (corr.).

Table 2.1 *Suggested melting-point standards for calibration of thermometers*

Compound	Melting point (°C)
naphthalene	80.6
1,3-dinitrobenzene	90.0
acetanilide	114.3
benzoic acid	121.7
benzamide	130
urea	132.7
sulfanilamide	166
p-toluic acid	181
succinic acid	188
3,5-dinitrobenzoic acid	205
anthracene	217
p-nitrobenzoic acid	242

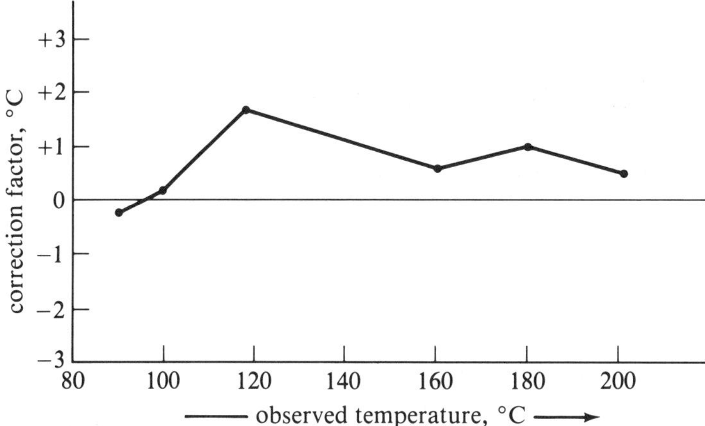

Figure 2.4 *A typical thermometer-calibration graph.*

Problems

2.1 Using the heating curves in Figure 2.2 (page 23), what should be your rheostat settings for a compound expected to melt at (a) 150°; (b) 110°?

2.2 Using the *Handbook of Chemistry and Physics*, find the expected melting points of the following compounds:

(a) indophenol (b) quinine
(c) acetophenone, 2,3-dihydroxy-

2.3 Using Table 2.1, construct a thermometer calibration graph from the following observed melting points:

acetanilide, mp 113°
benzamide, mp 130°
succinic acid, mp 188°
p-nitrobenzoic acid, mp 245°

2.4 (a) Using the following table, draw a melting-point diagram and estimate the eutectic temperature and composition:

Per cent composition	Melting range
100% A	125–126°
75% A–25% B	115–120°
65% A–35% B	128–131°
50% A–50% B	135–140°
100% B	151–152°

(b) If you wished to obtain a more accurate graph for this melting-point curve, suggest the best mixtures for additional mixed melting points.

2.5 A compound is observed to melt sharply at 111° with the vigorous evolution of a gas. Then the compound solidifies and does not melt until 155°, at which temperature it again melts sharply. Explain.

EXPERIMENT 2.1

Melting Points of Benzoic Acid and an Unknown

Discussion

Steps in Determining a Melting Point

1) *Preparation of the sample.* Pulverize the sample for a melting-point determination by placing 0.1–0.2 g of *dry* crystals on a watch glass and crushing them with a metal spatula or the end of a test tube. If the sample is for a mixed melting point, grind a 50:50 mixture of the two compounds (approximated, not necessarily weighed) *thoroughly* with a mortar and pestle to ensure a homogeneous and intimate mixture. Alternatively, dissolve the mixture of the two compounds in dichloromethane (CH_2Cl_2) on a watch glass; allow the solvent to evaporate; and pulverize the residue.

2) *Loading the capillary.* Mound up the pulverized sample and press the open end of the capillary into the sample against the surface of the watch glass or mortar. A small plug of sample will be pushed into the open end of the capillary tube. The ideal amount of sample is a plug about 1 mm in length. Tapping the sealed end of the capillary on the surface of the laboratory bench may knock the sample down to the desired position at the sealed end of the capillary. (CAUTION: Capillaries are fragile.) A safer and more efficient technique for driving the plug of sample to the sealed end of the capillary tube is to drop the tube (sealed end down) through 2–3 feet of ordinary glass tubing

vertically onto the bench top. The impact of the capillary with the hard surface knocks the sample down. It may be necessary to repeat this dropping two or three times.

It is important that the height of the sample in the capillary be only 1–2 mm, and that it be packed firmly. A larger sample takes longer to melt and may exhibit an erroneously large melting range. If the sample is loosely packed, it is difficult to determine the start of the melt.

3) Preliminary melting point. If the approximate melting point of a sample is not known, it saves time to take a preliminary melting point with a second capillary tube. The approximate melting point of the sample is determined by heating the melting-point apparatus fairly quickly, with a temperature increase of about 10° per minute. This preliminary melting point will be on the low side of the true melting point, but will indicate where the rate of temperature increase should be slowed in the determination of the exact melting point.

You need not determine a preliminary melting point if you know the name of the compound and can find its melting point in a reference book or journal. For example, the *Handbook of Chemistry and Physics* (CRC Press, Inc., Boca Raton, Florida) contains an extensive section called "Physical Constants of Organic Compounds," along with instructions for finding compounds in the listing.

4) Taking the melting point. Insert the capillary tube containing the sample into the melting-point apparatus, along with the thermometer, if necessary. Heat the apparatus rapidly to a temperature about 10° below the expected melting point, then slow the rate of temperature increase to about 1° per minute. If the temperature is increased too rapidly at the melting point, the thermometer, sample, and block (or bath) will not be at thermal equilibrium, and erroneous readings will result.

As we have mentioned, the initial, or lower, end of the melting range is the temperature at which the first drop of liquid is noted in the solid sample. The final, or upper, end of the range is the temperature at which the sample becomes a clear liquid containing no solid material. The determination of the final value of the melting range usually presents no problems. The initial value, however, may require judgment. Many organic compounds undergo changes in crystal structure prior to melting, and these phase changes may be mistaken for the first sign of melting. Sample sag, shrinking, changes in texture, and the appearance of droplets *outside the bulk of the sample* are not the start of the melt. The initial temperature of the melting range is taken as the first appearance of liquid *within the bulk of the sample.*

Another problem that may arise is *decomposition.* In a simple case, a sample may change color, effervesce, or otherwise change in appearance at its melting point. If the melting point is reasonably sharp, this type of behavior does not affect the value of the melting point as a criterion of purity or as an identification tool, but the decomposition should be reported along with the melting point—for example: mp 150.3–151.5° d; 150.3–151.5° decomp.; or 150.3–151.5° with darkening. If a sample decomposes over a large temperature range, the melting point cannot be used for identification purposes. The decomposition range should be reported, even if only approximate temperatures can be given. For example: decomp. 127–~150°.

After taking a melting point, discard the used, cooled capillary tube in the waste crock; the tube cannot be cleaned for reuse.

Experimental

EQUIPMENT:

10-mm glass tubing, about 1 meter long
melting-point apparatus
melting-point capillary tube (commercially available)
small mortar and pestle
spatula
watch glass

50-mL beaker (optional)
dropper or disposable pipet (optional)

CHEMICALS:

benzoic acid, 0.1 g
unknown, 0.1 g*

dichloromethane (methylene chloride), 1.0 mL (optional)

TIME REQUIRED: 1 hour

PROCEDURE

A. Melting Point of Benzoic Acid

Find the melting point of benzoic acid in the *Handbook of Chemistry and Physics* and record the value in your notebook.

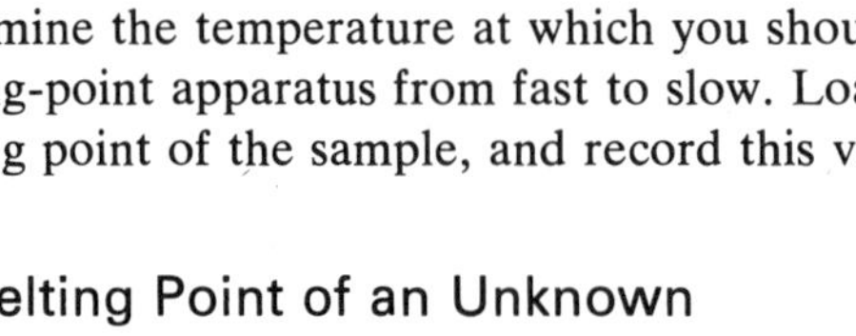

```
  H     H
   \   /
    C—C    O
   //  \\  ||
H—C     C—COH   or   (C6H5)—COH
   \   /                   ||
    C=C                    O
   /   \
  H     H
```

benzoic acid

Determine the temperature at which you should change the rate of heating the melting-point apparatus from fast to slow. Load a capillary tube, determine the melting point of the sample, and record this value in your notebook.

B. Melting Point of an Unknown

Obtain an unknown from your instructor and enter its number in your notebook. This unknown will be one of three compounds labeled A, B, or C in

* The instructor's guide contains a list of suggested unknowns.

bottles in your laboratory. Determine the melting point. By mixed melting points, determine which of the three unknowns your sample is. Enter your data and conclusion in your notebook.

Problems

2.6 What criteria are used to determine: (a) the lower value of a melting range; (b) the upper value of the range?

2.7 (a) What might you observe when a compound melts with decomposition?
(b) Suggest a way to verify that a sample has melted with decomposition.

2.8 Describe two methods of preparing samples for mixed melting-point determinations.

2.9 Suppose that your sample melts before you are ready to record the melting point. Should you (a) cool the capillary and redetermine the melting point, or (b) begin with a fresh sample? Explain.

»»» EXPERIMENT 2.2 «««

(SUPPLEMENTAL)

Mixed Melting Points of Benzoic Acid and 2-Naphthol

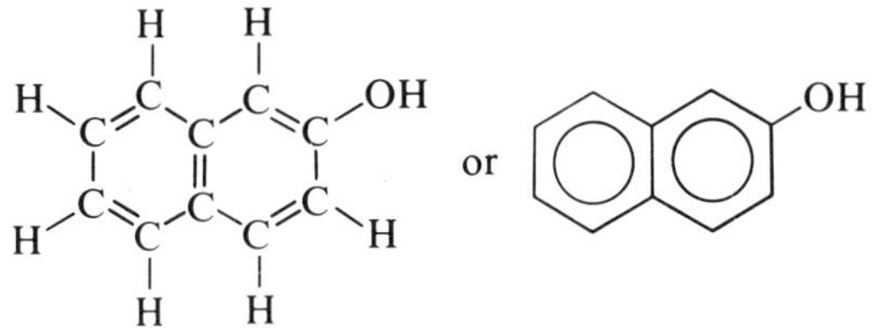

2-naphthol
(β-naphthol)

Experimental

EQUIPMENT: See Experiment 2.1.

CHEMICALS:

benzoic acid, 1.5 g
2-naphthol, 0.9 g

dichloromethane, 1–4 mL (optional)

TIME REQUIRED: 2–3 hours

Table 2.2 Mixtures for mixed melting points

Composition	***Weight of benzoic acid (g)***	***Weight of 2-naphthol (g)***
99% benzoic acid– 1% 2-naphthol[a]	0.495	0.005
80% benzoic acid–20% 2-naphthol	0.400	0.100
50% benzoic acid–50% 2-naphthol	0.250	0.250
20% benzoic acid–80% 2-naphthol	0.100	0.400

[a] In order to obtain accurate amounts in this first mixture, use *ten times* the amounts stated (4.95 g benzoic acid and 0.05 g 2-naphthol). To keep waste to a minimum, a group of students should use this particular mixture.

PROCEDURE

Find and enter the melting points of benzoic acid and 2-naphthol in your notebook. (2-Naphthol is listed as "Naphthalene, 2-hydroxy-" in the *Handbook of Chemistry and Physics.*) Load one capillary with pure benzoic acid and another capillary with pure 2-naphthol.

Table 2.2 lists four mixtures, the melting points of which are to be determined. Grind each mixture with a mortar and pestle until it is *homogeneous.* (It is surprisingly difficult to obtain homogeneity.) Alternatively, dissolve each mixture in 1–2 mL of dichloromethane in a 50-mL beaker with swirling, and then allow the solvent to evaporate (in a fume hood).

Load each mixture into a capillary. (Keep track of the identity of each capillary!) Run melting-point determinations on three samples simultaneously, in the following two groups:

1) benzoic acid
2-naphthol
99% benzoic acid–1% 2-naphthol

2) the remaining three mixtures

Record the melting points in your notebook. In the first group, you should be able to detect a small depression (about 0.5°) in the melting point of the 99% benzoic acid compared to that of pure benzoic acid. The mixtures in the second group will show wide melting ranges. Plan on starting your slow temperature rise at least 30° below the melting point of the pure materials, then increase the temperature at the rate of about 2° per minute.

Experiment 4.2 is the separation of a mixture of benzoic acid and 2-naphthol by extraction. From the data obtained in Experiment 2.2, determine (a) if you can report accurately the per cent composition of a mixture of benzoic acid and 2-naphthol by melting point alone, and (b) what is the smallest percentage of 2-naphthol detectable in benzoic acid by the melting point.

Problems

2.10 Why is it *not* good practice to mix two components by simply stirring them together and then sampling them for a mixed melting point?

2.11 In Experiment 2.2, why is it recommended that the melting point of 99% benzoic acid–1% 2-naphthol be determined simultaneously with the melting points of the pure compounds?

2.12 What conclusions would you draw from the following student observations? The melting point of unknown sample A is 83.0–89.5°. The melting point of an authentic sample of 1-naphthol is 93.5–94.0°. A mixed melting point of these two samples is observed to be 88.0–91.5°.

EXPERIMENT 2.3

(SUPPLEMENTAL)

A Temperature-Dependent Liquid Crystal Display

Discussion

When most solid organic compounds are heated, they undergo a phase change (usually from solid to liquid) at a specific temperature or else they decompose. Let us consider only those compounds that melt when heated. Most pure organic compounds melt over a 1–2° temperature range. A wide melting range usually arises from impurities, but this is not always the case. About 0.5% of pure organic compounds do not melt sharply, but form a **mesophase** before melting (Greek *meso*, "middle" or "intermediate"). This mesophase, which is intermediate between the solid phase and the liquid phase, is called a **liquid crystal.***

In a melting-point apparatus, a compound forming a liquid crystal phase does not show a sharp melting point. Instead, we observe two ill-defined transitions, each at its own characteristic temperature. In most cases, a transition from the solid to a turbid liquid is observed at one temperature and, at a higher temperature, another transition to a clear liquid is seen. This melting-point behavior is that usually associated with an impure organic compound.

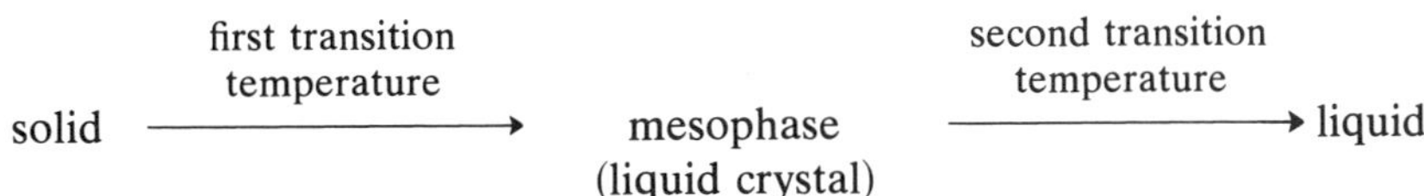

In the liquid crystal mesophase, the molecules are more orderly than those in a liquid, but less orderly than those in a solid. Because of the unusual

* We will discuss only the liquid crystals that are formed by heat (*thermotropic liquid crystals*) and not those formed by solvent action (*lyotropic liquid crystals*). For more extensive discussions, see the references given at the end of this section.

arrangement of their molecules, liquid crystals have some very interesting optical properties and some find application in watch displays, digital thermometers, and so forth.

Types of Liquid Crystals

Thermally produced liquid crystals are classified according to the molecular associations found in the mesophase. The types of molecules that can form liquid crystals are generally large organic molecules that are relatively flat and elongated, shaped somewhat like surfboards, or flattened rods. Many of these compounds are polar. In the liquid crystal mesophase, these rod-like molecules are aligned parallel to one another, much like a bundle of pencils. Further interactions or lack of them determine the classification of a particular liquid crystal. The three main classes of liquid crystals are:

1) *Nematic liquid crystals,* in which the parallel rod-like molecules move relatively freely, much like parallel pencils rolling around one another in a pencil box. (The word *nematic* is taken from the Greek word for "thread.")

2) *Smectic liquid crystals,* in which the parallel molecules are found in layers, or planes, with the molecules perpendicular to the planes so that each plane resembles a field of cornstalks. Each molecule can move freely within its plane, but not from plane to plane. (*Smectic* is from a Greek word meaning "grease" or "slime.")

3) *Cholesteric liquid crystals,* in which the parallel molecules are also in layers, but with the layers slightly displaced from one another. Because of the displacement, the layers form a helix, so that the liquid crystal resembles a spiral staircase. The displacement of the layers arises from side chains that "stick out" from the relatively flat molecules. (*Cholesteric* is derived from the chemical name "cholesterol" because this type of liquid crystal is commonly formed by derivatives of cholesterol.)

Of these three classes, the cholesteric liquid crystals (one of which you will prepare in this experiment) exhibit the most intriguing optical properties. One unique optical property of cholesteric liquid crystals is **circular dichroism,** which we will describe only briefly. When white light (which contains all visible wavelengths) strikes the surface of a cholesteric liquid crystal, the light is separated into two components. One of these components (colored, because it arises from only part of the original white light) is reflected from the surface. The other component (of another color) is transmitted through the liquid crystal (see Figure 2.5). The result is a vivid display of iridescence.

The circular dichroism exhibited by cholesteric liquid crystals is *temperature-dependent.* The temperature-dependency arises from the helical arrangement of the layers of rod-like molecules in the mesophase. A small change in temperature results in a small change in the pitch of the helix (distance required for one complete revolution of the helix). This change in pitch, in turn, alters the colors of light that are reflected and transmitted. The result is that the colors change with changing temperature, a property that is useful in liquid crystal thermometers and other novelty items.

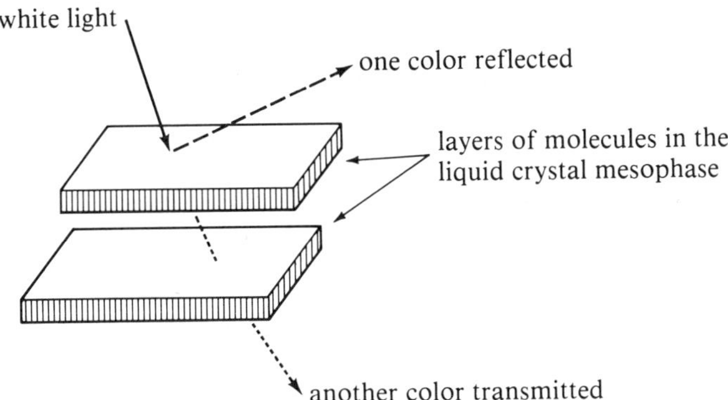

Figure 2.5 The dichroism characteristic of a cholesteric liquid crystal. (The helical structure of the liquid crystal is not depicted here.)

In this experiment you will prepare a simple thermal liquid crystal display; a pair of cholesteric liquid crystals are used so that the temperature changes between 0°C and body temperature can drive the display.

Suggested Readings

Brown, G. H. "Liquid Crystals." *Chemistry* **1967**, *40* (9), 10.

Brown, G. H. "Liquid Crystals and Some of Their Applications in Chemistry." *Anal. Chem.* **1969**, *41* (13), 26A.

Goodman, L. A. "Liquid Crystal Displays." *J. Vac. Sci. Tech.* **1973**, *10* (5), 804.

Verbit, L. "Liquid Crystals—Synthesis and Properties," *J. Chem. Ed.* **1972**, *49*, 36.

Experimental

EQUIPMENT:

hot plate
ice bath
laboratory tissues
melting-point apparatus
two microscope slides
spatula

CHEMICALS:

cholesteryl oleyl carbonate, 0.2 g (see Experimental Note)
cholesteryl pelargonate, 0.2 g (see Experimental Note)

TIME REQUIRED: 15 minutes for each part

PROCEDURE

A. Melting Point of Cholesteryl Pelargonate

Determine the melting point of cholesteryl pelargonate (approximate melting point of 75°). Note any unusual characteristics as the compound melts. Contrast the melting-point behavior of this compound with that of benzoic acid, a compound that does not form a mesophase upon melting.

B. Preparation of Thermal Liquid Crystal Display

On a clean microscope slide, place approximately 0.2 g cholesteryl pelargonate and approximately 0.2 g cholesteryl oleyl carbonate. The melting point of the oleyl carbonate is near room temperature; therefore, this compound will be either liquid or semisolid. Mix the two cholesteryl esters thoroughly on the microscope slide with a spatula. Carefully warm the microscope slide on a hot plate so that the mixture forms a thin film when sandwiched between two slides. Do not overheat the slide or allow the compounds to begin to smoke. Place a second slide on top of the first, sandwiching the liquid mixture, and press firmly. The mixture will spread out into a thin film, and the excess will ooze out the edges of the slides. Clean the edges with a laboratory tissue.

To observe the iridescent display, place the slides onto ice, then warm them slowly with your fingers. The color changes will be most evident if the slides are allowed to warm on a dark surface, such as a black laboratory bench. At the proper temperature (variable, depending on the per cent composition of the mixture), the colors will remain visible almost indefinitely.

EXPERIMENTAL NOTE

Both compounds are available from Aldrich Chemical Co., Inc. (The pelargonate catalog number is C7, 880–1 and the oleyl carbonate number is 15, 115–7.) The structures of these two compounds follow. (Each "angle," or "corner," of the polygons represents a carbon atom with the appropriate number of hydrogen atoms.)

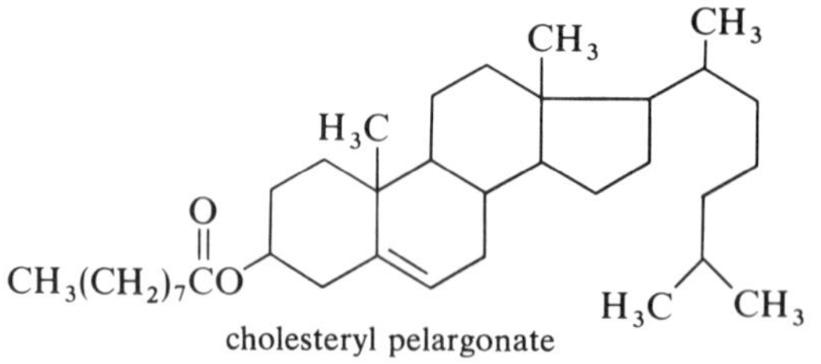

CH_3 CH_3 H_3C O $CH_3(CH_2)_7\overset{\|}{C}O$ H_3C CH_3

cholesteryl pelargonate

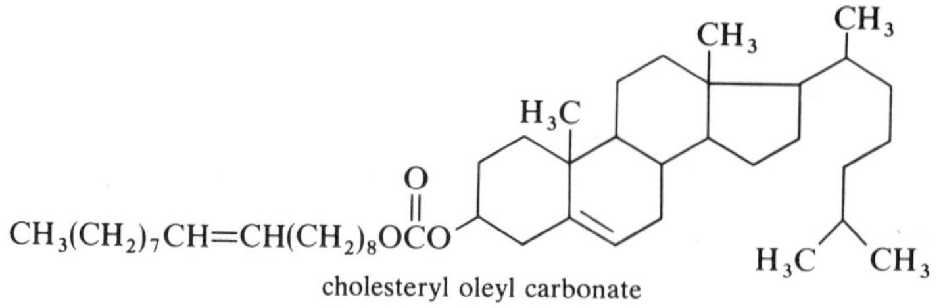

CH_3 CH_3 H_3C O $CH_3(CH_2)_7CH{=}CH(CH_2)_8O\overset{\|}{C}O$ H_3C CH_3

cholesteryl oleyl carbonate

Problems

2.13 Suggest a reason why some liquid crystal displays deteriorate if they are exposed to moist air.

2.14 Although pure soaps themselves are crystalline solids that do not form liquid crystals, a 50:50 mixture of some soaps with water shows liquid crystal properties. Suggest a reason for this behavior.

2.15 The colors reflected and transmitted by a cholesteric liquid crystal are partially dependent on the angle at which light strikes the liquid crystal. Explain.

2.16 Certain liquid crystals change from opaque to transparent when a small electrical potential is applied. Suggest a reason for this behavior.

2.17 If your liquid crystal mixture in Experiment 2.3 is iridescent at about 15°, explain (qualitatively) how you would modify the procedure to raise the temperature at which you would observe iridescence to about 25°.

»»» CHAPTER 3 «««

Crystallization

When a solid organic compound is prepared in the laboratory or isolated from some natural source, such as leaves, it is almost always impure. A simple technique for the purification of such a solid compound is **crystallization**. The compound is first dissolved in a minimum amount of hot solvent. If insoluble impurities are present, the hot solution is filtered. If the solution is contaminated with colored impurities, it may be treated with decolorizing charcoal and filtered. The hot, saturated solution is finally allowed to cool slowly so that the desired compound crystallizes at a moderate rate. When the crystals are fully formed, they are isolated from the **mother liquor** (the solution) by filtration.

Crystallization is not the same as precipitation. Precipitation is the *rapid* formation of solid material, while crystallization is the *slow* formation of a crystalline solid. If a hot saturated solution is cooled too quickly, then the compound may precipitate instead of crystallizing. A precipitated solute may contain many impurities trapped in the rapidly formed crystal structure. On the other hand, when a solution is allowed to crystallize slowly, impurities tend to be excluded from the crystal structure because the molecules in the crystal lattice are in equilibrium with the molecules in solution. Molecules that are unsuitable for the crystal lattice are likely to return to solution, and only the most suitable molecules are retained in the crystal structure. Because impurities are usually present in low concentration, they remain in solution even when the solution cools.

If an extremely pure compound is desired, the filtered crystals may be subjected to **recrystallization**. Of course, each crystallization results in some loss of the desired compound, which remains dissolved in the mother liquor along with the impurities.

3.1 Solvents for Crystallization

The ideal solvent for the crystallization of a particular compound is one that: (1) does not react with the compound; (2) boils at a temperature that is below the compound's melting point; (3) dissolves a moderately large amount of the compound when hot; (4) dissolves only a small amount of compound when cool; (5) is moderately volatile so that the final crystals can be dried readily; and (6) is nontoxic, nonflammable, and inexpensive. In addition, impurities should be either highly insoluble in the solvent (so that they can be filtered from the hot solution) or else highly soluble, so that they remain in solution during the crystallization. As you might guess, a solvent possessing *all* these attributes does not exist.

The primary consideration in choosing a solvent for crystallizing a compound is that the compound be moderately soluble. Unfortunately, the solubility of a compound in a solvent cannot be predicted with accuracy. Most commonly, the solubility of a specific compound in various solvents is determined by trial and error. General guidelines for predicting solubilities based upon the structures of organic compounds do exist. For example, an *alcohol*, a compound containing the hydroxyl (—OH) group as its functional group, may be soluble in water because it can form hydrogen bonds with water molecules. If an alcohol's molecules are largely hydrocarbon, the alcohol may be insoluble in water, but will probably be soluble in other alcohols, such as ethanol (CH_3CH_2OH). *Carboxylic acids* (compounds containing $—CO_2H$ groups) and *amines* (compounds containing NH_2, $>$NH, or $\geqslant$N groups) also can form hydrogen bonds and are also generally soluble in polar solvents such as alcohols.

Compounds that are largely hydrocarbon in structure are not soluble in polar solvents because C—C and C—H bonds are not polar. For these compounds, we would choose a nonpolar solvent—for example, low-boiling petroleum ether, which is a mixture of alkanes such as pentane, $CH_3(CH_2)_3CH_3$, and hexane, $CH_3(CH_2)_4CH_3$. Thus, in choosing crystallization solvents, chemists generally follow the rule of thumb: like dissolves like.

If the best solvent for crystallizing a compound is not known, small portions of the compound can be tested with a variety of likely solvents. The test samples can be recovered by boiling away the solvents. For well-known compounds, however, the best crystallization solvent has already been determined. This information can be found in textbooks, handbooks, and chemical journals. In this book, the crystallization solvent to be used is often specified. Table 3.1 lists some common crystallization solvents.

A. Solvent Pairs

Ideally, a compound to be crystallized should be soluble in the hot solvent, but insoluble in the cold solvent. When a proper solvent cannot be found, a chemist may use a **solvent pair**. A solvent pair is simply two miscible liquids, chosen so that one liquid dissolves the compound readily and the other does not.

For example, many polar organic compounds are highly soluble in ethanol, but insoluble in water. Such a compound is dissolved in a moderate amount of hot ethanol; then water is added dropwise until the solution becomes turbid, or cloudy. A few drops of ethanol are then added to redissolve the precipitating

Table 3.1 Some common crystallization solvents

Name	Structure	Bp (°C)	Miscibility with water	Comments
water	H_2O	100	—	—
methanol	CH_3OH	65	yes	flammable, toxic
ethanol (95%)	CH_3CH_2OH	78	yes	flammable
acetone (propanone)	$(CH_3)_2C{=}O$	56	yes	flammable
ethyl acetate	$CH_3CO_2CH_2CH_3$	77	no	flammable
carbon tetrachloride (tetrachloro-methane)	CCl_4	77	no	toxic
chloroform (trichloromethane)	$CHCl_3$	61	no	toxic
methylene chloride (dichloromethane)	CH_2Cl_2	41	no	—
diethyl ether (ethyl ether, ether)	$(CH_3CH_2)_2O$	35	no	highly flammable
benzene	(benzene ring)	80	no	flammable, toxic, carcinogenic, freezes at 5°
cyclohexane	(cyclohexane ring)	81	no	flammable, freezes at 6.5°
petroleum ether[a]	C_nH_{2n+2}	30–60	no	flammable
ligroin[b]	C_nH_{2n+2}	—	no	flammable

[a] A mixture of alkanes and not a true ether.

[b] A mixture of alkanes boiling at 60–90°, 90–150°, or other specified temperature range. Ligroin is occasionally referred to as "high-boiling petroleum ether."

compound. The resulting ethanol–water solution is a saturated solution and may be allowed to cool slowly for crystallization to occur. Table 3.2 lists some common solvent pairs.

3.2 Safety Using Solvents

Most organic solvents are flammable, toxic, or both. In Table 3.1, we have indicated the most highly flammable and the most highly toxic ones. Because crystallization involves heating solvents to their boiling points, caution must always

Table 3.2 Some common solvent pairs for crystallization

methanol–water	diethyl ether–methanol
ethanol–water	diethyl ether–acetone
acetone–water	diethyl ether–petroleum ether
benzene–ligroin	methanol–dichloromethane

be exercised. Solvent fumes should not be sniffed. Toxic solvents should be boiled only in a hood.

A Bunsen burner should be used only for aqueous solutions—and only when no flammable solvents are being used in the vicinity. Some hot plate surfaces are also capable of igniting flammable solvents. The safest heat source for heating a solvent, especially good for low-boiling solvents, is the steam bath. Section 1.3C describes its use.

Boiling chips, or **boiling stones**, should be used when a solvent is brought to a boil, unless the liquid can be constantly stirred or swirled. Boiling chips are small porous stones of calcium carbonate or silicon carbide that contain trapped air. When the chips are heated in a solvent, they release tiny air bubbles, which ensure even boiling. Without boiling chips, part of the solvent may become superheated and boil in spurts, a process called **bumping**. (Bumping is likely to result in part of the solution splashing out of the flask.)

In most cases, two or three boiling chips are all that are necessary to prevent bumping. *Boiling chips should never be added to a hot solution*! If a solution is at or near its boiling point when boiling chips are added, it is almost a certainty that the solution will boil out of the flask.

Boiling chips will maintain their function throughout a long boiling period, but, once they are used and cooled, the pores fill with liquid and lose their ability to release air bubbles. Therefore, fresh boiling chips should be used each time a solution is heated.

Problems

3.1 The proper procedure in crystallization is to allow the hot solution to cool to room temperature, then to chill the solution in an ice bath. Why do we not simply chill the hot solution in an ice bath initially?

3.2 Each of the following compounds, A–C, is equally soluble in the three solvents listed. In each case, which solvent would you choose? Give reasons for your answer. (More than one answer may be correct.)

(a) Compound A: benzene, acetone, or chloroform

(b) Compound B: carbon tetrachloride, dichloromethane, or ethyl acetate

(c) Compound C: methanol, ethanol, or water

3.3 Which of the following solvents could *not* be used as solvent pairs for crystallization?

(a) ligroin and water

(b) chloroform and diethyl ether

(c) acetone and methanol

3.4 From the following lists of organic compounds (on the left) and solvents (on the right), choose the solvent in which each solid is likely to be *most soluble* (not necessarily the best crystallization solvent).

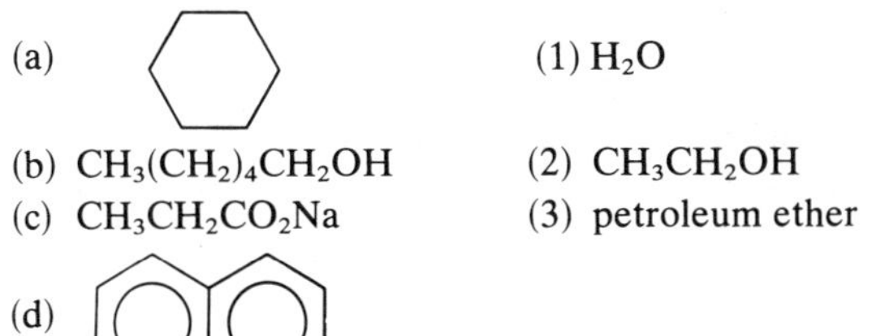

3.5 Using Tables 3.1 and 3.2 (page 38), suggest possible crystallization solvents for the following compounds:

(a) naphthalene, [structure] (mp 80°)

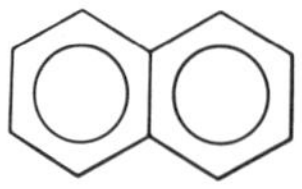

(b) succinic acid, $HOC(=O)(CH_2)_2C(=O)OH$ (mp 188°)

(c) *p*-iodophenol, 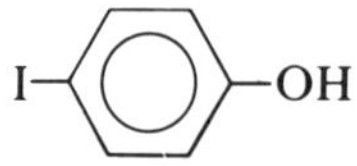 I–C_6H_4–OH (mp 94°)

(d) *p*-nitroaniline, O_2N–C_6H_4–NH_2 (mp 148°)

3.6 Draw structures that illustrate the two ways hydrogen bonding can occur between (a) water and methanol, CH_3OH; (b) water and diethylamine, $(CH_3CH_2)_2NH$; (c) water and acetic acid, CH_3CO_2H. (There are more than two correct answers in c.)

⟫⟫ EXPERIMENT 3.1 ⟪⟪

Crystallization of Benzoic Acid and an Unknown

Discussion

The basic steps in the crystallization of an organic solid were summarized in the brief description at the start of this chapter. In this section, we will discuss each of the steps in more detail, along with the techniques that are used to carry out each procedure.

Steps in Crystallization

1) Dissolving the compound. The first step in crystallization is dissolving the compound in a *minimum amount* of the appropriate solvent in an Erlenmeyer flask. An Erlenmeyer flask is used instead of a beaker or other container because the solution is less likely to slosh out and dust is less likely to get in; also, an Erlenmeyer flask can be corked and stored in your locker.

Pulverize a lumpy solid with a spatula to speed its dissolving. To ensure that a minimum amount of solvent is used, add the solvent in small portions, and heat the mixture with constant stirring or swirling. As soon as the compound dissolves, examine the solution for insoluble impurities. If these are discovered, filter the hot solution before the compound begins to crystallize.

This filtration is not necessary and is, in fact, undesirable if the solution looks clear and clean. If the solution appears to be contaminated with colored impurities, decolorizing charcoal may be added at this time. The use of decolorizing charcoal is discussed on page 50.

2) *Filtering insoluble impurities.* Filtering a hot, saturated solution inevitably results in cooling and in evaporation of some of the solvent. Therefore, a premature crystallization of the compound on the filter paper and in the funnel may be observed. A few precautions can minimize this premature crystallization.

Before filtering, add a little extra solvent (about 5–10% of the total volume) to the hot solution, and keep the solution hot while the filtration apparatus is being prepared. Preheat the funnel by resting it on the neck of the flask containing the hot solution. A stemless or short-stemmed funnel is convenient for filtering the hot solution because premature crystals may clog the stem of a long-stemmed funnel.

Filter the hot solution through either a plug of glass wool or filter paper. **Fluted filter paper** is preferred to folded filter paper because the increased surface area of the fluted paper allows the filtration to proceed more rapidly. Figure 3.1 shows how to prepare a piece of fluted filter paper. Filter paper is rated by its porosity. For hot filtration, a porous paper, such as Whatman's No. 1 or No. 4, should be used. (Do not use Whatman's No. 5 or No. 6 papers, which have slow filter speeds.)

Heat the empty receiving flask (another Erlenmeyer flask), which will be the actual flask used for the crystallization, on a steam bath or with hot water, and then place the funnel and fluted filter paper in the neck. Figure 3.2 shows two ways in which the funnel can be supported slightly away from the lip of the filtration flask to create an air space, which prevents a liquid "seal" from blocking the flow of air and solvent fumes.

During the filtration, keep both flasks hot on a steam bath or hot plate. To keep the solution hot, pour only small amounts into the filter paper (instead of filling the filter paper to the brim). If a flammable solvent and a hot plate are used, move the flasks away from the hot plate when pouring so that solvent vapors do not flow over the heating element. Wrap the flask containing the hot solution in a towel or hold it in a clamp while pouring, and avoid the hot fumes that flow out of the flask.

If crystallization occurs in the funnel, the crystals can often be removed by heating the receiving flask to boiling with the funnel still on it; solvent condensing in the funnel may dissolve the crystals and carry them back to the filtered solution. Alternatively, the solid can be washed into the flask with a little hot solvent.

After all of the hot solution has been filtered, wash the original flask with a small amount of hot solvent and pour this solvent through the filter paper into the receiving flask to transfer the final traces of the desired compound. Two washings may be necessary; however, use a minimum amount of solvent.

The crystallization flask should contain a hot, clear, saturated solution of the compound. Boil away excess solvent at this time. (Remember to use the hood for a toxic solvent and stay away from flames with a flammable solvent.)

Step 1. Fold the paper in half, then in quarters, creasing the folds as you proceed. However, do not crease the very center of the paper (the point), which might become weakened.

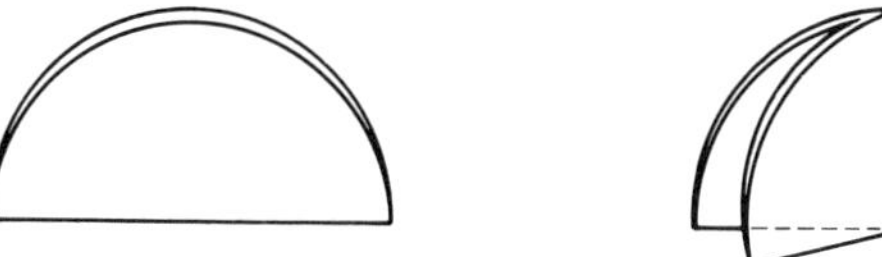

Step 2. Open the quarters to a half-sized piece, and then fold the edges in to the centerfold.

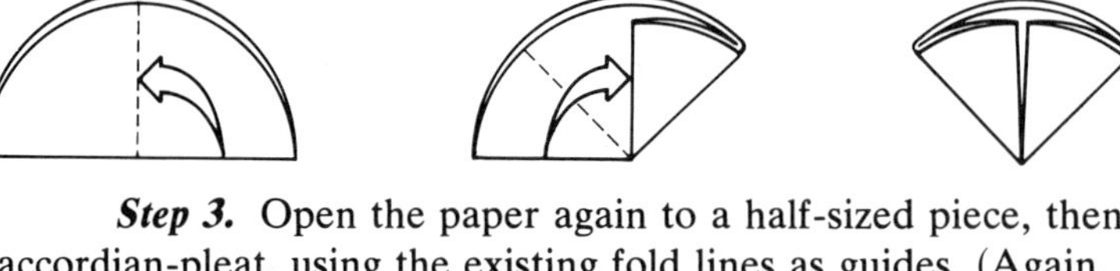

Step 3. Open the paper again to a half-sized piece, then accordian-pleat, using the existing fold lines as guides. (Again, do not crease the center of the paper.)

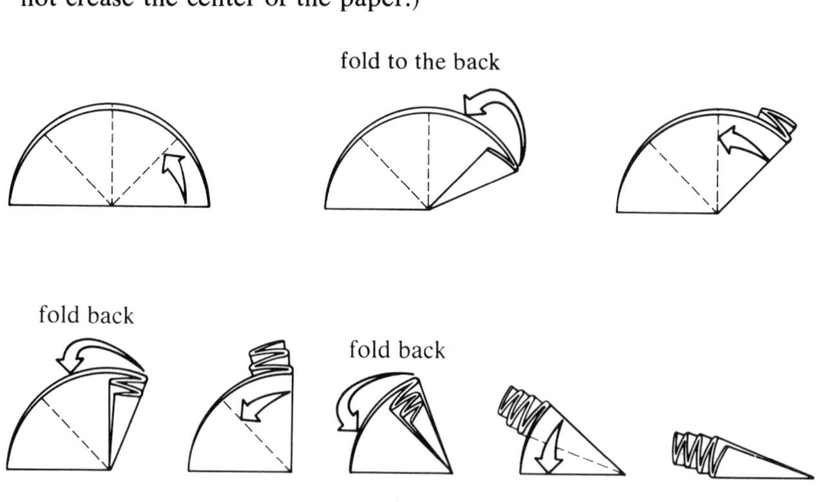

Step 4. Crease the folds (except at the point), then open the filter paper and place it in an appropriately sized funnel.

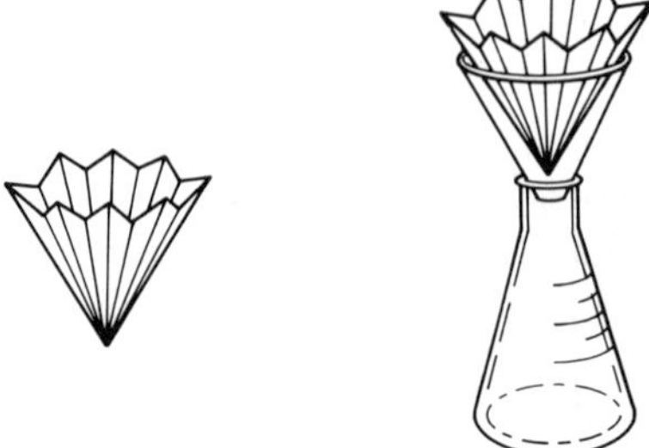

Figure 3.1 *How to prepare fluted filter paper.*

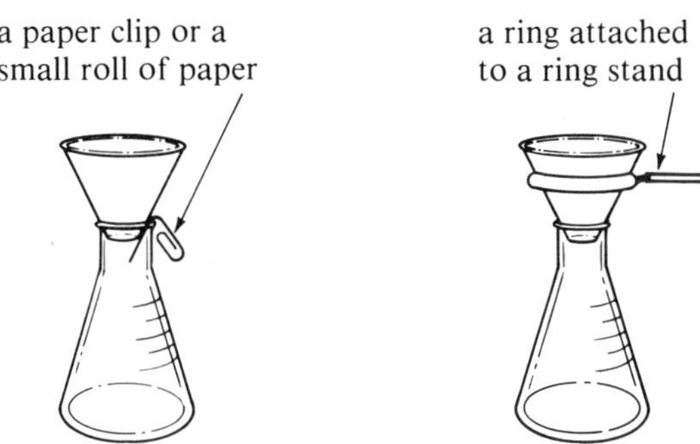

Figure 3.2 *Ways to create an air space between a funnel and an Erlenmeyer flask.*

If the hot solution starts to crystallize, reheat it to dissolve the crystals. If so much solvent has evaporated that the crystals will not dissolve, add a small additional amount of solvent to the flask.

3) *Crystallizing the compound.* Cover the flask containing the hot, saturated solution with a watch glass or inverted beaker to prevent solvent evaporation and dust contamination. Then, set the flask aside where it can remain undisturbed (no jostling or bumping, which will induce precipitation rather than crystallization) for an hour or several hours. If the flask must sit for several days, allow it to cool to room temperature, then stopper it with a cork (not a rubber stopper if an organic solvent was used) to prevent solvent evaporation.

Sometimes a hot solution cools to room temperature with no crystallization occurring. In such a case, your first question should be, "Is the solution *supersaturated*?" Often, crystallization can be induced in a supersaturated solution by scratching the inside of the flask up and down at the surface of the solution with a glass rod. The scratching of the glass on glass is thought to release microcrystals of glass, which serve as a template for crystal growth. If scratching the flask does not start the crystallization, a **seed crystal** may be added. This is a small crystal of the original material set aside for just this purpose; it provides a nucleus, or template, upon which other crystals can grow. Sometimes a seed crystal can be obtained from the glass rod used for scratching, after the solvent has evaporated from it. Allowing a few drops of solution to evaporate on a watch glass may also produce seed crystals. After addition of the seed crystal, set the flask aside for crystallization to proceed.

If scratching and seeding do not produce crystals, your next question should be, "Did I use too much solvent?" If more than the minimum amount of solvent was used in the earlier steps, then the excess must be boiled away and the flask set aside again to crystallize. Unless the solid begins to separate, reducing the volume of the solution by one third may result in the desired saturated solution.

Another problem encountered in crystallization is **oiling out**: instead of crystals appearing, an oily liquid separates from solution. A compound may oil out if its melting point is lower than the boiling point of the solvent. A very impure compound may oil out because the impurities depress its melting point.

The formation of an oil is not selective, as is crystallization; therefore, the oil (even if it solidifies) is probably not a pure compound. Reheat a mixture that has oiled out to dissolve the oil (add more solvent if necessary) and then allow the solution to cool slowly, perhaps adding a seed crystal or scratching with a glass rod.

If these techniques do not prevent oiling out, allow the oil to solidify (a seed crystal may be necessary), filter the solid or decant (pour) the solvent away, and attempt to crystallize the solid with fresh solvent. Enough impurities may have been removed in the first attempted crystallization that the second one will proceed smoothly. If the substance has a low melting point, a lower-boiling solvent may be necessary. Alternatively, keeping the temperature of the solvent below the melting point of the solute may help prevent oiling out.

4) *Isolating the crystals.* Crystals are separated from their mother liquor by filtration. Chilling the mixture in an ice-water bath before filtration will increase the yield of crystals. The standard filtration process used is called **vacuum** or **suction filtration.**

Vacuum filtration apparatus. Vacuum filtration has the advantage of being much faster than gravity filtration (simple filtration through filter paper and funnel). It has the disadvantage of requiring more equipment. Figure 3.3 shows the physical setup required for vacuum filtration. The trap is necessary regardless of whether a water aspirator or a centralized vacuum system is used.

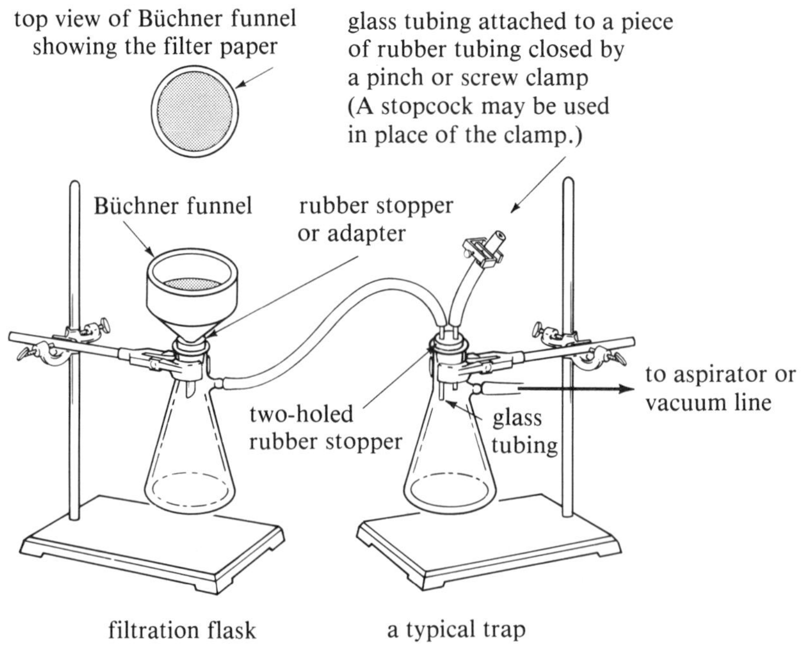

Figure 3.3 *A vacuum filtration apparatus.*

The purpose of this trap is two-fold: (a) Any solution accidentally sucked into the vacuum line will be caught in the trap. (b) Water from an aspirator sometimes backs up into the vacuum line. With a trap, this water will be caught before it contaminates the mother liquor in the filtration flask.

Heavy-walled vacuum tubing is used for connections because ordinary tubing collapses when vacuum is applied. All flasks should be clamped to ring stands. The filter flask, especially, should be firmly clamped because it usually becomes top-heavy when a Büchner funnel and vacuum line are connected to it.

Attach the Büchner funnel or Hirsch funnel to the filter flask with a rubber adapter or a one-holed rubber stopper (best) so that the connection will be airtight when vacuum is applied. Place a medium- or slow-speed filter paper (such as Whatman's No. 2, 5, or 6) on the perforated surface of the funnel. (A fast-speed, porous filter paper allows finely divided solids to pass through under vacuum.) The filter paper must lie flat and not curl up at the sides, yet it must cover all the holes. When the vacuum is applied, the filter paper is pulled snugly to the flat surface of the funnel by suction. To ensure no leakage around the edges, moisten the filter paper with the solvent before applying vacuum.

Water aspirator. Many laboratories are equipped with **water aspirators**, devices that attach to faucets and develop a vacuum through a side tube when water flows through the main tube. A large beaker in the sink under the water outlet minimizes splashing. Because aspirators are easily plugged, they should be checked before each use. Turn the water on *full force* and hold your finger on the vacuum hole to feel the suction before attaching your filtration apparatus.

As we have mentioned, an aspirator can "back up." A slight decrease in water pressure can cause the vacuum in the filtration apparatus to pull water into the apparatus. If you see water entering the trap, break the vaccum by opening the stopcock or clamp on the trap to save your mother liquor from the influx of tap water.

The actual filtration. For nonvolatile solvents like water, apply the vacuum and pour the crystallization mixture into the Büchner funnel at such a rate that the bottom of the funnel is always covered with some solution. If the solvent is volatile, keep the vacuum off when pouring the first portion of the mixture into the funnel; then apply the vacuum. When the vacuum is applied, the mother liquor is literally sucked through the filter paper into the filter flask while the crystals remain on the filter paper. When the mother liquor ceases to flow from the funnel stem, release the vacuum by opening the stopcock on the trap, then turn off the aspirator or vacuum line.

Washing. To wash the contaminating mother liquor from the crystals, transfer the crystal mass, or filter cake, to a small beaker, using a spatula to loosen, remove, and scrape the filter paper. Place fresh filter paper in the Büchner funnel, stir the crystals with a small amount of *chilled* solvent, then immediately refilter. Small amounts of crystals may be washed right in the funnel on the original filter paper. This procedure is not recommended because the wet filter paper may tear when you stir the wash solvent into the crystals and because this type of washing is not as thorough as a beaker washing.

Remove excess solvent from the crystals by putting a fresh piece of filter paper *on top of the crystals* still in the funnel and pressing this filter paper down firmly and all over with a cork. Keep the vacuum on during this pressing. When as much solvent has been pressed out of the filter cake as possible, leave the vacuum running for another minute or so. The air pulled through the filter cake will remove even more solvent. Then, open the trap clamp, turn off the vacuum, remove the Büchner funnel, and disconnect the filter-flask assembly from the vacuum line. Using a spatula, pry the filter cake from the funnel for drying.

Do not discard the mother liquor (in the filter flask), but place it in a corked Erlenmeyer flask until the completion of the experiment. The reason for saving the mother liquor is that it may still contain a substantial amount of the desired compound.

5) *Drying the crystals.* The filter cake removed from the Büchner funnel still contains an appreciable amount of solvent. The crystals must be dried thoroughly before they can be weighed or before a melting point can be taken. (Why?)

There are many methods for drying crystals. The simplest is *air-drying*, in which the crystals (with any lumps crushed) are spread out on a watch glass or large piece of filter paper and allowed to dry. Air-drying is sometimes slow, especially if water or some other high-boiling solvent was used. Unless the crystals are partially covered, they can collect dust. Another watch glass or a beaker, propped on corks to allow air to get to the crystals, may be used as a cover. For a melting-point determination, a few crystals may be removed from the mass and allowed to air-dry on a separate uncovered watch glass. If a compound is hygroscopic (attracts water from the air), it cannot be air-dried.

A **dessicator** may be used for drying a water-crystallized or hygroscopic compound. A dessicant (drying agent) such as anhydrous calcium chloride is placed in the bottom of the dessicator; the porcelain shelf is inserted; then a watch glass or beaker holding the crystals is placed on the shelf. When the cover is in place, the dessicant attracts water from the atmosphere in the dessicator as the water evaporates from the crystals. Some dessicators can be evacuated, thus speeding up the evaporation of any solvent.

6) *The second crop.* The mother liquor from a crystallization may still contain in solution a large amount of the desired compound. In many cases, another batch of crystals, called the **second crop**, can be obtained from this mother liquor. This can be accomplished by boiling away some of the solvent from the mother liquor and allowing it to cool and crystallize as for the first crop of crystals. A seed crystal added to the cooled solution may help start the crystallization process. The second crop of crystals is rarely as pure as the first crop because the impurities have been concentrated in the mother liquor. The purity of the second crop can be improved by recrystallization of these impure crystals with fresh solvent.

Additional crops can sometimes be harvested from each mother liquor by repeating the process described above. Each successive crop will be less pure than the preceding crop. For this reason, only one or two crops are usually taken.

Experimental

EQUIPMENT:

250-mL (or larger) beaker
two 250-mL Erlenmeyer flasks
filter paper (Whatman's No. 1 or 4 and No. 2, 5, or 6)
hot plate and water bath
ice bath
melting-point apparatus
short-stemmed funnel
spatula
6–8 test tubes
vacuum filtration assembly

CHEMICALS:

impure benzoic acid, 5 g
individual unknown,* 2.0 g
solvents: 95% ethanol, acetone, dichloromethane (methylene chloride), petroleum ether (30–60°), ligroin (60–90°)

TIME REQUIRED: 2 hours for Part A, $1\frac{1}{2}$ hours for Part B, plus overnight drying of the crystals and melting-point determinations. (The time required will be shorter if a burner is used in Part A instead of a hot plate.)

STOPPING POINTS: while any solution is crystallizing or while the crystals are drying

>>>> ***SAFETY NOTE*** Part A of this experiment includes the heating of water, while Part B includes the use of flammable solvents. It is recommended that a *hot plate* be used for the crystallizations from water. If a burner is used, it is imperative that you and your neighbors coordinate your activities so that burners and flammable solvents are *not* used at the same time.

PROCEDURE

A. Crystallization of Benzoic Acid from Water

Weigh approximately 5.0 g of crude benzoic acid into a 250-mL Erlenmeyer flask. Record the weight of the benzoic acid to the nearest 0.1 g.

Heat 150–200 mL of distilled water to boiling in a beaker on a hot plate. Because you will later need a warmed flask and funnel, add 10 mL of distilled water to another 250-mL flask, place a short-stemmed funnel in the neck of the flask, and heat to a gentle boil (adding more water, if necessary).

Flute a piece of porous filter paper as shown in Figure 3.1, page 42. Add 100 mL of distilled water and 2–3 boiling chips to the benzoic acid, and

* Suggestions for unknowns are listed in the instructor's guide.

bring the mixture to a boil. Since the benzoic acid used in this experiment contains insoluble impurities, most, but not all, of the solid should go into solution. If necessary, add more water (up to 25 mL) to dissolve the benzoic acid. Filter the hot solution through the fluted filter paper into the heated flask (emptied of the hot water).

After the filtration is complete, wash the original flask with 5 mL of boiling water and pour this wash water through the filter paper. Then wash the filter paper with an additional 5 mL of boiling water.

To remove excess water from the filtrate, add 2–3 fresh boiling chips, and boil the solution until its volume is about 100 mL. Then, cover the flask with an inverted beaker and place the flask on the bench top or in your locker to cool slowly to room temperature.

While the solution is cooling, equip a filter flask with a Büchner funnel and a water trap (see Figure 3.3, page 44). Then, vacuum-filter the crystals, wash the filter cake with ice-cold water as described in the discussion, and refilter. Transfer the filter paper with the crystals onto a piece of fresh filter paper and allow the crystals to air-dry overnight or longer.

Weigh the dry crystals and calculate the per cent recovery. (In your calculation, ignore the presence of impurities; assume that all the material placed in the Erlenmeyer flask was benzoic acid.) A 65% recovery is typical.

Determine the melting point of the crystallized benzoic acid. Enter all pertinent data in your notebook, place the benzoic acid in a properly labeled vial, and turn in the vial to your instructor.

If desired, a second crop (about 5% recovery) of benzoic acid can be obtained from the water filtrate. To obtain a second crop, transfer the mother liquor to an Erlenmeyer flask, boil on a hot plate to reduce the volume by about one third, cool, filter the crystals with vacuum, and air-dry them. Do not combine the second crop with the first crop (why not?), but determine its weight and melting point, then hand it in to your instructor in a vial labeled as the "second crop." Report the per cent recovery as two values: per cent recovery for the first crop and total per cent recovery (first and second crops).

B. Crystallization of an Unknown

Obtain an unknown (CAUTION: possibly toxic) from your instructor, and record its number in your notebook. Use the following procedure to choose a suitable crystallization solvent.

Weigh exactly 0.10 g of the unknown into a small test tube. Pipet exactly 1.0 mL of water into another test tube. Use these two test tubes as references for estimating 0.10 g of unknown and 1.0 mL of solvent for the other test tubes.

Place an estimated 0.10 g of the unknown in each of six test tubes, and add 1.0 mL of solvent to each. The following solvents should be tried: water, 95% ethanol, acetone, dichloromethane, petroleum ether (30–60°), and ligroin (60–90°). The boiling points of these solvents are listed in Table 3.1, page 38. (CAUTION: Most of these solvents are flammable; all burners should be extinguished!) Stir each sample, and determine the solubility of your unknown in each solvent at room temperature. Place the test tubes in a beaker of boiling water on a hot plate, and again note the solubilities. (Some of the unknowns may contain insoluble impurities even though the unknown compound itself dissolves.)

Return the test tubes to a rack and allow them to cool to room temperature. Finally, place the test tubes in an ice bath. Record your observations.

From the preceding tests, choose the best solvent (or solvent pair) for the crystallization of the unknown, then weigh and crystallize the remainder of your unknown sample. Air-dry the crystals overnight, then determine the per cent recovery and melting point. Hand in the dried crystals in a labeled vial to your instructor.

Problems

3.7 Give reasons for each of the following experimental techniques used in crystallization:

(a) A hot crystallization solution is not filtered unless absolutely necessary.

(b) An Erlenmeyer flask containing a hot solution is not tightly stoppered to prevent solvent loss during cooling.

(c) The suction of a vacuum filtration apparatus is broken before the vacuum is turned off.

(d) Vacuum filtration is avoided when crystals are isolated from a very volatile solvent.

(e) Carborundum (silicon carbide) boiling chips are better than calcium carbonate chips in the crystallization of an unknown.

3.8 A procedure suggests using 25 mL of solvent for crystallizing 5.0 g of a compound. What size flask would you choose to crystallize 30 g?

3.9 How would you recover the solid unknown that was used to determine the correct solvent for crystallization?

3.10 A chemist crystallizes 17.5 g of a solid and isolates 10.2 g as the first crop and 3.2 g as the second crop.

(a) What is the per cent recovery in the first crop?

(b) What is the total per cent recovery?

3.11 A student carried out a crystallization from water and obtained a 120% recovery. How could this happen?

3.12 A student crystallized a compound from benzene and observed only a few crystals when the solution cooled to room temperature. To increase the yield of crystals, the student chilled the mixture in an ice-water bath. The chilling greatly increased the quantity of solid material in the flask. Yet, when the student filtered these crystals with vacuum, only a few crystals remained on the filter paper. Explain this student's observations.

3.13 At 0°, the solubility of benzoic acid is 0.02 g/100 mL of water. Assume that you carry out Experiment 3.1A using 5.00 g of pure benzoic acid, boiling the solution down to exactly 100.0 mL, then chilling the solution to 0°.

(a) What is the theoretical maximum per cent recovery in the first crop?

(b) If you boiled the mother liquor down to 50 mL and collected a second crop, what is the theoretical maximum per cent recovery (total)?

EXPERIMENT 3.2

(SUPPLEMENTAL)

Crystallization of Acetanilide Using Decolorizing Charcoal

Discussion

Frequently, small amounts of colored compounds and tarry (long-chain, or polymeric) compounds are found as colored impurities in colorless organic compounds. These colored impurities can cause the crystallization solution and even the final crystals to have a tinge of color. These impurities can be removed with **decolorizing charcoal** (also called *activated charcoal* or *activated carbon*). The fine particles of carbon in decolorizing charcoal have a large surface area and adsorb organic compounds, especially colored and polymeric compounds.

The decolorizing charcoal is added to the initial crystallization solution after the impure solid has been dissolved. *Never* add decolorizing charcoal to a solution near its boiling point! The fine particles act like thousands of tiny boiling chips and will cause a hot solution to boil over. Therefore, always allow a hot solution to cool for 3–4 minutes before adding the charcoal.

Only a small amount of decolorizing charcoal should be used: a "pinch" (0.1–0.5 g) on the end of a spatula is sufficient for most purposes. It is always better to err by adding too little charcoal than too much, because the particles of carbon can adsorb the desired compound as well as impurities.

After the carbon is added, the mixture should be swirled a few times, then *carefully* reheated. (Boiling solutions containing decolorizing charcoal have a tendency to froth.) The hot mixture is then filtered through fluted filter paper as was described on page 42.

Because decolorizing charcoal is messy, and because its use always results in the loss of some of the compound being crystallized, decolorizing is carried out only when necessary and not as a routine procedure. If the use of decolorizing charcoal is being considered, it is generally advisable to first test an aliquot of the solution.

In this experiment, you will use decolorizing charcoal to decolorize acetanilide contaminated with the dye methylene blue.

$$\bigcirc\!\!-NH\overset{\overset{O}{\|}}{C}CH_3$$

acetanilide

Experimental

EQUIPMENT:

three 250-mL Erlenmeyer flasks
filter paper (Whatman's No. 1 or 4, and No. 2, 5, or 6)
hot plate
ice bath
melting-point apparatus
short-stemmed funnel
spatula
vacuum filtration assembly

CHEMICALS:

acetanilide contaminated with 0.1% methylene blue, 10 g
decolorizing charcoal, 0.1 g (Two 0.1-g quantities may be needed.)

TIME REQUIRED: $2\frac{1}{2}$ hours plus overnight drying and melting-point determination.

STOPPING POINTS: While any solution is crystallizing or while the crystals are drying.

»»> ***SAFETY NOTE*** Do *not* add decolorizing charcoal to a hot solution. Avoid getting decolorizing charcoal on your skin or clothing because it is difficult to wash way.

PROCEDURE

Weigh 5.0 g of contaminated acetanilide (see Experimental Note) in a 250-mL Erlenmeyer flask, add 100 mL of water, add 2–3 boiling chips, and boil the mixture until the acetanilide dissolves. (Be sure that it has dissolved and not simply melted.) Allow the acetanilide to crystallize, chill the flask in an ice bath, and filter the crystals with vacuum.

Weigh another 5.0-g sample of contaminated acetanilide into a clean flask, and dissolve this sample in 100 mL of boiling water as before. Remove the flask from the burner, then add 30 mL of cold water and about 0.1 g of decolorizing charcoal. Bring the mixture to a boil, but be prepared to remove the flask from the hot plate if it froths. Boil the mixture for 1–2 minutes.

Filter the hot mixture by gravity, using an additional 20 mL of hot water to wash the Erlenmeyer flask and the filter paper. The total volume of the filtrate should be around 150 mL. (If the solution is still blue, repeat the treatment with decolorizing charcoal.) Add 2–3 boiling chips and boil the solution down to 100 mL. Allow the flask to stand until crystals have formed, then cool it in an ice bath. Filter the crystals with vacuum, and allow both sets of crystals to air-dry at least overnight. Weigh the samples and determine their melting points simultaneously after looking up the melting point in the *Handbook of Chemistry and Physics*. Compare the appearance and melting points of the two samples, and record your observations and conclusions in your notebook. Turn in both samples (in properly labeled vials) to your instructor.

EXPERIMENTAL NOTE

The success of this experiment depends upon the amount of methylene blue used as an impurity and the activity of the decolorizing charcoal. It is assumed that the storeroom has balanced the two variables.

Problems

3.14 Explain why: (a) excess solvent is used when decolorizing a solution with decolorizing charcoal; (b) a crystallization solution is heated *after* the decolorizing charcoal is added.

3.15 How could you determine the minimum amount of decolorizing charcoal needed to decolorize a crystallization solution?

3.16 The solubility of acetanilide in hot water (5.5 g/100 mL at 100°) is not very great, and its solubility in cold water (0.53 g/100 mL at 0°) is significant. What would be the maximum theoretical per cent recovery (first crop only) from the crystallization of 5.0 g of acetanilide from 100 mL of water (assuming the solution is chilled to 0°)?

CHAPTER 4

Extraction

Extraction is the "pulling out" of a substance from one phase by another phase. The term is usually used to describe removal of a desired compound from a solid or liquid mixture by a solvent. In a coffee pot, caffeine and other compounds are extracted from the ground coffee beans by hot water. (Experiment 4.3 is the similar extraction of caffeine from tea leaves.) Vanilla extract is made by extracting the compound vanillin from vanilla beans. In the laboratory, the extraction of an organic compound from one liquid phase by another liquid, a technique called **liquid–liquid extraction**, or simply "extraction," is a common manipulation.

Extraction is often used as one of the steps in isolating a product of an organic reaction. After an organic reaction has been carried out, the reaction mixture usually consists of the reaction solvent and inorganic compounds, as well as organic products and by-products. In most cases, water is added to the reaction mixture to dissolve the inorganic compounds. The organic compounds are then separated from the aqueous mixture by extraction with an organic solvent that is immiscible with water. The organic compounds dissolve in the extraction solvent while the inorganic impurities remain dissolved in the water.

Various devices for separating the phases in a liquid–liquid extraction have been invented; the most commonly used device is the **separatory funnel**. The aqueous mixture to be extracted is poured into the funnel first, then the appropriate extraction solvent is added. The mixture is shaken (with the proper precautions) to mix the extraction solvent and the aqueous mixture, and then is set aside for a minute or two until the aqueous and organic layers have separated. The stopcock at the bottom of the separatory funnel allows the bottom layer to be drawn off

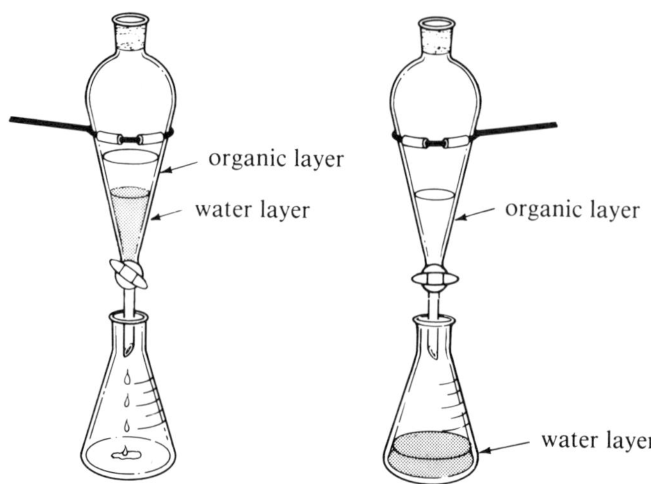

Figure 4.1 Two immiscible solutions can be separated with a separatory funnel. (The organic layer may be the upper or lower layer, depending on the relative densities of the two solutions.)

into a flask and makes possible the separation of the two layers. Figure 4.1 shows how this is accomplished. The result (ideally) is two separate solutions: an organic solution (organic compounds dissolved in the organic extraction solvent), and an inorganic solution (inorganic compounds dissolved in water). Unfortunately, often the water layer still contains some dissolved organic material. For this reason, the water layer is usually extracted one or two more times with fresh solvent to remove more of the organic compound.

After one or more extractions and separations, the combined organic solutions are usually treated with a solid drying agent to remove traces of water, and then they are filtered to remove the hydrated drying agent. Finally, the solvent is evaporated or distilled away. The resulting organic product can then be purified by a technique such as crystallization or distillation.

4.1 Distribution Coefficients

When a compound is shaken in a separatory funnel with two immiscible solvents, such as water and diethyl ether ($CH_3CH_2OCH_2CH_3$), the compound distributes itself between the two solvents. Some dissolves in the water and some in the ether. How much solute dissolves in each phase depends on the solubility of the solute in each solvent. The ratio of the concentrations of the solute in each solvent at a particular temperature is a constant called the **distribution coefficient** or **partition coefficient (*K*)**. (In calculations of distribution coefficients, we assume that the solute neither ionizes in nor reacts with either solvent.) Because a ratio is involved, the concentrations may be in any units, as long as the two concentrations are in the *same* units:

$$K = \frac{\text{concentration in solvent}_2}{\text{concentration in solvent}_1}$$

where solvent_1 and solvent_2 are immiscible liquids.

To a rough approximation, the ratio of concentrations in this equation is the same as the ratio of the *solubilities* of the compound in the two solvents, measured independently. For example, consider a compound, which we will call compound A, that is soluble in diethyl ether to the extent of 20 g/100 mL at 20°C and soluble in water to the extent of 5.0 g/100 mL at the same temperature. We can approximate the distribution coefficient of compound A in diethyl ether and water to be 4.0, where diethyl ether is solvent_2 in the following equation:

$$\begin{aligned} K &= \frac{\text{solubility in solvent}_2}{\text{solubility in solvent}_1} \\ &= \frac{20\ \text{g}/100\ \text{mL}}{5.0\ \text{g}/100\ \text{mL}} \\ &= 4.0 \end{aligned}$$

Given the distribution coefficient for a particular system, we can calculate approximately how much compound can be extracted. Suppose that we have a solution containing 5.0 g of compound A in 100 mL of water. If we shake this solution with 100 mL of diethyl ether, how much A will be extracted by the ether?

To solve this problem, we will let x be equal to the number of grams of compound A in the diethyl ether. Since we started with 5.0 g of A in the water, the number of grams remaining in the water is $5.0 - x$. The respective concentrations are the number of grams in each layer divided by each volume.

$$\begin{aligned} \text{concentration of A in diethyl ether} &= \frac{x\ \text{g}}{100\ \text{mL}} \\ \text{concentration of A in water} &= \frac{(5.0 - x)\ \text{g}}{100\ \text{mL}} \end{aligned}$$

Substituting,

$$\begin{aligned} K &= \frac{\text{concentration in diethyl ether}}{\text{concentration in water}} \\ 4.0 &= \frac{x\ \text{g}/100\ \text{mL}}{(5.0 - x)\ \text{g}/100\ \text{mL}} \end{aligned}$$

Solving,

$$\begin{aligned} 4.0 &= \frac{x}{5.0 - x} \\ 20 - 4.0\,x &= x \\ 5.0\,x &= 20 \\ x &= 4.0\ \text{g in 100 mL of diethyl ether} \end{aligned}$$

From this calculation we see that we will extract only 80% of A from the water solution. Therefore, 20%, or 1.0 g, of A will remain in the water layer.

Because a solute distributes itself between two solvents, a single extraction may not be very efficient. Considerable amounts of material may remain behind in the original solvent. Is it possible to make liquid–liquid extraction more efficient? Using the same amount of diethyl ether, can we extract more than 4.0 g of A from the original water solution? Yes, we can. Let us divide the 100 mL of ether into three portions of approximately 33 mL each. Then, let us extract the water solution three separate times, using 33 mL of fresh diethyl ether for each extraction. If you carry out the calculation for each extraction as before (allowing for the different volumes of solvent), you will find that a total of 4.5 g of A can be extracted.

1st extraction:	2.8 g
2nd extraction:	1.2 g
3rd extraction:	0.5 g
Total:	4.5 g

Our conclusion is that it is a more efficient use of solvent to perform three small extractions than one large one. A greater number of small extractions would remove an even greater quantity of the solute from water. If compound A were valuable, we would perform the extra extractions; otherwise, we would not bother.

Most organic compounds have distribution coefficients between organic solvents and water greater than 4. Therefore, a double or triple extraction generally removes almost all of the organic compound from the water. However, for water-soluble compounds, where K may be less than 1, we can calculate that little of such a compound will be extracted.*

The distribution coefficient of an organic compound between an organic solvent and water can be changed by adding sodium chloride to the water. (Other inorganic salts would have the same effect as sodium chloride, but the latter is the least expensive salt available.) Organic compounds are less soluble in salt water than in plain water. Sometimes, the solubility difference is dramatic. Therefore, by simply dissolving sodium chloride in the water layer, we can increase the distribution of an organic compound in the organic solvent. This effect is commonly referred to as **salting out** the organic compound.

4.2 Extraction Solvents

The preceding discussion has provided some clues for the choice of an extraction solvent. The extraction solvent must be immiscible with the first solvent, which is generally water. The compound to be extracted should be soluble in the extraction solvent and, of course, not undergo reaction with it. (Section 3.1 contains a brief discussion of solubilities.) Major impurities should not be soluble in the extraction solvent. In addition, the extraction solvent should be sufficiently volatile that it can be removed from the extracted material later.

It is also preferable that the solvent be nontoxic and nonflammable. Unfortunately, these last two criteria are not met by many organic solvents. Diethyl ether, the most common extraction solvent, is both volatile (boiling point only a few degrees above room temperature) and flammable. Benzene is toxic and flammable. Halogenated hydrocarbons are not all flammable, but most are toxic. When using a solvent for extraction, always proceed with caution. Table

* Another technique, **continuous liquid–liquid extraction**, could be used in this case. See L. C. Craig and D. Craig, "Extraction and Distribution," *Technique of Organic Chemistry*, Vol. III, Arnold Weissberger, Ed. New York: Interscience Publishing Co., Inc., 1950, p. 247.

Table 4.1 Some Common Extraction Solvents

Name	Formula	Density (g/mL)[a]	Bp (°C)	Comments
lighter than water:				
diethyl ether	$(CH_3CH_2)_2O$	0.7	35	highly flammable
petroleum ether	—[b]	~0.7	30–60	flammable
ligroin	—[b]	~0.7	>60[b]	flammable
benzene	C_6H_6	0.9	80	flammable, toxic, carcinogenic
heavier than water:				
methylene chloride (dichloromethane)	CH_2Cl_2	1.3	41	—
chloroform (trichloromethane)	$CHCl_3$	1.5	61	toxic
carbon tetrachloride (tetrachloromethane)	CCl_4	1.6	77	toxic, carcinogenic

[a] The density of water is 1.0 g/mL; that of saturated aqueous sodium chloride solution is 1.2 g/mL.
[b] See Table 3.1, page 38.

4.1 lists a few common extraction solvents, their densities, and their potential hazards.

Note from Table 4.1 that the chlorinated hydrocarbon solvents are more dense than water; these solvents sink to the bottom in a separatory funnel containing water. The other solvents listed usually float on water. An exception would be a solvent containing a high concentration of a dense solute, which can increase the density of the organic layer so that it becomes heavier than water. Pouring the wrong solution down the drain is a common error. For this reason, it is generally wise to test the two layers if there is any question as to which is the organic layer and which is the aqueous layer. A simple test is to add a few drops of each layer to a small amount of water in a pair of test tubes. The layer that is immiscible with water is the organic layer. Also, it is generally wise to save all layers until an experiment is complete.

4.3 Chemically Active Extraction

The type of extraction procedure that we have been discussing might be considered a "passive" process—the extraction of a compound by virtue of its distribution coefficient in a pair of solvents. A less common, but very powerful, extraction technique is **chemically active extraction**. In this type of extraction, a compound is altered chemically to change its distribution coefficient in a pair of solvents.

To illustrate how this technique works, let us consider a general example. Assume that we have a mixture of two compounds, A and B. Further, assume that A and B are both soluble in diethyl ether and insoluble in water. These two compounds could not be separated from each other by an ordinary passive extraction. However, if B could be changed into a salt (soluble in water, but insoluble in diethyl ether), then we could effect a clean separation of B (in the water layer) and A (in the ether layer). For example, if B is a base and A is neutral, we can treat the mixture with an acid to change the relative solubilities of A and B.

$$\underset{\text{water-insoluble and ether-soluble}}{\text{A + B:}} \xrightarrow{\text{HCl}} \begin{cases} \text{A in the ether layer} \\ \text{BH}^+\ \text{Cl}^- \text{ in the water layer} \end{cases}$$

This type of reaction and subsequent extraction is the principle behind chemically active extraction.

Most organic compounds are neutral, neither acidic nor basic to any great extent. The major exceptions are carboxylic acids and phenols, which are weak acids, and amines, which are weak bases. Compounds belonging to these classes can often be separated from compounds of other classes by chemically active acid or base extraction.

weak acids*:

$$\mathrm{R\overset{\overset{O}{\|}}{C}OH} \qquad \mathrm{Ar\overset{\overset{O}{\|}}{C}OH} \qquad \mathrm{ArOH}$$

carboxylic acids $pK_a \cong 5$ — phenols $pK_a \cong 10$ (*—OH attached to the C of an aromatic ring*)

weak bases*:

$$RNH_2 \qquad R_2NH \qquad R_3N$$

amines $pK_b \cong 4$

Carboxylic acids and phenols (but not alcohols, ROH) are strong enough acids to undergo reaction with a strong aqueous base such as sodium hydroxide to yield water-soluble salts. In this form, these compounds can be extracted from neutral organic compounds.

$$\mathrm{R\overset{\overset{O}{\|}}{C}OH + Na^+ + OH^-} \rightarrow \underbrace{\mathrm{R\overset{\overset{O}{\|}}{C}O^- + Na^+}}_{\text{a water-soluble salt}} + H_2O$$

$$\mathrm{ArOH + Na^+ + OH^-} \rightarrow \underbrace{\mathrm{ArO^- + Na^+}}_{\text{a water-soluble salt}} + H_2O$$

* In these formulas, R represents an alkyl group, such as $CH_3—$, $CH_3CH_2—$, etc. Ar represents an aryl group, such as $C_6H_5—$, $CH_3—C_6H_4—$ etc. Acid strength is shown by pK_a, the negative logarithm of the acidity constant. Base strength is shown by pK_b, the negative logarithm of the basicity constant for the reaction $B + H_2O \rightleftarrows BH^+ + OH^-$.

Carboxylic acids, with pK_a values typically around 5, undergo an acid–base reaction with the weak base sodium bicarbonate ($NaHCO_3$). A typical phenol, with a pK_a value of around 10, is only 1/100,000th as strong an acid as a carboxylic acid. Most phenols are too weakly acidic to undergo reaction with sodium bicarbonate. In Experiment 4.2, this difference in reactivity is used to separate a mixture of benzoic acid and 2-naphthol (a phenol).

$$\underset{\text{water-insoluble}}{R\overset{O}{\overset{\|}{C}}OH} + Na^+ + HCO_3^- \rightarrow \underbrace{R\overset{O}{\overset{\|}{C}}O^- + Na^+}_{\text{a water-soluble salt}} + H_2O + CO_2\uparrow$$

$$\underset{\text{water-insoluble}}{ArOH} + Na^+ + HCO_3^- \rightarrow \text{no appreciable reaction}$$

A carboxylic acid or a phenol may be regenerated from its salt by treatment with an aqueous mineral acid.

$$R\overset{O}{\overset{\|}{C}}O^- + H_3O^+ \rightarrow R\overset{O}{\overset{\|}{C}}OH + H_2O$$

Amines are bases for the same reason that ammonia is a base: an amine contains a nitrogen atom with an unshared pair of electrons and thus can accept a proton. Treatment of an amine with aqueous acid yields a water-soluble salt that can be separated from other organic compounds.

$$\underset{\text{water-insoluble}}{R\ddot{N}H_2} + H^+ + Cl^- \rightarrow \underbrace{R\overset{H}{\overset{|}{N}}H_2^+ + Cl^-}_{\text{a water-soluble salt}}$$

Because different compounds form water-soluble salts under different conditions, the acid–base reactions that we have presented can form the basis of a number of types of chemically active extractions. In the laboratory, the acids and bases usually used to effect these reactions are:

5–10% aqueous HCl to remove an amine as a water-soluble salt

5% aqueous $NaHCO_3$ to remove RCO_2H as a water-soluble salt

5% aqueous NaOH to remove ArOH as a water-soluble salt

In any of these acid–base reactions, hydrocarbons, halogenated hydrocarbons, alcohols, and other neutral organic compounds are usually unaffected.

Problems

4.1 What is *wrong* with the following procedure?

The reaction mixture, consisting of NaBr and an ethanol solution of the product, is diluted with an equal volume of water, then extracted once with an equal volume of diethyl ether. The lower aqueous layer is discarded.

4.2 Suppose you added an additional 50 mL of water to a separatory funnel containing a compound distributed between 50 mL of ether and 50 mL of water. How will this addition affect (a) the distribution coefficient of the compound? (b) the actual distribution of the compound?

4.3 The pain reliever phenacetin (Figure 6.5, page 113) is soluble in cold water to the extent of 1.0 g/1310 mL and soluble in diethyl ether to the extent of 1.0 g/90 mL.

(a) Determine the approximate distribution coefficient for phenacetin in these two solvents.

(b) If 50 mg of phenacetin were dissolved in 100 mL of water, how much ether would be required to extract 90% of the phenacetin in a single extraction?

(c) What per cent of the phenacetin would be extracted from the aqueous solution in (b) by two 25-mL portions of ether?

4.4 Complete the following equations. If no appreciable reaction occurs, write "no reaction."

(a) 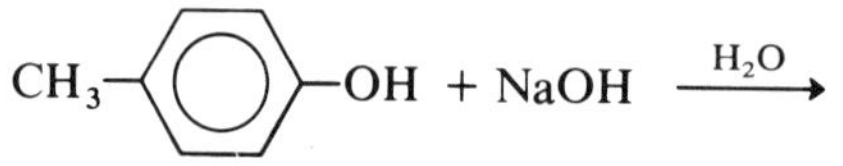 $CH_3-C_6H_4-OH + NaOH \xrightarrow{H_2O}$

(b) $C_6H_5-CH_2CH_2OH + NaOH \xrightarrow{H_2O}$

(c) $CH_3-C_6H_4-OH + NaHCO_3 \xrightarrow{H_2O}$

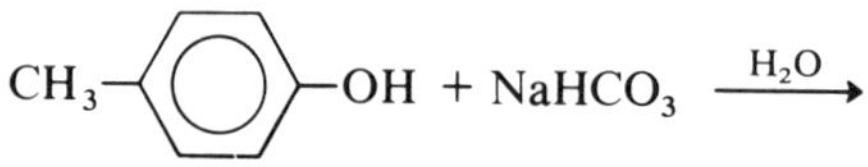

(d) $CH_3CH_2CH_2\overset{O}{\overset{\|}{C}}OH + NaHCO_3 \xrightarrow{H_2O}$

(e) $CH_3NHCH_2CH_2NH_2$ + excess HCl $\xrightarrow{H_2O}$

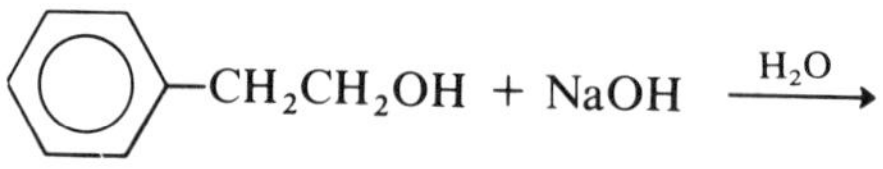

4.5 Which of the following pairs of compounds *could* be separated by chemically active extraction? What reagent would you use?

(a) $CH_3CH_2\overset{O}{\overset{\|}{C}}OH$ $ClCH_2CH_2\overset{O}{\overset{\|}{C}}OH$

(b) 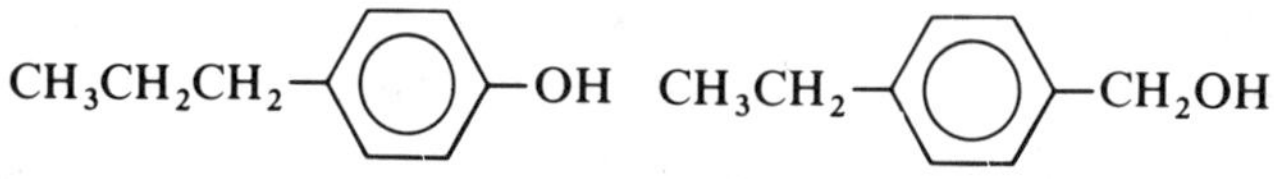 $CH_3CH_2CH_2-C_6H_4-OH$ $CH_3CH_2-C_6H_4-CH_2OH$

(c) [benzo-fused ring with NH] [benzo-fused ring with O]

NH O

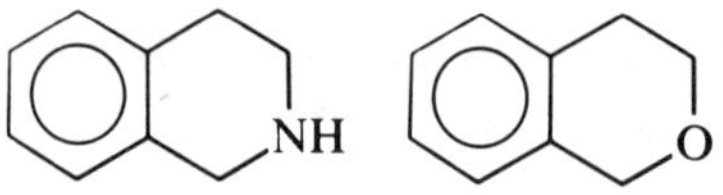

4.6 Draw a flow diagram for the separation of $CH_3CH_2CH_2Br$, $(CH_3CH_2CH_2)_3N$, and $CH_3CH_2CO_2H$.

4.7 At a pH of 8, what per cent of phenol (pK_a = 10) exists as an anion?

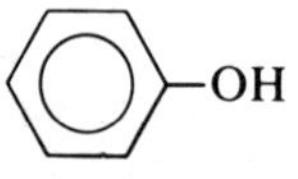

C_6H_5-OH

phenol

4.8 An aqueous solution containing 5.0 g of solute in 100 mL is extracted with three 25-mL portions of diethyl ether. What is the total amount of solute that will be extracted by the ether in each of the following cases?

(a) distribution coefficient (ether/water), $K = 0.10$

(b) $K = 1.0$

(c) $K = 10$

4.9 If the compound in Problem 4.8(b) were extracted with three 50-mL portions of diethyl ether, how much would be extracted?

4.10 In experiments later in this book, acetic acid ($K = 0.3$; ether/water) must be removed from diethyl ether. If 100 mL of ether contained 10 g of acetic acid, and if this solution were extracted with a series of 50-mL portions of water, it would take 63 aqueous extractions to remove 99.99% of the acetic acid (assuming that diethyl ether and water are immiscible).

(a) Diethyl ether is partially soluble in water (at 25°, approx. 6 g/100 mL). How would this solubility affect the efficiency of such an extraction procedure?

(b) Suggest an alternative technique for removing the acetic acid.

»»» EXPERIMENT 4.1 «««

Extraction of Adipic Acid and an Unknown from Water—A Quantitative Study

Discussion

In part A of this experiment, adipic acid is extracted from an aqueous solution by diethyl ether. In part B, an unknown carboxylic acid is extracted using the same procedure. In both parts of this experiment, only a single extraction by the ether is used. The quantities of both the adipic acid and the unknown acid in the aqueous solution before and after the extraction are determined by titration. These quantities are then used to calculate the distribution coefficient for these ether–water systems.

in the titration:

$$\underset{\text{hexanedioic acid (adipic acid)}}{HO\overset{O}{\overset{\|}{C}}CH_2CH_2CH_2CH_2\overset{O}{\overset{\|}{C}}OH} + 2Na^{+}\,{}^{-}OH \rightarrow \underset{\text{disodium hexanedioate (disodium adipate)}}{Na^{+}\,{}^{-}O\overset{O}{\overset{\|}{C}}CH_2CH_2CH_2CH_2\overset{O}{\overset{\|}{C}}O^{-}\,Na^{+}} + 2H_2O$$

A single extraction is adequate to illustrate the technique involved and to provide sufficient data for the calculation of a distribution coefficient. However, a single extraction is usually inadequate in synthetic experiments where yield is the goal. Toward the end of this discussion, we will consider multiple extractions, along with washing and drying an organic solution and solvent removal.

Steps in a Single Extraction

1) Preparation of the separatory funnel. Lightly grease the stopcock and stopper of the separatory funnel with stopcock grease. Use just enough grease to cover the frosted look of the clean ground glass when the joint is rotated. Too much grease will contaminate the organic solution and may even clog the stopcock. (If grease does get into the stopcock hole, it can be removed with a pipe cleaner or a rolled piece of paper.)

With the stopcock in place, set the separatory funnel in an iron ring attached to a rack or sturdy ring stand for support. The iron ring should have on it three short lengths of rubber tubing, slit lengthwise and slipped over the ring, to form a cushion for the funnel (see Figure 4.2).

2) Adding the liquids. Be sure the stopcock is closed. Using an ordinary long-stem funnel, pour the solution to be extracted into the separatory funnel, followed by a measured amount of extraction solvent. Do not fill the separatory funnel more than three-quarters full; otherwise, there will not be enough room for mixing the liquids.

>>>> ***SAFETY NOTE*** If you are using a flammable solvent, make sure that there are no flames in the vicinity!

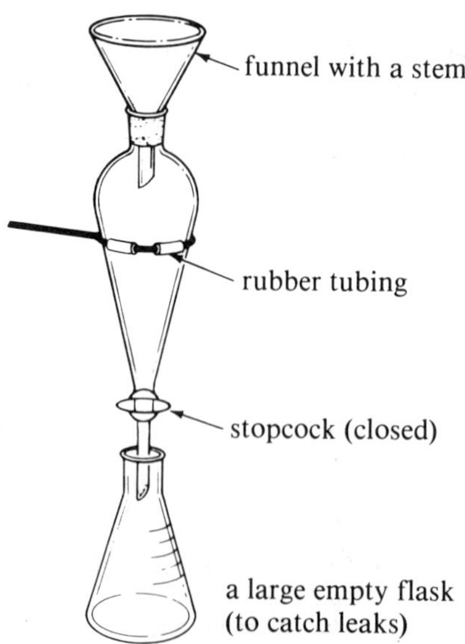

Figure 4.2 The appearance of the separatory funnel before liquids are poured into it.

3) Mixing the layers. Before inserting the stopper, swirl the separatory funnel gently. The swirling is especially important if an acid and sodium bicarbonate are present because carbon dioxide is given off by their reaction. The swirling will drive off some of the carbon dioxide and minimize pressure build-up during the shaking process.

Insert the lightly greased stopper, then *holding the stopper in place with one hand,* pick up the separatory funnel and invert it. Immediately, open the stopcock with your other hand to vent solvent fumes or carbon dioxide.

>>>> **SAFETY NOTE** Always aim the stem of the separatory funnel away from your neighbors when venting. Better, aim the stem into the hood.

Figure 4.3 shows the proper method for holding a separatory funnel. Hold the stopper and stopcock firmly in place throughout the entire shaking process to prevent their falling out.

After venting, close the stopcock, gently shake or swirl the mixture in the inverted funnel, then re-vent the fumes. If excessive pressure build-up is not observed, the separatory funnel and its contents may be shaken up and down vigorously in a somewhat circular motion for 2–3 minutes so that the layers are thoroughly mixed. Vent the stopcock several times during the shaking period. After completing the shaking, vent the stopcock one last time. With the stopcock closed, place the separatory funnel back in the iron ring and remove the stopper. If you are extracting a small amount of material, wash the stopper into the separatory funnel with a few drops of extraction solvent, using a dropper. Place a large Erlenmeyer flask under the stem of the separatory funnel in case the stopcock should develop a leak. Allow the separatory funnel to sit until the layers have separated.

4) Separating the layers. Before proceeding, make sure the stopper has been removed. (It is difficult to drain the lower layer from a stoppered funnel

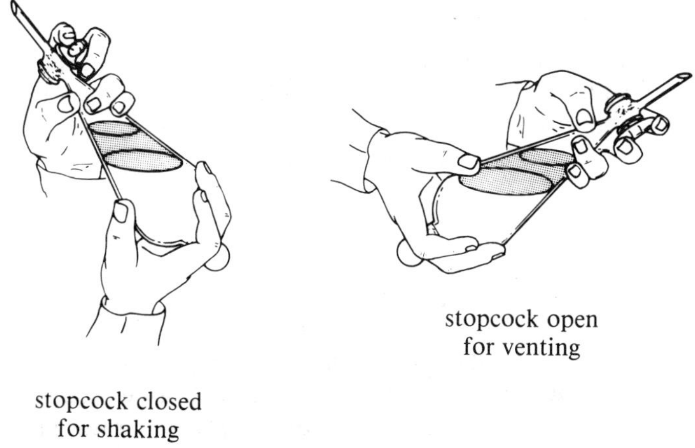

Figure 4.3 How to hold and vent a separatory funnel.

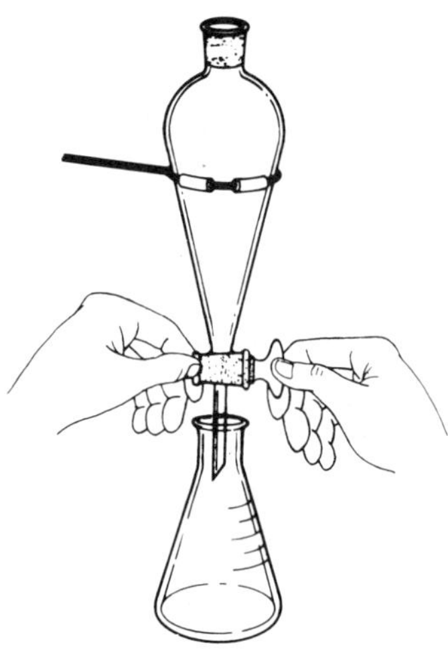

Figure 4.4 How to hold a separatory funnel while draining the lower layer.

because a vacuum is created in the top portion of the funnel.) Partially open the stopcock to drain the lower layer into the flask, holding the stopcock in place and bracing the separatory funnel with both hands so that the stopcock cannot accidentally slip out of place (see Figure 4.4). Splashing is minimized if the tip of the stem touches the side of the flask so that the liquid can run down its side.

When the lower layer is almost, but not quite, drained into the flask, close the stopcock and gently swirl the separatory funnel. This swirling knocks drops clinging to the sides of the funnel to the bottom, where they can be removed. Carefully and slowly drain the last of the lower layer into the flask. Finally, tap the stem of the funnel to knock any clinging drops into the flask.

At this point in a synthesis experiment, the organic layer containing the desired product would be washed and poured from the top of the funnel into a clean Erlenmeyer flask. In this way, the upper layer is not contaminated with drops remaining in the stem.

5) Cleaning the separatory funnel. Wash the separatory funnel as soon as you are finished with it, because the organic solvent can dissolve stopcock grease and cause the stopper or the stopcock to freeze. Then regrease the ground-glass joints or else store them separately.

Other Techniques Used in Extractions

1) Multiple extraction. In practice, a single extraction is rarely used; multiple extractions are the rule. A typical double-extraction sequence of an aqueous solution by diethyl ether is diagrammed in Figure 4.5.

The original aqueous solution that has already been extracted once and drained into a flask is returned to the dirty, but empty, separatory funnel, along with a fresh portion of extraction solvent. If a second separatory funnel is available, the aqueous layer can be drained directly into it for the second

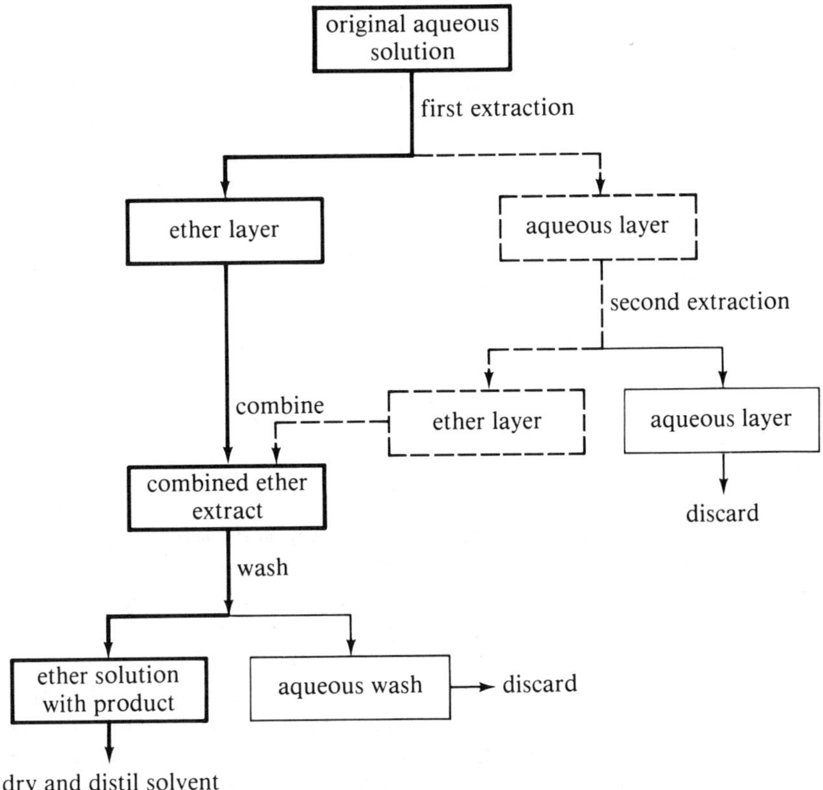

Figure 4.5 *Diagram of a two-stage diethyl ether extraction of an aqueous solution. (The location of the organic product is indicated by the heavy black lines.)*

extraction. (If the solvent forms a *lower* layer, it is drained off first. In this case, the original solution to be re-extracted can simply remain in the separatory funnel, and fresh solvent added.) The second (and possibly a third) extraction is carried out just as the first one was. After the multiple extraction, the organic layers are combined in one flask. The aqueous layer (if that is the layer that will be discarded) should be labeled and saved until the entire experiment is finished.

2) *Washing.* If the original solution was an aqueous one, the organic layers, separately or combined, may be *washed* (extracted with fresh water) to remove any water-soluble impurities. Use a volume of wash water about 10% that of the organic layer. A 5% sodium bicarbonate wash may be used to remove traces of acid from the organic layer. A sodium chloride wash (saturated aqueous solution) may be used to remove water from the organic solution.

Washing is accomplished by placing the organic solution in the separatory funnel along with the wash solution, then shaking and separating in the usual manner. Aqueous washes should be combined with the original water solution. $NaHCO_3$ or NaCl washes should be saved separately until the experiment is complete.

3) Drying organic solutions. No two liquids are completely immiscible. In any liquid–liquid extraction, the desired layer (usually the organic layer) contains some of the other solvent (usually water). Before the organic solvent is removed, the organic solution is usually dried so that water will not contaminate the product.

The amount of water that dissolves in an organic solvent can be substantial and is sometimes evidenced by a cloudy solution. In some cases, the water dissolved in the organic solution can be largely removed by shaking the solution in the separatory funnel with a small amount of saturated aqueous sodium chloride. Your instructor may recommend that you do this routinely with diethyl ether extracts.

Drying agents are commonly used to remove the last traces of water from organic solutions. These drying agents are generally anhydrous inorganic salts (insoluble in organic liquids) that attract water and are converted to hydrated salts. Then, the hydrated salts are removed by filtering or careful decanting (pouring off the solution) prior to distillation of the organic solvent. Table 4.2 lists a few common drying agents and comments about their use.

A moderate amount of drying agent (about 1–5 grams, or just enough to cover the bottom of the flask) is sufficient for drying most solutions. If the drying agent becomes wet-looking or clumpy, the solution should be filtered or decanted into a clean, dry flask and a fresh portion of drying agent added.

Most drying agents are fairly swift (15 minutes) in removing the bulk of the water from an organic solution; however, most are quite slow in trapping the last vestiges of water. For this reason, an overnight drying is always preferable to a 15-minute drying. If a trace of water is not particularly undesirable, however, the 15-minute drying may allow you to continue the experiment and thus save laboratory time.

Table 4.2 Some drying agents for organic solutions

Name	*Formula*	*Comments*
magnesium sulfate	$MgSO_4$	fast and generally useful; best filtered, not decanted (because of fine particle size); Lewis acid
sodium sulfate	Na_2SO_4	slow, but inexpensive, easy to use, and generally useful; neutral
calcium chloride	$CaCl_2$	slow; forms complexes with O and N compounds, such as alcohols, amines, ketones, and carboxylic acids; may contain CaO as an impurity
calcium sulfate	$CaSO_4$	fast; neutral
potassium carbonate	K_2CO_3	moderately fast; basic—reacts with acidic compounds such as phenols and carboxylic acids
saturated sodium chloride solution	$NaCl + H_2O$	fast; used to remove bulk of water from an organic solution so that less solid drying agent is needed

Table 4.3 Composition and boiling points of some low-boiling binary azeotropes containing water

Composition	*Bp of azeotrope (°C)*
91% benzene (bp 80.1°) 9% water (bp 100.0°)	69.4
96% carbon tetrachloride (CCl_4; bp 76.8°) 4% water	66.8
97% chloroform ($CHCl_3$; bp 61.7°) 3% water	56.3
99% dichloromethane (CH_2Cl_2; bp 40°) 1% water	38.8

Some solvents form **low-boiling azeotropes** with water—that is, when they are distilled, they produce a mixed distillate of constant composition with a lower boiling point than that of either water or the pure solvent (see also Section 5.1D). For example, benzene (bp 80.1°) and water (bp 100.0°) form an azeotrope composed of 91% benzene and 9% water that boils at 69.4°. Table 4.3 lists some solvents that form low-boiling azeotropes with water.

Solutions in these solvents may be dried by simply boiling off the azeotropic mixture until the distillate is clear and no longer a two-phase mixture of water plus solvent. Once this low-boiling azeotrope is boiled off, the solution contains no more water. This is the procedure used to dry the organic extract in Experiment 4.3.

4) *Removing solvents.* After an extraction has been completed and the extract dried, there are a number of ways to remove a solvent from an organic compound. Small amounts of solvent may be removed by simple evaporation or by boiling the solvent and using a stream of clean, dry air or a vacuum to remove the vapors from the flask. This is the technique used in Experiment 4.3. (When heating such a mixture, take care that the residue in the flask does not become overheated.)

Large amounts of solvent should not be boiled away into the atmosphere, even in a fume hood; instead, the solvent should be distilled and collected. (See Chapter 5 for a discussion of distillation.) This is the technique used in the synthesis experiments later in this book.

›››› **SAFETY NOTE** Organic extraction solvents should *never* be boiled or evaporated except under a hood. Flammable solvents should *never* be heated with a burner. A steam bath is the safest source of heat, or a spark-free hot plate may be used.

When the experiment is completed, the extracted aqueous layer may be poured down the drain and the drain flushed with water. A small amount of organic solvent may also be flushed down the drain. In most cases, however, the large amounts of solvent used in a classroom should be stored in waste jugs. This is particularly true of diethyl ether because of its volatility and flammability.

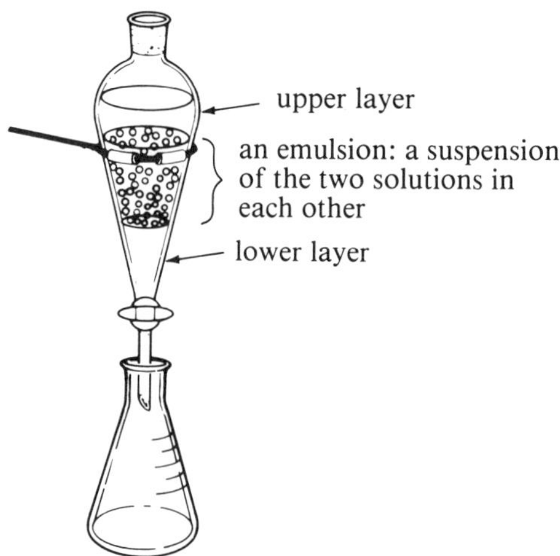

Figure 4.6 *An emulsion in the separatory funnel.*

5) *Emulsions.* An **emulsion** is a suspension of one material in another that does not separate quickly by gravity. In liquid–liquid extractions, an emulsion refers to the suspension of one of the solutions in the other (see Figure 4.6). The result is that the two layers do not separate completely, a very annoying situation.

If it is known that a particular extraction might lead to an emulsion, a few preventive steps might save time later. The *addition of sodium chloride* to the aqueous layer decreases the solubility of water in the organic solvent and vice versa, and therefore may prevent an emulsion. *Swirling* the separatory funnel, instead of shaking it vigorously, when mixing the two layers may also prevent an emulsion from forming. However, swirling is a less efficient (and thus slower) way of reaching the equilibrium distribution of solute.

Once an emulsion has formed, it can sometimes be broken up by one or more of the following techniques:

1) allowing the separatory funnel to sit in an iron ring for a few minutes with periodic gentle swirling;

2) adding a few drops of a saturated solution of aqueous sodium chloride with a dropper;

3) squirting a few drops of 95% ethanol or a commercial antifoam agent on the emulsion "bubbles."

If none of the preceding techniques break up the emulsion, one of the following techniques might work:

4) filtering the mixture with vacuum (which removes solid particles that help stabilize the emulsion), then proceeding with the separation;

5) drawing off as much of the lower layer in the separatory funnel as possible, adding fresh solvent to the top layer remaining in the funnel to dilute it, and swirling gently;

6) pouring the mixture into a flask, corking the flask, and allowing the mixture to sit overnight or until the next laboratory period. (Do not let the mixture stand for any length of time in the separatory funnel; the stopcock will eventually leak under the pressure of liquid.)

Experimental

EQUIPMENT:

50-mL buret
dropper or disposable pipet
three 250-mL Erlenmeyer flasks
funnel
two 50-mL and one 100-mL graduated cylinders
hot plate
ice bath
250-mL separatory funnel

magnetic stirrer and stir bar (optional)

CHEMICALS:

adipic acid, 1.1 g
diethyl ether, 50 mL for each part (100 mL total)
phenolphthalein indicator, 0.2 mL per titration (0.8 mL total)
0.200*N* sodium hydroxide, 30–50 mL per titration (120–200 mL total)
unknown,* 1.1 g

TIME REQUIRED: 1–2 hours

STOPPING POINTS: after the titration in each part of the experiment

›››› ***SAFETY NOTE*** Diethyl ether (bp 34.6°) is extremely volatile and extremely flammable. Under no circumstances should any burners be used in the laboratory. (See Sections 1.1I and 1.1L.)

It is *imperative* that the solution to be extracted with diethyl ether be *chilled*. If the solution is warm, the diethyl ether vapors will pressurize the separatory funnel and, in all probability, pop the stopper and stopcock out of the funnel.

PROCEDURE

A. Extraction of Adipic Acid

Place 1.1 g of adipic acid in a 250-mL Erlenmeyer flask. Add exactly 100.0 mL of distilled water and heat the flask on a hot plate with occasional swirling. When all the adipic acid has dissolved, cool the flask in an ice bath until the temperature of the solution is less than 20° (see Safety Note).

* Recommended unknowns are given in the instructor's guide.

Using a 50-mL graduated cylinder and a disposable pipet, transfer exactly 50.0 mL of the cool adipic acid solution to a 250-mL Erlenmeyer flask. Add 0.2 mL of phenolphthalein indicator and titrate to the end point with 0.200*N* sodium hydroxide solution (see Experimental Note 1).

Measure the volume of the remainder of the adipic acid solution in a 50-mL graduated cylinder and carefully transfer it to a 250-mL separatory funnel (not a dropping funnel; see Figure 1.3, page 11). Using a clean, dry, 50-mL graduated cylinder, measure 50 mL of diethyl ether (CAUTION: flammable! See Safety Note) and transfer it to the separatory funnel. Stopper the separatory funnel and shake it for 2–3 minutes as described in the discussion. (CAUTION: To prevent pressure build-up, vent the funnel after inverting it, and vent it periodically during the shaking.)

After the layers have separated, drain the lower aqueous layer into a clean 100-mL graduated cylinder. Dispose of the ether layer remaining in the separatory funnel in a jug labeled "waste ether" in the hood. (Do *not* pour it down the sink, where it would be a fire hazard.)

Record the volume of the aqueous solution in the graduated cylinder, then transfer the solution with washing into an Erlenmeyer flask for titration. (After the volume has been measured, extra water will not affect the results of the titration.) Titrate this aqueous layer as you did the original adipic acid solution and record the data in your notebook.

From your data, calculate the distribution coefficient for adipic acid in diethyl ether and water and the per cent of the adipic acid that was extracted (see Experimental Note 2). Record these calculations in your notebook.

B. Determination of the Distribution Coefficient for an Unknown

Obtain an unknown from your instructor and record its number in your notebook. The unknown will be a water-soluble carboxylic acid that can be titrated with 0.200*N* sodium hydroxide to the phenolphthalein end point. Carry out the extraction procedure described in part A, substituting your unknown for the adipic acid. Calculate the distribution coefficient as well as the percentage of the unknown acid extracted by the ether.

EXPERIMENTAL NOTES

1) *Procedure for titration.* It is assumed that you have previously carried out an acid–base titration; the following description is intended only as a review.

Clamp a buret to a ring stand with a buret holder (not a standard flask clamp). Rinse the buret two or three times with standardized NaOH solution. (This solution should be pre-mixed and stored in a plastic container in the laboratory. The exact concentration will be marked on the container.) Fill the buret to the zero mark, using the lower level of the meniscus for measuring. Be sure there are no air bubbles in the stem of the buret.

Measure the volume of solution to be titrated, and transfer it to a 250-mL Erlenmeyer flask. Add a few drops of phenolphthalein solution and a magnetic stir bar, then place the flask on a magnetic stirrer. (If you do not use a magnetic stirrer, swirl the flask frequently during the titration.)

With the magnetic stirrer set at slow speed, add the NaOH solution from the buret until the solution turns from colorless to pink. Add the NaOH solution rapidly at first, then more slowly as you approach the end point. At the end point, one drop should change the color. (Do not be fooled by a premature pink color that fades back to colorless.) Near the end of the titration, wash down the sides of the Erlenmeyer flask with water so that any adipic acid solution that has splashed onto the sides of the flask will be returned to the bulk of the solution.

At the end of the titration, record the volume of NaOH added, along with its concentration.

2) *Sample calculations.* The following sample calculations are based upon the use of a solution containing 2.40 g of propanoic acid ($CH_3CH_2CO_2H$) dissolved in 100.0 mL of water and titrated with 0.200*N* NaOH. You will be using adipic acid, not propanoic acid. Although propanoic acid is a monobasic acid and adipic acid is a dibasic acid, the method of calculation is the same.

(a) ***Distribution coefficient for propanoic acid.*** Titration of a 50.0-mL aliquot of the propanoic acid solution required 80.2 mL of 0.200*N* NaOH. At the end point of the acid–base titration:

$$\begin{aligned}
\text{milliequivalents (mEq)} &= (N)(\text{mL}) \\
\text{mEq of acid} &= \text{mEq of base} \\
\text{mEq of acid before extraction} &= (N_{\text{base}})(\text{mL}_{\text{base}}) \\
&= (0.200)(80.2) \\
&= 16.0
\end{aligned}$$

After extraction of 50.0 mL of propanoic acid with 150.0 mL of ether, titration of the aqueous solution (measured to be 50.0 mL) required 28.9 mL of 0.200*N* NaOH.

$$\begin{aligned}
\text{mEq of acid in water after extraction} &= (N_{\text{base}})(\text{mL}_{\text{base}}) \\
&= (0.200)(28.9) \\
&= 5.78
\end{aligned}$$

Therefore,

$$\begin{aligned}
\text{mEq of acid in ether} &= 16.0 - 5.78 \\
&= 10.2
\end{aligned}$$

$$\begin{aligned}
K &= \frac{\text{concentration of propanoic acid in diethyl ether}}{\text{concentration of propanoic acid in water}} \\
&= \frac{10.2\ \text{mEq}/150\ \text{mL}}{5.78\ \text{mEq}/50\ \text{mL}} \\
&= 0.59
\end{aligned}$$

(b) ***Per cent of propanoic acid extracted.*** From (a), we extracted 10.2 mEq of the acid from a total of 16.0 mEq. Therefore,

$$\frac{10.2}{16.0} \times 100 = 64\%$$

Problems

4.11 Why should you not use a dropping funnel for an extraction procedure?

4.12 Diagram a two-stage dichloromethane (d = 1.3) extraction of an aqueous solution, showing how this procedure differs from the diethyl ether extraction diagrammed in Figure 4.5.

4.13 Which drying agent(s) in Table 4.2 could be used to dry an ether solution of each of the following compounds?

(a) $CH_3CH_2CH_2CH_2OH$ (b) $CH_3CH_2CH_2CH_2Br$

(c) $CH_3CH_2\overset{O}{\overset{\|}{C}}OH$ (d) $CH_3CH_2CH_2CH_2NH_2$

4.14 Propanoic acid, 7.30 g, is diluted to 300 mL with water. How much 0.500N NaOH is required to neutralize a 50.0-mL aliquot?

4.15 A 50.0-mL aliquot of a 0.300M solution of compound A (a monobasic carboxylic acid) is extracted with 50.0 mL of diethyl ether. Titration of the extracted aqueous solution requires 23.5 mL of 0.500N NaOH. Calculate the distribution constant of compound A in this extraction.

EXPERIMENT 4.2

(SUPPLEMENTAL)

Chemically Active Extraction

Discussion

Benzoic acid and 2-naphthol are both ether-soluble and water-insoluble. They are both weak acids and react with strong bases to form water-soluble salts. Benzoic acid and 2-naphthol are not acids of the same strength; benzoic acid is about 100,000 times stronger. Consequently, it is possible to select a weak base that will react with benzoic acid but not with 2-naphthol. Sodium bicarbonate ($NaHCO_3$) is just such a weak base. When an ether solution containing a mixture of benzoic acid and 2-naphthol is treated with an aqueous solution of $NaHCO_3$, the benzoic acid forms a water-soluble, ether-insoluble salt. The 2-naphthol does not react and remains in the ether layer.

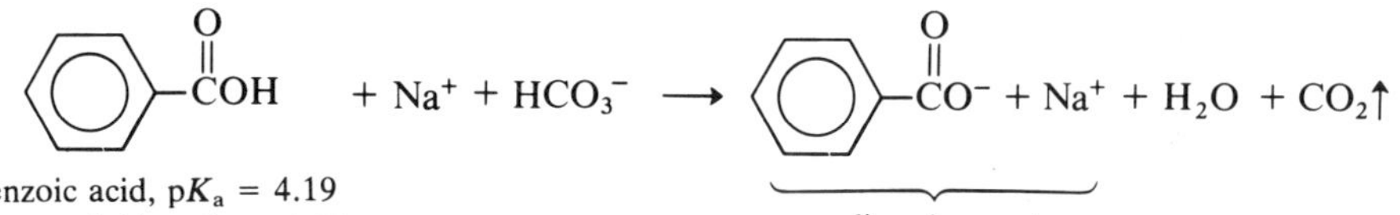

$$C_6H_5COOH + Na^+ + HCO_3^- \longrightarrow C_6H_5COO^- + Na^+ + H_2O + CO_2\uparrow$$

benzoic acid, $pK_a = 4.19$
water-insoluble, ether-soluble

sodium benzoate
water-soluble, ether-insoluble

$$C_{10}H_7OH + Na^+ + HCO_3^- \longrightarrow \text{no appreciable reaction}$$

2-naphthol, $pK_a = 9.51$
water-insoluble, ether-soluble

When the ether and water phases are separated, the aqueous solution of sodium benzoate is drained into a flask. Two extractions are needed to remove the bulk of the benzoic acid.

After the benzoic acid has been extracted, we could evaporate the ether to obtain the 2-naphthol; however, it is better to extract the 2-naphthol so that the neutral impurities remain in the ether. The 2-naphthol is extracted from the ether layer with an aqueous solution of NaOH (a strong base). The water-soluble sodium salt of 2-naphthol is formed, and its solution is drained into a flask.

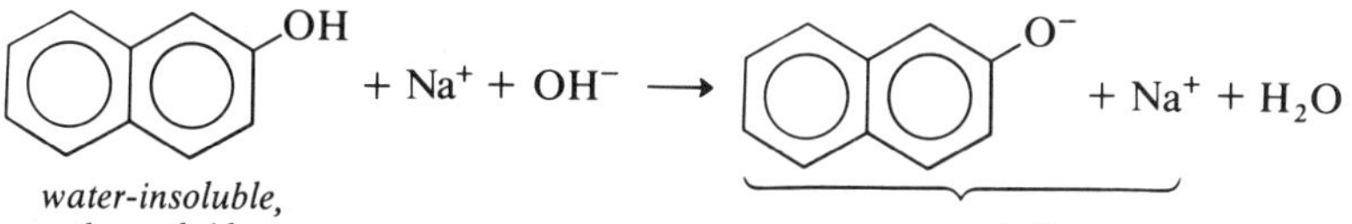

$$C_{10}H_7OH + Na^+ + OH^- \longrightarrow C_{10}H_7O^- + Na^+ + H_2O$$

water-insoluble, ether-soluble

water-soluble, ether-insoluble

The complete extraction procedure results in two flasks that contain aqueous solutions of the sodium salts of the desired compounds. When these solutions are acidified, the sodium salts are converted back to their water-insoluble, acidic forms, which precipitate. The crude materials are collected by vacuum filtration. The degree of purity of each is estimated by its melting point. If desired, the two compounds can be further purified by crystallization from water.

Experimental

EQUIPMENT:

dropper or disposable pipet
125-mL and two 250-mL Erlenmeyer flasks
graduated cylinder
ice bath
250-mL separatory funnel
vacuum filtration assembly

CHEMICALS:

benzoic acid, 3.0 g
diethyl ether, 50 mL
conc. hydrochloric acid, 10–20 mL
litmus paper or pH paper
2-naphthol, 3.0 g
10% sodium bicarbonate solution, 100 mL
10% sodium hydroxide solution, 50 mL

TIME REQUIRED: $1\frac{1}{2}$ hours plus overnight drying and time for melting-point determinations

STOPPING POINTS: at any point, if all liquids are kept in well-stoppered flasks

›››› ***SAFETY NOTE 1*** Diethyl ether is volatile and flammable. There must be no flames in the laboratory! (See Sections 1.1I and 1.1L.)

›››› ***SAFETY NOTE 2*** In the first extraction, CO_2 gas is generated. Be sure to: (a) hold both the stopcock and the stopper of the separatory funnel tightly; (b) swirl before inverting the separatory funnel and vent the gas through the stopcock immediately after inverting; (c) vent the funnel frequently. Do not point the stem of the funnel at anyone when you vent it; the outrushing gas can cause liquid to be spewed out.

›››› ***SAFETY NOTE 3*** Concentrated hydrochloric acid is a strong acid and emits irritating vapors. Use it only in the fume hood. Wash any splashes on your skin with copious amounts of water. (See Section 1.1D.)

PROCEDURE

Weigh 3.0 g of benzoic acid and 3.0 g of 2-naphthol into a 125-mL Erlenmeyer flask. Add 50 mL of diethyl ether (CAUTION: *flammable*!), and swirl the mixture until solution is complete. Transfer the ether solution to a 250-mL separatory funnel and add 50 mL of 10% sodium bicarbonate solution. (See Safety Notes 1 and 2.) Swirl the funnel until frothing has subsided, then shake the funnel with frequent venting. Drain the lower, aqueous layer into a clean 250-mL Erlenmeyer flask; label the flask "bicarbonate extract." Keep the ether layer in the separatory funnel.

Repeat the extraction with another 50-mL portion of 10% sodium bicarbonate solution. Combine the lower, aqueous layer with the first bicarbonate extract. Again, leave the ether layer in the separatory funnel.

Add 50 mL of ice-cold 10% sodium hydroxide solution (CAUTION: *caustic*!) to the ether solution in the separatory funnel. Shake the funnel with frequent venting. Drain the lower, aqueous layer into a clean 250-mL

Erlenmeyer flask labeled "hydroxide extract." Discard the ether in the appropriate waste jug. (Do not pour it down the drain.)

Add concentrated hydrochloric acid to each flask *dropwise* until each solution is acidic; about 5–10 mL will be required. (CAUTION: See Safety Note 3; also see Experimental Note.) During the addition, the bicarbonate extract froths because of the release of CO_2 from excess $NaHCO_3$ in the solution; therefore, add the acid slowly. A precipitate forms in each flask as the acid is added. The 2-naphthol (in the "hydroxide extract" flask) has a tendency to oil out (see Experiment 3.1) when acidified; therefore, chill the extract in an ice bath before and during the acidification step.

When both solutions are acidic, cool each flask in an ice bath, then filter the solids by vacuum (see Section 3.3). Air-dry the two products, then weigh them, calculate the per cent recovery, and determine the melting points. If you performed Experiment 2.2, compare the melting points with those that you obtained in that experiment. If desired, the two products can be further purified by crystallization from water.

EXPERIMENTAL NOTE

It is often difficult to judge with pH paper when these mixtures are acidic. To verify that they are indeed acidic, test the mother liquor after filtration with either pH paper or with conc. HCl (to see if additional precipitate forms).

Problems

4.16 Write a flow diagram of the chemically active extraction in Experiment 4.2.

4.17 In Experiment 4.2, can the extraction with NaOH be carried out first? Explain.

4.18 Write equations that show how the sodium salts of (a) benzoic acid, and (b) 2-naphthol are converted to their free acidic forms.

4.19 If 1.0 liter of an aqueous solution is sufficiently basic that 50% of 1.0 g of dissolved benzoic acid is in the ionic form, approximately what per cent of 1.0 g of 2-naphthol in the same solution will be ionized?

4.20 Suppose that you wish to extract 10 g of benzoic acid from an ether solution with aqueous $NaHCO_3$.

(a) What is the minimum amount of $NaHCO_3$ that must be added to 50 mL of water to carry out the extraction?

(b) What is the minimum amount of conc. HCl (37%) that would be needed to convert the sodium benzoate back to its free acid?

EXPERIMENT 4.3
(SUPPLEMENTAL)

Extraction of Caffeine from Tea

Discussion

An **alkaloid** ("like an alkali") is a physiologically active nitrogen compound that can be extracted from plants by aqueous acid. Alkaloids are amines, and therefore yield water-soluble salts when treated with acids. Caffeine is an example of an alkaloid.

O
CH_3N NCH_3
O N N
CH_3

caffeine

According to legend, caffeine was discovered by Arabian priests who were told that the local goats became frisky after eating coffee beans. The priests subsequently learned to consume coffee beans to stay awake during lengthy prayer sessions. Tea, on the other hand, is native to Indochina and India. It became a popular beverage in China about 700 A.D.

Caffeine is water-soluble, and thus can be extracted from ground coffee beans or tea leaves by hot water, as any coffee- or tea-drinker knows. Tea leaves contain a surprisingly large amount of caffeine; an ordinary cup of brewed tea contains about 50 mg of this compound (almost as much as a cup of coffee). In this experiment, a larger amount of tea leaves than the amount required for a cup of tea is used.

Along with caffeine, numerous other organic compounds are extracted from tea leaves by hot water. While these other compounds lend body and aroma to a cup of tea, their presence interferes with filtration of the tea leaves from the water extract and leads to emulsions when the caffeine is later extracted from the water. To suppress the extraction of impurities from the tea leaves, a solution of sodium carbonate, rather than pure water, is used for the boiling-water extraction. To circumvent a filtration step, tea bags, rather than free tea leaves, are used. Because the alkaline sodium carbonate solution causes the paper of some types of tea bags to disintegrate, the tea bags are first wrapped tightly with cheese cloth.

To remove the caffeine from the water extract, the black solution is extracted with dichloromethane. Caffeine is very soluble in this solvent, but the other water-soluble components of tea are not. Dichloromethane is quite volatile, and is easily removed from the caffeine by boiling it off on a hot plate or steam bath in the fume hood.

Experimental

EQUIPMENT:

50-mL, 200-mL, 250-mL, and 400-mL (or better, 600-mL) beakers
cheese cloth
hot plate
ice bath
250-mL separatory funnel
string

CHEMICALS:

dichloromethane (methylene chloride), 160 mL
sodium carbonate monohydrate, 35 g (or 30 g anhydrous)
20 tea bags, 2.5 g each (not loose tea)

TIME REQUIRED: $2\frac{1}{2}$ hours plus melting-point determination

STOPPING POINTS: after boiling the tea leaves; after the dichloromethane extraction

PROCEDURE

Obtain from your instructor the net weight of tea in one tea bag, then roll 20 tea bags (about 50 g) into a tight pack with cheese cloth, using four or more layers of cloth. Tie the pack tightly with string in at least two directions. The roll should be small enough to fit loosely in a 400-mL or 600-mL beaker. Dissolve 35 g of sodium carbonate monohydrate in 250 mL of water and pour this solution over the tea package. The water solution should nearly cover the tea package when the package is pushed to the bottom of the beaker.

Bring the contents of the beaker to a boil on a hot plate, and boil *gently*. (If boiled vigorously, the mixture will froth and boil over.) Using a stirring rod, push the tea package into the boiling liquid occasionally. After 20–30 minutes, cool the black solution in the beaker to below 20° in a cold-water bath or ice bath. Decant the thick liquid into a 250-mL separatory funnel. Use a small beaker to press (firmly but gently) the residual liquid from the tea package, and combine this liquid with that in the separatory funnel. Pour 50 mL of water over the tea package. Again, press the liquid out of the tea package, and combine it with the material in the separatory funnel (see Experimental Note 1).

Extract the black aqueous solution with 30 mL of dichloromethane. (CAUTION: *Do not shake the separatory funnel*! See Experimental Note 2.) Drain the dichloromethane solution (the clear, lower layer) into a clean, dry, 250-mL beaker. Re-extract the black liquid with four more 30-mL portions of dichloromethane and combine the extracts.

Add 2–3 boiling chips, then boil the extracts down to 20–30 mL on a hot plate in the hood. Transfer the residue to a tared 50-mL beaker, using 5–10 mL of fresh dichloromethane to rinse the last of the caffeine into the beaker. Boil the final solution to dryness, being careful not to overheat the residue. A typical recovery is 0.2–0.4 g of crude caffeine, mp 224–228° (see Experimental Note 3). Calculate the per cent of caffeine recovered from the tea. Turn in the product (in a labeled vial) to your instructor.

EXPERIMENTAL NOTES

1) If the combined volume of extract exceeds 200 mL, return the liquid to a suitably sized beaker, boil to reduce its volume, and chill in an ice bath. Return the liquid to the separatory funnel and continue with the extraction.

2) Emulsions are inevitable in this extraction. To minimize their formation, gently swirl the inverted separatory funnel instead of shaking it. Then, allow the funnel to sit in an iron ring until the emulsion that does form breaks up. Place an empty beaker beneath the separatory funnel while you are waiting because the dichloromethane solution may penetrate the stopcock grease and leak.

3) The melting point of pure caffeine is 238°, but its sublimation temperature is 178°. You may be able to obtain a melting point by heating the capillary tube fairly rapidly. Otherwise, seal the open end of the capillary with a small flame before taking the melting point (see page 411).

Problems

4.21 Carbon tetrachloride, chloroform, and dichloromethane are all suitable solvents for the extraction of caffeine. Why was dichloromethane chosen?

4.22 In Experiment 4.3, why is it not necessary to dry the dichloromethane with an inorganic drying agent prior to its removal on a hot plate?

4.23 In some other experiment, you observe frothing while boiling an aqueous solution. If you are to extract that solution in a subsequent step, would you expect to encounter emulsions? Explain.

4.24 Which of the following compounds is an alkaloid?

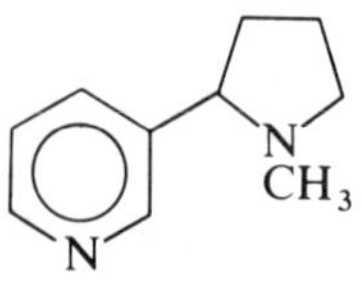

nicotine
in tobacco

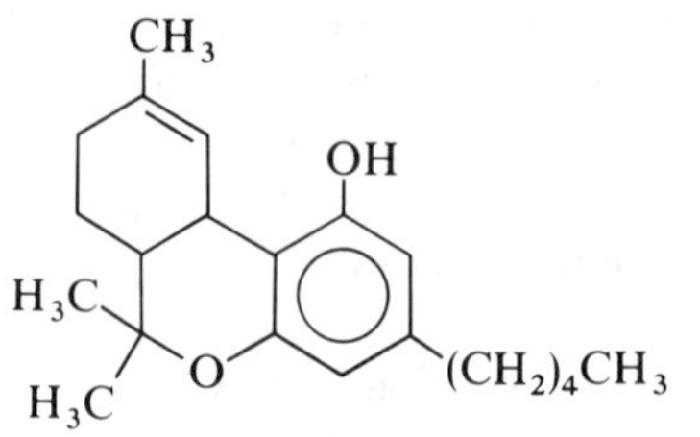

tetrahydrocannabinol
in marijuana

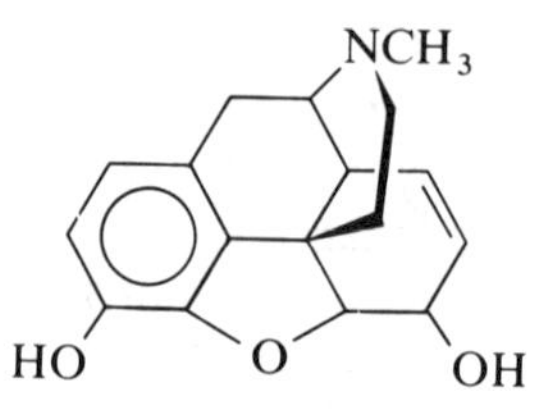

morphine
in opium poppies

4.25 The two nitrogens in the five-membered ring of caffeine are basic (although not to the same degree). Write an equation that illustrates the basic character of each.

»»» CHAPTER 5 «««

Distillation

Distillation is a general technique used for removing a solvent, purifying a liquid, or separating the components of a liquid mixture. In distillation, a liquid is vaporized by boiling, then condensed back to a liquid, called the **distillate** or **condensate**, and collected in a separate flask (the **receiving flask**). In an ideal situation, a low-boiling component can be collected in one flask, then a higher-boiling component can be collected in another flask, while the highest-boiling components remain in the original distillation flask as the **residue**.

Several types of distillation procedures have been developed. The most common are *simple distillation*, *fractional distillation*, *vacuum distillation*, and *steam distillation*. In this chapter, we will first discuss the theory of distillation, then consider briefly each type of distillation procedure. The experiments in this chapter deal only with simple and fractional distillation. Experiments involving the other two types of distillation appear elsewhere in this book.

5.1 Characteristics of Distillation

A. The Boiling Point

The **boiling point** of a liquid is defined as the temperature at which its vapor pressure equals the atmospheric pressure, and is characterized by vigorous bubbling and churning of the liquid as it vaporizes. The temperature of a boiling liquid is seldom reproducible because of the presence of impurities and the

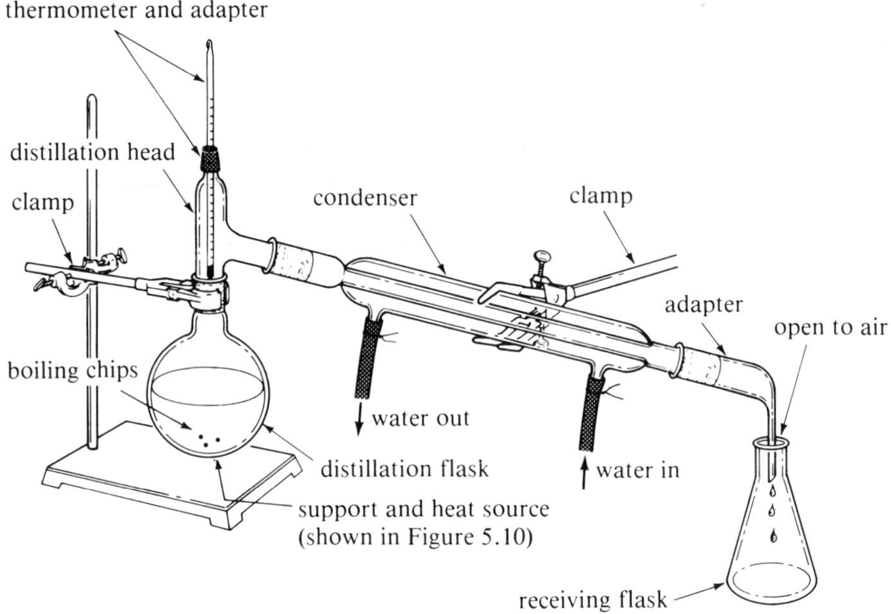

Figure 5.1 The apparatus for a simple distillation.

possibility of superheating the liquid. The actual temperature of a boiling liquid is invariably *higher* than the boiling point of the material distilling. Therefore, the boiling point that we record is measured above the surface of the liquid, where vapor and liquid are in equilibrium with each other.

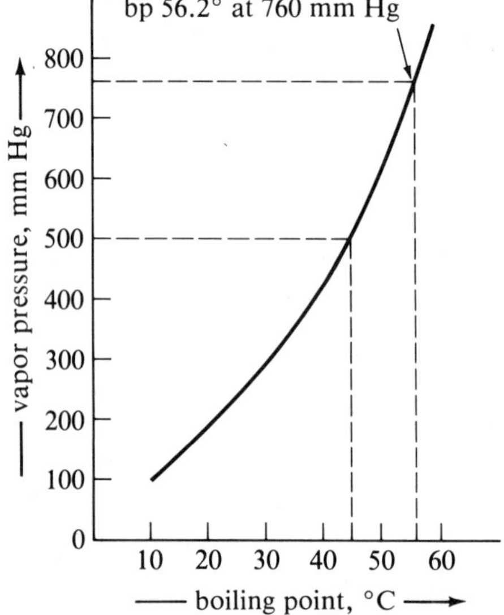

Figure 5.2 Vapor pressure–boiling point diagram for acetone, $(CH_3)_2C{=}O$.

Figure 5.1 shows the apparatus for a **simple distillation**. When the liquid in the distillation flask boils, vapor rises to the top of the flask, through the distillation head, past the thermometer, and out the side arm into the condenser. Some of the vapor condenses on the walls of the flask and head and on the thermometer. As long as vapor is flowing out the side arm, the condensed liquid on the tip of the thermometer is in equilibrium with the vapor, and an accurate boiling point can be determined. Like the melting point, a boiling point is usually reported as a range. A pure compound exhibits a range of 1–2° or less (but not all liquids with constant boiling points are single pure compounds; see page 85).

The boiling point is often reported at a particular pressure—for example, bp 83.5–84.5° (752 mm Hg). The reason for this is that boiling points vary significantly with changes in pressure. Figure 5.2 is a graph of the vapor pressure of acetone versus its boiling point. Because a liquid boils when its vapor pressure equals the pressure above it, the term "vapor pressure" in the figure can be replaced equally well by "atmospheric pressure" or "applied pressure." Note that acetone boils at about 56° at 1.0 atmosphere of pressure (760 mm Hg or 760 Torr). At 500 mm Hg pressure, the boiling point of acetone is only about 45°.

B. Distillation of a Single Volatile Liquid

When a solvent is removed by heating from a solution containing a nonvolatile component, or when a reasonably pure liquid is distilled, the observed temperature rises rapidly to the boiling point. Once the distillation apparatus has reached thermal equilibrium, the boiling point remains relatively constant, not changing more than 1–2° during the course of the distillation. A significant drop in temperature is the signal that the distillation is complete (see Figure 5.3).

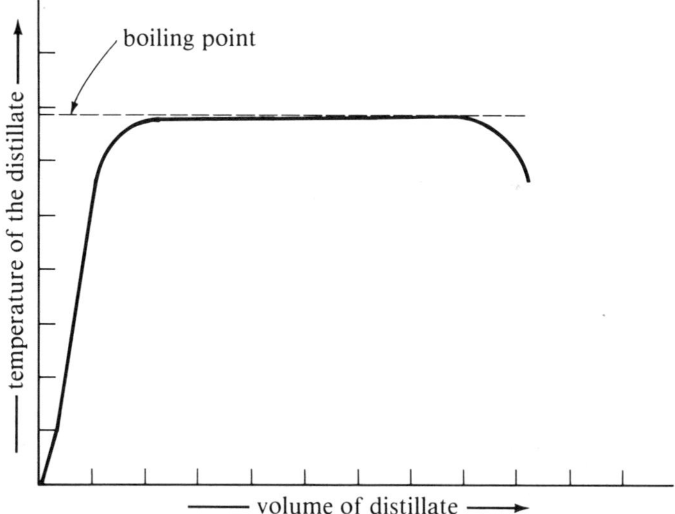

Figure 5.3 *Volume versus temperature of the distillate in the distillation of a pure liquid.*

C. Distillation of Mixtures

Because most distillations are performed with mixtures, let us consider the theory of distillation of an ideal (noninteracting) mixture of two miscible liquids A and B, where A is the lower-boiling component. If the difference in boiling points between A and B is large (100° or more), then the distillation temperature will rise to the boiling point of compound A, the lower-boiling component, and remain constant while A is distilling. When all of compound A has distilled, the temperature will rise to the boiling point of compound B, the higher-boiling component, and remain constant while B is distilling. Unfortunately, this type of distillation is rarely encountered in laboratory practice because separations usually involve mixtures of compounds with boiling points closer together than 100°.

A more common experience is that the distillation temperature rises more or less steadily during the distillation because *mixtures* of A and B distil. The first, lower-boiling portions of the distillate contain more of A than of B, but this distillate will not be pure A. Similarly, the last, higher-boiling portions of the distillate will be predominantly B, but will also contain some A. Although enrichment of A and B in the first and last portions of the distillate can be accomplished, neither pure A nor pure B will be obtained. The reason for this is that B has a significant vapor pressure, even at temperatures below its boiling point. The vapor pressure of B in a boiling mixture of A and B is a function of a number of factors; to discuss these, we must consider Dalton's law and Raoult's law.

Dalton's law and Raoult's law. A liquid boils when its vapor pressure equals the atmospheric pressure. **Dalton's law of partial pressures** states that the total pressure of a gas is the sum of the partial pressures of its individual components. Thus, the total vapor pressure of a liquid (P_{total}) is the sum of the partial vapor pressures of its components. In our example, the mixture of A and B boils when the sum of the two partial pressures (P_{A} and P_{B}) equals the atmospheric pressure.

$$P_{\text{total}} = P_{\text{A}} + P_{\text{B}}$$

The **mole fraction *X***, a measure of concentration, is the number of moles of one particular component in a mixture divided by the total number of moles (all components) present.

$$X_{\text{A}} = \frac{\text{moles of A}}{\text{moles of A} + \text{moles of B}}$$

$$X_{\text{B}} = \frac{\text{moles of B}}{\text{moles of A} + \text{moles of B}}$$

where X_{A} and X_{B} are the mole fractions of A and B in the liquid

Raoult's Law states that, at a given temperature and pressure, the partial vapor pressure of a compound in an ideal solution is equal to the vapor pressure of that pure compound multiplied by its mole fraction in the liquid. By Raoult's law, if P^0_{A} represents the vapor pressure of *pure* A at a given temperature, then the partial vapor pressure of A (P_{A}) in a solution is equal to $X_{\text{A}}P^0_{\text{A}}$. Similarly, the partial vapor pressure of B (P_{B}) at that temperature is $X_{\text{B}}P^0_{\text{B}}$.

for the liquid mixture:

$$P_A = X_A P_A^0 \qquad \text{and} \qquad P_B = X_B P_B^0$$

and, because $P_{total} = P_A + P_B$,

$$P_{total} = X_A P_A^0 + X_B P_B^0$$

From Dalton's law and Raoult's law, we conclude that the total vapor pressure of the liquid is a function of the vapor pressures of the pure components and their mole fractions in the mixture.

Composition of the liquid. Distillation is a dynamic process. Vapor is removed as condensed distillate, while more vapor is generated by the boiling liquid. During the course of distillation, the liquid mixture of A and B becomes progressively poorer in the lower-boiling component A and richer in B. The combination of Dalton's law and Raoult's law shows this mathematically.

To show the changing composition, let us begin with a mixture of A (bp 80.1°) and B (bp 110.6°) in a molar ratio of 1:1 ($X_A = 0.50$ and $X_B = 0.50$). Compound A boils at a lower temperature than does compound B; therefore, P_A^0 is larger than P_B^0. For this reason, compound A contributes a larger partial vapor pressure ($P_A = X_A P_A^0$). At the start of the distillation, the vapor from the boiling mixture contains more A than B. As the distillation proceeds, the boiling liquid contains progressively less A: X_A decreases and X_B increases.

At the start, greater value for P_A^0 means more A distils.

$$P_{total} = X_A P_A^0 + X_B P_B^0$$

Later, greater value for X_B means an increasing amount of B distils.

Because P_B^0 is lower than P_A^0, the temperature of the liquid must be increased to maintain a boil as the liquid accumulates a greater concentration of B. Figure 5.4 is a plot of the relative concentrations of A and B versus the temperature of the boiling liquid. In a distillation, we would start at one point on the curve and move to the right.

Composition of the vapor. The mole fraction of a compound such as A in the vapor above a boiling mixture (not in the boiling liquid itself) is equal to the ratio of its partial vapor pressure (P_A) to the total pressure.

for the vapor:

$$X'_A = \frac{P_A}{P_{total}} \qquad X'_B = \frac{P_B}{P_{total}}$$

where X'_A and X'_B are the mole fractions of A and B in the vapor

At the start of the distillation of a 1:1 mixture of A and B, the vapor contains a greater mole fraction of A because P_A is greater than P_B. As the distillation proceeds, P_A decreases because of the decreasing amount of A in the liquid, and thus X'_A decreases.

Figure 5.5 is a plot of the vapor composition versus the temperature of the vapor. The temperature of the vapor gradually rises as the mole fraction of B that is distilling increases.

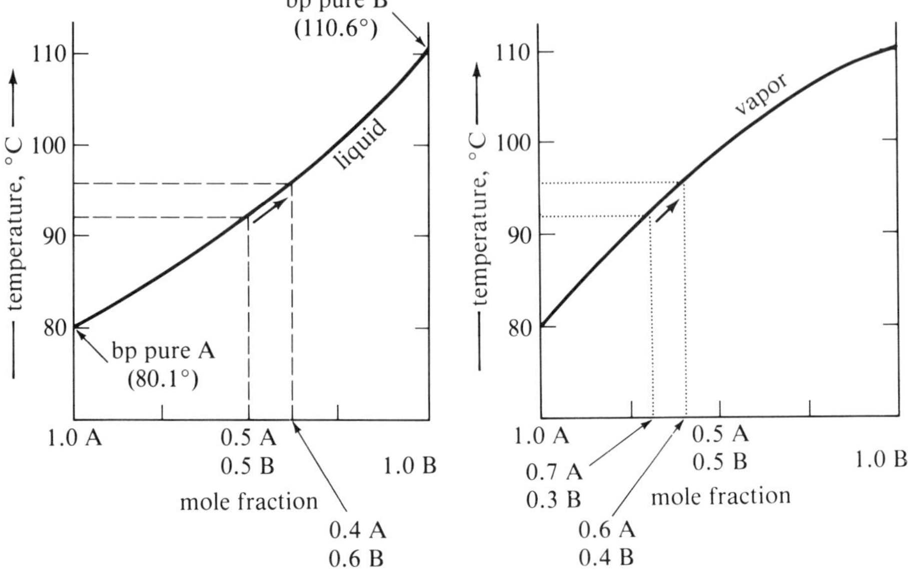

Figure 5.4 *The boiling point of an ideal binary liquid solution versus mole fractions of components in the liquid. (The arrow on the curve shows the direction of change in temperature and composition as the distillation proceeds.)*

Figure 5.5 *The boiling point of an ideal binary liquid solution versus mole fractions of components in the vapor.*

Relating liquid and vapor compositions. Generally, the curves for liquid and vapor compositions versus temperature are combined into a single diagram called a *boiling point–composition diagram,* or *phase diagram,* as shown in Figure 5.6 (a combination of the two curves previously presented in Figures 5.4 and 5.5). In Figure 5.6, we show the same diagram twice to show what happens in the distillation of a 1 : 1 mixture of A and B. From the first diagram, we see that a 1 : 1 mixture of A and B boils at 92°, and that the vapor at 92° contains approximately 70 mole % A and 30 mole % B (mole fractions 0.7 and 0.3, respectively). (If we know the vapor pressures of pure A and B, we could calculate this composition from the equations on page 83.)

As the distillation continues, more A is distilled than B; therefore, the liquid contains a progressively larger percentage of B. When the boiling liquid contains 40% A and 60% B (mole fractions 0.40 and 0.60, respectively) its temperature is 95°. At 95°, however, the vapor contains 60% A–40% B (mole fractions 0.6 and 0.4). These compositions are marked in the second diagram in Figure 5.6. As the distillation continues, we move farther to the right on each curve. The net results of the changing compositions are (1) a steady increase in the boiling point, and (2) a distillate containing progressively less A and more B.

In our example, a distillate of pure A could never be obtained. However, the early portion of the distillate (containing 80% or 90% A, for example) could be redistilled, yielding somewhat purer A at the start of the distillation. Then, *this* distillate could be redistilled, yielding even purer A. Because these distillations

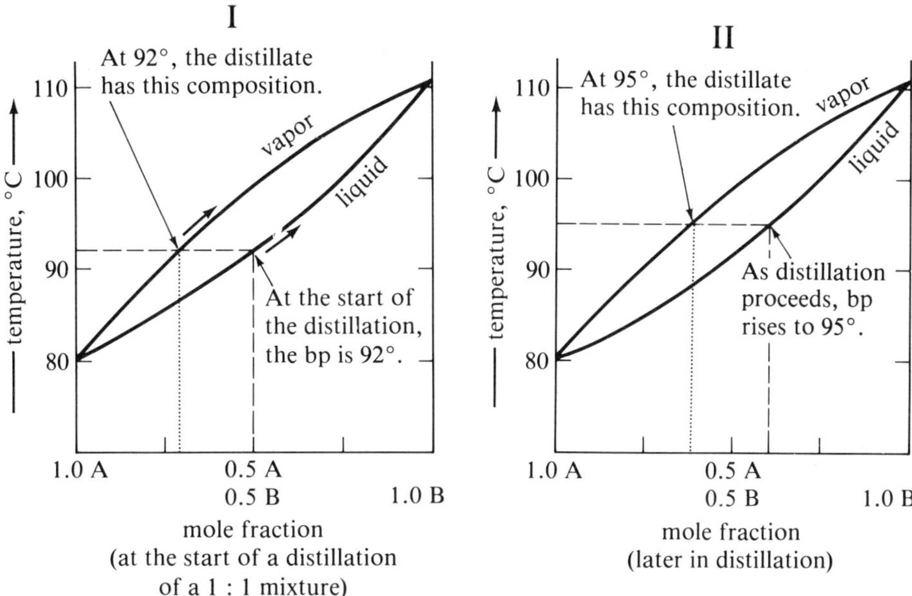

Figure 5.6 Boiling point–composition diagram, or phase diagram, for an ideal binary solution. (A distillation of a mixture of A and B would begin at the liquid of that composition and progress from there to the right along the lower curve. The composition of the vapor at each temperature is read from the upper curve.)

would be tedious and time-consuming, *fractional distillation* was developed. We will discuss fractional distillation in Section 5.2.

D. Azeotropes

Because of intermolecular interactions, such as hydrogen bonding, many binary mixtures do not follow Raoult's Law and do not give rise to the idealized phase diagram shown in Figure 5.6. For example, a pair of liquids that form an **azeotrope** (a mixture that distils at a constant boiling point and with a constant composition) might show the type of phase diagram depicted in Figure 5.7. The low point on the graph represents the azeotropic mixture, which boils at a constant boiling point just as a pure compound does. Note that the boiling point of the azeotrope shown in Figure 5.7 is *lower* than that of either pure component. If this mixture has a greater percentage of one component than will distil with the azeotrope, this compound will distil as a pure component only after the lower-boiling azeotrope has distilled.

Although most azeotropes have boiling points lower than those of their components, some azeotropes have boiling points intermediate between those of their components, while other azeotropes have higher boiling points than those of the components. For example, the azeotrope of α-bromotoluene (bp 183.7°) and n-octanol (bp 195°) boils at an intermediate temperature of 184.1°, and the azeotrope of acetone (bp 56°) and chloroform (bp 61°) boils at a higher temperature of 64.7°. Azeotropes of three or more components, as well as binary azeotropes, are also well-known.

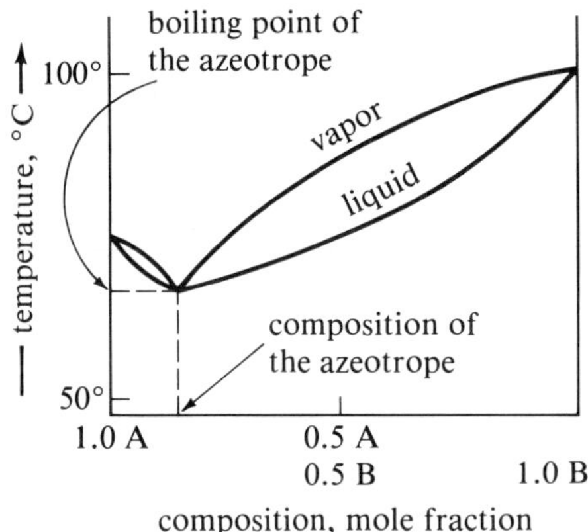

Figure 5.7 *Phase diagram for a typical binary mixture that forms a low-boiling azeotrope.*

A common azeotropic mixture encountered in the laboratory (but by no means the *only* common one; see Table 4.3, page 67) is the binary azeotrope of ethanol (CH_3CH_2OH) and water. Ethanol for beverage purposes is made by the fermentation of aqueous sugars and starches, a series of reactions catalyzed by enzymes found in certain types of yeast. The fermentation of the carbohydrate mixture stops when the alcoholic content of the mixture reaches 12–14% because ethanol at this concentration inhibits the action of the enzymes. To yield a higher concentration of alcohol, the fermentation mixture is distilled. Because ethanol (bp 78.3°) and water (bp 100°) form a low-boiling azeotrope (bp 78.15°), the first major fraction to distil is the azeotrope. This particular azeotrope contains 95% ethanol and 5% water. After all the ethanol has distilled as part of the azeotrope, the remaining water distils at its normal boiling point. Pure ethanol cannot be distilled from an ethanol–water solution that contains 5% or more water.

5.2 Types of Distillation Procedures

A. Simple Distillation

Simple distillation involves boiling a liquid and collecting the condensed vapors of successively higher boiling points in one or more portions, or fractions, in a receiving flask. Simple distillation may be used to separate compounds boiling more than 100° apart. More commonly, it is used to remove a low-boiling solvent from high-boiling organic compounds or simply to purify a compound or a solvent. A typical apparatus for simple distillation is shown in Figure 5.1, page 80.

B. Fractional Distillation

The technique of **fractional distillation** differs from simple distillation in only one respect. The vapor and condensate from the boiling liquid are passed through a **fractionation column** (Figure 5.8) before they reach the distillation head. This

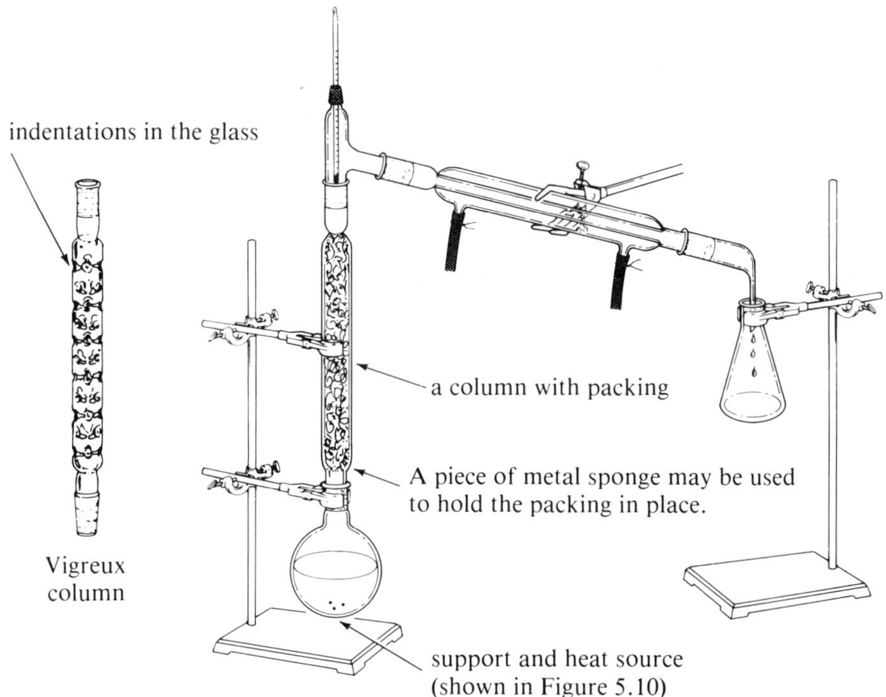

Figure 5.8 *Apparatus for a fractional distillation.*

column contains a packing such as metal turnings or glass beads. As vapor rises through this column, it condenses on the packing and revaporizes continuously. Each revaporization of condensed liquid is equivalent to another simple distillation. Each of these "distillations" leads to a distillate successively richer in the lower-boiling component. Substantial enrichment of the vapor in the lower-boiling component occurs by the time the vapor reaches the head.

With a good fractionation column and proper operation, compounds with boiling points only a few degrees apart may be separated successfully. (Of course, an azeotropic mixture can *not* be separated into its components by fractional distillation, because an azeotrope is a mixture of constant composition with a constant boiling point.)

Efficiency of the fractionation column. How well a fractionation column can separate a pair of compounds with a known boiling-point difference is called its **efficiency**. The efficiency of a column depends on both its length and its packing. For columns of the same length, increasing the surface area and the heat conductivity of the packing results in an increased efficiency.

Vigreux glass helices glass beads metal turnings metal sponge
→
increasing efficiency

The efficiency of a column is reported in **theoretical plates**, where one theoretical plate is equivalent to one simple distillation. A distillation assembly with an efficiency of two theoretical plates is therefore equivalent to two simple distillations.

A column with one theoretical plate (a simple distillation head) separates compounds with a difference in boiling points of about 100° or more, but does not give good separation of compounds boiling closer together. At the other extreme, a column of 100 theoretical plates can separate a pair of compounds boiling as close as 2° apart. Most laboratory fractionation columns vary from 2–15 theoretical plates. For example, a 25-cm column packed with glass beads or metal turnings has an efficiency of approximately 8 theoretical plates and, at best, can separate a binary mixture with a boiling-point difference of 30–40°.

The number of theoretical plates is only one consideration in the choice of a column. Some fractionation columns have a greater *hold-up* than others. This term refers to the quantity of condensed vapor that remains on the packing when the distillation is stopped. If a very small amount of liquid is to be distilled, a column with less hold-up, even though it is less efficient, might be the column of choice. One such column, which contains no packing, is the **Vigreux column** (Figure 5.8).

C. Vacuum Distillation

As its name implies, **vacuum distillation** (simple or fractional) is distillation under reduced pressure, as low as 1 mm Hg or even less. Because the reduction in pressure lowers the boiling point, vacuum distillation is useful for distilling high-boiling or heat-sensitive compounds. The procedures used in simple vacuum distillation will be discussed in Experiment 22.1, page 396.

D. Steam Distillation

Steam distillation is the distillation of an immiscible mixture of organic compounds and water (steam). The experimental techniques for steam distillation will be discussed in Experiment 10.4.

Immiscible mixtures do not behave like solutions. The components of an immiscible mixture boil at *lower temperatures than the boiling points of any of the components*. Thus, a mixture of high-boiling organic compounds and water can be distilled at a temperature less than 100°, the boiling point of water. Natural products, such as flavorings and perfumes from leaves and flowers, are sometimes separated from their sources by steam distillation. This technique is also used occasionally in the organic laboratory in lieu of vacuum distillation or other separation techniques.

The principle that gives rise to steam distillation is that the total vapor pressure of a mixture of immiscible liquids is equal to the *sum of the vapor pressures of the pure individual components*. The total vapor pressure of the mixture thus equals atmospheric pressure (and the mixture boils) at a lower temperature than the boiling point of any one of the components singly.

for immiscible liquids:

$$P_{\text{total}} = P_A^0 + P_B^0$$

vapor pressures of pure components

Note that this boiling behavior is different from that of miscible liquids, where the total vapor pressure is the sum of the *partial* vapor pressures of the components.

for miscible liquids:

$$P_{\text{total}} = \underbrace{X_A P_A^0}_{} + \underbrace{X_B P_B^0}_{}$$

partial vapor pressures

5.3 Refractive Index

Two analytical tools used in conjunction with distillation are refractive index and gas–liquid chromatography. Gas–liquid chromatography is discussed in detail in Chapter 7. In this section, we will consider the refractive index.

Refraction is the bending of a ray of light as it passes obliquely from one medium to another of different density. Refraction arises from the fact that light travels more slowly through a more dense substance. In organic chemistry, we are interested in the refraction of light as it passes from air through a liquid layer. Refraction is useful because the degree of refraction varies, depending on the liquid compound.

The **refractive index (*n*)** of a particular substance is defined as the ratio of the speed of light in a vacuum or in air (the values are very close) to the speed of light in the substance.

$$\text{refractive index, } n = \frac{\text{speed of light in air}}{\text{speed of light in the substance}}$$

The refractive index is measured in an instrument called a **refractometer**, which measures the angle of refraction of light between the liquid and a prism.

Different wavelengths of light are refracted different amounts. This is the reason that sunlight can be split into the visible spectrum (a rainbow) by droplets of water. When the refractive index is used as a physical constant, only one wavelength of light is used, usually the sodium D line, at 589.3 nm. This single wavelength can be obtained from either a sodium lamp or an ordinary white light with a prism system.

Besides being dependent on wavelength, the refractive index is temperature-dependent. For this reason, the temperature is always specified when a refractive index is reported. A typical refractive index for a compound, such as for benzene or ethanol, is reported as in the following examples:

benzene, ${n_D}^{20}$ 1.5011 ethanol, ${n_D}^{20}$ 1.3611

where D refers to the wavelength (the sodium D line) and the superscript number is °C

The refractive index may be used as one piece of evidence to establish the identity of a compound. More often, the refractive index is used as a criterion of purity, especially for a series of distillation fractions. The refractive index is a very sensitive physical property. Unless a compound is extremely pure, it is almost impossible to duplicate a literature value exactly. For example, a sample of benzene from a fractional distillation might exhibit n_D^{20} 1.4990 instead of the reported value of n_D^{20} 1.5011. However, the closer the observed refractive index is to the reported value, the more pure the compound is likely to be.

A. Correcting for Temperature Differences

Refractive index varies inversely with temperature. An increase in temperature causes a liquid to become less dense, and almost always causes a decrease in refractive index. It has been determined experimentally that the average temperature-correction factor for a wide variety of compounds is 0.00045 units for each degree Celsius. The following examples show how to adjust a refractive index to a different temperature.

Example 1 (adjusting to a higher temperature):

To adjust n_D^{18} 1.4370 to n_D^{20},

calculate factor:

$$2°C \times 0.00045/°C = 0.00090$$

subtract from measured value:

$$\begin{aligned} n_D^{20} &= 1.4370 - 0.0009 \\ &= 1.4361 \end{aligned}$$

Example 2 (adjusting to a lower temperature):

To adjust n_D^{23} 1.5066 to n_D^{20},

calculate factor:

$$3°C \times 0.00045/°C = 0.00135 \quad \text{(round to 0.0014)}$$

add to measured value:

$$\begin{aligned} n_D^{20} &= 1.5066 + 0.0014 \\ &= 1.5080 \end{aligned}$$

B. Steps in Using a Refractometer

Your instructor will give you specific instructions on the use of the refractometer in your laboratory. The following general steps, however, apply to most refractometers.

1) The prism section can be opened mechanically. If the prism surfaces are not absolutely clean, clean them with a few drops of ethanol and a lens-cleaning tissue. (CAUTION: The prism surfaces are very easily scratched. Do not touch these surfaces with anything hard, such as an eye dropper, the end of a glass rod, or a metal spatula. Clean them gently by dabbing with the tissue.)

2) Place 1–3 drops of the liquid to be analyzed on the lower prism surface, then close the prism section. The prism surfaces fit closely together; when the section is closed, the liquid is forced to spread out as a thin film.

3) Position the lamp to shine directly into the glass prism of the instrument.

4) Look through the eyepiece and use the appropriate knobs to focus the cross hairs and to bring a light area and a dark area into view. The appearance of the field is illustrated in Figure 5.9. If the dividing line between the light and dark areas is fringed with red or blue, the prism system is not directing the sodium

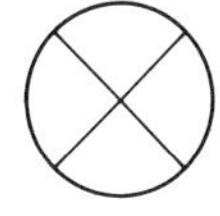
the cross hairs

the light and dark areas

properly adjusted for reading the refractive index

Figure 5.9 The field of view through the eyepiece of a refractometer.

D line exactly through the sample. The prisms may be adjusted by a dispersion correction knob. The field is in proper adjustment when there is a sharp dividing line, with no color, between the light and dark areas.

5) To determine the refractive index, move the dividing line between the light and dark areas *exactly* to the center of the cross hairs, read the value for the refractive index, and record the temperature.

6) Clean the instrument as was described in Step (1).

Problems

5.1 A mixture of two miscible liquids with widely different boiling points is distilled. The temperature of the distilling liquid is observed to plateau, then drop before rising again. Explain the temperature drop.

5.2 If a mixture is fractionally distilled rapidly, the separation of its components is poorer than if the mixture is distilled slowly. Explain.

5.3 Why is a fractionation column packed with glass helices more efficient than a Vigreux column of the same length and same diameter?

5.4 From the graph in Figure 5.2 (page 80), estimate:
- (a) the boiling point of acetone at 1000 mm Hg;
- (b) the vapor pressure of acetone at room temperature (23°C);
- (c) the per cent of the total pressure contributed by acetone above an open beaker of this compound at room temperature (23°) and 760 mm.

5.5 Near atmospheric pressure, a 10-mm drop in pressure usually lowers a boiling point by about 0.5°. Using this rule of thumb, predict the boiling points of the following compounds at the pressure indicated:
- (a) water at 745 mm Hg
- (b) dichloromethane (bp 41° at 760 mm) at 737 mm
- (c) benzene (bp 80° at 760 mm) at 765 mm

5.6 At low pressures such as used in vacuum distillation, reducing the pressure by half reduces a boiling point by about 10°. If a compound boils at 180° at 10 mm Hg, what would be its approximate boiling point at:

(a) 15 mm Hg; (b) 2.0 mm Hg?

5.7 Correct the following refractive indexes to 20°C.

(a) n_D^{22} 1.4398 (b) n_D^{30} 1.4702 (c) n_D^{16} 1.3962 (d) n_D^{18} 1.4022

5.8 A 50% aqueous solution of ethanol (50 mL total) is distilled and collected in 10-mL fractions. Predict the boiling range of each fraction.

5.9 Calculate the mole fraction of each compound in the following mixtures:

(a) 95.0 g CH_3CH_2OH and 5.0 g H_2O

(b) 10.0 g CH_3OH, 10.0 g CH_3CH_2OH, and 10.0 g $CH_3CH_2CH_2OH$

5.10 (a) Using the graph in Figure 5.6, determine the composition of the liquid solution boiling at 100°.

(b) What is the composition of the vapor being given off?

5.11 A mixture of immiscible liquids (both water-insoluble) is subjected to steam distillation. At 90°, the vapor pressure of pure water is 526 mm Hg. If the vapor pressure of compound C is 127 mm Hg and that of D is 246 mm Hg at 90°:

(a) What is the total vapor pressure of the mixture at 90°?

(b) Would this mixture boil at a temperature above or below 90°?

(c) What would be the effect on the vapor pressure and boiling temperature by doubling the amount of water used?

5.12 A mixture of ideal miscible liquids E and F is distilled at 760 mm Hg pressure. At the start of the distillation, the mole per cent of E in the mixture is 90.0, while that of F is 10.0. The vapor is condensed and found to contain 15.0 mole per cent of E and 85.0 mole per cent of F. Calculate:

(a) the partial vapor pressures (P) of E and F in this mixture;

(b) the vapor pressures (P^0) of pure E and F.

5.13 Given the following mole fractions and vapor pressures for miscible liquids G and H, calculate the composition (in mole per cent) of the vapor from a distilling ideal binary solution at 150° and 760 mm Hg.

for the solution:

$X_G = 0.40$ $\qquad X_H = 0.60$

$P^0_G = 1710$ mm Hg $\qquad P^0_H = 127$ mm Hg

5.14 Compound I has a refractive index of 1.4577 at 20°, while compound J (miscible with compound I) has a refractive index of 1.5000 at the same temperature.

(a) Experimentally, how could you determine if the refractive indexes of mixtures of I and J follow a linear relationship to the mole per cent compositions of the mixtures?

(b) If the relationship is linear, what is the composition (in mole per cent) of a mixture that has a refractive index of 1.4678 at 20°?

EXPERIMENT 5.1

Simple Distillation

Discussion

This experiment is a simple distillation of a mixture of two esters (RCO_2R'): ethyl acetate and *n*-propyl acetate.

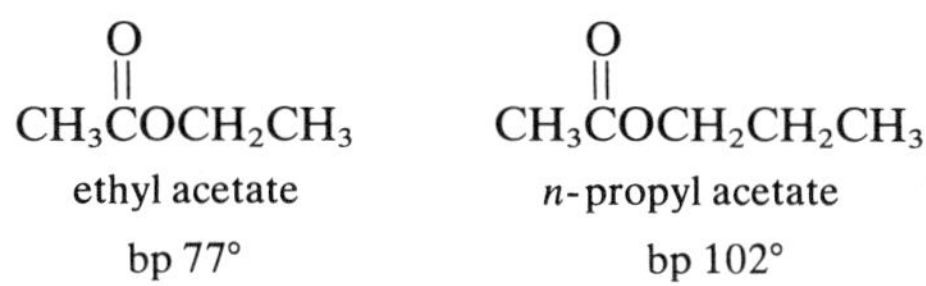

We will first describe the general steps used in any simple distillation, then mention some specific features of this experiment.

Steps in a Simple Distillation

The apparatus for a simple distillation is shown in Figure 5.1, page 80. Study this figure carefully, noting the placement and clamping of the distillation flask, the distillation head, and the condenser. Note that water flows into the bottom of the condenser's cooling jacket and out the top. If the water inlet were at the top, the condenser would not fill. Also, note the placement of the thermometer bulb, *just below the level of the side arm* of the distillation head. If the bulb were placed higher than this position, it would not be in the vapor path and consequently would show an erroneously low reading for the boiling point.

>>>> ***SAFETY NOTE*** Distillation of noxious or toxic substances should always be carried out in a fume hood. Special precautions should also be taken with distillations of highly flammable substances, such as most solvents. Never use a burner in these cases, and avoid allowing an excess of uncondensed vapors to flow into the room.

1) *The distillation flask.* Use only a round-bottom flask, never an Erlenmeyer flask, for distillation. The flask should be large enough that the material to be distilled fills $\frac{1}{2}$–$\frac{2}{3}$ of its volume. If the flask is overly large, a substantial amount of distillate will be lost as vapor filling the flask at the end of the distillation. If the flask is too small, boiling material may foam, splash, or boil up into the distillation head, thus ruining the separation.

Grease the ground-glass joint of the flask lightly, then securely clamp the flask to a ring stand or rack. Before adding liquid, support the bottom of the flask with an iron ring and a heating mantle or wire gauze (if a burner must be used), as shown in Figure 5.10. The heating mantle should fit snugly around the flask. Introduce the liquid into the distillation flask, using a funnel with a stem to prevent the liquid from contaminating the ground-glass joint. Finally, add three or four boiling chips. (CAUTION: Never add boiling chips to a hot liquid!)

2) *The distillation head.* Grease the ground-glass joints of the distillation head lightly and place the head on the flask. It is generally not necessary to clamp the head. (Do *not* attach the thermometer at this time.)

3) *The condenser.* Grease the ground-glass joints of the condenser lightly and attach rubber tubing firmly to the jacket inlet and outlet (which should *not* be greased). A strong clamp (oversized, if available) is needed to

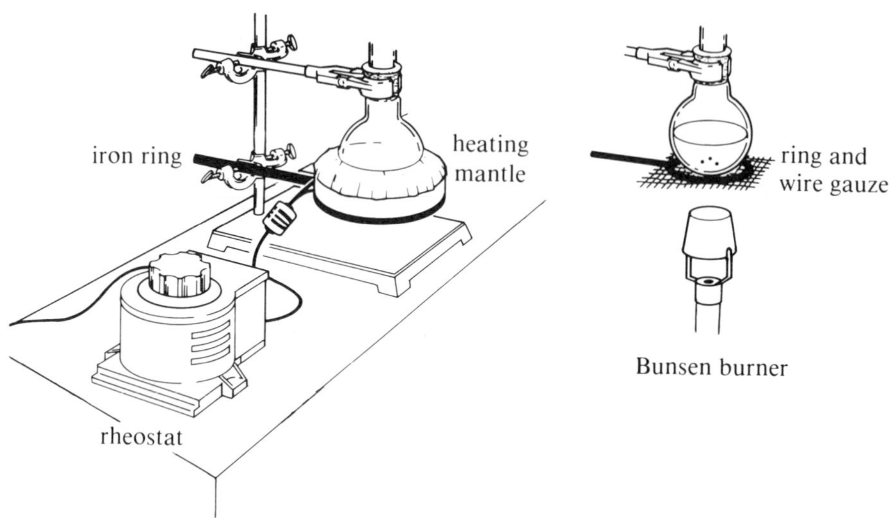

Figure 5.10 *Heat sources and supports for distillation flasks.*

hold the condenser in place. Because of the weight and angle of the condenser, it will tend to pull away from the distillation head. For this reason, check the tightness of this joint frequently before and during a distillation.

Attach the rubber tubing from the lower end of the condenser to an adapter on the water faucet. Place the end of the upper outlet tubing from the condenser in the sink or drainage trough. Turn on the water cautiously; after it fills the condenser and flows out, adjust the flow to a "heavy trickle." Water should flow, not drip, from the outlet tubing; however, a forceful flow of water is likely to cause the tubing to pop off the condenser. If you leave a distillation unattended for even a short while, twist short pieces of wire around the tubing on the condenser inlet and outlet to secure them. Because of variations in water pressure and because many faucets tend to tighten gradually, check the flow of water frequently during the distillation.

4) The adapter. The adapter directs the flow of distillate plus uncondensed vapors into the receiving flask. Figure 5.1 (page 80) shows the usual type of adapter, but a vacuum adapter (Figure 5.11) may also be used. If desired, a piece of rubber tubing attached to the vacuum adapter can be used to carry fumes to the floor. (Rubber tubing is no substitute for a fume hood, however.) Whichever type of adapter is used, grease its joint lightly before attaching it to the condenser. A rubber band may be used to secure the adapter, as shown in Figure 5.11.

5) The receiving flask. Almost any container can be used as a receiver, as long as it is large enough to receive the expected quantity of distillate. An Erlenmeyer flask is recommended. A beaker is not recommended because its wide top allows vapors and splashes to escape and allows dirt to get into the distillate. Either set the receiving flask on the bench top or clamp it in place. (It is not good practice to prop up a receiving flask on a stack of books.) If you are collecting several fractions, prepare a series of clean, dry, tared (weighed

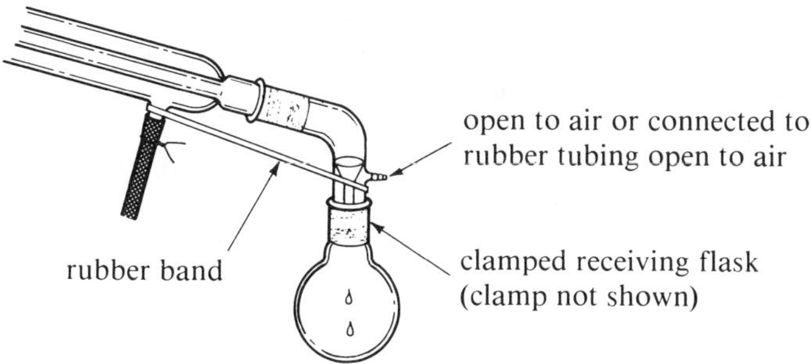

Figure 5.11 *A vacuum adapter.*

empty) flasks. If the volume, rather than the weight, of distillate is to be determined, you may use a clean, dry graduated cylinder as the receiver. A round-bottom flask with a ground-glass joint is also a good receiver. This type of flask will fit onto a vacuum adapter, but must be clamped in place.

>>>> ***SAFETY NOTE*** When distilling at atmospheric pressure, *always* leave the apparatus open to the air at the adapter–receiver end. If you attempt a distillation with a closed system, the pressure build-up inside the apparatus may cause it to explode.

6) The thermometer. Attach the thermometer last (and remove it first) because thermometers are expensive and easily broken. The easiest type of thermometer to insert is one with a ground-glass joint that fits a joint at the top of the distillation head. Neoprene adapters are available for attaching ordinary thermometers. Alternatively, a short piece of rubber tubing used as a sleeve can be used to hold the thermometer in place. A one-hole rubber stopper is not recommended because the hot vapors and condensate of the distilling liquid may dissolve the rubber, which can discolor the distillate. When attaching the thermometer, be sure to place the bulb just below the level of the side arm, as shown in Figure 5.1.

7) The actual distillation. Before proceeding, check the water flow through the condenser and make sure that all ground-glass joints are snug. Plug the heating mantle into a rheostat, then plug the rheostat into the wall socket. If you use a burner, check the vicinity for flammable solvents. (Do not use a burner when distilling a flammable compound!)

Slowly heat the mixture in the distilling flask to a gentle boil. You will then see the **reflux level** (the ring of condensate, or upper level of vapor condensing and running back into the flask) rise up the walls of the flask to the thermometer and side arm. At this time, the temperature reading on the thermometer will rise rapidly until it registers the *initial boiling point,* which should be recorded. The vapors and condensate will pass through the side arm and into the condenser, where most of the vapor will condense to liquid, and will finally drip from the adapter into the receiving flask.

The proper *rate of distillation* is one drop of distillate every 1–2 seconds. This rate is achieved by controlling the amount of heat supplied to the distillation flask. A too slow rate means that not enough vapor is passing the thermometer to give an accurate boiling point. A too rapid rate leads to a lack of separation of components and also to uncondensed vapor being carried through the condenser and into the room. It is generally necessary to increase the amount of heat applied to the distillation flask (by increasing the rheostat setting) during the course of a distillation.

8) Collecting the fractions. Volatile impurities are the first compounds to distil. This first fraction, called the **fore-run**, is generally collected separately. When the temperature has risen to the desired level and has been recorded, place a fresh receiver under the adapter to collect the main fraction. In some cases, the main fraction can be collected in a single receiver. In other cases, it should be collected as a series of smaller fractions. Each time you change a receiver, note the temperature reading and record the boiling range of the fraction. After checking the purity of a group of fractions, you may want to combine some of these fractions later. Figure 5.12 shows typical notebook records of distillations collected in several fractions.

Impurities that are higher boiling than the desired material are generally not distilled, but are left in the distillation flask as the residue. If higher-boiling impurities are present in large quantities, the temperature may rise from the desired level as the impurities begin to distil. However, the temperature frequently *drops* after the main fractions have distilled. This happens because not enough vapor and condensate are present in the head to keep the thermometer bulb hot. If the temperature drops at the end of a distillation, the last temperature to record is the *highest* temperature, before the drop occurred.

At the conclusion of a distillation, remove the heat source. Turn off a heating mantle and lower it from the flask immediately. Allow the entire apparatus to cool before dismantling it.

>>>> ***SAFETY NOTE*** Never carry out a distillation to dryness, but always leave a small amount of residue in the distillation flask. The presence of boiling residue will prevent the flask from overheating and breaking and will also prevent the formation of pyrolytic tars (difficult to wash out).

In the simple distillation of the mixture of ethyl acetate and *n*-propyl acetate, you will collect the distillate in ten fractions and measure the refractive index of each. Using this information, you will construct two graphs: (1) boiling point versus total volume of distillate; and (2) refractive index versus total volume of distillate. (Sample graphs are shown in Figure 5.13.)

If you also carry out Experiment 5.2 (fractional distillation), you will be able to compare the graphs of each experiment to see how the two types of distillation differ in their ability to separate mixtures. If you are to analyze the distillation fractions using gas–liquid chromatography (Experiment 7.3), it is important that you store the distillation fractions in *tightly stoppered containers* so that their compositions do not change by evaporation.

Date: 10/15/83

	tare weight	Bp	gross weight	net weight
Fraction 1	25.93 g	35-98°	30.32 g	4.39 g
2	25.64	98-100	35.21	9.57
3	25.99	100-100	37.82	11.83
4	26.02	100-101	36.59	10.57
5	25.24	101-102	31.01	5.77
6	26.15	102-110	28.76	2.61
residue	—	—	—	approx. 3 g

Date: 10/17/83

	Bp °C (749 mm)	Volume, mL	n_D^{20}
Fraction 1	50-110	0.5	1.3372
2	110-112	1.0	1.4952
3	112-114	15.5	1.4960
4	114-125	6.0	1.4982
residue	–	2.0	1.5047

Figure 5.12 Typical notebook records of distillations.

Experimental

EQUIPMENT:

condenser
condenser adapter
distillation head with thermometer adapter
droppers or disposable pipets
10-mL and 50-mL graduated cylinders
heating mantle for 100-mL flask, with rheostat
long-stemmed funnel
refractometer
100-mL round-bottom flask
eleven test tubes with corks
test-tube holder
thermometer

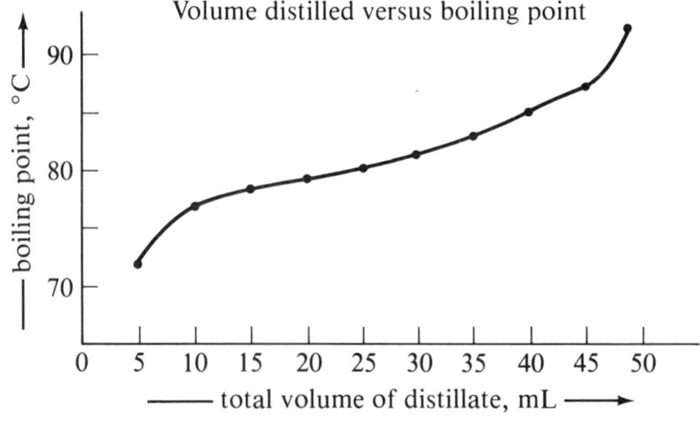

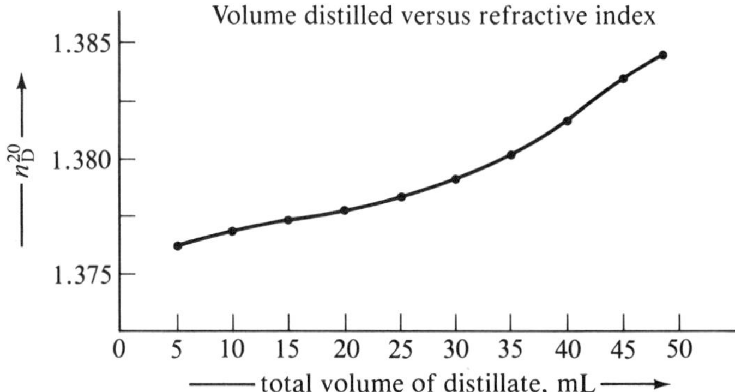

Figure 5.13 Sample graphs for the simple distillation of a mixture of ethyl acetate and n-propyl acetate.

CHEMICALS:

ethyl acetate, 25 mL
n-propyl acetate, 25 mL

TIME REQUIRED: 2 hours plus $\frac{1}{2}$ hour for refractive index measurement

STOPPING POINT: after the distillation

>>>> ***SAFETY NOTE*** Ethyl acetate and *n*-propyl acetate are volatile and flammable. There must be no flames in the laboratory.

PROCEDURE

Add 25 mL of ethyl acetate, 25 mL of *n*-propyl acetate, and 3–4 boiling chips to a 100-mL round-bottom flask, clamped to a ring stand and supported by a 100-mL heating mantle as shown in Figure 5.10. Assemble the distillation apparatus as shown in Figure 5.1 (page 80), following the instructions in the discussion of this experiment.

Assemble eleven clean, dry test tubes of the same size: add 5.0 mL of water to one and set it in a test tube rack. The volume of liquid in this test tube is used to estimate (by comparison) the volumes of the distillation fractions.

Plug the heating mantle into a rheostat, then plug the rheostat into the wall socket. A setting of 40 volts should bring this particular mixture to a boil. Increase or decrease the voltage setting to achieve a steady boil that maintains a drip rate of distillate of about 1 drop every 2 seconds.

Collect the distillate in 5-mL fractions in the test tubes. Record the temperature at the start and at the end of each fraction. Cork each test tube after the fraction has been collected to prevent evaporation.

The distillation is complete when the distillation flask is almost empty and the temperature starts to drop or fluctuate. (Because of hold-up, your last fraction will not be 5 mL.) When the apparatus is cool, transfer the residue to a small graduated cylinder and record the volume.

Measure the refractive index of each fraction, along with the refractive indexes of the starting ethyl acetate and *n*-propyl acetate. Record these data in your notebook in the format shown in Figure 5.12. Construct two graphs: one of boiling point versus mL distilled and the other of refractive index versus mL distilled. Use the upper value of the boiling range for each fraction as the boiling point in your graph. The graphs that you obtain should look similar to those in Figure 5.13.

Problems

5.15 As a liquid begins to distil, a student notes that the boiling chips are missing. The student removes the thermometer and drops a few chips into the flask. What will probably happen? What is the correct procedure?

5.16 *n*-Propyl acetate (bp 102°) evaporates rapidly when exposed to air, but water (bp 100°) does not. Explain.

5.17 During a distillation, you observe that the thermometer bulb is dry. List at least three possible causes, and state what you should do in each case.

5.18 When a solvent is used to extract a small amount of a high-boiling product from a reaction mixture, it is common practice to first distil the solvent by simple distillation, transfer the residue to a smaller flask, and isolate the product in a second distillation. Why not just continue the first distillation to isolate the product?

5.19 Refer to the graph of n_D^{20} versus volume of distillate in Figure 5.13. Assuming a linear relationship between per cent composition (in mole per cent) and the change in refractive index, calculate the per cent composition of the distillate when:

(a) 10 mL has distilled; (b) 40 mL has distilled.

EXPERIMENT 5.2

(SUPPLEMENTAL)

Fractional Distillation

Discussion

Like Experiment 5.1, this experiment is the distillation of a mixture of ethyl acetate and *n*-propyl acetate. Unlike Experiment 5.1, you will use a fractionation column so that you can compare the efficiency of fractional distillation to that of simple distillation in separating mixtures.

Steps in Fractional Distillation

The following general procedure applies to fractional distillations in general, not only the distillation in this experiment.

1) Packing the fractionation column. The technique for packing a fractionation column depends on the packing material. Metal turnings or sponges are best pulled up into the column with a hook on the end of a wire. If the packing is glass beads, glass helices, or small metal turnings, first place a piece of metal sponge at the bottom of the column to support the packing (see Figure 5.8, page 87). Then pour or drop the pieces of packing in. Regardless of the type of packing used, it should be loosely, but uniformly, packed. "Holes" in the packing will decrease efficiency, while spots of very tight packing may plug the column.

2) Setting up the apparatus. Assemble the apparatus shown in Figure 5.8 with the fractionation column clamped in a vertical position. When high-boiling compounds are distilled, the column should be insulated around the outside with glass cloth, dry rags, or a double layer of loosely wrapped aluminum foil. Whenever practical, however, it is preferable to leave the column uncovered so that the behavior of the liquid–vapor mixture in the column can be observed. In Experiment 5.2, low-boiling compounds are distilled; therefore, the column needs no insulation.

Clamp the distillation flask ($\frac{1}{2}$–$\frac{2}{3}$ full, containing boiling chips, and with its joint lightly greased) to the fractionation column. Clamp the receiving flask in position, then insert the thermometer into the distillation head.

>>>> ***SAFETY NOTE*** Before heating, check that all joints are snug, that fresh boiling chips have been added, and that the system is open to the atmosphere at the receiver.

3) The fractional distillation. Heat the distillation flask *slowly*. When the solution boils, you will observe the ring of condensate rising up the fractionation column. If heating is too rapid and the condensate is pushed up too rapidly, equilibration between liquid and vapor will not occur and separation of the components will not be satisfactory.

If you heat the distillation flask too strongly before the column has been warmed by hot vapors and condensate, the column may *flood*, or show an excessive amount of liquid in one or more portions of the packing. Flooding is due to lack of equilibration between condensate and vapor and is more likely to occur if the packing has not been inserted uniformly. Ideally, the packing should appear wet throughout, but no portion of it should be clogged with liquid.

Flooding can be stopped by lowering the heat source. As the boiling of the liquid diminishes, the excess liquid in the column flows back into the distillation flask. At this time, resume heating, but more slowly than before. If flooding recurs, insulate the column as described in Step (2) so that the vapors will have less tendency to condense. If the flooding is due to an incorrectly packed column, cool the apparatus, repack the column, and begin again.

4) Collecting the fractions. In a fractional distillation, read the boiling points and collect the fractions just as in a simple distillation. It is always better to collect a large number of small fractions than a few large ones. Small fractions of the same composition can always be combined, but a fraction that contains too many components must be redistilled.

Experimental

EQUIPMENT:

same as in Experiment 5.1, plus:
additional condenser or fractionation column
copper turnings or other column packing
wire with hook (to prepare the fractionation column)

CHEMICALS:

same as in Experiment 5.1

TIME REQUIRED:

fractional distillation: $2\frac{1}{2}$ hours (plus 10–30 min to pack the fractionation column)
refractive indexes: $\frac{1}{2}$ hour

STOPPING POINT: after the distillation

>>>> **SAFETY NOTE** Ethyl acetate and *n*-propyl acetate are volatile and flammable. There must be no flames in the laboratory.

PROCEDURE

Assemble the fractional distillation apparatus as described in the discussion, using a 100-mL distillation flask containing 25 mL of ethyl acetate, 25 mL of *n*-propyl acetate, and 3–4 boiling chips (see Experimental Note).

Distil the mixture and collect the fractions as described in Experiment 5.1. Because of the hold-up in the fractionation column, you will collect only nine fractions. Construct a pair of graphs similar to those in Experiment 5.1. Compare the graphs from the simple distillation and the fractional distillation. Record your conclusions about the relative efficiencies of the two distillations and the relative purities of the fractions.

EXPERIMENTAL NOTE

Although copper turnings are recommended as the packing, any good packing material (such as coarse stainless steel scouring pads, glass beads, or glass helices) may be used. A Vigreux column may also be used.

»»» EXPERIMENT 5.3 «««

(SUPPLEMENTAL)

Fractional Distillation of an Unknown Mixture

Experimental

EQUIPMENT:

same as in Experiment 5.2

CHEMICALS:

unknown mixture, 50 mL*

TIME REQUIRED: same as Experiment 5.2

STOPPING POINT: after the distillation

PROCEDURE

Fractionally distil the mixture of unknown compounds (CAUTION: flammable!) using the procedure described for Experiment 5.2. From the graphs that you obtain, estimate the boiling point of each unknown and the amount (as a volume per cent) present in the unknown mixture before distillation.

* Recommended unknowns are listed in the instructor's guide.

Problems

5.20 Which of the following circumstances might contribute to column flooding and why?

(a) "holes" in column packing

(b) packing too tight

(c) heating too rapidly

(d) column too cold

5.21 Explain why flooding in the fractionation column can lead to a poor separation of distilling components.

5.22 Refer to Figure 5.12 (page 97). In each of the distillations recorded in the figure, which fractions could be combined to yield reasonably pure compounds?

5.23 A chemist has a small amount of a compound (bp 65°) that must be fractionally distilled. Yet, the chemist does not want to lose any of the compound to hold-up on the column. What can the chemist do?

»»» CHAPTER 6 «««

Introduction to Chromatography; Thin-Layer Chromatography

Chromatography is a general term that refers to a number of related techniques used for analyzing, identifying, or separating mixtures of compounds. The use of chromatographic techniques is not limited to organic chemistry—these techniques find wide use in a variety of scientific areas. For example, gas–liquid chromatography is used in criminology laboratories for blood alcohol tests; thin-layer chromatography and gas–liquid chromatography are used in environmental and biology laboratories; and all types of chromatography are used in medical laboratories for both research and routine analyses.

All chromatographic techniques have one principle in common: a liquid or gaseous solution of the sample, called the **moving phase**, is passed (moved) through an adsorbent, called the **stationary phase**. The different compounds in the sample move through the adsorbent at different rates because of physical differences (such as vapor pressure) and because of different interactions (adsorptivities, solubilities, etc.) with the stationary phase. Thus, the individual compounds in a sample become separated from one another as they pass through the adsorbent and can be either collected or detected, depending on the chromatographic technique and the quantity of sample used.

In this chapter, we will introduce column chromatography, thin-layer chromatography, and paper chromatography. One experiment, the thin-layer chromatography of some analgesic drugs, is included here, while Experiment 22.2 is a column chromatography experiment.

6.1 Column Chromatography

Column chromatography, also called **elution chromatography**, is used to separate mixtures of compounds. In this technique, a vertical glass column is packed with a polar adsorbent along with a solvent. The sample is added to the top of the column; then, additional solvent is passed through the column to wash the components of the sample, one by one (ideally), down through the adsorbent to the outlet. Figure 6.1 illustrates the column and the technique.

The sample on the column is subjected to two opposing forces: the solvent dissolving it and the adsorbent adsorbing it. The dissolving and the adsorption constitute an equilibrium process, with some sample molecules being adsorbed and others leaving the adsorbent to be moved along with the solvent, only to be re-adsorbed farther down the column. A compound (usually a nonpolar one) that is highly soluble in the solvent, but not adsorbed very strongly, moves through the column relatively rapidly. On the other hand, a compound (usually a more polar compound) that is more highly attracted to the adsorbent moves through the column more slowly.

Because of the differences in the rates at which compounds move through the column of adsorbent, a mixture of compounds is separated into *bands*, each compound forming its own band that moves through the column at its own rate. Figure 6.1 shows the formation of a pair of bands (which are not always visible).

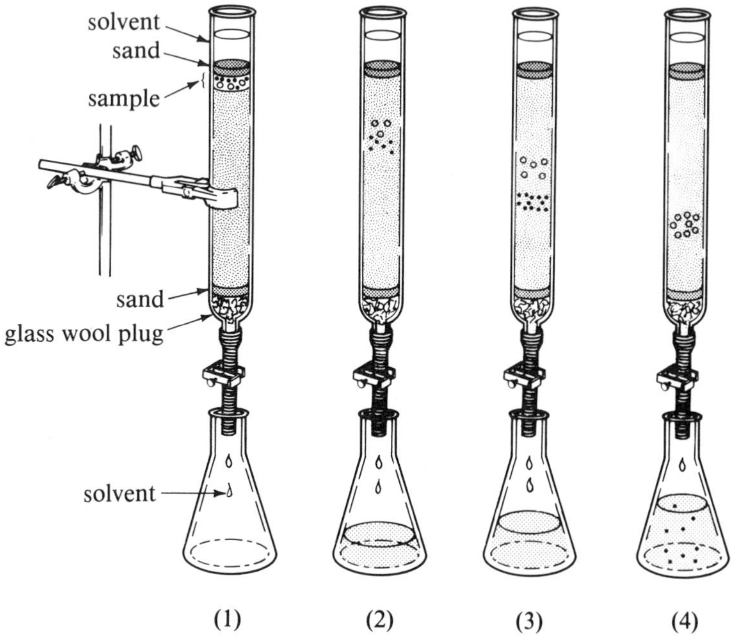

Figure 6.1 Column chromatography. (1) The sample has just started to move into the column of adsorbent. (2) and (3) As more solvent is passed through the column, the sample moves down and begins to separate into its components because of differences in attraction to the adsorbent and solvent. (4) The faster-moving compound is eluted into a flask.

Table 6.1 Order of elution of the classes of organic compounds

	Name of class	General formula
fast	alkanes	RH
	alkenes	$R_2C{=}CR_2$
	ethers	R_2O
	halogenated hydrocarbons	RX
increasing polarity	aromatic hydrocarbons	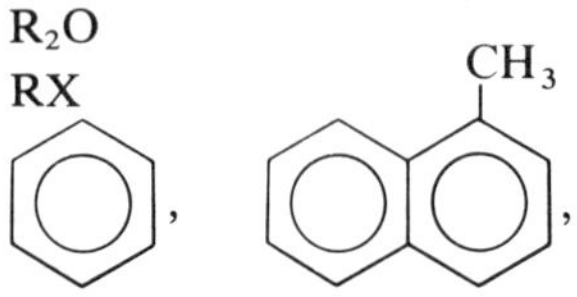 (benzene), (methylnaphthalene, CH_3), etc.
	aldehydes and ketones	$RCH{=}O$ and $R_2C{=}O$
	esters	$RC(=O)OR$
	alcohols	ROH
	amines	RNH_2, R_2NH, R_3N
slow	carboxylic acids	$RC(=O)OH$

The bands are finally washed out of, or *eluted* from, the bottom of the column, each to be collected in a separate flask. Table 6.1 lists the usual order of elution of different types of compounds.

A. The Adsorbent

The selective action of an adsorbent is very similar to the action of decolorizing charcoal (page 50), which selectively adsorbs colored compounds. In fact, activated carbon is sometimes used as an adsorbent in column chromatography. The adsorption process is due to intermolecular attractions, such as dipole–dipole attractions or hydrogen bonding.

$$\overset{\delta-}{X}-\overset{\delta+}{R} \quad \cdots \quad \delta+Al_2\,O_3\delta- \qquad\qquad RO-H\cdots Al_2O_3$$

dipole–dipole attractions — a hydrogen bond

Different adsorbents attract different types of molecules. A highly polar adsorbent adsorbs polar molecules strongly, but does not have much attraction for nonpolar molecules such as hydrocarbons. This is why nonpolar compounds are usually eluted first and more-polar compounds are eluted later. Because adsorbents differ in their adsorbing power, an adsorbent chosen for a particular chromatographic separation depends in part on the types of compounds being separated. Table 6.2 lists a few of the common adsorbents used to pack chromatography columns.

B. The Solvent

Generally, solvents are organic compounds. They, too, can be adsorbed by the packing in a chromatography column and compete with the sample components

Table 6.2 Typical adsorbents used in column chromatography[a]

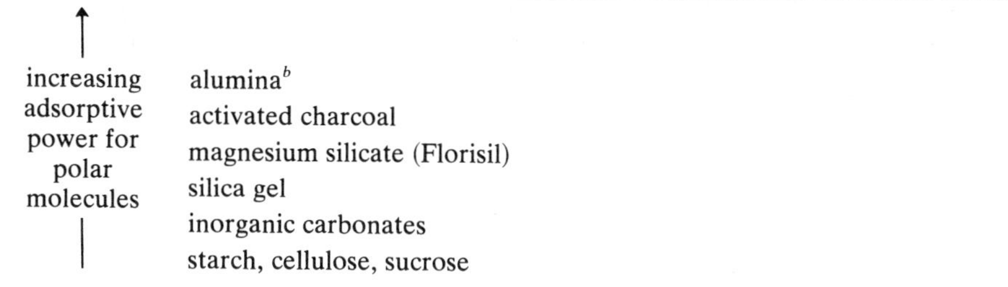

↑ increasing adsorptive power for polar molecules	alumina[b]
	activated charcoal
	magnesium silicate (Florisil)
	silica gel
	inorganic carbonates
	starch, cellulose, sucrose

[a] Generally, these adsorbents are "activated," usually by heating, to drive off adsorbed water and other volatile substances.
[b] "Basic alumina" contains hydroxide ions. "Acid-washed alumina," with no hydroxide ions, is used for base-sensitive compounds. The adsorptive power of either type of alumina can be adjusted by the addition of a small amount of water.

for positions on the adsorbent. However, nonpolar solvents are not as highly attracted to an adsorbent as are other organic compounds.

The action of a solvent, or a series of solvents, can be used to advantage in effecting a separation. Assume that you have a mixture of polar and nonpolar compounds and wish to separate the compounds by column chromatography. After adding the sample to the top of the column, you would begin by dripping a nonpolar hydrocarbon solvent like petroleum ether (or a mixture of petroleum ether containing about 1% diethyl ether) through the column. These solvent molecules are not adsorbed to any degree. Polar molecules in the sample are more attracted to the adsorbent than to the nonpolar solvent; therefore, the polar molecules are selectively held back on the column. The nonpolar components of the mixture are not strongly adsorbed and are highly soluble in the nonpolar solvent. These compounds move down the column at a relatively rapid rate to be eluted and collected.

Table 6.3 Some solvents used in column chromatography

	Name	***Structure***
increasing polarity ↓	alkanes (petroleum ether, ligroin, hexane)	—[a]
	benzene	
	halogenated hydrocarbons (dichloromethane, chloroform)	CH_2Cl_2, $CHCl_3$
	diethyl ether	$(CH_3CH_2)_2O$
	ethyl acetate	$CH_3C(=O)OCH_2CH_3$
	acetone	$(CH_3)_2C{=}O$
	alcohols (methanol, ethanol)	CH_3OH, CH_3CH_2OH
	acetic acid	$CH_3C(=O)OH$

[a] Petroleum ether and ligroin are mixtures of alkanes; see Table 3.1, page 38.

To remove the more polar compounds from the column, a more polar solvent is used. If you were using a petroleum ether–diethyl ether solvent system, you would gradually increase the per cent of diethyl ether so that the polarity of the eluting solvent system would be gradually increased. Table 6.3 lists some common solvents in order of their polarity. Note the similarities between this list and the list in Table 6.1.

6.2 Thin-Layer Chromatography

Thin-layer chromatography (abbreviated **tlc**) is a variation of column chromatography. Instead of a column, a strip of glass or plastic is coated on one side with a thin layer of alumina or silica gel (sometimes mixed with plaster of Paris, $CaSO_4$, a binder) as the adsorbent. Other adsorbents can also be used.

As an analytical tool, tlc has a number of advantages; it is simple, reasonably quick, inexpensive, and requires only small amounts of sample. Tlc is generally used as a qualitative analytical technique, such as checking the purity of a compound or determining the number of components in a mixture. Tlc is also useful for determining the best solvents for a column chromatographic separation or as an initial check on the identity of a compound (by spotting the plate with a known as well as with the sample). With calibration, tlc can be used as a quantitative technique. Preparative work can be carried out with special thick-layered tlc plates.

In a tlc analysis, about 10 μL (where 1 μL = 10^{-6} liter, or 10^{-3} mL) of a solution of the substance to be tested is placed ("spotted") in a single spot near one end of the plate, using a microcapillary. The plate is "developed" by placing it in a jar with a small amount of solvent. Figure 6.2 shows a tlc plate in a developing jar. The solvent rises up the plate by capillary action, carrying the components of the sample with it. Different compounds in the sample are carried different distances up the plate because of variations in their adsorption on the adsorbent coating. If several components are present in a sample, a column of spots is seen on the developed plate, with the more polar compounds toward the bottom of the plate and the less polar compounds toward the top.

A. The R_f Value

The distance that the spot of a particular compound moves up the plate relative to the distance moved by the solvent front is called the **retention factor**, or **R_f value**.

$$R_f = \frac{\text{distance travelled by the compound}}{\text{distance travelled by the solvent}}$$

Figure 6.3 shows how these distances are measured. When the developed tlc plate is removed from the developing jar, the solvent front is marked immediately with a pencil before the solvent evaporates. Assuming that the compound spots are colored, the spots are outlined with a pencil in case the color fades. The distance that a compound has travelled is measured from the original spot to the center of the new spot. If the spot is elongated, the "center" is estimated (usually closer to the leading edge). The distance that the solvent has travelled is measured from the original spot to the solvent front.

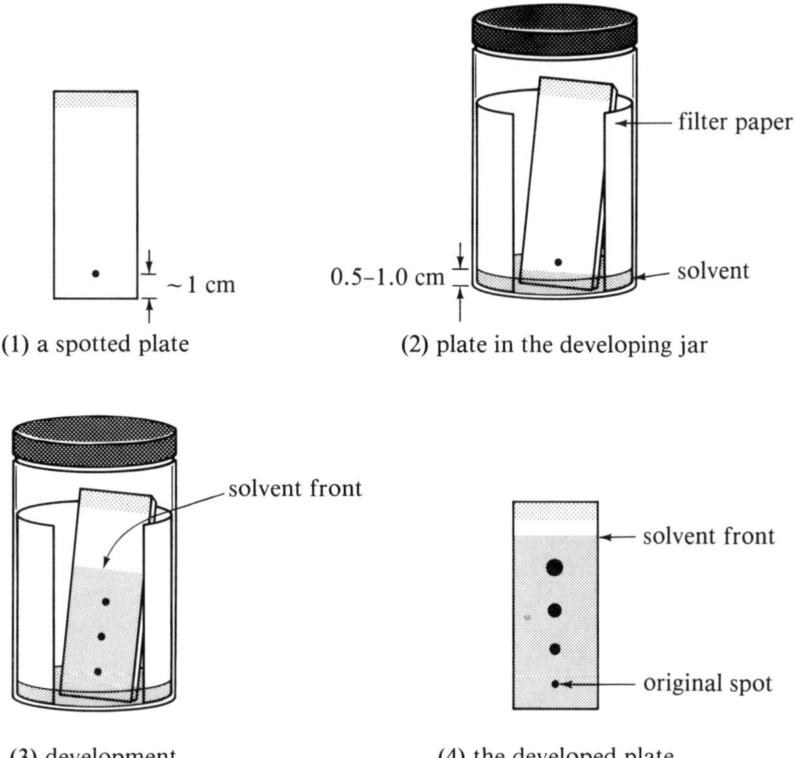

Figure 6.2 *Thin-layer chromatography. The spotted plate is placed in the developing jar with a piece of filter paper, which acts as a wick to saturate the atmosphere with solvent. Different compounds move up the plate at different rates: the less polar compounds move the fastest and are found closer to the solvent front.*

The R_f value for a compound is a constant, but only if all variables are also held constant: temperature, solvent, adsorbent, thickness of adsorbent, and amount of compound on the plate. Because it is difficult to duplicate all these factors exactly, an unknown sample is usually compared with a known compound

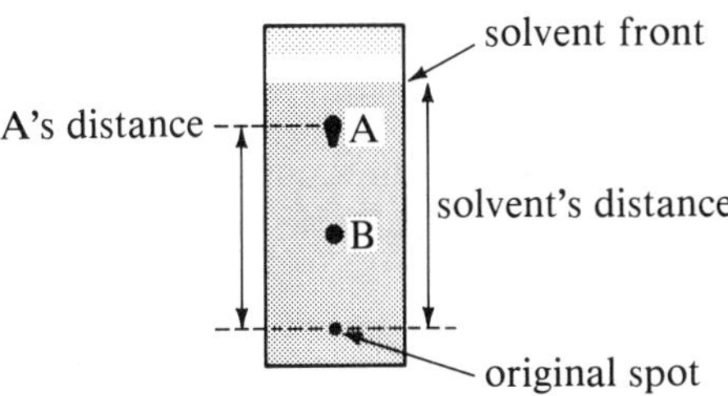

Figure 6.3 *The R_f value for compound A is the ratio of the distance it has travelled to the distance the solvent has travelled. The spot for A is not circular here, but shows "tailing"; therefore, the center of the spot is estimated. (Tailing is sometimes caused by too much sample in the original spot.)*

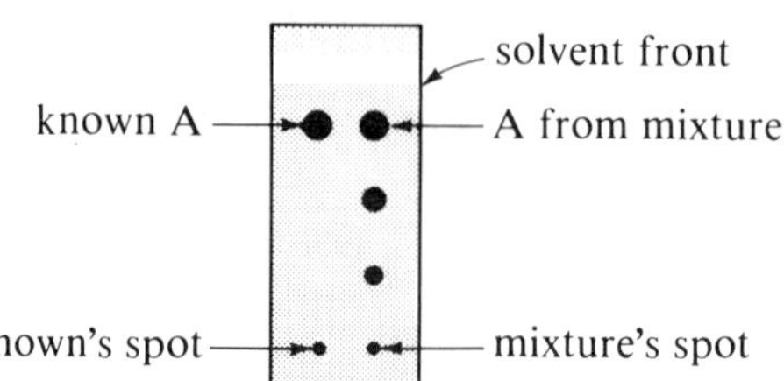

Figure 6.4 *A known and an unknown sample may be analyzed on the same plate at the same time.*

on the same plate. Figure 6.4 shows how a mixture containing compound A compares with pure A on the same plate.

If two substances have the same R_f value, they are likely (but not necessarily) the same compound. A second tlc comparison with a different solvent may result in different R_f values, in which case the substances are not the same. If the second tlc analysis results in the same R_f value for the pair, the likelihood that the samples are identical increases. Even in this case, some other type of corroborating evidence should be sought.

6.3 Paper Chromatography

In many respects, **paper chromatography** is similar to thin-layer chromatography. Instead of an adsorbent-coated plate, a strip of paper is used. Instead of a solid adsorbent, a thin film of water on the paper constitutes the adsorbent. Therefore, paper chromatography is a *liquid–liquid* partition technique, rather than a liquid–solid technique such as column chromatography and tlc.

For exacting analyses, commercial chromatographic paper strips equilibrated in a humid atmosphere should be used. In many analyses, however, filter paper can be used for paper chromatography because it is almost pure cellulose with few impurities. Under most atmospheric conditions, filter paper adsorbs moisture from the air. This adsorbed water makes up about 20% by weight of the filter paper and is usually sufficient for successful paper chromatography. Often, a damp solvent is used to develop the paper chromatogram to ensure the presence of sufficient water. A solvent that is fairly immiscible with water does not disturb the film of water adsorbed onto the cellulose.

Because very polar water molecules form the adsorbent layer, paper chromatography is most successful with very polar organic compounds. (Nonpolar compounds, which are not attracted to water, are usually carried with the solvent front.) Paper chromatography is commonly used for the identification of amino acids, which exist as highly polar *dipolar ions*, species containing full positive and negative ionic charges.

$$H_3\overset{+}{N}-\underset{\displaystyle R}{\overset{\displaystyle H}{\underset{|}{\overset{|}{C}}}}-\overset{\displaystyle O}{\overset{\|}{C}}O^-$$

the dipolar ion
of an amino acid

Suggested Readings

Chromatography: A Laboratory Handbook of Chromatographic and Electrophoretic Methods, E. Heftmann, Ed. New York: Wiley-Interscience, 1971.

Stock, R., and Rice, C. B. F. *Chromatographic Methods*, 3rd ed. London: Chapman and Hall, 1974.

Snyder, L. R., and Kirkland, J. J. *Introduction to Modern Liquid Chromatography*, 2nd ed. New York: Wiley-Interscience, 1979.

Thin-Layer Chromatography. A Laboratory Handbook, 2nd ed. E. Stahl, Ed. New York: Springer, 1969.

Paper Chromatography, I. M. Hais and K. Macek, Eds. New York: Academic Press, Inc., 1963.

Problems

6.1 A chemist wishes to carry out an elution chromatographic separation using diethyl ether and ethanol as the eluting solvents.

(a) With which solvent should the chemist begin the elution?

(b) What would happen if the chemist started with the other solvent?

6.2 A highly polar compound is moving through an elution chromatography column too slowly. What can the chromatographer do to increase its rate of movement?

6.3 Which packing would *not* be suitable for elution chromatography of a mixture of carboxylic acids? Explain your answer.

(a) basic alumina

(b) a magnesium silicate

(c) limestone ($CaCO_3$)

6.4 List the following compounds in order of expected elution from a chromatography column packed with acid-washed alumina, using a petroleum ether–diethyl ether solvent system.

(a) OH

(b) $\overset{O}{\parallel}$COH

(c) CH_3

(d) OCH_3

6.5 Calculate the R_f values for the following compounds:

(a) spot, 5.0 cm; solvent front, 20.0 cm

(b) spot, 3.0 cm; solvent front, 12.0 cm

(c) spot, 9.8 cm; solvent front, 12.0 cm

6.6 If two compounds have R_f values of 0.50 and 0.61, how far will they be separated from each other on a plate when the solvent front is developed to:

(a) 5 cm? (b) 15 cm?

6.7 As a separation-and-detection method, would tlc or paper chromatography yield better results in the analysis of each of the following pairs of compounds?

(a) $CH_3CH_2\overset{O}{\underset{O}{\overset{\parallel}{\underset{\parallel}{S}}}}OH$ and $CH_3CH_2\overset{O}{\underset{O}{\overset{\parallel}{\underset{\parallel}{S}}}}NH_2$

(b) 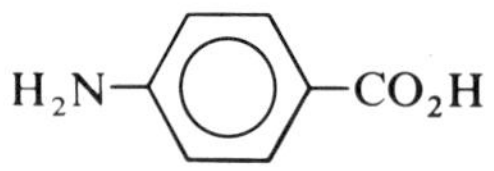 $H_2N-C_6H_4-CO_2H$ (para) and $H_2N-C_6H_4-CO_2H$ (meta)

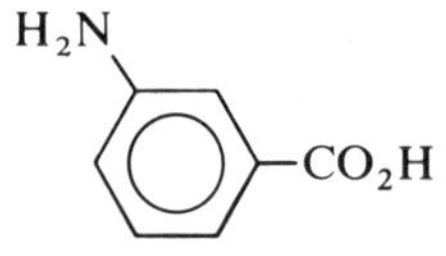

(c) $CH_3(CH_2)_6C(=O)CH_3$ and $CH_3(CH_2)_6CH(OH)CH_3$

6.8 When a mixture of colorless compounds is subjected to elution chromatography, the chromatographer cannot see whether a compound or only solvent is being eluted. Suggest ways in which the following devices can be used to determine when one compound is being eluted or when a second compound is following the first:

(a) analytical balance

(b) refractometer

(c) melting-point apparatus

»»› EXPERIMENT 6.1 ‹««

Thin-Layer Chromatography of Analgesic Drugs

Discussion

Figure 6.5 shows the names and structures of the four most common over-the-counter analgesics (pain killers). Of these, aspirin is the most widely sold and has the most potent biological activity. Besides being an analgesic, aspirin is an antipyretic (reduces fever), an anti-inflammatory agent (frequently used to treat rheumatoid arthritis), and an anticlotting agent. Acetaminophen and phenacetin are also mild antipyretics. How aspirin works is only partially understood. It is thought that aspirin inhibits the biosynthesis of *prostaglandins*: biological compounds that moderate hormone activity and also stimulate immune responses in the body.

Commercial headache remedies contain one or more of the drugs shown in Figure 6.5 plus an inert binder, such as starch. Caffeine, which is often included in these formulations, is a vasodilator (dilates the veins) and, as such, can also help relieve headaches.

In this experiment, the active ingredients of two brands of analgesic tablets are extracted from their binders by methanol. The methanol solutions are then subjected to tlc analysis on a single plate. To provide tentative identifications of two of the components, methanol solutions of pure aspirin and caffeine (obtained from the extraction of tea in Experiment 4.3) are spotted on the same plate. After development, the spots on the plate are "visualized" using a black lamp or iodine vapor and the R_f values are determined.

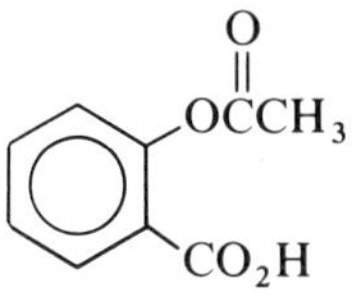

acetylsalicylic acid (aspirin)

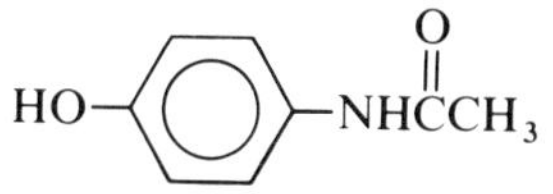

N-acetyl-*p*-aminophenol ("acetaminophen"; found in Tylenol)

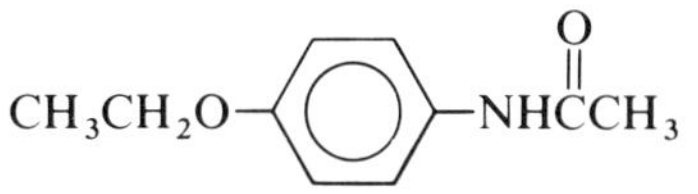

O-ethyl-*N*-acetyl-*p*-aminophenol ("acetophenetidin"; "phenacetin"; found in Empirin, along with aspirin and caffeine)

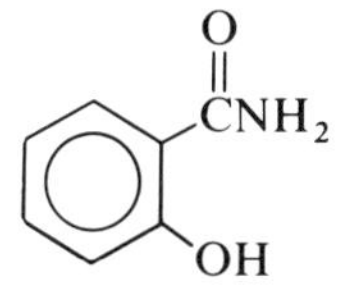

salicylamide (found in Excedrin, along with other ingredients)

Figure 6.5 *Some common analgesics.*

Steps in a Tlc Analysis

1) Spotting the plate. Dissolve about 1 mg of the solid or liquid sample in a few drops of a volatile solvent such as methanol, CH_3OH, or acetone, $(CH_3)_2C{=}O$. Dip the end of a fresh micropipet into this solution, which rises into the pipet by capillary action.

Touch the end of the pipet *gently* and *briefly* to the adsorbent about 1.5 cm from the end of the plate, so that the solution runs out of the pipet onto the adsorbent (Figure 6.6). It is desirable to have the spot as small as possible (1–3 mm in diameter); therefore, allow only a small amount of liquid to run out before lifting the pipet. As soon as the solvent evaporates, more sample may be added to the same spot. Depending upon the concentration of the sample solution, one to three applications are usually sufficient.

It is important to spot the compounds high enough on the plate that they will be *above the solvent level* in the developing jar. If the spots are below the solvent level, they will be dissolved away from the plate by the solvent and you will have to prepare a fresh plate.

If more than one sample is being analyzed on the same plate, space the spots well apart and at the same distance from the bottom of the plate. Samples that are spotted too close together may spread out and run together as

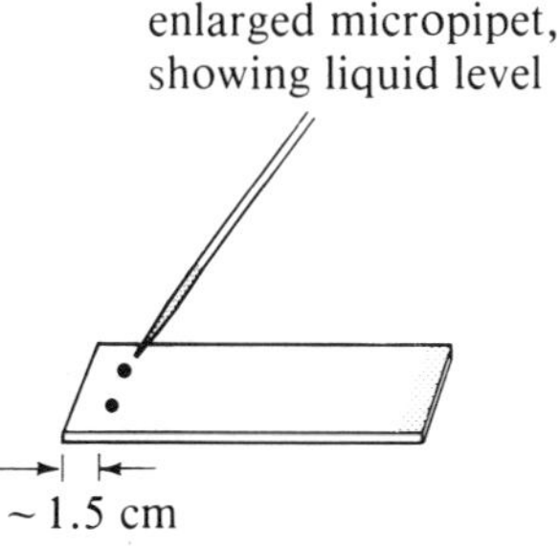

Figure 6.6 *Spotting a tlc plate with solution from a micropipet.*

they are developed; therefore, a maximum of 3–4 sample spots per 5-cm-wide plate is advised. Use a fresh micropipet for each sample and discard it after use.

2) Preparing the developing jar. Any type of tall jar, such as an instant coffee jar or mason jar, with a lid or screw top, may be used for developing a tlc plate. The jar should be narrow enough that the plate can be propped upright inside without danger of its falling over (Figure 6.2, page 109).

Line the inside of the jar half-way around with a piece of filter paper, which will act as a wick to saturate the atmosphere in the jar with solvent vapor. Before inserting the tlc plate, pour a small amount of the developing solvent into the jar to soak the filter paper and to cover the bottom of the jar to a depth of about 0.5–1.0 cm. The solvent level should cover the edge of the adsorbent on the plate, yet not reach the spots. Cap the jar and allow it to sit for at least 15 minutes to reach liquid–vapor equilibrium. Check that the solvent level is still about 0.5–1.0 cm, and add more solvent if necessary before inserting the plate.

3) Developing the plate. Prop the plate upright in the center of the jar (spots at the bottom) in such a way that the adsorbent side of the plate is visible through the side of the jar. Cap the jar and do not move it during the development.

The solvent will rise rapidly up the adsorbent on the plate by capillary action. When the solvent front has risen almost to the top of the plate (about 1–2 cm from the end), open the jar, remove the plate, and quickly mark a line across the plate at the solvent front with a pencil. Check the plate for visible spots. Outline these carefully with the pencil, keeping your lines at the perimeters of the spots. A spot from a colorless organic compound will not be visible on the plate. Therefore, one or more visualization procedures may be followed.

4) Visualizing the spots. The most convenient method of visualizing is with an ultraviolet (black) lamp. Some compounds show up in ultraviolet light as bright spots that then can be outlined. Another technique consists of using a black lamp to visualize spots on a tlc plate containing an inorganic fluorescent compound, such as ZnS, in its coating. When such a plate is placed under a black lamp, the entire plate glows. An organic compound capable of absorbing ultraviolet light, and thus quenching the fluorescence, will show up as a dark spot.

A third simple technique for visualizing spots is to place the dry, developed plate in a dry, covered jar with a few crystals of iodine (I_2). (CAUTION: Iodine is a strong irritant!) Most organic compounds form colored complexes with iodine. In the jar, vapors of iodine are selectively adsorbed onto the tlc plate wherever there is a concentration of organic compound. The spots or organic compounds will consequently turn colored (usually yellow-brown to purple). If spots do not appear, warm the bottom of the jar gently (with your hand or briefly on a steam bath) to vaporize the iodine crystals. As soon as the spots are well-defined, remove the plate from the jar and circle the spots with a pencil, because the iodine will rapidly sublime from the plate, leaving the spots colorless again.

If a plate is left in the iodine vapor too long, the entire plate will become dark, and the spots will no longer be distinguishable. In such a case, remove the plate and observe it carefully as the iodine sublimes. Generally, the iodine will evaporate more quickly from the bulk of the plate than from the organic compounds. The spots should thus become apparent, and they can be circled with a pencil.

In this experiment, I_2 is used as a visualizing agent. If a black lamp is available, your plate should also be examined under that.

Experimental

EQUIPMENT:

two developing jars
disposable micropipets
mortar and pestle
scissors (to cut the tlc sheets)
eight test tubes
commercial, plastic-backed, alumina thin-layer chromatography sheets or plates (see Experimental Note 1)
visualization jar

black lamp (optional)

CHEMICALS:

glacial acetic acid, about 18 mL
acetone, about 50 mL
acetylsalicylic acid (reagent grade), 0.5 g
plain aspirin tablet
caffeine (from Experiment 4.3), 0.5 g
chloroform, about 100 mL
one Excedrin tablet (or tablet of unknown composition)
iodine crystals, 0.2 g
methanol, about 11 mL
toluene, about 120 mL

TIME REQUIRED: about 3 hours (To save time, prepare the developing jars first.)

STOPPING POINT: after extraction of the active ingredients from the tablets

>>>> ***SAFETY NOTE*** The solvent mixtures used in the developing jars are toxic and irritating. Handle glacial acetic acid as you would conc. HCl. Work in the fume hood!

PROCEDURE

Using a clean mortar and pestle, crush one aspirin tablet and place the powder in a test tube. Then crush one Excedrin tablet and place its powder in another test tube. Estimate the volume in each of these two test tubes and place approximately equal volumes of reagent-grade acetylsalicylic acid (aspirin) and caffeine in two additional test tubes. Add 2.5 mL of methanol to each test tube. Mix the contents of each with a stirring rod for 3–5 minutes, then filter each mixture into a clean test tube and cork. Discard the solid residues.

Cut two 5×10-cm strips from a plastic-backed, alumina tlc sheet (see Experimental Note 1). Using a fresh micropipet for each sample (see Experimental Note 2), spot each of the four solutions 1.5 cm from one end of the strip, as described in the discussion.

Prepare the following two solvent systems (CAUTION: See Safety Note!) in Erlenmeyer flasks: (1) a solution of 50 mL acetone and 50 mL chloroform, and (2) a solution of 120 mL of toluene, 50 mL of chloroform, 18 mL of glacial acetic acid, and 1 mL of methanol.

Prepare two developing jars, one for each solvent system, as described in the discussion (see Experimental Note 3). Place the plates in the developing jars as described in the discussion. When each plate is developed, mark the solvent front and then set the plates in the hood for 5–10 minutes to allow the solvent to evaporate (see Experimental Note 4).

To visualize the spots, place about 0.2 g of iodine crystals in a clean, dry developing jar (without solvent or filter paper), place the plate in the jar, and cover the jar. (CAUTION: Iodine is a strong irritant. Do not handle the crystals with your fingers or inhale the vapors.) In 1–2 minutes, the compounds on the plate will appear as yellow-brown or purple spots (or an intermediate color). When you can see the spots clearly, remove the plate and circle the spots with a pencil. Because visualization with I_2 does not always yield satisfactory results, also visualize the plates with a black lamp if one is available and compare the results.

In your notebook, record the total number of components in each sample. (For Excedrin, how does this number compare with the listed ingredients?) Calculate the R_f value for each spot in each chromatogram as described in Section 6.2A. Compare the relative abilities of the two solvent systems to separate the components. If time permits, try some solvent systems of your own and compare them with those suggested in this experiment.

EXPERIMENTAL NOTES

1) Tlc sheets and plates. Commercial tlc sheets are coated with silica gel (SiO_2) or alumina (Al_2O_3). In this experiment, alumina sheets should be used because silica gel sheets give poor results. Kodak Chromatogram Sheets (alumina adsorbent without fluorescent indicator) #13253 and Kodak Chromatogram Sheets (alumina adsorbent with fluorescent indicator) #13252 are both satisfactory.

Preparation of tlc plates. If commercial tlc sheets are unavailable, plates can be made from microscope slides and a slurry of 1 g aluminum oxide G and 2 mL chloroform. Dip two slides, back-to-back, in the slurry; allow the excess slurry to drain; separate the slides; and allow them to dry in a fume hood. Making satisfactory plates requires practice;

therefore, prepare a number of plates and select the most evenly coated ones for the experiment. Microscope slide plates are shorter than those cut from commercial sheets; consequently, the separation of components is not as clean.

2) *Pipets.* Commercial 10-μL disposable pipets are best. If commercial pipets are not available, draw out some soft glass tubing in a flame. The diameter of the pipet should be about $\frac{1}{4}$ that of a melting-point capillary.

3) *Developing jars.* Developing chambers with the proper solvent systems may be prepared for the class as a whole and kept in the fume hood. To prepare your own chamber, you can use any long cylindrical container with a top, as described in the discussion. Beakers are unsuitable; they are too shallow for their volume and cannot be satisfactorily capped.

4) The odor of acetic acid can still be detected after a 10-minute solvent evaporation period; however, the acetic acid does not interfere with the visualization using I_2. It is possible that acetic acid would interfere with other visualization techniques; in such a case, a longer evaporation time is recommended.

Problems

6.9 List the toxic characteristics of the solvents used in this experiment.

6.10 A wick of filter paper is placed in a tlc developing jar and the atmosphere in the jar is saturated with solvent before a plate is developed. What would happen if a plate were developed in a jar with an atmosphere not saturated with solvent vapor?

6.11 Would you expect Br_2 or Cl_2 to be as suitable as I_2 for a visualizing agent in tlc? Explain.

6.12 Which of the following reactions could be used as the basis for a visualization technique?

(a) $R_2C{=}CR_2 + I_2 \rightleftarrows R_2CI{-}CIR_2$

(b) $RCO_2H + NaOH \rightarrow RCO_2Na + H_2O$

(c) $RH + H_2SO_4 \rightarrow C + CO_2 + H_2SO_4{\cdot}H_2O$

6.13 Like ZnS, some organic compounds are fluorescent. Could one of these compounds be used in place of ZnS as a visualizing agent in tlc? Explain.

6.14 What other solvents might be suitable to extract the active agents from the analgesic tablets used in Experiment 6.1? What types of solvents would be unsuitable?

6.15 If an analgesic tablet contained the sugar sucrose:

(a) Would you expect the sucrose to dissolve in the methanol or remain behind with the binder?

(b) If the sucrose did dissolve, would you expect it to exhibit a larger or smaller R_f value than phenacetin? (The structure of phenacetin is shown in Figure 6.5, page 113. Consult your lecture text for the structure of sucrose.)

»»» CHAPTER 7 «««

Gas–Liquid Chromatography

Gas–liquid chromatography (glc) is known by a variety of other names: *gas chromatography* (*gc*), *gas–liquid phase chromatography* or *gas–liquid partition chromatography* (*glpc*), and *vapor phase chromatography* (*vpc*). All these terms refer to the same technique, which is an instrumental method of analyzing the components of a liquid or reasonably volatile solid. The instrument, called a **gas chromatograph**, is diagrammed in Figure 7.1.

About 1–10 μL of sample is injected with a small hypodermic syringe into the gas chromatograph, through a rubber septum on the heated injection port of the instrument. The sample is vaporized and carried through a heated column by an inert carrier gas (usually helium or nitrogen). The adsorbent in the column is a high-boiling liquid suspended on a solid inert carrier. Because of differing interactions with the adsorbent and differing vapor pressures, the components of the sample move through the column at different rates. At the end of the column, each component passes through a detector, which is connected to a recorder. The recorder produces a tracing that shows *when* each component of a mixture passes the detector and also indicates the *approximate relative quantity* of each component.

Glc is commonly used for checking the purity of volatile samples, such as distillation fractions; checking on the identity of a substance by comparison of its glc with that of a known; and analyzing a mixture for the presence or absence of a known compound (such as alcohol in blood). In addition, the relative quantities

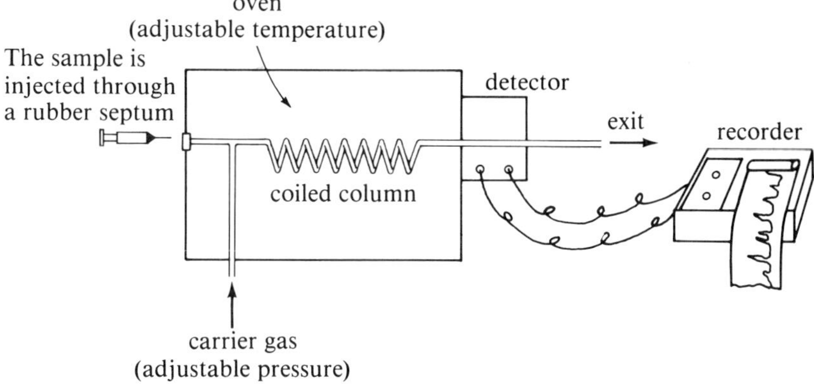

Figure 7.1 *Diagram of a gas chromatograph. The vaporized sample is carried through the column by an inert carrier gas. The passage of organic compounds is sensed by the detector and this information is graphed by the recorder.*

of the components of a mixture can be determined with about 5% accuracy; one way to do this is presented in Experiment 7.2.

In most glc units, small quantities of sample are used and no attempt is made to collect the material that has been chromatographed. However, special *preparative gas chromatographic columns* make possible the separation and collection of samples. Also, special techniques allow the effluent from a gas chromatograph to be further analyzed—for example, by a mass spectrometer.

7.1 The Gas Chromatograph

The column of a typical gas chromatograph is constructed of metal tubing, which is coiled so that a long column (3–10 feet) will fit into a small volume. The column contains an inert solid packing material (such as crushed firebrick or diatomaceous earth) coated with the liquid phase. A number of interchangeable columns can be purchased for a gas chromatograph so that the appropriate packing, liquid, and column length can be chosen for a particular sample.

In the instrument, the column is located in an insulated oven with adjustable temperature controls. The upper useful limit of temperature is determined by the vapor pressure of the liquid phase in the column. The usual operating temperature is 100–200°, but higher temperatures may be used with some types of columns. The operating temperature chosen depends on the boiling points of the compounds in the sample to be chromatographed.

A typical liquid phase used in a glc column is Silicone Oil DC-550, which is useful for a wide variety of compounds. For polar compounds, a more polar liquid phase, such as a Carbowax, may be used. Columns containing less polar liquids, such as mineral oil, are occasionally used for the glc analysis of nonpolar compounds.

typical liquid phases used in glc columns:

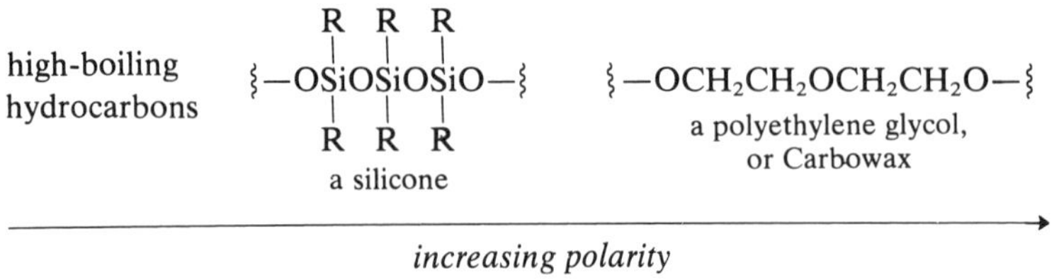

high-boiling hydrocarbons

$$-\mathrm{O}\overset{\mathrm{R}}{\underset{\mathrm{R}}{\mathrm{Si}}}\mathrm{O}\overset{\mathrm{R}}{\underset{\mathrm{R}}{\mathrm{Si}}}\mathrm{O}\overset{\mathrm{R}}{\underset{\mathrm{R}}{\mathrm{Si}}}\mathrm{O}-$$

a silicone

$$-OCH_2CH_2OCH_2CH_2O-$$

a polyethylene glycol, or Carbowax

→ *increasing polarity*

The liquid phase of a glc column does not truly adsorb the gaseous sample. Instead, the sample dissolves in the liquid phase. The dissolved compound is in equilibrium with its vapor, a process that depends primarily on the compound's vapor pressure. (Recall Henry's law: the solubility of a gas in a liquid solvent is directly proportional to the pressure of that gas above the solvent.)

As vapor of the sample is carried through the column by the carrier gas, it dissolves and revaporizes continuously. A nonpolar compound with a high vapor pressure moves along a nonpolar column at a fairly rapid rate. A less volatile compound moves more slowly. Thus, the compounds tend to concentrate in bands (similar to those in column chromatography) as they move through the column.

In nonpolar columns, the separation of compounds in glc is similar to their separation in distillation because both processes depend primarily on the vapor pressures of the components. In both glc and distillation, the low-boiling components of a mixture are eluted first, followed by successively higher-boiling components.

The detector at the end of the column signals when a compound is being eluted. A number of different types of detectors have been invented, but only two types are in common use: **thermal conductivity (hot wire) detectors** and **flame ionization detectors**.

In a thermal conductivity detector, an electric current is passed through a wire located directly in the flow of gaseous effluent from the column. The electrical resistance of a hot wire varies with the heat conductivity of a gas passing over it. Helium, with a high heat conductivity, absorbs heat from the wire and keeps it relatively cool. Organic compounds have lower heat conductivities; therefore, when the vapors of an organic compound pass over the wire, the wire becomes hotter and its resistance changes. The hot wire is actually one arm (the sample arm) of a *Wheatstone bridge* circuit (see Figure 7.2). The other arm (the reference arm) is a similar hot wire with pure helium (no sample) passing over it. When no sample is passing, the bridge is in balance and the recorder pen rides on the base line. The passage of a gaseous organic compound changes the resistance of the sample arm. Thus, the bridge becomes unbalanced, and an electric current passes to the recorder.

In a flame ionization detector (Figure 7.2), part of the gaseous effluent from the column is combined with a separate gas stream of hydrogen and oxygen, and this gaseous mixture is burned. When organic compounds are present in the effluent from the column, they are oxidized and converted to ions by the high temperature of the flame. The ions then pass through a metal ring, creating an electrical potential difference between the ring and the barrel of the burner. The small potential difference is amplified, then sent to the recorder.

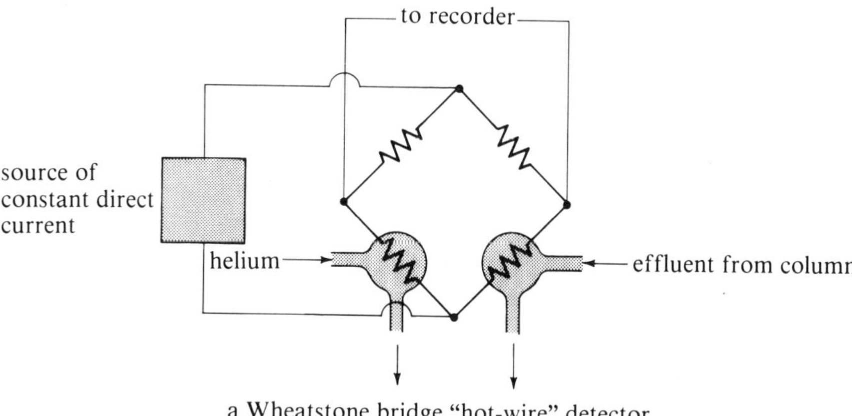

a Wheatstone bridge "hot-wire" detector

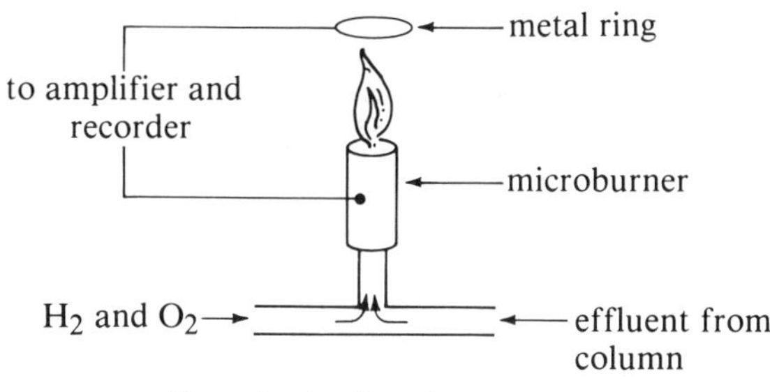

a flame ionization detector

Figure 7.2 *Two common types of glc detection devices.*

The **recorder** receives information from either type of detector and produces a graph (the **chromatogram**) of the components passing through the detector. Figure 7.3 is a representation of a typical gas chromatogram. When the sample is first injected, the start of the chromatogram is indicated by a small reference peak. This reference peak can arise from the operator moving the pen slightly or marking the position of the pen on the paper with a pencil at the time of injection. Alternatively, with some instruments, a small bubble of air can be injected from the hypodermic syringe along with the sample. This air passes through the column rapidly; when it passes the detector, the recorder pen automatically traces a small peak. A small air peak is shown on the chromatogram in Figure 7.3.

7.2 Retention Time

The time it takes for a particular compound to pass through the column after its injection and reach the detector is called the compound's **retention time**. The retention time is a function of the vapor pressure of the compound; the rate of gas flow; the temperature; and the composition, length, and diameter of the column.

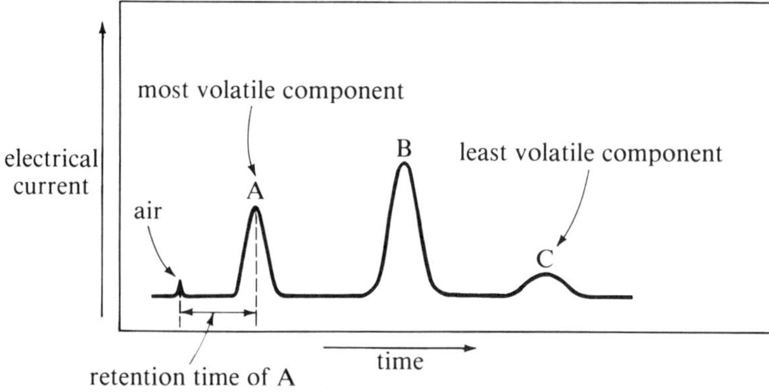

Figure 7.3 A gas chromatogram. The air peak signals the time of injection. The retention time of a compound is measured as the distance between the air peak or other reference mark and the top of the compound peak, converted to minutes.

The retention time is measured from the reference mark to the top of the compound peak. Retention times may be reported in minutes (which are the correct units) or as centimeters. Because the recorder paper moves at a constant number of cm/min, the units of distance can easily be converted to units of time. For example, if the paper moves at 1.27 cm/min, a measured retention time of 5.2 cm equals 4.1 min.

$$\text{retention time} = \frac{\text{number of cm from injection to peak}}{\text{speed of the paper (cm/min)}}$$

Under carefully controlled conditions, retention times can be duplicated. If accuracy is desired, an inert compound such as a hydrocarbon can be added to the sample and *relative retention times* measured from the internal standard's peak instead of from the operator's mark or an air peak. With this technique, relative retention times can be reproduced with an accuracy of about 1%.

7.3 Glc Analyses

As we have mentioned, glc is commonly used for checking the purity of a volatile sample. Determining the actual identity of a sample component from a glc analysis cannot be done directly. In a research laboratory, it may be necessary to isolate a compound, such as by preparative glc, and to determine its structure by physical constants, spectral data, and chemical analysis. In routine laboratory analyses, when we already have a good idea of the components in a mixture, such a lengthy identification procedure is usually not necessary.

A common problem faced in a simple glc analysis is that the components of a mixture are known, but the identity of each peak in the chromatogram is in question. In this circumstance, the retention times of components of the mixture can be compared to those of known pure compounds and each peak assigned to a particular compound. The procedure for doing this is described in the discussion in Experiment 7.1.

In some cases, the structure of a compound giving rise to a peak in a glc chromatogram can be determined indirectly. If the compound is known to belong to a particular homologous series, it is possible to identify the compound by comparing the log of its retention time to those of three (or more) other members of the same homologous series. For example, suppose that the sample compound is a methyl ester of a carboxylic acid with a continuous, saturated chain of unknown length. We would select three known methyl esters of continuous-chain carboxylic acids and plot the logs of their retention times versus the number of carbons in their chains. The result is a straight line. The number of carbons in the chain of the unknown ester can then be determined from the graph and the log of its retention time.

Besides showing the number of components in a sample and their retention times, a gas chromatogram obtained from a nonpolar column can show the approximate relative boiling points. If the instrument is calibrated with similar compounds of known boiling points, we can closely estimate the actual boiling points of the components in the sample.

The relative quantities of components of a mixture can be estimated from relative peak sizes (height and width, or area). However, the exact quantities cannot be determined from this information alone because detector response varies with different types of compounds. Several techniques have been devised to circumvent this problem; one of these techniques is described in Experiment 7.2.

Suggested Readings

Crippen, R. C. *Identification of Organic Compounds with the Aid of Gas Chromatography*. New York: McGraw-Hill, 1973.

Grant, D. W. *Gas–Liquid Chromatography*. New York: Van Nostrand Reinhold, 1971.

Littlewood, A. B. *Gas Chromatography, Techniques and Applications*. New York: Academic Press, Inc., 1962.

Purnell, H. *New Developments in Gas Chromatography*. New York: Wiley-Interscience, 1973.

Problems

7.1 Calculate the retention times of peaks at the following distances from the reference mark if the glc recorder paper is moving at 3 in./min:

(a) 12.0 in. (b) 3.0 cm (c) 20 mm

7.2 Alcohols have very long retention times on Carbowax columns. Why?

7.3 Suppose an organic compound has a *higher* heat conductivity than the carrier gas. How would its glc signal appear on a chromatogram run on an instrument with: (a) a thermal conductivity detector? (b) a flame ionization detector?

7.4 Why should a thermal conductivity detector exhibit different degrees of response to different compounds?

7.5 Why are the following procedures *invalid*?

(a) To minimize the time required for an analysis, the temperature of a glc instrument is increased above the boiling points of the components in the mixture.

(b) To increase the retention times and thus maximize the separation between peaks, the gas flow is adjusted to a very low value.

7.6 Between sample injections, some students clean a microsyringe with acetone rather than with the mixture to be analyzed. Why is this a poor technique?

7.7 Helium is the carrier gas of choice for a gas chromatograph containing a thermal conductivity detector, and nitrogen gas is preferred for a gas chromatograph having a flame ionization detector. Suggest a reason for this.

7.8 Suppose that, on a glc column, methyl hexanoate eluted in 1.87 min, methyl heptanoate eluted in 2.35 min, and methyl octanoate eluted in 3.0 min. What is the structure of an unknown methyl ester, known to belong to the same homologous series, that elutes at 4.7 min?

⟫⟫ EXPERIMENT 7.1 ⟪⟪

Glc Analysis of a Mixture

Discussion

This experiment is intended to provide familiarization with the operation of a gas–liquid chromatograph by the analysis of a mixture of ethyl acetate and *n*-propyl acetate. Experiment 7.3 (supplemental) is the glc analysis of the distillation fractions obtained in Experiments 5.1 and 5.2, which also contain these two compounds.

ethyl acetate   *n*-propyl acetate

Because of differences in procedure required for each model of gas chromatograph, the following description is generalized. You should obtain specific instructions for your instrument from your instructor.

Steps in a Glc Analysis

1) *Preparing the instrument.* A gas chromatograph cannot be turned on and used immediately. The injection port, the detector, and possibly the oven must be allowed to heat and come to thermal equilibrium. For this reason, one or more of the heat sources (depending on the instrument) and a slow flow of carrier gas should be started well in advance of instrument use. Your instructor will probably prepare the instrument before class. Each time you make a run on the instrument, make sure that the gas flow is adjusted properly; the temperature is at the proper level; the auxiliary equipment (detector, recorder) is switched on; and the recorder pen is tracing a base line.

2) *Injecting the sample*. Use a microhypodermic syringe to inject a sample through the silicone rubber septum into the vaporizing chamber. The amount to inject depends on the capacity and sensitivity of the instrument; a few microliters is usually sufficient. (CAUTION: Microsyringes are delicate and expensive. The metal plunger and needle are easily bent. Dirt in the glass barrel of the syringe will cause the plunger to stick; if the plunger is forced, the barrel will split. Your instructor will demonstrate the proper handling and cleaning procedures for microsyringes.)

Inject the sample quickly, so that it vaporizes all at once. With a larger microsyringe (100 μL), hold the plunger before and after the injection. (The carrier gas in some instruments is under enough pressure that it can cause the plunger to fly out.) Some instruments are inconsistent in showing an automatic injection signal; therefore, mark the position of the pen on the recorder paper with a soft pencil or felt pen when you make an injection.

If a *solution* (instead of an undiluted sample mixture) is injected, a large solvent peak will be observed at the start of the chromatogram. Do not worry if the recorder pen runs to the top of the recorder paper and whines; the pen will return to the base line when the solvent has all passed the detector.

3) *Obtaining the chromatogram*. Once an injection is made, an operator has little to do but watch the tracing of the pen. If it is evident that you have injected too little or too much sample, prepare another sample. After the chromatogram has been recorded, stop the flow of paper from the recorder and cut or tear off your chromatogram. Mark it with your name; the date; the name of the sample; the sample size; the column type (silicone, Carbowax, etc.); the temperature of the column; the flow rate of the carrier gas; the speed of the paper; and any other germane comments.

4) *Calibration of gas chromatograms*. There are two ways to calibrate a gas chromatogram to identify components. The first is to run a chromatogram of a pure sample of known composition immediately before or after the chromatogram of the unknown sample. The conditions (temperature, gas flow, etc.) should not be changed during the calibration run and actual run.

If a sample has already been subjected to glc analysis so that the appearance of the gas chromatogram is known, a drop of the known can be mixed with a few drops of the sample. The chromatogram of this mixture is then compared with the original chromatogram. If the addition of the known has increased the size of one of the existing peaks, the compound giving rise to that peak may be identical with the known. If the addition of the known compound to the mixture results in a new peak, then that known compound is not present in the sample. While such identification techniques are useful in the laboratory, they are not considered an absolute proof of structure. Additional corroborating evidence is needed.

Problems Encountered in Glc

Most of the problems encountered in glc analysis are easily correctable. *Poor resolution*, in which peaks overlap each other, may be the result of too large a sample, too high a gas flow, or too high a temperature. In some cases, a different type of column (for example, Carbowax instead of silicone oil) or a longer column of narrower diameter is necessary for good resolution of peaks.

A peak that runs off the top of the recorder paper is usually the result of either too large a sample or too much sensitivity. Most instruments have a sensitivity control, which may be adjusted if necessary.

A peak that is very broad or unsymmetrical can result from the injection of too much material. A compound with an unusually long retention time may also be broad and sometimes very flat. In this case, the gas flow should be rechecked. An increase in temperature or switching to another type of column usually solves this problem.

If an injected sample gives rise to no recorder signal and if the instrument is otherwise operating normally, the sample may have been retained in the column. A retained sample can result in contamination of future analyses, a change in the character of the column, and even a plugged column. For these reasons, avoid injecting tars or polymers into a glc column. (In fact, your instructor may ask you to analyze only *distilled* samples by glc.) Notify your instructor if your sample does not pass through the column.

Experimental

EQUIPMENT:

gas chromatograph
microsyringe
three test tubes

CHEMICALS:

ethyl acetate, about 1.0 mL
n-propyl acetate, about 1.0 mL

TIME REQUIRED: $\frac{1}{2}$–1 hour of laboratory work, depending upon the instrument

STOPPING POINT: after the instrument work

PROCEDURE

Place about 0.3 mL of ethyl acetate in a test tube, 0.3 mL of *n*-propyl acetate in a second test tube, and a mixture containing 0.3 mL each of ethyl acetate and *n*-propyl acetate in a third test tube. Label and cork each test tube (see Experimental Notes 1 and 2).

Obtain a microsyringe from your instructor. (See the precautionary note on page 125.) Clean the syringe with the mixture by pulling up and then ejecting a few μL into the hood sink or into a tissue (which should be placed in the hood to dry). Draw up 2.0 μL (or the volume you are instructed to use with your instrument) into the syringe. If there is an air bubble in the syringe, adjust the *liquid* level (not the plunger level) to 2.0 μL.

Following the instructions on pages 124–125, obtain the gas chromatogram of the sample. The two peaks on the chromatogram should be well-separated, with the valley between them dropping to the base line. The heights of the peaks should be about one half the height of the paper. If the peaks are not this height, increase or decrease the sensitivity of the instrument (or plan to inject more or less sample), make any other necessary adjustments, and re-inject another sample of the mixture.

When the instrument and your injection size are adjusted correctly, inject each of the three samples (and also a sample of the pure internal standard if you are using one; see Experimental Note 1). Clean the syringe with the sample before each injection and wait for a previous injection to be recorded before making a new injection. On the recorder paper, mark each point of injection and the name of the sample.

After all the samples have been analyzed, stop the recorder paper flow and note on the recorder paper the information outlined in Step (3), page 125. Inspect the tracings, noting any small peaks representing impurities. Determine the retention times of ethyl acetate and *n*-propyl acetate in each sample.

EXPERIMENTAL NOTES

1) Your instructor may wish to have you measure *relative retention times* rather than uncorrected retention times. If so, add about 5 drops of methanol or other internal standard to your sample. (The instrument settings may need readjustment so that the peaks are separated.)

2) Some glc instruments are too sensitive to analyze pure liquid samples correctly. In such a case, your instructor will supply directions for dilution with a solvent.

Problems

7.9 If the difference in retention times of two components of a mixture is 0.75 min, by how many cm will the signals for the two components be separated on a chromatogram if the chart speed is 2 in./min?

7.10 How would you prepare a solution of a sample if you want to inject 0.5 μL of sample as a 3-μL injection?

(SUPPLEMENTAL)

Quantitative Glc Analysis of a Mixture

Discussion

In this experiment, you will analyze a mixture of unknown composition to determine the weight per cents of the components. These weight per cents can be determined from the *areas under the peaks* and *correction factors* that compensate for the different detector responses arising from the different compounds in the mixture. Although there are several ways to convert area ratios to corrected weight ratios, we will present only one: the **internal-standard technique**.

Determining Peak Areas

Many gas chromatographic recorders are equipped with mechanical or electronic **integrators** that automatically provide peak areas. If your recorder has one of these devices, your instructor will show you how it is used.

If a gas chromatograph does not have an integrator, the area can be approximated with an accuracy of about 3% by considering the peak to be a *triangle* (area = $\frac{1}{2}$base × height). Instead of measuring the base of a glc peak and dividing by 2, the width of the peak at $\frac{1}{2}$ peak height is used because it yields more reproducible values (see Figure 7.4). Determining areas by triangulation works well only if the peaks are *well-resolved* and *symmetrical.* (The Suggested Readings, page 123, list sources that describe determination of the areas of overlapping peaks.)

Another technique for determining the relative areas under peaks, including unsymmetrical peaks, is by carefully cutting out the peaks (or tracings of the peaks on plain, uniform bond paper) and weighing them. The weights of the peaks are proportional to the areas.

With either technique, the recorder paper should be set to a fast speed to maximize peak width and minimize errors.

Correcting Areas with an Internal Standard

A detector produces a signal that is directly proportional to the quantity of compound injected. Because different compounds exhibit different proportionality constants, it is necessary to obtain correction factors for quantitative determinations of mixture composition. The internal-standard technique utilizes weight and area ratios for these corrections.

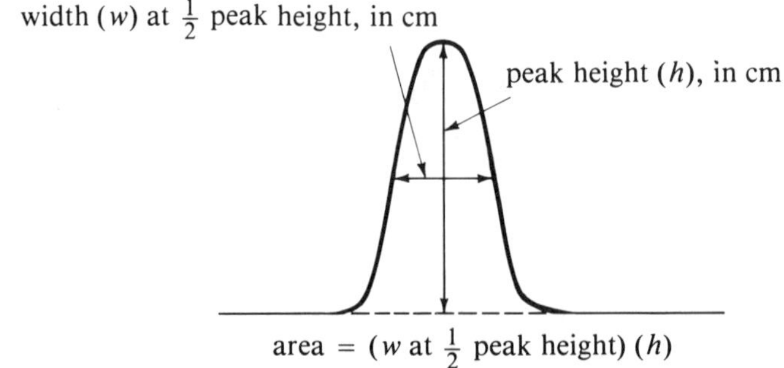

Figure 7.4 The area of a glc peak is approximately equal to the width of the peak at its half height multiplied by its full height.

In the internal-standard technique, a reference compound, or standard, is added to both a calibrating mixture and the mixture to be analyzed. Any compound can be used as a standard. However, for best results, the standard should be structurally similar to the components being analyzed and should be present in the mixture in approximately the same concentration. Also, the instrument should be adjusted so that the peak of the standard is well-separated from the other peaks in the chromatogram.

The internal-standard technique involves three general steps. First, the calibrating mixture of *known weights* of the components to be analyzed and the standard is prepared; the chromatogram of this known mixture is then obtained. Next, the weights and peak areas for this known mixture are used to calculate a correction factor for each component, which can be applied to a chromatogram of a mixture of unknown per cent composition. Finally, the chromatogram of the mixture of unknown per cent composition (also containing the reference compound) is obtained, and the per cent composition of this mixture is calculated.

Let us consider the procedure in more detail. Assume that we wish to analyze a mixture of compounds A, B, and C to determine its per cent composition. The specific steps in the procedure follow.

For the known mixture:

1) Add a weighed amount of standard to a mixture containing known weights of compounds A, B, and C.

2) Obtain the gas chromatogram and determine the areas of the peaks.

3) Calculate the weight ratio (component/standard) and the area ratio (component/standard) for each component. For example:

for A,

$$\text{wt. ratio (A/standard)} = \frac{\text{wt. of A}}{\text{wt. of standard}}$$

$$\text{area ratio (A/standard)} = \frac{\text{area of A's peak}}{\text{area of standard peak}}$$

Similar ratios are calculated for B and C.

4) For each component, calculate the weight/area correction factor.

$$\text{correction factor, } F = \frac{\text{wt. ratio}}{\text{area ratio}}$$

For A,

$$\text{correction factor } F_A = \frac{\text{wt. ratio (A/standard)}}{\text{area ratio (A/standard)}}$$

Similar ratios are calculated for B and C.

For the unknown mixture:

5) Add a sufficient amount of the standard to the unknown to give a peak approximately equivalent in area to the other peaks in the chromatogram. It is not necessary to know the exact amount of the standard added.

6) Obtain a gas chromatogram and determine the area ratio (component/standard) for each component.

For A,

$$\text{area ratio (A/standard)} = \frac{\text{area of A's peak}}{\text{area of standard peak}}$$

7) Using the correction factor for each component, determine the weight ratio (component/standard).

For A,

$$\text{wt. ratio (A/standard)} = F_A \times \text{area ratio (A/standard)}$$

8) Calculate the weight per cent for each component.

$$\text{wt. per cent A} = \frac{\text{wt. ratio (A/standard)}}{\text{sum of wt. ratios of components (A + B + C) to standard}} \times 100$$

Because the internal standard technique involves so many steps, let us go through the procedure with a specific example.

Example:

For the known:

1) Add 0.25 g of internal standard to a mixture of 0.22 g of compound A, 0.26 g of compound B, and 0.19 g of compound C.

2) *The areas of the glc peaks* are listed in Table 7.1.

3) The *wt. ratios and area ratios* (also listed in Table 7.1) for each component/standard are calculated as follows for A:

$$\text{wt. ratio (A/standard)} = \frac{0.22\text{ g}}{0.25\text{ g}} = 0.88$$

$$\text{area ratio (A/standard)} = \frac{8.8\text{ cm}^2}{6.6\text{ cm}^2} = 1.25$$

4) *The correction factors* (also shown in Table 7.1) for each component

Table 7.1 Data for the known sample in the example

Component	***Weight, g***	***Area, cm²***	***Wt. ratio (component/ standard)***	***Area ratio (component/ standard)***	***Correction factor F (wt. ratio/ area ratio)***
standard	0.25	6.6	—	—	—
A	0.22	8.30	0.88	1.25	0.70
B	0.26	10.4	1.04	1.57	0.66
C	0.19	5.00	0.76	0.75	1.0

in the mixture are calculated as follows for A:

$$F_A = \frac{\text{wt. ratio for A}}{\text{area ratio for A}} = \frac{0.88}{1.25} = 0.70$$

For the unknown:

5) Add about 0.20 g of internal standard to the mixture of unknown composition.

6) The *areas of the glc peaks* and the *area ratios* (components/standard) are listed in Table 7.2. The area ratios are calculated as follows for A:

$$\text{area ratio (A/standard)} = \frac{8.0\ \text{cm}^2}{4.0\ \text{cm}^2} = 2.0$$

7) Using the correction factors, determine the wt. ratio of each component to the standard. The values obtained are shown in Table 7.2. The calculation for component A follows:

$$\begin{aligned}\text{wt. ratio (A/standard)} &= F_A \times \text{area ratio (A/standard)} \\ &= (0.70)(2.0) \\ &= 1.4\end{aligned}$$

8) Calculate the *weight per cents* from each weight ratio and the sum of the weight ratios.

$$\begin{aligned}\text{sum of wt. ratios of A, B, and C} &= 1.4 + 0.12 + 1.8 \\ &= 3.32\end{aligned}$$

$$\text{wt. per cent A} = \frac{1.4}{3.32} \times 100 = 42\%$$

$$\text{wt. per cent B} = \frac{0.12}{3.32} \times 100 = 4\%$$

$$\text{wt. per cent C} = \frac{1.8}{3.32} \times 100 = 54\%$$

Had the internal-standard technique not been used in the example, and only the relative glc *areas* of A, B, and C been used to determine the weight per cent values, the erroneous values of 50% A, 5% B, and 45% C would have been obtained.

Table 7.2 Data for the unknown sample in the example

Component	***Area, cm²***	***Area ratio (component/ standard)***	***Correction factor F[a]***	***Wt. ratio (component/ standard)***
standard	4.0	—	—	—
A	8.0	2.0	0.70	1.4
B	0.75	0.19	0.66	0.12
C	7.2	1.8	1.0	1.8

[a] from Table 7.1

This experiment is the analysis of three-component mixtures (1-propanol, 1-butanol, and *n*-butyl acetate), using 1-bromobutane as the internal standard.

$CH_3CH_2CH_2OH$ — 1-propanol

$CH_3CH_2CH_2CH_2OH$ — 1-butanol

$CH_3\overset{O}{\overset{\|}{C}}OCH_2CH_2CH_2CH_3$ — *n*-butyl acetate

$CH_3CH_2CH_2CH_2Br$ — 1-bromobutane

You will prepare two mixtures of known composition, then use the glc data from the first mixture to calculate the weight per cents of the components in the second mixture. Because you can calculate the weight per cents in the second mixture independently, you will have a check on your calculations. Finally, you will analyze a mixture of unknown composition.

Experimental

EQUIPMENT:

analytical balance
disposable pipets
gas chromatograph
microsyringe
two test tubes

CHEMICALS:

1-bromobutane, 0.3–0.8 g
1-butanol, 0.2–0.6 g
n-butyl acetate, 0.2–0.6 g
1-propanol, 0.2–0.6 g
mixture of 1-propanol, 1-butanol, and *n*-butyl acetate for unknown

TIME REQUIRED: 1–2 laboratory hours, depending upon the instrument

STOPPING POINT: after the instrument work

PROCEDURE

A. Analysis of Known Mixture

Using an analytical balance and two test tubes, prepare two mixtures containing 0.1–0.3 g each (weighed to the nearest mg) of 1-propanol, 1-butanol, 1-bromobutane, and *n*-butyl acetate. Cork the test tubes to prevent evaporation.

Select instrument settings so that each component in one of the mixtures is well-resolved. Inject each component as a pure compound to identify each peak in the mixture. Select an instrument sensitivity that allows the tallest peak in the mixture to trace about 80% of the paper width. Increase the recorder-paper flow to fast speed, then inject a sample of each mixture. On each chromatogram, mark the composition and the instrument settings.

Using the procedure outlined in the discussion, use one mixture to obtain correction factors, then calculate the weight-per cent composition of the second mixture. Also, calculate the weight-per cent composition of the second sample from the actual weights. The calculated and actual values should not differ by more than 5%.

B. Analysis of the Unknown

Obtain a mixture of unknown composition from your instructor. Run an initial chromatogram, then add enough 1-bromobutane to the unknown so that its peak is about the same height as the peaks in the mixture. Run a final chromatogram and calculate the weight per cents of 1-propanol, 1-butanol, and *n*-butyl acetate in the unknown mixture.

EXPERIMENT 7.3

(SUPPLEMENTAL)

Glc Analysis of Distillation Fractions From Experiments 5.1 and 5.2

Experimental

EQUIPMENT:

microsyringe
gas chromatograph

CHEMICALS:

distillation fractions obtained from the simple and fractional distillations in Experiments 5.1 and 5.2

TIME REQUIRED: about 2 hours of laboratory time

STOPPING POINT: after the instrument work

PROCEDURE

Using the procedure outlined in Experiment 7.2, determine the weight per cents of ethyl acetate and *n*-propyl acetate in the distillation fractions obtained in Experiments 5.1 and 5.2. For each distillation, plot the weight per cent of each component versus mL distilled. Compare these graphs with the graphs of refractive index and boiling point versus volume that you prepared in Experiments 5.1 and 5.2. Also, compare the efficiencies of the two distillations if you performed both.

Problems

7.11 In determining the area of a glc peak, why is it more accurate to use the width at half peak height instead of one half the base of the peak?

7.12 Using the data given in Table 7.1 (page 130), calculate the weight-per cent composition of a mixture with glc areas of: internal standard, 4.3 cm^2; A, 6.2 cm^2; B, 7.0 cm^2; and C, 3.2 cm^2.

7.13 Determine the weight percentages of the components in mixture I.

Calibration:

Component	*Weight, g*	*Area, cm²*
standard	0.60	70
A	0.20	25
B	0.45	60
C	0.60	88

Mixture:

Component	*Area, cm²*
standard	30
A	22
B	36
C	18

7.14 In Problem 7.13, if mixture I contains 0.20 g of the internal standard and 0.80 g of the mixture of A, B, and C, what are the actual weights of A, B, and C in the mixture?

»»» CHAPTER 8 «««

Substitution Reactions of Alkyl Halides

The majority of alkyl halide (RX) reactions fall into one of two categories: **substitution reactions**, in which another atom or group is substituted for the halogen, or **elimination reactions**, in which HX is lost and one or more alkenes result. The experiments in this chapter will deal with only substitution reactions.

$$\underset{\text{2-bromopropane}}{CH_3\overset{\overset{\displaystyle Br}{|}}{C}HCH_3} + OH^- \begin{cases} \xrightarrow{\text{substitution}} \underset{\text{2-propanol}}{CH_3\overset{\overset{\displaystyle OH}{|}}{C}HCH_3} + Br^- \\ \xrightarrow{\text{elimination}} \underset{\text{propene}}{CH_3CH{=}CH_2} + H_2O + Br^- \end{cases}$$

Because the same reactants can undergo either type of reaction, mixtures of products are often obtained. By the choice of reactants and reaction conditions (solvent, temperature, etc.), a chemist can control, to an extent, which type of products are obtained.

8.1 S_N2 Reactions of Primary Alkyl Halides

Primary alkyl halides (RCH_2X) undergo substitution reactions much more readily than they undergo elimination reactions. In Experiment 8.1 you will treat 1-bromobutane, a primary alkyl halide, with sodium methoxide ($Na^+ \ {}^-OCH_3$) to synthesize *n*-butyl methyl ether.

The reaction of a methyl halide or primary alkyl halide with an alkoxide (RO^-) to yield an ether is quite general. (Secondary alkyl halides give lower yields of ethers). This reaction is called the **Williamson ether synthesis** after the English chemist A. W. Williamson, who developed the reaction about 1850.

$$\underset{\text{methoxide ion}}{CH_3O^-} + \underset{\text{1-bromobutane}}{CH_3CH_2CH_2CH_2Br} \rightarrow \underset{n\text{-butyl methyl ether}}{CH_3CH_2CH_2CH_2OCH_3} + \underset{\text{bromide ion}}{Br^-}$$

The substitution reaction of a primary alkyl halide with an alkoxide is an example of a reaction that proceeds by an **S_N2 mechanism**: backside attack of the nucleophile ("positive charge seeker") RO^- on the carbon joined to X, accompanied by the simultaneous leaving of X^-.

$$R\ddot{O}:^- + \overset{\delta+}{C}H_2(R')-\overset{\delta-}{\ddot{B}r}: \longrightarrow \left[R\ddot{O}\cdots C(H)(H)(R')\cdots\ddot{B}r: \right]^- \longrightarrow R\ddot{O}-CH_2(R') + :\ddot{B}r:^-$$

transition state,
or activated complex

S_N2 reactions typically exhibit **second-order kinetics**, which means that, under the same reaction conditions (solvent, temperature, etc.), the rate of the reaction is directly proportional to the concentrations of *two* reactants. Because of this proportionality, the rate of product formation can be varied by varying either the concentration of the alkyl halide (RCH_2X) or the concentration of the nucleophilic reagent (RONa). For example, doubling the concentration of *either* reactant would double the rate of the reaction, while doubling *both* concentrations would quadruple the rate. The proportionality may be expressed by a **rate equation**, using k as the proportionality constant, or **rate constant**.

$$S_N2 \text{ rate} = k[\text{RONa}][\text{RX}]$$

8.2 S_N1 Reactions of Tertiary Alkyl Halides

Although tertiary alkyl halides (R_3CX) are too sterically hindered to undergo reaction by a backside attack, they can undergo substitution by a different path, called an **S_N1 mechanism.** S_N1 reactions of tertiary alkyl halides are possible only if the nucleophile is an extremely weak base, because a strong base causes a tertiary halide to undergo elimination. Water and alcohols are sufficiently weak bases that they can be used as nucleophiles in S_N1 reactions of tertiary alkyl halides. Because the water or alcohol is also usually used as the solvent, these reactions are called **solvolysis** ("cleavage by solvent") reactions.

Experiment 8.2 is a study of the solvolysis of *t*-butyl chloride by water and an alcohol. The mechanism of the reaction with water involves three steps: (1) ionization of the alkyl halide to yield a **carbocation** (an ion containing a positively charged carbon atom), followed by (2) reaction of the carbocation with the solvent. In the last step, (3) a proton is lost from the adduct of the carbocation and solvent to yield the final products of the reaction. Figure 8.1 shows these steps. The reaction of a tertiary alkyl halide with an alcohol follows a similar path.

overall reaction:

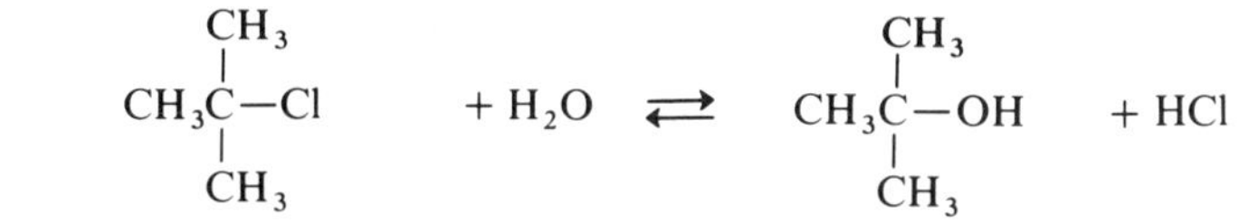

$$\underset{\substack{\text{2-chloro-2-methylpropane}\\(t\text{-butyl chloride})}}{(CH_3)_2CCl\text{-}CH_3} + H_2O \rightleftharpoons \underset{\substack{\text{2-methyl-2-propanol}\\(t\text{-butyl alcohol})}}{(CH_3)_3C\text{-}OH} + HCl$$

mechanism:

Step 1, ionization (rate-limiting):

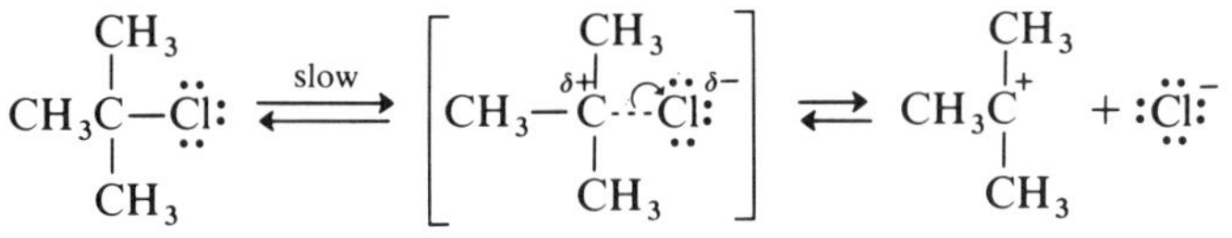

$$(CH_3)_3C\text{-}\ddot{\underset{..}{Cl}}: \xrightleftharpoons{\text{slow}} \left[CH_3\text{-}\overset{\delta+}{C}(CH_3)_2\cdots\overset{\delta-}{Cl}:\right] \rightleftharpoons (CH_3)_3C^+ + :\ddot{\underset{..}{Cl}}:^-$$

Step 2, reaction with solvent:

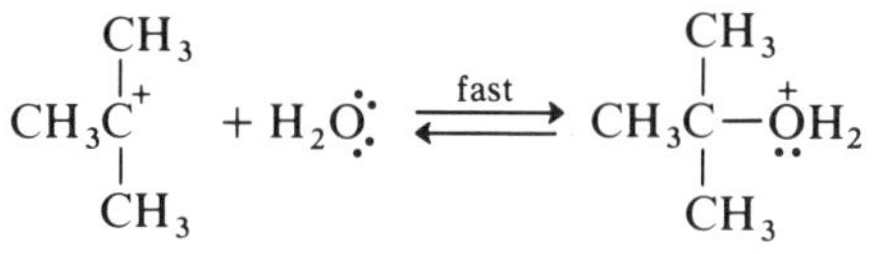

$$(CH_3)_3C^+ + H_2\ddot{O}: \xrightleftharpoons{\text{fast}} (CH_3)_3C\text{-}\overset{+}{\underset{..}{O}}H_2$$

Step 3, loss of H^+ to solvent:

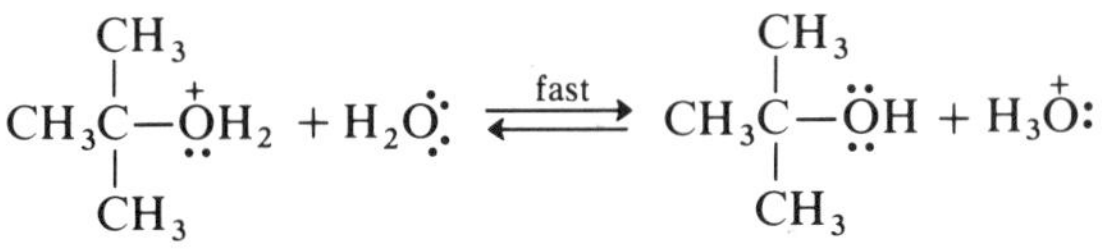

$$(CH_3)_3C\text{-}\overset{+}{\underset{..}{O}}H_2 + H_2\ddot{O}: \xrightleftharpoons{\text{fast}} (CH_3)_3C\text{-}\ddot{\underset{..}{O}}H + H_3\overset{+}{O}:$$

Figure 8.1. The mechanism of the solvolysis of t-butyl chloride with water, an S_N1 reaction.

Note that the first step in the S_N1 mechanism, the breaking apart of *t*-butyl chloride into ions, is *slow* relative to Steps 2 and 3. In any multistep mechanism, the slowest step generally determines the overall rate of the entire reaction sequence. Thus, the slow step is called the **rate-limiting**, or **rate-determining, step**.

Because the slowest step determines the overall rate, the rate of a typical solvolysis reaction of an alkyl halide in water depends on the concentration of *only the alkyl halide*. The rate is independent of the concentration of the water because water is not involved in the slow step. (Of course, other factors, such as a very low concentration of water, can change this rate dependence by changing the rate-limiting step or by changing the character of the reaction mechanism.) A reaction with a rate that is dependent on the concentration of a single reactant is said to follow **first-order kinetics**.

Solvent systems that can support ions in solution (water is the best) enhance the rate of an S_N1 reaction. Conversely, the rate is diminished by solvents such as acetone that are not polar enough to support ions. The rates of reaction in mixtures of the two types of solvents (acetone–water, for example) are intermediate. Experiment 8.2 is a study of the effects of such solvent systems.

8.3 Kinetics

Chemical kinetics, the study of rates of reactions, is extremely important in biochemistry and chemical engineering, as well as in the study of reaction mechanisms. For example, many clinical assays are based upon enzyme kinetics. Questions of chemical manufacturing processes are often answered by kinetic studies. Let us consider briefly two techniques by which rate equations can be determined.

A. Changes with Time in the Concentrations of Reactants and Products

The mathematical form of a kinetics equation cannot be predicted from the balanced chemical equation, but must be determined from experimental data, such as measurements of the concentrations of reactants or products at various time intervals. Let us consider a hypothetical reaction in which 1.0 mol of A reacts with 1.0 mol of B to yield 1.0 mol of C.

$$A + B \rightarrow C$$

In this example, when A and B are mixed, their initial concentrations $[A]_0$ and $[B]_0$ are equal and at their maximum, while the initial concentration of C, $[C]_0$, is zero. As the reaction proceeds, the concentrations of A and B decrease, while that of C increases. Figure 8.2 shows graphs of the concentrations of A, B, and C versus time.

In determining the progress of the reaction between A and B, we need to know the initial concentrations plus the concentration of only *one* of the three compounds (A, B, and C) at time t. The other two concentrations are easily calculated. Let us define the concentrations at a specific time t as $[A]_t$, $[B]_t$, and $[C]_t$. If $[A]_0 = [B]_0$, $[A]_t$ and $[B]_t$ must be equal because they react with each other to yield the product. Since one molecule of C is formed for every molecule of A (or B) that reacts, $[C]_t$ is equal to $[A]_0 - [A]_t$ (or to $[B]_0 - [B]_t$).

If $[A]_0 = [B]_0$,

then $[A]_t = [B]_t$

and $[C]_t = [A]_0 - [A]_t = [B]_0 - [B]_t$

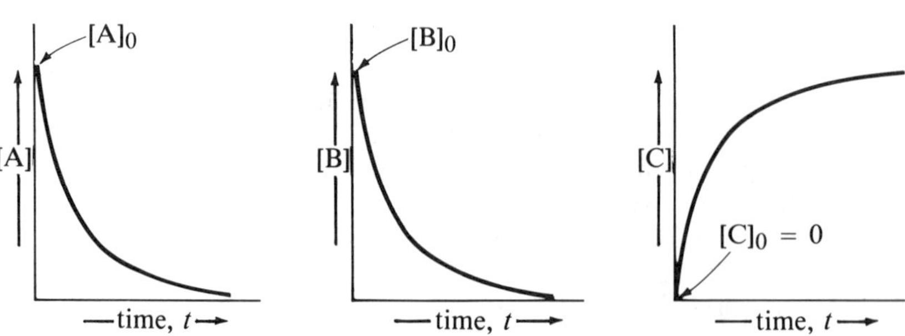

Figure 8.2. Graphs of the concentrations of A, B, and C versus time for the reaction A + B → C.

As an example, let us start with 1.00 mol of A and 1.00 mol of B in 1.00 L of solution. In 15 minutes, we determine experimentally the concentration of C in the reaction mixture to be 0.25*M*. With this value, we can calculate the concentrations of A and B at $t = 15$ minutes:

$$[C]_t = [A]_0 - [A]_t$$
$$[A]_t = [A]_0 - [C]_t$$
$$= 1.00M - 0.25M$$
$$= 0.75M$$

Since $[B]_t = [A]_t,$

$$[B]_t = 0.75M$$

B. Rate Equations

The concentrations of reactants determined at various time intervals can be used to determine the rate constant k. The best technique for calculating k uses a combination of differential and integral calculus.

Because of concentration relationships, the rate of a reaction exhibiting first-order or second-order kinetics can be represented by one of the following sets of equalities:

$$S_N1 \text{ rate} = k[A] = \frac{-d[A]}{dt} = \frac{-d[B]}{dt} = \frac{d[C]}{dt}$$

$$S_N2 \text{ rate} = k[A][B] = \frac{-d[A]}{dt} = \frac{-d[B]}{dt} = \frac{d[C]}{dt}$$

In order to relate actual experimental data to a kinetics equation and to determine k, such as in the first-order equation, it is necessary to integrate the differential equation. The mathematics for integrating the equation for a first-order reaction are shown in Figure 8.3.

C. Half-Life of a Reaction

For reactions that exhibit first-order kinetics, another approach for obtaining the rate constant k is to use the **half-life of the reaction** $t_{1/2}$, the time required for one-half of the reactants to be converted to products. At $t_{1/2}$, a reaction of a 1.0*M* solution of reactant has been converted to a solution that is 0.5*M* in that reactant. The integrated form of the first-order equation at the half-life is

$$\ln 0.5 = -kt_{1/2}$$

or using $\log_{10}$,

$$2.303 \log 0.5 = -kt_{1/2}$$

The rate constant can easily be calculated:

$$k = \frac{2.3 \log 0.5}{t_{1/2}} = \frac{0.69}{t_{1/2}}$$

$$\frac{-d[\mathrm{A}]}{dt} = k[\mathrm{A}]$$

$$\frac{-d[\mathrm{A}]}{[\mathrm{A}]} = k\,dt \quad \text{or} \quad \frac{d[\mathrm{A}]}{[\mathrm{A}]} = -k\,dt$$

Let $[\mathrm{A}]_0$ = initial concentration

and $[\mathrm{A}]_t$ = concentration at time t

Then:

$$\int_{[\mathrm{A}]_0}^{[\mathrm{A}]_t} \frac{d[\mathrm{A}]}{[\mathrm{A}]} = -k\int_0^t dt$$

$$\ln[\mathrm{A}]_t - \ln[\mathrm{A}]_0 = -kt$$

or

$$\ln\left(\frac{[\mathrm{A}]_t}{[\mathrm{A}]_0}\right) = -kt$$

Therefore: A plot of $-\ln\left(\frac{[\mathrm{A}]_t}{[\mathrm{A}]_0}\right)$ versus t is a straight line with a slope equal to the rate constant k.

Figure 8.3 *Solution of the first-order kinetics equation to show how k can be determined from experimental data—in this case, the concentration of A at various time intervals.*

Note that the half-life of a reaction that follows first-order kinetics is independent of concentration. This fact simplifies the experimental determination of the rate constant.

Problems

8.1 Write an equation (not balanced) to show the substitution and elimination products that could be obtained from each of the following pairs of reactants:

(a) $CH_3CH_2CHBrCH_3$ and $NaOCH_3$

(b) 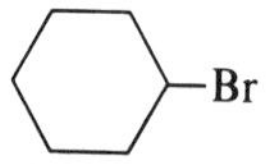 (cyclohexyl)—Br and $NaSCH_2CH_3$

(c) 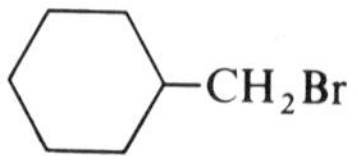 (cyclohexyl)—CH_2Br and KOH

8.2 Write equations for two different Williamson syntheses for each of the following ethers:

(a) 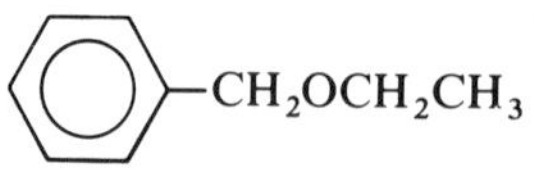 (phenyl)—$CH_2OCH_2CH_3$ (b) $CH_3CH_2CH_2OCH(CH_3)_2$

8.3 Write the mechanism for the following solvolysis reaction. Label the carbocation, the leaving group, the nucleophile, the transition state(s), and the rate-limiting step.

$(CH_3CH_2)_3CCl + CH_3CH_2OH \rightarrow$

8.4 What solvolysis products would be observed for each of the following sets of reactants?

(a) $(CH_3CH_2)_3CCl + H_2O + CH_3OH \rightarrow$

(b) 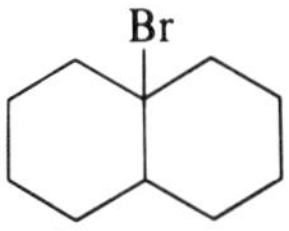 $+ H_2O + CH_3CH_2OH \longrightarrow$

8.5 Ethyl *n*-pentyl ether is prepared from sodium ethoxide and 1-bromopentane in two different Williamson syntheses. In the first, 0.10 mol of each reactant and 100 mL of ethanol (as solvent) are used. In the second, 0.10 mol of each reactant and 200 mL of ethanol are used. If the temperature is the same in each reaction, (a) which reaction will proceed faster? (b) How much faster? Explain your answer.

8.6 *t*-Butyl ethyl ether is prepared by two different solvolysis reactions of *t*-butyl chloride at 25°. In the first reaction, 0.10 mol of the alkyl halide is mixed with 100 mL of ethanol. In the second, 0.10 mol is mixed with 200 mL of ethanol. How much faster is the first reaction? Explain.

8.7 Which of the following changes in reaction conditions would change the rate constant (k) of a reaction? Explain.

(a) concentration of the reactants

(b) temperature

(c) solvent

8.8 (a) Would the following reaction proceed by backside attack or by way of a carbocation?

(b) Would the reaction exhibit first-order or second-order kinetics? Explain.

$^-OCH_2CH_2CH_2CH_2CH_2Cl \longrightarrow$ 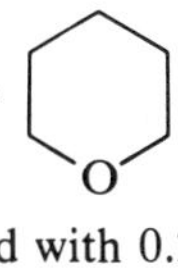 $+ Cl^-$

8.9 1-Bromobutane (0.50 mol) is treated with 0.20 mol of sodium methoxide in methanol. The mixture is diluted with methanol to 100 mL.

(a) What is the maximum number of grams of *n*-butyl methyl ether that can be formed?

(b) If 10.3 g of NaBr is formed in 20 minutes, how many grams of 1-bromobutane remain?

(c) What are the initial concentrations of 1-bromobutane and sodium methoxide?

(d) What are the concentrations of these two reactants after 20 minutes?

8.10 If the $t_{1/2}$ of a first-order reaction is 15 minutes, what is the rate constant for that reaction?

8.11 If $t_{1/2}$ for a first-order reaction is 10 minutes and if the initial concentration is $0.20M$, what is the concentration after (a) 30 minutes? (b) one hour?

»»» EXPERIMENT 8.1 «««

Synthesis of n-Butyl Methyl Ether

Step 1, Preparation of Sodium Methoxide

	CH_3OH	+	Na	→	$CH_3O^- Na^+$	+	$\frac{1}{2}H_2\uparrow$
	methanol		sodium		sodium methoxide (not isolated)		hydrogen
MW:	32.04		22.99		—		—
weight:	158 g (200 mL)		7.0 g		—		—
moles:	4.93		0.30		0.30 (theory)		—

Step 2, Reaction of Sodium Methoxide and 1-Bromobutane

	$CH_3O^- Na^+$	+	$CH_3CH_2CH_2CH_2Br$	→	$CH_3CH_2CH_2CH_2OCH_3$	+	$Na^+ Br^-$
	sodium methoxide		1-bromobutane		*n*-butyl methyl ether		
MW:	—		137.03		88.15		
weight:	—		27.5 g		17.6 g (theory)		
moles:	0.30 (theory)		0.20		0.20 (theory)		

Table 8.1 Physical properties of reactants and products

Name	***Bp (°C)***	n_D^{20}	***Density (g/mL)***	***Solubility***
methanol	65.0	1.3288	0.79	sol. H_2O and alcohols
sodium metal	—	—	3.9	—[a]
1-bromobutane	101.6	1.4401	1.28	insol. H_2O, sol. CH_3OH
n-butyl methyl ether	71	1.3736	0.74	insol. H_2O, sol. CH_3OH

[a] Reacts vigorously with methanol, violently with water.

Discussion

A Williamson ether synthesis consists of two separate reactions: the preparation of an alkoxide, and the reaction of this reagent with the alkyl halide.

Sodium methoxide is prepared by the addition of sodium metal to methanol. In the reaction, sodium metal is oxidized to sodium cations and the hydrogen atoms of the —OH groups are reduced to hydrogen gas. A large excess of methanol is used to act as a solvent for the sodium methoxide.

The exothermic nature of the reaction causes the methanol to boil. To prevent the methanol from boiling away, an upright condenser, called a **reflux condenser**, is attached to the reaction flask. Methanol vapors condense in the condenser and the liquid runs back into the flask. Experimental Note 1 and Figure 8.4 (page 147) describe the process of refluxing.

Because of the vigor of the reaction, the sodium must be added *slowly*; otherwise, the methanol will boil violently, overwhelm the capacity of the reflux condenser, and spew out the top of the condenser. An uncontrolled reaction of this type, called a **runaway reaction**, may erupt like a volcano, throwing flammable solvent and corrosive chemicals over laboratory workers and the work area.

There are a number of techniques by which a solid can be added to a reaction mixture. In this experiment, small pieces of sodium are added to the flask through the reflux condenser tube. Most solids (generally powders) should *not* be added to reaction vessels in this manner because they tend to stick to the sides of the condenser. Should the sodium stick to the side of the condenser, a long glass rod or tubing can be used to push it down into the methanol. (It is prudent to bend an L at one end of the glass rod so that it cannot drop completely through the condenser and puncture the flask.)

After all the sodium has reacted, excess methanol is removed by distillation. Decreasing the volume of solvent increases the rate of reaction of sodium methoxide with 1-bromobutane and permits the entire reaction sequence to be carried out in a single laboratory period.

It is possible to stop the synthesis at this point until the next laboratory period. However, it is not desirable to do this because sodium methoxide is strongly basic. It absorbs moisture from the air and is converted to methanol and NaOH.

$$CH_3O^- + H_2O \rightleftharpoons CH_3OH + OH^-$$

If it is necessary to store the sodium methoxide, the flask must be sealed with a well-greased glass stopper.

After the excess methanol has been distilled, the flask containing the sodium methoxide is refitted with the reflux condenser, and 1-bromobutane is added in small portions through the condenser. The vigorous spontaneous reaction should be allowed to subside before each addition. As the reaction proceeds, sodium bromide precipitates. After all the 1-bromobutane has been added, the reaction mixture is heated at reflux (heated to boiling with a reflux condenser attached to the flask) to complete the reaction.

After a reaction has been carried out, the product must be isolated from the reaction mixture and purified. The general procedure is termed a **work-up**. A specific work-up procedure is dictated by the physical and chemical properties of the products and by-products in the mixture. In this experiment, the first step in the work-up is the addition of water to dissolve the sodium bromide. Two phases result: an aqueous–methanol phase and an organic phase. These two phases could be separated in a separatory funnel; however, the separation would be poor because the product *n*-butyl methyl ether is partly soluble in the methanol–water phase. Because *n*-butyl methyl ether and methanol form an azeotrope that boils at 56°, they are easily separated from the mixture by a distillation. The distillate contains *n*-butyl methyl ether and methanol, while the residue contains water, sodium bromide, and the bulk of the methanol.

The next step in the work-up procedure is to remove the methanol from the product ether. Methanol is water-soluble, but the ether is not. An aqueous extraction of the methanol can be used at this point because most of the solvent methanol was left behind in the distillation step. An aqueous solution of calcium chloride instead of pure water is used to extract the methanol from the ether. The presence of the salt in the aqueous layer "salts out" the ether (page 56) so that it is not carried into the water solution by the methanol. After extraction, the ether is dried with calcium chloride. Calcium chloride forms solid molecular complexes with both methanol and water; therefore, any residual methanol and water are removed from the product.

Finally, the product is purified by distillation. If the earlier distillation and drying steps were carried out carefully, only the clear product (bp 65–68°) will distil. If the distillate is cloudy, it contains water and must be redried and redistilled. If part of the material boils at 56°, the boiling point of the ether–methanol azeotrope, then all the methanol was not removed. In this case, the product must be rewashed with aqueous calcium chloride, redried, and redistilled.

Experimental

EQUIPMENT:

condenser
distillation assembly
droppers
two 50-mL Erlenmeyer flasks
glass rod or tubing with an L at one end
100-mL graduated cylinder
heating mantle and rheostat
ice bath
50-mL, 100-mL, and 250-mL round-bottom flasks
125-mL separatory funnel

ground-glass stopper (optional)

CHEMICALS:

1-bromobutane, 27.5 g
anhydrous calcium chloride, 3 g
25% aqueous calcium chloride, 45 mL
methanol, 200 mL
sodium metal, 7 g

TIME REQUIRED: 4–5 hours

STOPPING POINTS: after preparation of sodium methoxide (if necessary); after the reaction with 1-bromobutane; while the ether is being dried with $CaCl_2$.

›››› ***SAFETY NOTE 1*** Methanol is toxic and flammable. Ingestion or excessive inhalation of the vapors can cause blindness or death. Use an efficient condenser when distilling it. (If possible, use the fume hood.) Wash any splashes on your skin with water.

›››› ***SAFETY NOTE 2*** *The sodium metal must not come into contact with water*! Do not throw sodium scraps down the sink or wash the work area with a wet towel or sponge. The reason is twofold: (a) The reaction of sodium with water is very exothermic and the hydrogen gas formed usually ignites and explodes. (b) The other product of the reaction is concentrated NaOH, which is corrosive to both clothing and skin.

Follow the procedure outlined in Experimental Note 2 for handling sodium.

›››› ***SAFETY NOTE 3*** Hydrogen gas is given off in this experiment. Flames cannot be used in the laboratory, and adequate ventilation must be provided. In some laboratories, it may be advisable to carry out the preparation of sodium methoxide in the fume hood.

›››› ***SAFETY NOTE 4*** Sodium methoxide is a strong base and very caustic. Wash any spills with copious amounts of water (see Sections 1.1D and 1.1E).

PROCEDURE

Step 1, Preparation of Sodium Methoxide. Place 200 mL of methanol in a 250-mL round-bottom flask and fit the flask with a reflux condenser (see Experimental Note 1). Add 7.0 g of diced sodium metal through the condenser, *one piece at a time,* allowing the reaction to subside before adding the next piece (see Safety Note 2 and Experimental Note 2). If the sodium sticks to the inside of the condenser tube, push it into the reaction flask with a long glass tube or rod.

After the sodium has completely reacted, fit the flask for a simple distillation, add 3–4 boiling chips, and distil 125 mL of methanol (bp 64°) into a graduated cylinder. (CAUTION: See Safety Notes 1 and 4.) If it is necessary to stop the experiment at this point, store the sodium methoxide in the round-bottom flask with a heavily greased glass stopper.

Step 2, Reaction of Sodium Methoxide and 1-Bromobutane. Fit the 250-mL round-bottom flask containing the sodium methoxide with a reflux condenser. Cool the flask to room temperature with an ice bath, if necessary. Weigh 27.5 g of 1-bromobutane into a 50-mL Erlenmeyer flask. Using a dropper, add the bromobutane to the reaction vessel through the top of the condenser in 1–2-mL aliquots over about a 10-minute period. Although warming may be required to start the reaction, the reaction is quite exothermic. Do *not* add all the bromobutane in one portion. Cool the reaction flask with an ice bath, if necessary, during the addition. Cork the Erlenmeyer flask between additions so that the 1-bromobutane does not evaporate.

After the addition has been completed, let the reaction vessel stand at room temperature until the exothermic reaction has subsided and the methanol ceases to reflux. Then heat the mixture at a gentle reflux, or "simmer," for one-half hour with a heating mantle. The mixture will bump because of the precipitated solid; therefore, do not attempt a vigorous reflux.

After the reflux, cool the reaction vessel with a water or ice bath and add 20–30 mL of water to the reaction mixture through the reflux condenser. If all the sodium bromide does not dissolve, add a few additional milliliters of water. Dissolving the sodium bromide prevents bumping in the next step, a distillation.

Equip the flask for simple distillation and add 2–3 boiling chips. Distil the two-layered mixture, collecting the material that boils up to 64°. The volume of distillate should be about 40 mL. Transfer the distillate to a 125-mL separatory funnel and extract it with three 15-mL portions of 25% aqueous calcium chloride. (The calcium chloride solution is the lower layer in this extraction, and the interface is difficult to see.) Pour the product from the separatory funnel into a 100-mL round-bottom flask, add about 3 g of anhydrous calcium chloride, stopper the flask, and let it stand overnight.

Decant the dried product into a 50-mL round-bottom flask. Use a dropper to transfer the final portion, being careful not to transfer any solid. Add 2–3 boiling chips to the flask; fit it for simple distillation; and distil the product, collecting the material boiling at 65–68°. You should obtain about 10 g (57%) of *n*-butyl methyl ether.

Measure the refractive index of the product, and calculate your per cent yield (Section 1.2C). Transfer the product to a correctly labeled vial and hand it in to your instructor.

EXPERIMENTAL NOTES

1) A **reflux condenser** is an ordinary condenser arranged in an upright position, as shown in Figure 8.4, so that vapors from a boiling liquid are condensed and returned to the flask. A reflux condenser can be used during spontaneous exothermic reactions or when a liquid is being boiled by a heat source. The purposes of reflux are twofold: the reaction temperature can be maintained, and solvent is not lost to the atmosphere.

2) Sodium is stored under mineral oil to protect it from air and moisture. If the sodium has been diced by the storeroom personnel, remove the pieces with tweezers, blot off excess mineral oil with a laboratory tissue, and weigh the sodium on a tared watch glass. Do not allow it to sit in the air for any length of time.

If the sodium has not been diced, use the following procedure. Remove a lump from the jar and blot off the oil. Dip your fingers in fresh mineral oil (not the sodium jar!) to provide a protective coating. (Using latex gloves is unwise. If a fire should occur, the latex can catch fire and cause severe burns.) Cut the crust from the sodium with a razor blade or knife to expose the shiny metal. Cut a wedge of the fresh metal and transfer it to a watch glass for dicing into pieces about the size of small peas, smaller than the inside diameter of the condenser.

After the sodium has been cut, return all small slivers of the metal back to the main jar. *Clean all utensils, desk, and watch glasses*

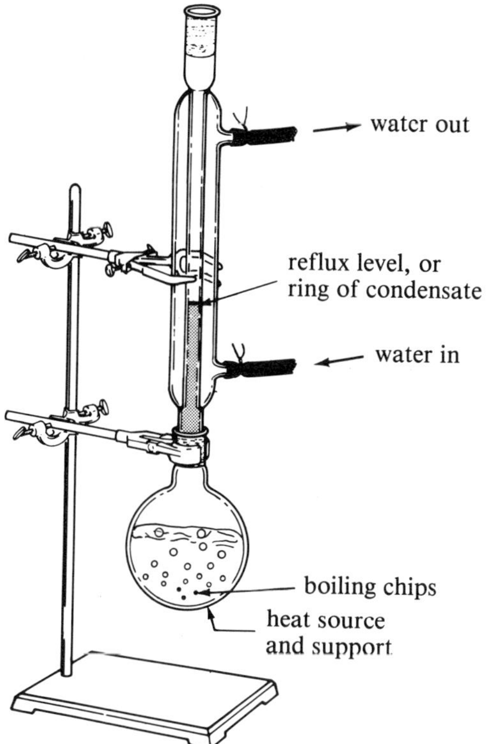

Figure 8.4 *A round-bottom flask fitted with a reflux condenser.*

with a tissue dampened with isopropyl alcohol or methanol. Soak the tissues used for cleaning in a beaker with methanol in the hood for one hour before disposing of them (or before allowing them to come into contact with water). Clean your hands with a tissue soaked in isopropyl alcohol, then wash them thoroughly with soap and water.

Problems

8.12 (a) Why is sodium metal stored under mineral oil?
(b) What is the crust that forms on sodium metal?

8.13 A chemist desires to wash the mineral oil off some sodium with a solvent. Which solvent or solvents would be appropriate?
(a) water (b) ethanol (c) pentane (d) petroleum ether

8.14 What would be the expected results of each of the following actions by a student while carrying out Experiment 8.1? [Use equations in your answers to (d) and (e).]
(a) Removing the reflux condenser to add Na to the methanol.
(b) Adding the Na rapidly through the condenser.

(c) Distilling 175 mL of methanol from the sodium methoxide instead of the 125 mL called for. (Give the immediate results and the results in subsequent reaction with 1-bromobutane.)

(d) Using 95% ethanol instead of methanol in the experiment.

(e) Leaving the sodium methoxide solution exposed to the air until the next laboratory period before adding the 1-bromobutane.

(f) Pouring the 1-bromobutane through the condenser in one portion.

8.15 Explain why the bulk of the methanol is removed before the sodium methoxide is treated with 1-bromobutane in Experiment 8.1. Use a kinetics equation in your answer.

8.16 By error, a student uses 1-bromopropane instead of 1-bromobutane in Experiment 8.1.

(a) What will the product be?

(b) What is the theoretical yield of this product?

8.17 Complete and balance the following equations:

(a) [cyclohexane ring]–OH + Na ⟶

(b) $CH_3CH_2CH_2CH_2OH + K \rightarrow$

(c) [the product from (b)] + $CH_3OH \rightarrow$

»»» EXPERIMENT 8.2 «««

(SUPPLEMENTAL)

Solvent Effects in an S_N1 Solvolysis Reaction: A Kinetics Study

Discussion

The mathematics for kinetics calculations are discussed in Section 8.3. In this experiment, you will not conduct a detailed quantitative kinetics study. Instead, you will determine the *relative rates* of the solvolysis of *t*-butyl chloride in three different solvent systems (methanol–water, ethanol–water, and acetone–water) and express the results in graphic form.

In the first two solvent systems, mixtures of organic products are formed because the alcohols contain hydroxyl (OH) groups. With the acetone–water mixture, only the water participates in the solvolysis reaction. Figure 8.5 shows flow equations for these reactions. Regardless of the solvent system, the inorganic product is HCl. Note that for each molecule of *t*-butyl chloride that undergoes reaction, one molecule of HCl is generated. Because of this 1 : 1

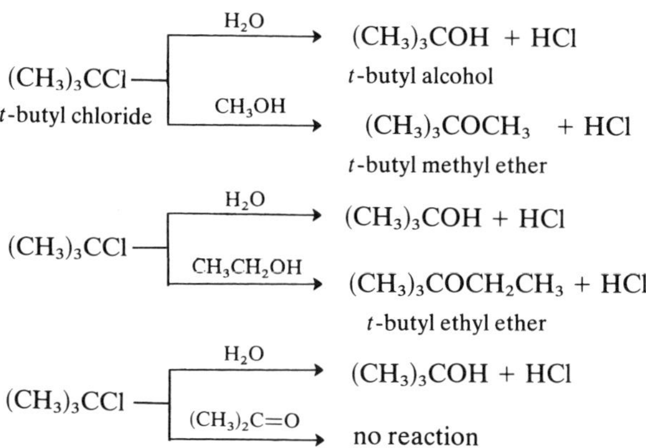

Figure 8.5 The solvolysis reactions of t-butyl chloride in various solvent systems.

correspondence, the course of the reaction can be conveniently followed by measuring the acidity of the reaction mixture.

In this experiment, you will be comparing the time it takes for each of several solvolysis reactions to reach the same per cent of completion. Because this is a study of *relative* rates, the exact per cent of completion does not matter as long as all the reactions are carried to the same point. (Your solvolysis reactions will be carried to only about 5% completion.) Detecting the per cent completion point is accomplished by adding a measured amount of NaOH to each reaction mixture. Under these conditions, each mixture remains alkaline until the NaOH has been neutralized by the HCl being generated.* Then, the solvolysis mixture will turn acidic as additional HCl is generated. We can detect the change from an alkaline solution to one that is acidic by including phenolphthalein in the solvolysis mixture. When the mixture becomes acidic, the solution changes from pink to colorless. From a plot of the per cent water in each solvent system versus the time required to reach the phenolphthalein end point, we can compare the effects of various solvent systems upon the rate of the S_N1 reaction.

The experiment as described is semiquantitative, and the results cannot be used to calculate the rate constant. Your instructor may wish to modify this experiment for quantitative measurements. In a quantitative procedure, a known weight of *t*-butyl chloride is mixed with the solvolysis solvent (without NaOH). Aliquots of the mixture are removed periodically, and the HCl in the aliquot is titrated with standardized base. From these data, the per cent reaction at specific time intervals can be calculated. From a plot of per cent reaction versus time elapsed, $t_{1/2}$ can be determined, and then the rate constant can be calculated.

* Under the conditions of this experiment, a tertiary alkyl halide can also undergo elimination by an E1 or E2 path. In this experiment, these side reactions are ignored. If you carry out titrations of aliquots of the reaction mixture (instead of adding NaOH directly to the reaction mixture), side reactions are minimized.

Experimental

EQUIPMENT:

two burets (or a 10-mL graduated cylinder and graduated pipets to deliver 2.0 mL)
clock with a second hand
dropper or disposable pipet
3–5 styrofoam cups
15 test tubes, 13 × 100 mm, with corks
thermometer

CHEMICALS:

acetone, 6–10 mL
t-butyl chloride, less than 5 mL
95% ethanol, 6–10 mL
methanol, 6–10 mL
phenolphthalein indicator, a few drops
0.5*N* sodium hydroxide, less than 5 mL

TIME REQUIRED: 2–3 hours

STOPPING POINTS: Although the experiment could be stopped after any batch has been run, it is preferable to run all the solvolysis reactions in the same laboratory period. This will ensure that the same droppers and the same NaOH solution are used.

PROCEDURE

The solvent systems to be tested are listed in Table 8.2. Because 15 separate mixtures will be tested, plan to run 3–5 reactions simultaneously. Each reaction requires 5–30 minutes; therefore, your various runs should be planned to overlap.

Place 2.0 mL of the appropriate solvent mixture in a clean, *labeled* test tube. Use a buret to add the proper volume of solvent and a second buret to add the distilled water (see Experimental Note 1). Cork the test tubes and place them in a constant-temperature bath for about 5 minutes to come to

Table 8.2 *Solvent mixtures for Experiment 8.2*

Composition (per cent by volume) (solvent : water)	*Volumes for 2.0-mL of mixture*[a]	
	Solvent[b]	*Water*
50:50	1.0 mL	1.0 mL
55:45	1.1	0.9
60:40	1.2	0.8
65:35	1.3	0.7
70:30	1.4	0.6

[a] See Experimental Note 1.
[b] The solvents to be mixed with water are methanol, 95% ethanol, and acetone.

thermal equilibrium. A styrofoam coffee cup is a convenient insulating container for a constant-temperature bath. Place water at 30° ± 1° in the cup, along with a thermometer. During the course of the experiment, add a few milliliters of warm water to the bath to maintain the temperature at 30° ± 1°.

To each test tube, add 3 drops of 0.5*N* sodium hydroxide solution and 1–2 drops of phenolphthalein indicator (see Experimental Note 2).

To one tube at a time, add 3 drops of *t*-butyl chloride. Mix or shake the contents of the tube immediately and record the time of the addition to the nearest second. When the pink color disappears, again record the time. Repeat this procedure for each of the solvent systems.

Calculate the elapsed time for reaction in each solvent system to the nearest 0.1 minute. Plot per cent water in each solvent system versus elapsed time. (Place all three plots on the same graph.) Compare the three plots and record your observations and conclusions.

EXPERIMENTAL NOTES

1) Your instructor may choose to set up communal burets containing the solvents and water for the experiment. If burets are not available, then mix ten times the volumes listed in Table 8.2, using a 10-mL graduated cylinder, and transfer a 2.0-mL aliquot to the test tube. (Several students can use each solvent mixture.)

2) The actual size of a drop of liquid varies, depending upon the dropper. For consistent results, use one dropper for all NaOH additions and another dropper for all *t*-butyl chloride additions.

Problems

8.18 Why is temperature control important in Experiment 8.2?

8.19 If 0.100 g of *t*-butyl chloride is present in the solvolysis mixture in Experiment 8.2, how much 0.500*N* NaOH solution must be added to the mixture in order to have the indicator change color at $t_{1/2}$?

8.20 A student adds four drops (rather than three) of *t*-butyl chloride to each of the test tubes in Experiment 8.2.

(a) Would this added reactant affect the times of reaction?

(b) How would this extra quantity of reactant affect the student's graphs?

8.21 Using equations, show why the rate of elimination of *t*-butyl chloride by way of a *t*-butyl cation is the same as the rate of solvolysis.

8.22 What are the solvolysis products of *t*-butyl chloride in a mixture of 1-propanol and 1-butanol?

8.23 Complete the following equations, showing the principal organic products only. (If no reaction occurs, write "no reaction.")

(a) $(CH_3CH_2)_3CBr$ + (cyclohexyl)–OH ⟶

(b) $(CH_3)_3CCl + NaOCH_2CH_3 \rightarrow$

(c) $(CH_3CH_2)_3CBr + CH_3CH_2OCH_2CH_3 \rightarrow$

»»» CHAPTER 9 «««

The Grignard Reaction

An organometallic compound is an organic compound that contains carbon bonded to a metal. Some organometallic compounds are highly reactive (dangerously so), while others are relatively stable.

The degree of reactivity of an organometallic compound depends on the degree of *ionic character* of the carbon–metal covalent bond. Because carbon is more electronegative than metals, a carbon atom bonded to a metallic atom withdraws electrons towards itself. Thus, the carbon atom attains a partial negative charge and the organic group can behave as a **carbanion** ($R_3C:^-$).

$$\overset{\delta-}{CH_3} \leftarrow \overset{\delta+}{Li} \qquad \overset{\delta-}{CH_3} \leftarrow \overset{\delta+}{Mg} \rightarrow \overset{\delta-}{I}$$

A more negative carbon is more reactive than a less negative carbon; therefore, the most reactive organometallic compounds are those containing the most electropositive metals (the alkali metals). An organometallic compound containing a less electropositive metal is less reactive.

$$\underset{\textit{increasing ionic character; increasing reactivity}}{\xrightarrow{CH_3CdCH_3 \qquad CH_3MgI \qquad CH_3Li}}$$

In 1901, the French mathematician–chemist Victor Grignard reported *organomagnesium halides* (RMgX),* now called **Grignard reagents**, in his doctoral

* The structure of a Grignard reagent is not as simple as we have shown it. There is evidence that a Grignard reagent is actually an equilibrium mixture:

$$2\ RMgX \rightleftarrows R_2Mg + MgX_2$$

dissertation. He received the 1912 Nobel prize in chemistry for his discovery and subsequent development of the reactions of these reagents.

$$CH_3CH_2Br + Mg \xrightarrow{\text{diethyl ether}} CH_3CH_2MgBr$$

ethylmagnesium bromide
a Grignard reagent

One useful reaction of a Grignard reagent is its reaction with a ketone to yield the magnesium salt of a tertiary alcohol, which can be converted to the alcohol by hydrolysis. Experiment 9.1 is such a synthesis.

$$\underset{\text{a ketone}}{R\overset{O}{\overset{\|}{C}}R} + \underset{\text{a Grignard reagent}}{R'MgX} \xrightarrow{\text{diethyl ether}} \underset{\text{a magnesium alkoxide}}{R\overset{O^{-}\,{}^{+}MgX}{\overset{|}{\underset{\underset{R'}{|}}{C}}}R} \xrightarrow{H_2O,\,H^+} \underset{\text{a 3° alcohol}}{R\overset{OH}{\overset{|}{\underset{\underset{R'}{|}}{C}}}R} + Mg^{2+} + X^-$$

9.1 Preparation of a Grignard Reagent

A Grignard reagent is prepared by the reaction of magnesium metal with an organohalogen compound in an ether solvent.

$$R{-}X + Mg \xrightarrow{\text{an ether}} R{-}Mg{-}X$$

where R = alkyl, aryl, or vinyl

An ether solvent is necessary for the formation of the Grignard reagent. It is thought that the unshared electrons of the oxygen coordinate with the Mg to stabilize the organometallic compound. Also, the alkyl portion of the ether provides a hydrocarbon-like solvent for the organic part of the Grignard reagent. Both these factors result in Grignard reagents being *soluble in ethers.*

$$\begin{array}{c} R'\diagdown\ddot{\underset{\cdot\cdot}{O}}\diagup R' \\ \downarrow \\ R{-}Mg{-}X \\ \uparrow \\ R'\diagup\ddot{\underset{\cdot\cdot}{O}}\diagdown R' \end{array}$$

Diethyl ether is the usual solvent because it is inexpensive. Other ethers, such as tetrahydrofuran, can also be used.

$CH_3CH_2OCH_2CH_3$
diethyl ether
("ether")

tetrahydrofuran
(THF)

9.2 Reactions of Grignard Reagents

Grignard reagents are *extremely strong bases* that react with acids, alcohols, amines, water, and even such weak acids as alkynes with a terminal triple bond

($RC{\equiv}CH$, $pK_a \cong 26$). In these reactions, the R of RMgX removes a proton from the acidic compound and becomes an alkane (RH).

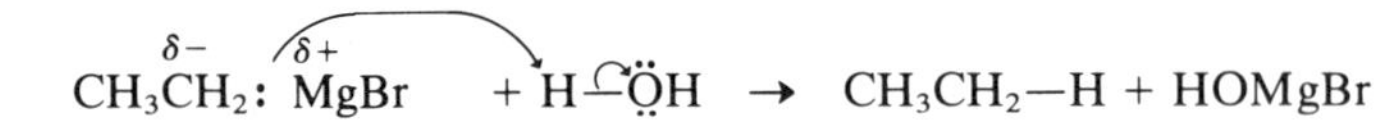

$$\overset{\delta-}{CH_3CH_2}\!:\overset{\delta+}{MgBr} + H{-}\ddot{\underset{..}{O}}H \rightarrow CH_3CH_2{-}H + HOMgBr$$

ethylmagnesium bromide ethane

Because acid–base reactions are very rapid compared to other organic reactions, a compound with a removable H^+ quickly destroys a Grignard reagent. Therefore, in carrying out a Grignard reaction, we must use scrupulously dry glassware, solvents, and reagents. In addition, the reagents must not contain a group with a reactive hydrogen unless an alkane is the desired product.

In the reaction with a carbonyl compound, the organic portion of a Grignard reagent acts as a *strong nucleophile* and attacks the partially positive carbonyl carbon to yield a magnesium alkoxide.

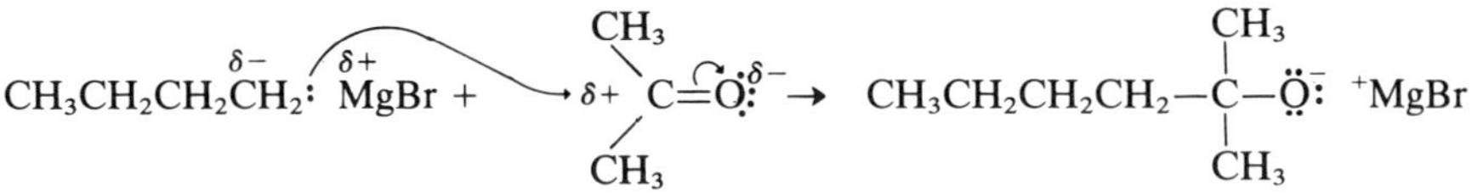

$$\overset{\delta-}{CH_3CH_2CH_2CH_2}\!:\overset{\delta+}{MgBr} + \overset{\delta+}{C}(CH_3)_2{=}\ddot{O}\!:^{\delta-} \rightarrow CH_3CH_2CH_2CH_2{-}C(CH_3)_2{-}\ddot{\underset{..}{O}}\!:^{-}\ {}^{+}MgBr$$

Subsequent treatment of the alkoxide with aqueous acid yields the alcohol itself.

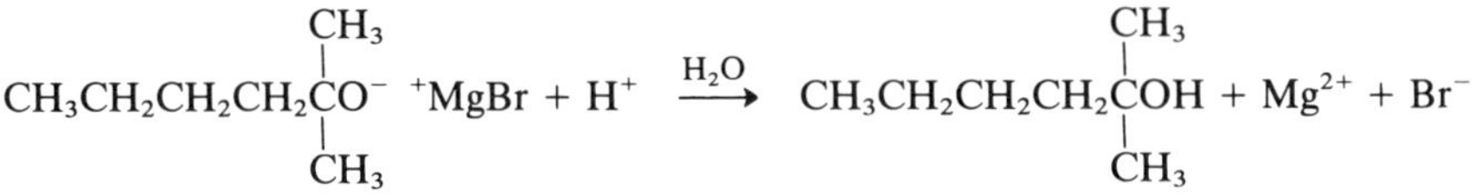

$$CH_3CH_2CH_2CH_2C(CH_3)_2O^-\ {}^{+}MgBr + H^+ \xrightarrow{H_2O} CH_3CH_2CH_2CH_2C(CH_3)_2OH + Mg^{2+} + Br^-$$

Problems

9.1 Which compound of each of the following pairs would be the more reactive?

(a) $CH_3CH_2CH_2Li$ or $(CH_3CH_2CH_2)_2Zn$

(b) $(CH_3CH_2)_2Hg$ or $CH_3CH_2CH_2CH_2Na$

(c) $(CH_3)_2Cd$ or $(CH_3)_4Si$

9.2 Circle the carbon in each of the following structures that has the greatest carbanion character:

(a) 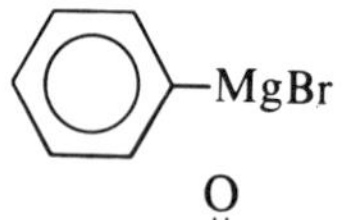 $C_6H_5{-}MgBr$ (b) $CH_3OCH_2CH_2CH_2Li$

(c) $BrZnCH_2\overset{O}{\overset{\|}{C}}OCH_2CH_3$

9.3 Complete the following equations:

(a) $CH_3CH_2CH_2Br + Mg \xrightarrow{\text{diethyl ether}}$

(b) 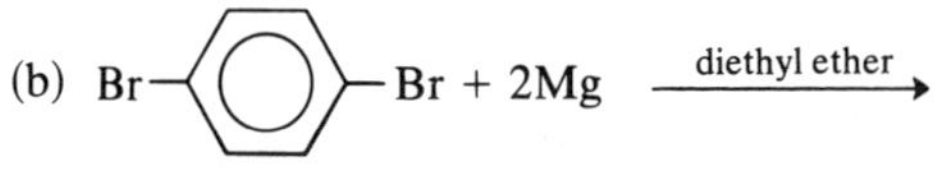 $Br{-}C_6H_4{-}Br + 2Mg \xrightarrow{\text{diethyl ether}}$

(c) $CH_3CH_2CH_2MgI + CH_3CH_2OH \xrightarrow{\text{diethyl ether}}$

(d) $CH_3CH_2MgBr + CH_3CH_2\underset{\displaystyle NH_2}{\underset{|}{C}}H\overset{\displaystyle O}{\overset{\|}{C}}CH_3 \xrightarrow{\text{diethyl ether}}$

(e) 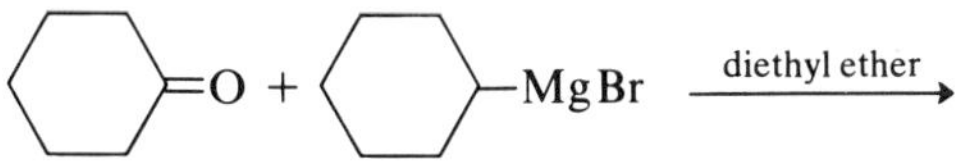 cyclohexanone (C_6H_{10}=O) + cyclohexyl–MgBr $\xrightarrow{\text{diethyl ether}}$

(f) [the product from (e)] + $H^+ \xrightarrow{H_2O}$

(g) 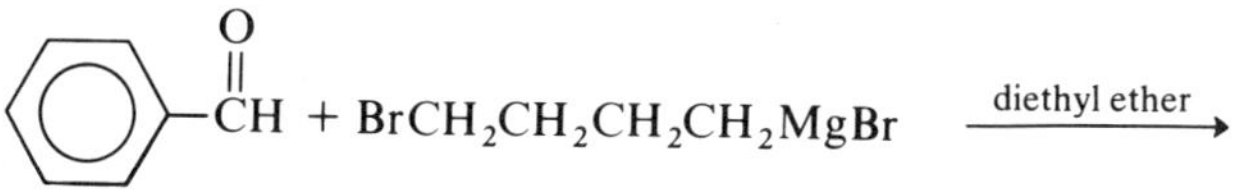 C_6H_5–$\overset{\displaystyle O}{\overset{\|}{C}}H + BrCH_2CH_2CH_2CH_2MgBr \xrightarrow{\text{diethyl ether}}$

(h) [the product from (g)] + $H^+ \xrightarrow{H_2O}$

(i) $CH_3CH_2MgBr + CH_3\overset{\displaystyle O}{\overset{\|}{C}}OH \xrightarrow{\text{diethyl ether}}$

(j) 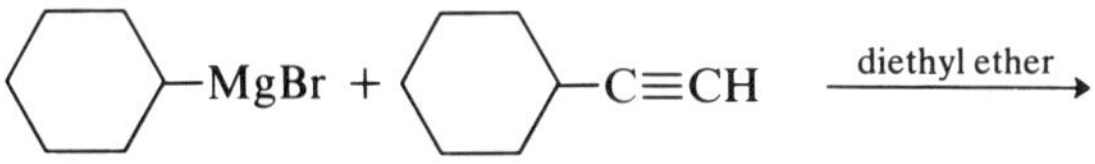 cyclohexyl–MgBr + cyclohexyl–C≡CH $\xrightarrow{\text{diethyl ether}}$

9.4 Which of the following compounds could be used as a solvent for a Grignard reaction? Explain.

(a) $CH_3OCH_2CH_2OCH_3$ (b) 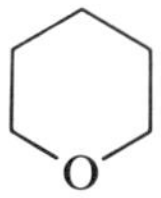 tetrahydropyran (six-membered ring containing O) (c) $HOCH_2CH_2OCH_3$

9.5 How many grams of water will completely destroy 10 g of *n*-hexylmagnesium bromide?

9.6 Would you expect the solvent-complexing of a Grignard reagent to increase or decrease the carbanion character of the carbon bonded to the magnesium, compared to that in a nonsolvated Grignard? Explain your answer.

9.7 Would you expect a Grignard reagent prepared in tetrahydrofuran to be more reactive or less reactive than the same Grignard reagent prepared in diethyl ether? Explain your answer.

9.8 Write flow equations for Grignard syntheses of the following compounds:

(a) 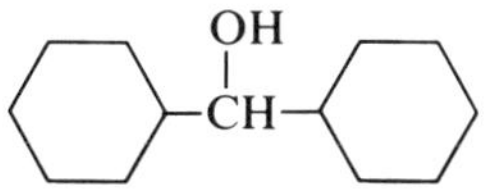 cyclohexyl–$\underset{}{\overset{\displaystyle OH}{\overset{|}{C}}}H$–cyclohexyl (b) $(CH_3CH_2)_3COH$ (c)  cyclohexane with H and D on the same carbon

9.9 Suggest a series of reactions to accomplish the following conversions:

(a) 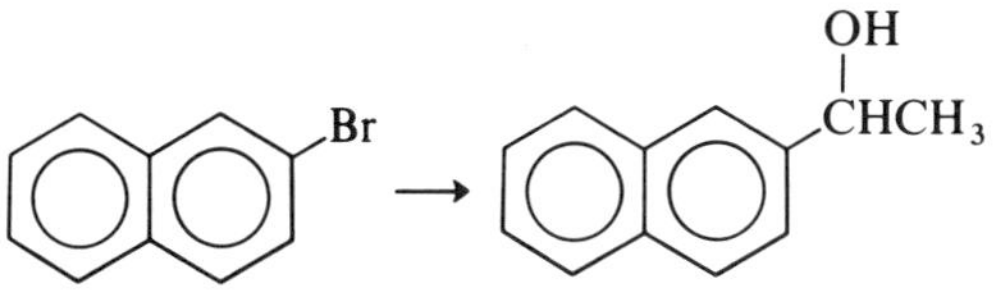 2-bromonaphthalene (naphthalene–Br) → naphthalene–$\overset{\displaystyle OH}{\overset{|}{C}}HCH_3$

(b) $CH_2{=}CHI \rightarrow (CH_3)_2\overset{\displaystyle OH}{\overset{|}{C}}CH{=}CH_2$

(c) $CH_3CH_2CH_2Cl \rightarrow CH_3CH_2CH_2CH_2OH$

»» EXPERIMENT 9.1 ««

Synthesis of 2-Methyl-2-Hexanol: A Grignard Reaction

Step 1, Preparation of n-Butylmagnesium Bromide

$$CH_3CH_2CH_2CH_2Br + Mg \xrightarrow{(CH_3CH_2)_2O} CH_3CH_2CH_2CH_2MgBr$$

	1-bromobutane (*n*-butyl bromide)	magnesium	*n*-butylmagnesium bromide (not isolated)
MW:	137.03	24.31	—
weight:	13.7 g	2.4 g	—
moles:	0.10	0.10	0.10 (theory)

Step 2, Reaction of n-Butylmagnesium Bromide with Acetone

$$CH_3CH_2CH_2CH_2MgBr + CH_3\overset{\overset{\displaystyle O}{\|}}{C}CH_3 \rightarrow CH_3CH_2CH_2CH_2\underset{\underset{\displaystyle CH_3}{|}}{\overset{\overset{\displaystyle O^- \; ^+MgBr}{|}}{C}}CH_3$$

		acetone	a magnesium alkoxide (not isolated)
MW:	—	58.08	—
weight:	—	5.8 g	—
moles:	0.10 (theory)	0.10	0.10 (theory)

Step 3, Hydrolysis of the Magnesium Alkoxide

$$CH_3CH_2CH_2CH_2\underset{\underset{\displaystyle CH_3}{|}}{\overset{\overset{\displaystyle O^- \; ^+MgBr}{|}}{C}}CH_3 + NH_4Cl \xrightarrow{H_2O} CH_3CH_2CH_2CH_2\underset{\underset{\displaystyle CH_3}{|}}{\overset{\overset{\displaystyle OH}{|}}{C}}CH_3 + MgClBr + NH_3$$

		2-methyl-2-hexanol
MW:	—	116.21
weight:	—	11.6 g (theory)
moles:	0.10 (theory)	0.10 (theory)

Discussion

A standard Grignard synthesis is carried out in three steps: (1) preparation of RMgX; (2) the reaction of RMgX with the carbonyl compound or other reactant; and (3) the acidic hydrolysis. The first two steps (and often all three steps) are generally carried out in the same reaction vessel. The intermediate products (the Grignard reagent and the alkoxide) are rarely isolated.

Table 9.1 Physical properties of reactants and products

Name	***Bp (°C)***	n_D^{20}	***Density (g/mL)***	***Solubility***
1-bromobutane	101.6	1.4401	1.28	insol. H_2O, sol. $(CH_3CH_2)_2O$
n-butylmagnesium bromide	—	—	—	decomposes in H_2O, sol. $(CH_3CH_2)_2O$
acetone	56.2	1.3588	0.79	sol. H_2O, sol. $(CH_3CH_2)_2O$
2-methyl-2-hexanol	143	1.4175	0.81	sl. sol. H_2O, sol. $(CH_3CH_2)_2O$

Preparation of the Grignard Reagent (Step 1)

In this experiment, the Grignard reagent is prepared by slowly adding a solution of 1-bromobutane in *anhydrous* diethyl ether (not solvent ether, which is wet) to Mg turnings.

Unfortunately, the reaction leading to the formation of a Grignard reagent is often difficult to initiate. Difficulties can usually be traced to contaminants, primarily water. Therefore, scrupulous care must be taken to dry all glassware and to use only dry reagents and solvents. The techniques used to start the reaction are discussed in Experimental Note 3, page 164.

Once started, the formation of a Grignard reagent is *exothermic*; therefore, excess 1-bromobutane should not be added to initiate the reaction. If a large excess of 1-bromobutane is present in the reaction vessel, the reaction may be difficult to control. Once the reaction has begun, the 1-bromobutane is added at a rate that will maintain a gentle reflux of the ether in the reaction flask.

At the end of the reaction, some of the magnesium may remain unconsumed. The reason for this is that some 1-bromobutane is destroyed by undergoing a coupling reaction with the Grignard reagent to yield octane.

$$CH_3CH_2CH_2CH_2Br + CH_3CH_2CH_2CH_2MgBr \rightarrow \underset{\text{octane}}{CH_3(CH_2)_6CH_3} + MgBr_2$$

Other than providing a mechanical inconvenience in the extraction steps, the residual magnesium metal does not interfere with the remainder of the experiment.

An ether solution of a Grignard reagent has a translucent gray-to-black tint. The color arises from impurities in the magnesium metal, rather than from the Grignard reagent itself.

Once formed, the Grignard reagent must be carried on to Step 2 (the reaction with acetone) immediately. It cannot be saved until the next laboratory period because it reacts with oxygen and moisture from the air.

$$RMgX \xrightarrow{O_2} ROOMgX \xrightarrow{RMgX} 2\,ROMgX$$

Reaction with Acetone (Step 2)

The reaction of *n*-butylmagnesium bromide with acetone is extremely vigorous. The acetone must be added very slowly; otherwise, the reaction mixture will boil over. The product is a magnesium alkoxide of an alcohol and thus insoluble in diethyl ether. This alkoxide sometimes forms a crusty precipitate that must be broken up by swirling the flask so that the unreacted acetone can become mixed with the Grignard reagent.

After the reaction of acetone and the Grignard reagent is completed, it is no longer necessary to protect the reaction mixture from air or moisture. This mixture can be stored until the next laboratory period.

Hydrolysis of the Alkoxide (Step 3)

The alkoxide product of the Grignard reaction is converted to 2-methyl-2-hexanol by treatment with aqueous NH_4Cl instead of with a dilute mineral acid. The reason is that the final product is a *tertiary alcohol* (R_3COH) and is easily dehydrated to an alkene by a strong acid.

When the magnesium alkoxide is poured into aqueous NH_4Cl, the alkoxide ion (a strong base) reacts with water or NH_4^+ to extract a proton.

$$\underset{\text{strong base}}{R\ddot{\underset{..}{O}}:^-} + NH_4^+ \rightarrow R\ddot{\underset{..}{O}}H + \underset{\text{weak base}}{:NH_3}$$

Water alone is not used as a hydrolyzing agent for two reasons. First, the product hydroxide ion is only a slightly weaker base than the alkoxide ion. The addition of an acid results in a more favorable equilibrium.

$$\underset{\text{slightly stronger base}}{R\ddot{\underset{..}{O}}:^-} + H_2\ddot{O}: \rightleftarrows R\ddot{\underset{..}{O}}H + \underset{\text{slightly weaker base}}{:\ddot{\underset{..}{O}}H^-}$$

Second, in alkaline solution, the magnesium ions are converted to a gelatinous precipitate of $Mg(OH)_2$, which is difficult to remove from the product. In a neutral or acidic medium, the magnesium ions remain in solution.

The product alcohol is extracted from the aqueous layer with diethyl ether (solvent grade). The aqueous layer, which contains the magnesium salts, is discarded. The ether solution is washed with sodium carbonate solution to ensure alkalinity prior to distillation. (Any acid remaining in the ether layer would cause dehydration of the alcohol during the distillation.) Because diethyl ether can dissolve a considerable amount of water (1.2 g H_2O in 100 g of ether), the ether extract is partially dried by extraction with saturated NaCl solution before an inorganic drying agent is used. The final drying is accomplished by allowing the ether solution to stand over anhydrous $MgSO_4$ or K_2CO_3.

The bulk of the ether is removed by simple distillation. Before the alcohol is distilled, the residue is transferred to a smaller distillation flask; otherwise, a considerable amount of product would be lost as vapor filling the large flask.

Experimental

EQUIPMENT:

400-mL beaker
calcium chloride drying tube
Claisen head
condenser
disposable pipet
25-mL or 50-mL tared distillation receiving flask
125-mL dropping funnel
50-mL, 125-mL, and two 250-mL Erlenmeyer flasks
heating mantle and rheostat (or steam bath)
ice bath
50-mL (or 100-mL) and 250-mL round-bottom flasks
250-mL or 400-mL separatory funnel
simple distillation apparatus
stirring rod
warm water bath

CHEMICALS:

ammonium chloride, 25 g
anhydrous acetone, 5.8 g
anhydrous diethyl ether, 50 mL
anhydrous magnesium sulfate or potassium carbonate, 5 g
10% aqueous sodium carbonate, 25 mL
1-bromobutane, 13.7 g
diethyl ether (for extraction), about 75 mL
magnesium turnings, 2.4 g
saturated aqueous NaCl, 25 mL

TIME REQUIRED: $3\frac{1}{2}$ hours, plus $1\frac{1}{2}$ hours for the distillation. The Grignard reagent must be used immediately after its formation. Therefore, enough time should be allotted to carry out Steps 1 and 2 (about 1 hour each) in a single laboratory period. IMPORTANT: If anhydrous acetone is not available, then reagent acetone must be dried with anhydrous $MgSO_4$ (5 g for each 50 mL) for at least 24 hours prior to the Grignard reaction (see Experimental Note 4).

STOPPING POINTS: after the acetone has been added to the Grignard reagent (and reaction has subsided); when the ether extracts are drying

>>>> ***SAFETY NOTE 1*** Diethyl ether (bp 34.6°) is used as a reaction solvent and as an extraction solvent. It is very volatile and extremely flammable. There must be *no flames* in the laboratory. An efficient condenser must be used for both the reaction and the distillation. The distillation should be carried out *slowly* to minimize ether vapors escaping into the room.

›››› ***SAFETY NOTE 2*** Because the formation of a Grignard reagent and a Grignard reaction are both exothermic, there is the danger of a runaway reaction. Keep an ice bath handy at all times in case the reaction flask needs rapid cooling.

›››› ***SAFETY NOTE 3*** The heavy caked precipitate that sometimes forms makes thorough mixing of acetone and the Grignard reagent difficult and can allow unreacted acetone to accumulate in one spot. When this acetone eventually contacts the Grignard reagent, the reaction may become impossible to control. Therefore, swirl the reaction flask gently, but *frequently* and *thoroughly*.

PROCEDURE

Step 1, Preparation of n-Butylmagnesium Bromide. Heat a 250-mL round-bottom flask, a Claisen head, a condenser, and a dropping funnel in a drying oven until they are hot to the hand. Then assemble them as shown in the left-hand illustration in Figure 9.1. Fit the reflux condenser with a drying tube containing anhydrous calcium chloride (see Figure 9.2 and Experimental Note 1). To prevent atmospheric moisture from condensing inside the condenser, do not turn on the condenser water until the reaction is initiated.

Place 2.4 g of oven-dried magnesium turnings in the round-bottom flask. To the dropping funnel, add a well-mixed solution of 13.7 g of 1-bromobutane and 50 mL of anhydrous diethyl ether. (CAUTION: flammable! See Safety Note 1; see also Experimental Note 2.)

To initiate the reaction, add 10–15 mL of the ether solution from the dropping funnel to the reaction flask. Loosen the clamp holding the round-bottom flask and gently swirl the flask to mix the contents. When the Grignard reagent begins to form, the ether solution will become cloudy and then begin to boil. Turn on the condenser water at this time. If your Grignard reagent does not start to form within 5–10 minutes, follow the procedure outlined in Experimental Note 3. Because the reaction is exothermic once initiated, do not add an excessive amount of 1-bromobutane to the magnesium at any one time (see Safety Note 2).

After the reaction has been initiated, add the remaining ether solution dropwise at a rate that maintains a gentle reflux. After all the solution has been added, close the stopcock of the dropping funnel and heat the mixture at a gentle reflux for 15 minutes in a warm (50°) water bath. As the magnesium is consumed, the mixture will become darker colored. At the end of the reflux period, proceed immediately to Step 2.

Step 2, Reaction of n-Butylmagnesium Bromide with Acetone. Chill the flask containing the Grignard reagent with an ice bath. Pour 5.8 g of anhydrous acetone (not ordinary reagent grade) in the dropping funnel, and add it *a few drops at a time* to the reaction mixture. After each addition, loosen the clamps to the reaction assembly and gently swirl the reaction flask. (CAUTION: See Safety Note 3.)

When the addition of the acetone is completed, allow the reaction mixture to stand at room temperature for 30 minutes or longer before going on

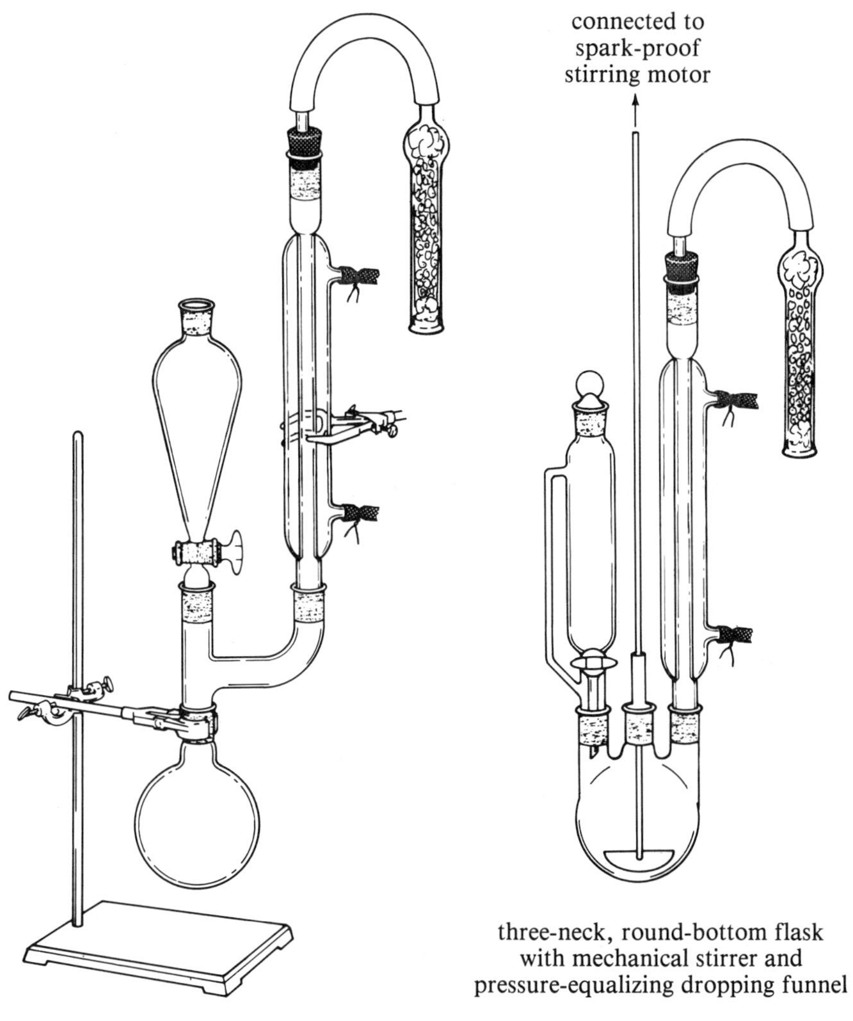

Figure 9.1 *Two reaction assemblies for a Grignard reaction. (With the right-hand apparatus, which is the type used in research laboratories, care should be taken that ether fumes do not flow over the stirring motor.)*

to the hydrolysis step. If the mixture will be standing for more than an hour, stopper the reaction flask with a glass stopper or a cork (not a rubber stopper) to prevent the solvent from evaporating.

Step 3, Hydrolysis and Purification. Prepare 100 mL of 25% aqueous ammonium chloride. Mix 75 mL of this solution with 50 g of crushed ice in a 400-mL beaker. Transfer the remaining 25 mL of the ammonium chloride solution to a 50-mL Erlenmeyer flask, and chill it in an ice bath. Slowly pour the Grignard reaction mixture into the ice mixture in the beaker, stirring

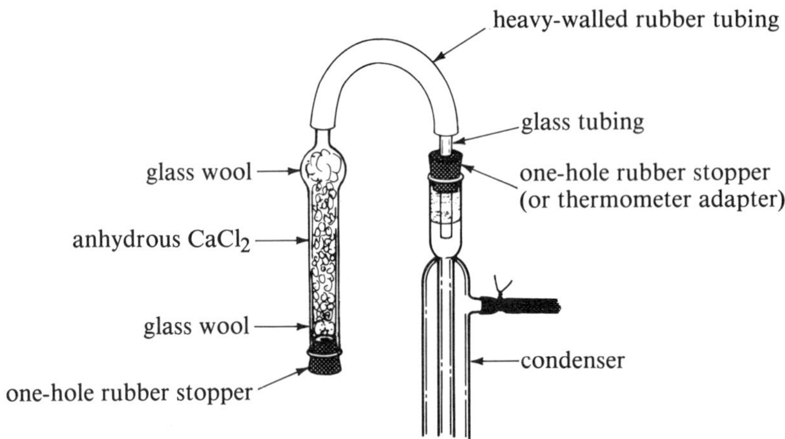

Figure 9.2 How to attach a straight-neck drying tube to the reaction assembly.

vigorously (see Experimental Note 4). Rinse the reaction vessel into the ice mixture, first with the 25 mL of chilled NH_4Cl solution, then with 25 mL of solvent ether.

Transfer the contents of the beaker to a 400-mL separatory funnel. (If a 250-mL separatory funnel must be used, divide the mixture into two batches and process each separately.) Add solvent ether to bring the volume of the upper ether layer to about 50 mL, shake the funnel, and allow the layers to separate. Drain the lower aqueous layer into a 250-mL Erlenmeyer flask or a second separatory funnel and drain the ether layer into a separate Erlenmeyer flask. (Draining instead of pouring minimizes evaporation of the ether.) Return the aqueous layer to the separatory funnel, add 25 mL of fresh solvent ether, and shake the mixture again. Drain and discard the lower, aqueous layer.

Add the first 50-mL ether extract to the second extract in the separatory funnel. Rinse the flask that contained the original extract into the separatory funnel with a few mL of ether. Wash the combined ether extracts by shaking them with 25 mL of water, then with 25 mL of 10% sodium carbonate solution. Finally, wash the ether solution with 20–25 mL of saturated sodium chloride solution.

Pour the ether solution into a clean, dry 250-mL Erlenmeyer flask, add 5 g of anhydrous magnesium sulfate or potassium carbonate, cork the flask tightly, and allow it to stand for at least one hour (overnight is better).

Carefully decant (or filter through a small plug of glass wool) the dried solution into a 250-mL round-bottom flask for distillation of the ether. Add a few boiling chips and *slowly* distil the bulk of the ether (bp 34.6°), using a steam bath or heating mantle and an efficient condenser. Stop the distillation when there is about 25 mL remaining in the distillation flask. Cool the flask and, using a disposable pipet, transfer the residue to a 50-mL round-bottom flask for distillation of the product. Wash the last traces of crude product from the 250-mL flask into the 50-mL flask with a few mL of *anhydrous* diethyl ether.

Add fresh boiling chips and distil the product, collecting the fraction boiling at 135–143° in a tared receiver. A typical yield is 5.0 g (43%). (The yield may vary considerably, depending on the degree of dryness of the anhydrous ether used in Step 1.) Determine the refractive index of the product. Place the distilled product in a correctly labeled vial, and hand it in to your instructor.

EXPERIMENTAL NOTES

1) The purpose of the drying tube is to prevent atmospheric moisture from entering the reaction vessel via the condenser and yet allow the reaction vessel to be open to the atmosphere so that gas pressure does not build up. There are two types of drying tubes: curved (better) and straight (less expensive). A straight drying tube must not be connected directly to the top of the condenser because the dessicant can liquefy and drain into the condenser. Connect the straight tube to the condenser by a short length of heavy-walled rubber tubing, as shown in Figure 9.2.

In either type of drying tube, the dessicant is held in place with loose plugs of glass wool. A one-hole rubber stopper may be used as a secondary plug at the wide end of the drying tube.

2) Solvent ether contains an appreciable amount of water (up to 1–2%) and is totally unsuitable as a Grignard solvent. Anesthesia ether contains ethanol, which makes it also unsuitable. Commercial anhydrous ether is adequate only if a *freshly opened* can is used.

Anhydrous ether must not be left open to the air because it absorbs both oxygen and moisture. (Oxygen and ethers yield peroxides, which can explode if the ether is distilled. Absorbed moisture will ruin a Grignard reagent.)

Your instructor will probably provide anhydrous ether for this experiment. In many laboratories, storeroom personnel prepare anhydrous ether by passing solvent ether through a column containing molecular sieves, which are adsorbents with pores that trap molecules of a certain size (in this case, H_2O molecules). Another procedure for the preparation of anhydrous ether from solvent ether and a procedure for the testing of peroxides in anhydrous ether follow. If you find it necessary to prepare your own anhydrous ether, allot an additional laboratory period.

Preparation of Anhydrous Diethyl Ether. With cooling, mix a 2:1 ratio of solvent ether and conc. H_2SO_4 in a large round-bottom flask, and distil about two-thirds of the ether. (Do not distil all the ether.) Any water and ethanol contaminating the ether will remain with the sulfuric acid. To discard the residue, pour it onto cracked ice, allow the residual ether to evaporate in the hood, then dilute the aqueous acid with water and pour it down the hood drain with additional water.

Add freshly prepared sodium wire or ribbon to the distilled ether, then allow the ether to stand at least overnight in the fume hood with the fan on. Stopper the container with a very *loose-fitting* cork or drying tube to allow the hydrogen gas to escape.

Sodium wire or ribbon is prepared by pressing sodium metal through a die, using a press. If a sodium press is not available, the ether can be dried with finely diced sodium; however, diced sodium is inferior to wire or ribbon. Another method is to add a few grams of CaH_2 to the ether and allow the mixture to stand until hydrogen has ceased to be evolved. The dried ether can then be decanted or (better) pipetted, using a rubber bulb, as needed. Commercial anhydrous ether can be further dried with sodium wire without the sulfuric acid purification step.

Peroxide Test. Shake 5 mL of ether with a solution of 1 mg of sodium dichromate and one drop of dilute H_2SO_4 in 1 mL of water in a corked test tube. If the ether layer turns blue (from perchromate ion), peroxides are present and must be removed.

Peroxide Removal. Shake the peroxide-contaminated ether with 5% aqueous ferrous sulfate ($FeSO_4$) solution acidified with H_2SO_4. The iron(II) ions are oxidized with concurrent reduction of the peroxide. Aqueous sodium sulfite (Na_2SO_3) can be substituted for the ferrous sulfate solution.

3) The most common cause of failure of initiation of the reaction leading to the Grignard reagent is moisture (in the apparatus, in the ether, or on the magnesium turnings). In addition, in a humid atmosphere, water will collect on the sides of a cold condenser. If the initial cloudiness becomes a white precipitate, then the Mg is being converted to $Mg(OH)_2$ by the water, and not to RMgX. If excessive moisture is present, it is best to begin anew with *dry* equipment and reagents.

Sometimes, Grignard reagents are reluctant to form because of a *magnesium oxide coating* on the metal turnings. The following procedure can often overcome this difficulty. First, warm the reaction flask with a pan of warm water (about 50–60°). This warming will cause the ether to boil (not a sign of initiation, in this case). Remove the warm water bath and watch for the signs of initiation (spontaneous boiling of the ether). This warming may be repeated if initiation does not occur.

If repeated warming does not initiate Grignard-reagent formation, add an additional 5 mL of the 1-bromobutane solution from the dropping funnel and warm the flask again.

As a last resort, another reagent may be added to activate the surface of the magnesium and/or indirectly complex with any water present. A number of reagents are useful: a crystal of I_2, a few drops of Br_2, 1.0 mL of iodomethane (methyl iodide) or dibromomethane (methylene bromide). (Only one, not all, of these should be added.) Add the reagent directly to the reaction mixture without swirling, then warm the flask in the water bath.

The two inorganic halogen compounds function by reacting with the magnesium to yield an anhydrous magnesium halide, which complexes with any water present. Iodomethane and dibromomethane are reactive organohalogen compounds that react with the magnesium in slightly different ways. For example, iodomethane first forms a Grignard reagent (even when a less reactive alkyl halide does not react), which then reacts with any water present and thus removes it from solution.

$$CH_3I \xrightarrow{Mg} CH_3MgI \xrightarrow{H_2O} CH_4 + HOMgX$$

4) The reaction mixture may contain small pieces of unreacted magnesium metal. If possible, avoid transferring these bits of metal to the ice mixture. However, a tiny piece of magnesium that cannot be removed easily from the ice mixture will do no harm.

Problems

9.10 Write equations for the three standard steps in a Grignard synthesis in which the principal reactants are cyclohexanone and ethylmagnesium bromide.

9.11 A student oven-dries the glassware needed for a Grignard reaction, then stores them in a locker until the next laboratory period. Will the glassware still be dry when the Grignard reaction is begun? Explain.

9.12 Suggest a reason for using magnesium turnings instead of magnesium powder or chunks in a Grignard reaction.

9.13 In which of the following steps in a Grignard synthesis is anhydrous ether (instead of solvent ether) necessary? Explain.

(a) Preparation of the Grignard reagent.

(b) Addition of an ether solution of a ketone (instead of pure ketone, as in Experiment 9.1).

(c) Extracting the product from the hydrolysis mixture.

(d) Washing the dried product into a distillation flask.

9.14 Are diethyl ether vapors lighter or heavier than air? What are the safety implications of your answer?

9.15 As an alternative to drying tubes to protect a Grignard reaction, some chemists carry out these reactions under a dry nitrogen atmosphere. Which of the following techniques could also be used to keep a Grignard reagent dry? Explain.

(a) a dry argon atmosphere

(b) a dry carbon dioxide atmosphere

(c) a gentle stream of dried air passed over the surface of the mixture

9.16 Which of the following reagents could be substituted for NH_4Cl in the hydrolysis step in Experiment 9.1? Explain.

(a) NH_4Br (b) HBr (c) NaH_2PO_4 (d) CH_3CO_2Na

9.17 (a) Write the equation for the side reaction that could occur if a strong acid were used, instead of ammonium chloride, to hydrolyze the magnesium alkoxide in Experiment 9.1.

(b) In this case, what would be observed when the product is purified by distillation?

9.18 A chemist carries out a Grignard reaction using *n*-butylmagnesium bromide and formaldehyde (methanal, $H_2C{=}O$). Which of the following reagents could be used to acidify the water for the hydrolysis step?

(a) NH_4Cl (b) HCl (c) H_2SO_4

9.19 Suppose your Grignard reagent did not completely react with the acetone in Experiment 9.1.

(a) What would be the by-product of the hydrolysis?

(b) Is this by-product a gas or a liquid?

(c) What are the safety implications?

9.20 Of the drying agents listed in Table 4.2 (page 66), which could *not* be used for drying the ether solution of the product in Experiment 9.1? Explain.

9.21 Assuming a 43% yield of 2-methyl-2-hexanol in the Grignard synthesis in Experiment 9.1, what quantities of starting materials would you use to *isolate* 11.6 g of product?

»» CHAPTER 10 ««

The Chemistry of Alcohols

Alcohols are used in beverages, as germicides, as solvents, and as industrial and chemical intermediates. Alcohols are of interest to the organic chemist primarily because they can be readily converted to a variety of other types of compounds. The experiments in this chapter were chosen to illustrate four typical reactions of alcohols: two substitution reactions (S_N2 and S_N1), an elimination reaction, and an oxidation reaction.

substitution:

$$ROH + HBr \rightarrow RBr + H_2O$$

elimination:

$$R_2\underset{}{\overset{OH}{\overset{|}{C}}}-CHR_2 \overset{H^+}{\rightleftharpoons} R_2C{=}CR_2 + H_2O$$

oxidation:

$$R\overset{OH}{\overset{|}{C}}HR \xrightarrow{[O]} R\overset{O}{\overset{\|}{C}}R$$ where [O] represents an oxidizing agent

Three alcohols will be used to carry out these four reactions:

$CH_3CH_2CH_2CH_2OH$	(cyclohexane ring)–OH	$CH_3\overset{CH_3}{\overset{\vert}{C}}OH$ with $\vert$ CH_3 below
1-butanol (*n*-butyl alcohol) *a primary (1°) alcohol*	cyclohexanol (cyclohexyl alcohol) *a secondary (2°) alcohol*	2-methyl-2-propanol (*t*-butyl alcohol) *a tertiary (3°) alcohol*

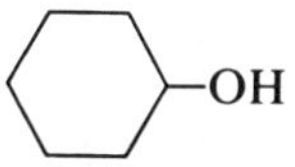

10.1 Substitution Reactions of Alcohols

The —OH group of an alcohol is a poor leaving group and does not undergo direct displacement by a nucleophile. Consequently, alcohols are inert toward nucleophiles such as Cl^- or Br^- if the reaction conditions are neutral or alkaline. In strongly acidic solution, however, substitution can occur. The reason is that, in acidic solutions, the —OH group can be *protonated*. The resulting oxonium ion contains the $—OH_2^+$ group, which is easily displaced as water.

Exactly how water leaves the molecule depends upon the class of alcohol undergoing reaction. In reactions of *primary alcohols*, water is displaced in an S_N2 process. In reactions of *secondary and tertiary alcohols*, the oxonium ion loses water to yield a carbocation, which then combines with a nucleophile in an S_N1 process. Figure 10.1 shows these two reaction mechanisms with the alcohols that you will be using in Experiments 10.1 and 10.2.

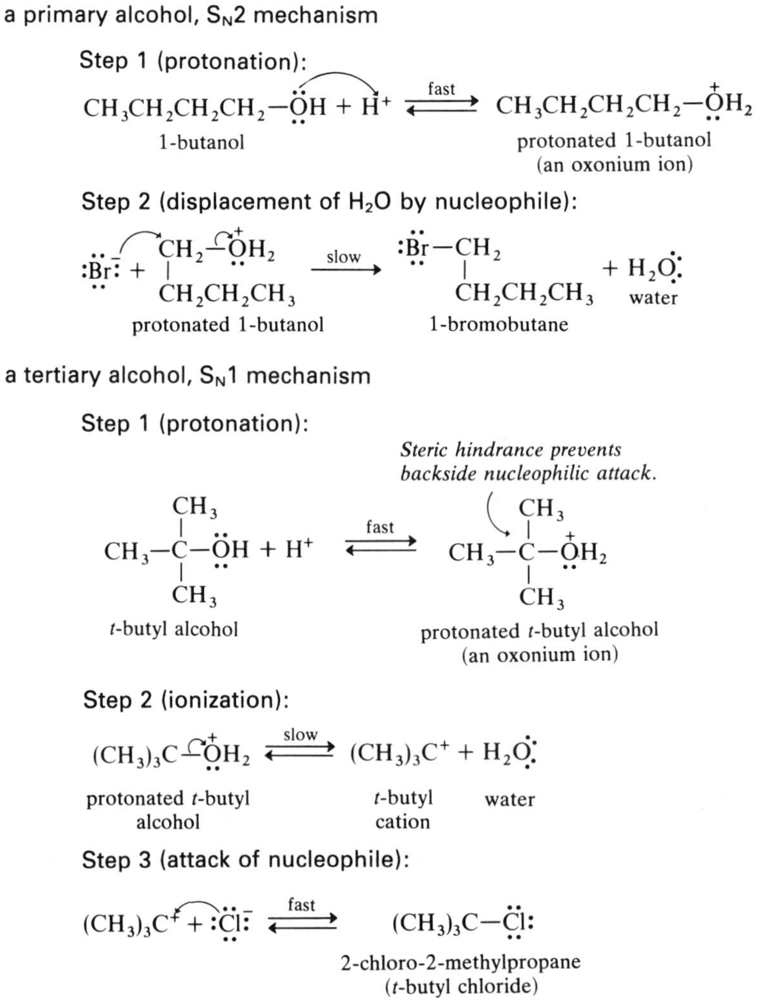

a primary alcohol, S_N2 mechanism

Step 1 (protonation):

$$\underset{\text{1-butanol}}{CH_3CH_2CH_2CH_2{-}\ddot{O}H} + H^+ \underset{}{\overset{\text{fast}}{\rightleftharpoons}} \underset{\substack{\text{protonated 1-butanol}\\ \text{(an oxonium ion)}}}{CH_3CH_2CH_2CH_2{-}\overset{+}{\ddot{O}}H_2}$$

Step 2 (displacement of H_2O by nucleophile):

$$:\ddot{Br}:^- + \underset{\text{protonated 1-butanol}}{CH_3CH_2CH_2{-}CH_2{-}\overset{+}{\ddot{O}}H_2} \xrightarrow{\text{slow}} \underset{\text{1-bromobutane}}{:\ddot{Br}{-}CH_2CH_2CH_2CH_3} + \underset{\text{water}}{H_2\ddot{O}:}$$

a tertiary alcohol, S_N1 mechanism

Step 1 (protonation):

Steric hindrance prevents backside nucleophilic attack.

$$\underset{t\text{-butyl alcohol}}{(CH_3)_3C{-}\ddot{O}H} + H^+ \overset{\text{fast}}{\rightleftharpoons} \underset{\substack{\text{protonated } t\text{-butyl alcohol}\\ \text{(an oxonium ion)}}}{(CH_3)_3C{-}\overset{+}{\ddot{O}}H_2}$$

Step 2 (ionization):

$$\underset{\text{protonated } t\text{-butyl alcohol}}{(CH_3)_3C{-}\overset{+}{\ddot{O}}H_2} \overset{\text{slow}}{\rightleftharpoons} \underset{t\text{-butyl cation}}{(CH_3)_3C^+} + \underset{\text{water}}{H_2\ddot{O}:}$$

Step 3 (attack of nucleophile):

$$(CH_3)_3C^+ + :\ddot{Cl}:^- \overset{\text{fast}}{\rightleftharpoons} \underset{\substack{\text{2-chloro-2-methylpropane}\\ (t\text{-butyl chloride})}}{(CH_3)_3C{-}\ddot{Cl}:}$$

Figure 10.1 The mechanisms of nucleophilic substitution of 1-butanol and t-butyl alcohol.

10.2 Dehydration of Alcohols

An elimination reaction of an alcohol yields an alkene and water. Since a molecule of water is lost from the alcohol molecule, elimination reactions are also called **dehydration reactions.** The dehydration of an alcohol usually proceeds by way of a carbocation (E1 reaction). For the alcohol to form a carbocation, the —OH group must first be protonated; therefore, a strong acid must be used as the dehydrating agent.

The rate of carbocation formation (the slow step in the reaction) determines the rate of the overall reaction. The rate of carbocation formation, in turn, is determined by the relative stability of the carbocation. Because the order of carbocation stability is 3° > 2° > 1°, the relative rates of dehydration reactions of alcohols are in the following order:

1° ROH 2° ROH 3° ROH
→
increasing rate of dehydration

The mechanism of an E1 reaction of an alcohol, such as cyclohexanol, is similar to that of an S_N1 reaction. Initial protonation of the —OH group (Step 1), followed by loss of water (Step 2), yields the carbocation. Instead of combining with a nucleophile as in the S_N1 reaction, the carbocation loses a proton to a base in the solution (Step 3). The proton acceptor can be water, the hydrogen sulfate ion (if H_2SO_4 is used as the dehydrating agent), or another molecule of the alcohol. The product of Step 3 is the alkene. (In the following equations, we show the hydrogen atoms on two of the carbons for clarity.)

Step 1 (protonation):

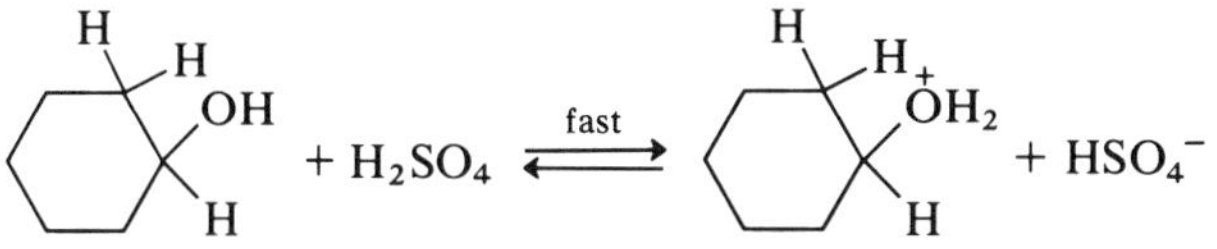

H H OH H + H_2SO_4 $\underset{}{\overset{\text{fast}}{\rightleftharpoons}}$ H H $\overset{+}{O}H_2$ H + HSO_4^-

Step 2 (carbocation formation):

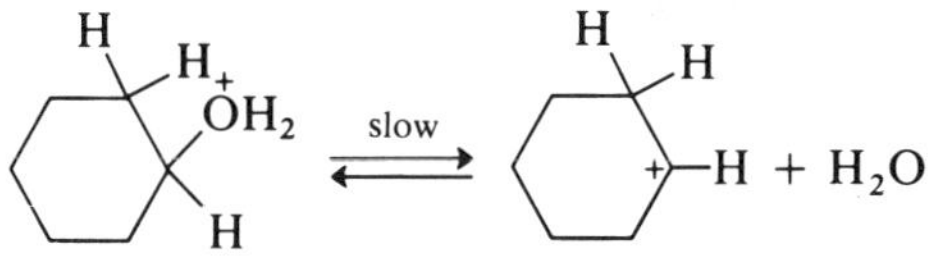

H H $\overset{+}{O}H_2$ H $\overset{\text{slow}}{\rightleftharpoons}$ H H +—H + H_2O

Step 3 (loss of H^+):

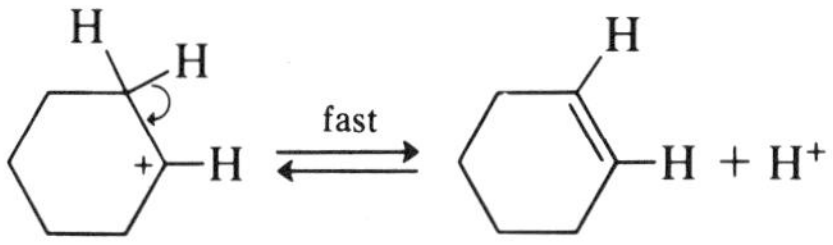

H H +—H $\overset{\text{fast}}{\rightleftharpoons}$ H —H + H^+

cyclohexene

Note that the acid in a dehydration reaction is a catalyst and not a reactant. In Step 1, the alcohol is protonated; however, in Step 3, another proton is produced.

A. Side Reactions

Either cyclohexanol or *t*-butyl alcohol can yield only one alkene upon dehydration. Some other alcohols can yield more than one alkene, either because of the presence of nonequivalent "losable" β hydrogens or because of carbocation rearrangements. Dehydrations that yield more than one product are generally avoided because of the practical problem of separating the alkene mixtures.

Because carbocations can undergo substitution as well as elimination, substitution and elimination reactions of alcohols are competing reactions. The actual product that predominates in the reaction mixture depends upon a number of factors, one of which is the nucleophile present in the mixture (the anion of the acid). To favor elimination, acids such as H_2SO_4 or H_3PO_4 are used because their anions (HSO_4^- and $H_2PO_4^-$) are relatively poor nucleophiles. To favor substitution, acids such as HCl or HBr are used because halide ions are relatively good nucleophiles.

Two other side reactions, **hydration** and **polymerization**, that can decrease the yields of alkenes, arise from the reactivity of the product alkenes in acidic solution. Alkenes can become protonated to reform carbocations. These carbocations can then react with water to yield alcohols in the reverse reaction of the dehydration. This dehydration reaction will be discussed in greater detail in Chapter 13 (the reactions of alkenes).

a typical hydration reaction of an alkene:

$$\text{cyclohexene} \underset{}{\overset{H^+}{\rightleftarrows}} \text{cyclohexyl}^+ \underset{}{\overset{H_2O}{\rightleftarrows}} \text{cyclohexyl}-\overset{+}{O}H_2 \underset{}{\overset{-H^+}{\rightleftarrows}} \text{cyclohexyl}-OH$$

Alkenes also undergo polymerization (and thus tar formation) in the presence of acid:

a typical acid-catalyzed polymerization:

$$\underset{\substack{\text{methylpropene}\\\text{(isobutylene)}}}{CH_3\overset{\overset{CH_2}{\|}}{\underset{\underset{CH_3}{|}}{C}}} \overset{H^+}{\rightleftarrows} CH_3\overset{\overset{CH_3}{|}}{\underset{\underset{CH_3}{|}}{C^+}} \xrightarrow{CH_2=\overset{\overset{CH_3}{|}}{C}CH_3} CH_3\overset{\overset{CH_3}{|}}{\underset{\underset{CH_3}{|}}{C}}-CH_2\overset{\overset{CH_3}{|}}{\underset{\underset{CH_3}{|}}{C^+}} \xrightarrow{\text{many steps}} \underset{\text{polyisobutylene}}{\left(-CH_2\overset{\overset{CH_3}{|}}{\underset{\underset{CH_3}{|}}{C}}-\right)_x}$$

In Experiment 10.3, these side reactions are minimized by distilling the alkene away from the reaction mixture as it is formed.

10.3 Oxidation of Alcohols

Primary and secondary alcohols are readily oxidized by dichromates ($Na_2Cr_2O_7$) and permanganates ($KMnO_4$). Secondary alcohols can be oxidized by reagents containing positive chlorine, such as *t*-butyl hypochlorite [$(CH_3)_3COCl$] or hypochlorous acid (HOCl, formed *in situ* from sodium hypochlorite and acetic acid). Under these strong oxidizing conditions, a *primary alcohol* is oxidized to a carboxylic acid (in acidic solution) or the carboxylate ion (in base).

A *secondary alcohol* is oxidized to a ketone—the highest oxidation state available without the breaking of carbon–carbon bonds. In Experiment 10.4, cyclohexanol is oxidized to cyclohexanone using either sodium dichromate or hypochlorous acid as the oxidizing agent.

$$C_6H_{11}{-}OH \xrightarrow{[O]} C_6H_{10}{=}O$$

cyclohexanol
a secondary alcohol

cyclohexanone

A. Balancing Organic Oxidation–Reduction Equations

The fact that an inorganic ion has undergone oxidation or reduction can usually be determined by inspection of the atom's oxidation state. For example, it is easy to see that the conversion of Fe^{3+} to Fe^{2+} is a reduction in which the iron ion has gained one electron.

When an organic compound is oxidized or reduced, the change in the number of electrons is not always readily apparent. For this reason, many organic oxidation–reduction equations cannot be balanced by inspection easily. A number of techniques have been developed to balance such equations. The technique that we shall present is called the **half-reaction method**. Like most techniques for balancing oxidation–reduction equations, the half-reaction method is artificial. It is not a mechanism, nor is it an attempt to explain how electrons are transferred between molecules; it is merely an aid for balancing equations.

In balancing an equation by half-reactions, the equation for the overall reaction is divided into two separate equations: one for the oxidation half-reaction and the other for the reduction half-reaction. These two equations are then balanced independently by "juggling" water molecules, electrons, and hydrogen ions (if the solution is acidic) or hydroxide ions (if the solution is basic). The final step in the technique is the addition of the two half-reaction equations to yield the complete, balanced equation.

To illustrate this technique, let us use, as an example, the oxidation of cyclohexanol to cyclohexanone by sodium dichromate in an acidic medium (Experiment 10.4B).

equation for the overall reaction (not balanced):

$$C_6H_{11}{-}OH + Cr_2O_7^{2-} \longrightarrow C_6H_{10}{=}O + Cr^{3+}$$

This unbalanced equation can be divided into the oxidation half-reaction and the reduction half-reaction.

oxidation half-reaction (not balanced):

$$C_6H_{11}{-}OH \longrightarrow C_6H_{10}{=}O$$

reduction half-reaction (not balanced):

$$Cr_2O_7^{2-} \rightarrow Cr^{3+}$$

Starting with the reduction half-reaction, we can deduce that one $Cr_2O_7^{2-}$ yields two Cr^{3+}. The seven oxygen atoms of the dichromate ion must be converted to 7 H_2O.

$$Cr_2O_7^{2-} \rightarrow 2\,Cr^{3+} + 7\,H_2O \qquad \text{(partially balanced)}$$

To balance the hydrogen atoms, we must add 14 H^+ to the left side of the equation.

$$Cr_2O_7^{2-} + 14\,H^+ \rightarrow 2\,Cr^{3+} + 7\,H_2O \qquad \text{(partially balanced)}$$

Now the numbers of atoms in the half-reaction equation are balanced, but not the charges. To balance the charges, we add 6 e^- to the left side of the equation.

balanced equation for the reduction half-reaction:

$$Cr_2O_7^{2-} + 14\,H^+ + 6\,e^- \rightarrow 2\,Cr^{3+} + 7\,H_2O$$

Turning to the oxidation half-reaction, we determine by inspection that the alcohol is oxidized to the ketone by the removal of 2 H^+. (If this is not apparent to you, write out the complete structures of cyclohexanol and cyclohexanone, showing all the carbon and hydrogen atoms.)

$$\text{C}_6\text{H}_{11}\text{–OH} \longrightarrow \text{C}_6\text{H}_{10}\text{=O} + 2H^+ \qquad \text{(partially balanced)}$$

To balance the charges, we add 2 e^- to the right side of the equation.

balanced equation for the oxidation half-reaction:

$$\text{C}_6\text{H}_{11}\text{–OH} \longrightarrow \text{C}_6\text{H}_{10}\text{=O} + 2H^+ + 2\,e^-$$

Before we combine the two half-reaction equations, we must equalize the number of electrons in each half-reaction. In the reduction half-reaction, six electrons are gained. However, in the oxidation half-reaction, only two electrons are lost. To equalize the numbers of electrons, we multiply the half-reaction equations by appropriate whole numbers. In this case, we multiply the oxidation half-reaction equation by 3.

$$3\,\text{C}_6\text{H}_{11}\text{–OH} \longrightarrow 3\,\text{C}_6\text{H}_{10}\text{=O} + 6H^+ + 6\,e^-$$

The two half-reaction equations may now be added.

$$3\,\text{C}_6\text{H}_{11}\text{–OH} + Cr_2O_7^{2-} + 14H^+ + 6\,e^- \rightarrow 3\,\text{C}_6\text{H}_{10}\text{=O} + 2Cr^{3+} + 6H^+ + 6\,e^- + 7H_2O$$

By canceling like terms, we arrive at the balanced equation.

the balanced net equation:

$$3\,\text{C}_6\text{H}_{11}\text{–OH} + Cr_2O_7^{2-} + 8H^+ \rightarrow 3\,\text{C}_6\text{H}_{10}\text{=O} + 2Cr^{3+} + 7H_2O$$

In the laboratory, we do not deal with isolated ions, even if their partner ions do not participate in the reaction. However, nonreactive ions are easily inserted into the net equation. If sodium dichromate dihydrate is the actual oxidizing agent, and if sulfuric acid is the acid, the complete balanced equation becomes:

$$3\ \text{C}_6\text{H}_{11}\text{—OH} + Na_2Cr_2O_7 \cdot 2H_2O + 4H_2SO_4 \rightarrow$$

$$3\ \text{C}_6\text{H}_{10}\text{=O} + Cr_2(SO_4)_3 + Na_2SO_4 + 9H_2O$$

Problems

10.1 List the following alcohols in order of increasing reactivity (least reactive first) toward (1) dehydration with concentrated H_2SO_4; (2) substitution with HCl:

(a) $\text{C}_6\text{H}_{11}\text{—OH}$ (b) $(CH_3CH_2)_3COH$ (c) $\text{C}_6\text{H}_{11}\text{—}CH_2OH$

10.2 Look up the pK_a of hydrogen fluoride in a handbook. Would you expect this reagent to be useful for converting a primary alcohol to an alkyl fluoride? Explain.

10.3 The following pairs of reagents can be used to prepare alkyl halides. (1) Which pair or pairs will probably require a reflux period? (2) Which pair would be most likely to yield substantial amounts of an alkene by-product?

(a) 1-propanol and HBr

(b) 3-methyl-3-pentanol and HBr

(c) 2-pentanol and HCl

10.4 State whether each of the following conversions is an oxidation, a reduction, or neither of these:

(a) $\text{C}_6\text{H}_{10}\text{=O} \longrightarrow HO\overset{O}{\overset{\|}{C}}(CH_2)_4\overset{O}{\overset{\|}{C}}OH$ (b) $CH_3CHCl_2 \rightarrow CH{\equiv}CH$

(c) cyclohexene $\longrightarrow \text{C}_6\text{H}_{11}\text{—OH}$ (d) 1,4-benzoquinone (O=C$_6$H$_4$=O) $\longrightarrow$ hydroquinone (HO—C$_6$H$_4$—OH)

(e) $CH_3\overset{O}{\overset{\|}{C}}H \rightarrow CH_3\overset{O}{\overset{\|}{C}}OH$ (f) $CH_3\overset{O}{\overset{\|}{C}}CH_3 \rightarrow CH_3\overset{O}{\overset{\|}{C}}CH_2Cl$

10.5 Predict the organic products of the chromate oxidation (acidic solution) of the following compounds:

(a) 1-hexanol (b) 2-hexanol (c) 1,4-hexanediol

(d) 4-hydroxyhexanoic acid

10.6 Suggest a synthetic route for each of the following conversions:

(a) 1-bromopropane to propanoic acid

(b) 2-bromopropane to acetone

10.7 Complete the following equations for dehydration reactions. List all the alkenes that could be formed. (If you have not yet covered rearrangement reactions in your lecture course, you may ignore rearranged alkenes.)

(a) $(CH_3CH_2)_2C(OH)CH_3 \xrightarrow{H^+}$

(b) [decahydronaphthalen-1-ol] $\xrightarrow{H^+}$

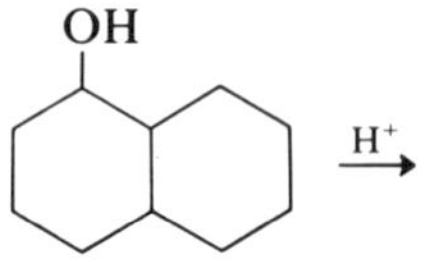

(c) $C_6H_5-CH_2CH_2CH(OH)CH_3 \xrightarrow{H^+}$

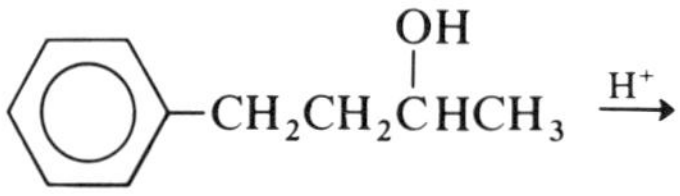

(d) [cyclohexyl]$-CH_2CH_2OH \xrightarrow{H^+}$

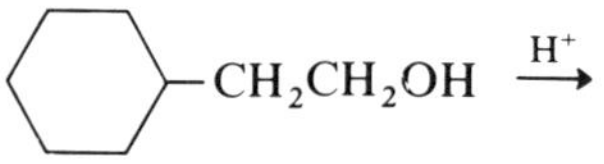

10.8 Complete and balance the following equations for oxidation–reduction reactions:

(a) $(CH_3)_2CHOH + MnO_4^- \xrightarrow{H^+}$

(b) $CH_3CH_2OH + Cr_2O_7^{2-} \xrightarrow{H^+}$

(c) [cyclohexene] $+ MnO_4^- \xrightarrow{OH^-}$ [cyclohexane-1,2-diol, OH, OH]

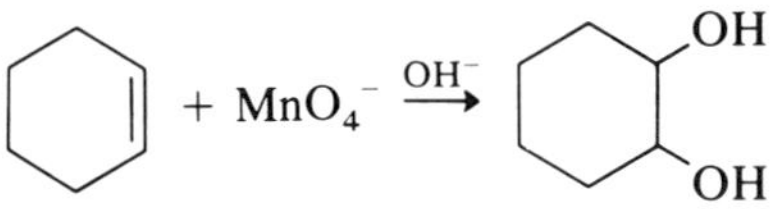

(d) $CH_3CH_2CH_2OH + CrO_4^{2-} \xrightarrow{OH^-}$

»»› EXPERIMENT 10.1 ‹««

Synthesis of 1-Bromobutane from 1-Butanol

	$CH_3CH_2CH_2CH_2OH$	+	NaBr	+	H_2SO_4	→	$CH_3CH_2CH_2CH_2Br$ + $NaHSO_4$ + H_2O
	1-butanol		sodium bromide		sulfuric acid		1-bromobutane
MW:	74.12		102.90		98.08		137.03
weight:	18.5 g		30.0 g		25 mL (46 g)		34.2 g (theory)
moles:	0.25		0.29		0.47		0.25 (theory)

Table 10.1 Physical properties of reactants and products

Name	***Bp (°C)***	n_D^{20}	***Density (g/mL)***	***Solubility***
1-butanol	117.2	1.3991	0.81	sl. sol. H_2O, sol. H_2SO_4
sulfuric acid	338	—	1.84	sol. H_2O
1-bromobutane	101.6	1.4401	1.28	insol. H_2O and H_2SO_4

Discussion

The treatment of a primary alcohol with a hydrogen halide yields a primary alkyl halide. The reaction proceeds by an S_N2 mechanism, and competing dehydration is minimal. The reaction requires a strong acid to protonate the hydroxyl group. Aqueous HBr, gaseous HBr, and "constant boiling" HI (57% aqueous solution) can all be employed, without additional catalyst, to prepare the alkyl halide. In this experiment, HBr is generated in the reaction mixture by treatment of NaBr with H_2SO_4.

$$\text{NaBr} + \text{H}_2\text{SO}_4 \rightleftharpoons \text{HBr} + \text{NaHSO}_4$$

When the mixture of alcohol, H_2SO_4, and NaBr is heated, gaseous HBr is given off; therefore, if the reaction is not carried out in a fume hood, a trap for the HBr must be arranged (Figure 10.2, page 179). In the trap, the HBr emitted from the reflux condenser is passed over aqueous sodium hydroxide and thus converted by an acid–base reaction to sodium bromide and water.

An excess of sulfuric acid is used in the experiment to provide a strongly acidic medium for the protonation of the alcohol. Sulfuric acid, which is a dehydrating agent, also combines with the water that is formed as a product of the substitution reaction.

A number of side reactions occur in this reaction. 1-Butanol can react with HSO_4^- ions present in solution to yield a hydrogen sulfate ester ($ROSO_3H$). This inorganic ester, in turn, can undergo elimination to yield 1-butene (a gas that is lost during the reflux and work-up) or substitution with 1-butanol to yield di-*n*-butyl ether (which must be removed during work-up).

$$\text{ROH} \xrightleftharpoons{\text{H}^+} \text{R}-\overset{+}{\text{O}}\text{H}_2 \xrightleftharpoons{^-\text{OSO}_3\text{H}} \text{ROSO}_3\text{H} + \text{H}_2\text{O}$$

$$\text{ROSO}_3\text{H} \xrightarrow{\text{heat}} \text{an alkene} + \text{H}_2\text{SO}_4$$

$$\text{ROSO}_3\text{H} \xrightarrow[\text{ROH}]{} \underset{\text{an ether}}{\text{ROR}} + \text{H}_2\text{SO}_4$$

Another side reaction that occurs is oxidation of the 1-butanol by either H_2SO_4 or Br_2 (formed by oxidation of Br^- by H_2SO_4).

$$\text{CH}_3\text{CH}_2\text{CH}_2\text{CH}_2\text{OH} \xrightarrow{[\text{O}]} \underset{\text{butanoic acid}}{\text{CH}_3\text{CH}_2\text{CH}_2\overset{\overset{\text{O}}{\|}}{\text{C}}\text{OH}}$$

At the end of the reaction, the mixture consists of two phases. The upper layer contains the desired 1-bromobutane plus organic by-products, and the lower layer contains the inorganic components. The work-up techniques in this experiment consist of four steps: (1) an initial steam distillation; (2) extraction; (3) drying; and (4) a final distillation to purify the product. In the steam distillation (see Section 5.2), the water and 1-bromobutane codistil, leaving the inorganic compounds behind in the distillation residue. Unfortunately, di-*n*-butyl ether, butanoic acid, and unreacted 1-butanol also codistil with water and must be removed from the distillate by extraction.

The first extraction, a water wash, removes some of the 1-butanol, which is slightly soluble in water. The second extraction is with cold, concentrated sulfuric acid. (If the acid is not cold, extensive charring of the organic material will occur.) Each of the two major impurities (1-butanol and di-*n*-butyl ether) and the minor impurity (butanoic acid) contains an oxygen atom. In strong acid, each of these compounds is protonated to yield a sulfuric acid-soluble salt. 1-Bromobutane does not form a salt with sulfuric acid; consequently, it remains in the separatory funnel as a separate layer. This extraction is thus an example of a chemically active extraction (Section 4.3).

$$CH_3CH_2CH_2CH_2OH + H_2SO_4 \rightleftharpoons \underset{\text{sulfuric acid-soluble}}{CH_3CH_2CH_2CH_2OH_2^{+}\ {}^{-}OSO_3H}$$

$$(CH_3CH_2CH_2CH_2)_2O + H_2SO_4 \rightleftharpoons \underset{\text{sulfuric acid-soluble}}{(CH_3CH_2CH_2CH_2)_2OH^{+}\ {}^{-}OSO_3H}$$

A subsequent extraction with aqueous sodium hydroxide solution removes any sulfuric acid clinging to the sides of the separatory funnel. The wet alkyl halide is then dried with anhydrous calcium chloride. Calcium chloride is the drying agent of choice in this reaction because it forms complexes with any residual alcohol, as well as with water. After drying, the 1-bromobutane is purified by distillation.

Experimental

EQUIPMENT:

condenser for reflux
distillation assembly
dropper or disposable pipet
three 50-mL and one 125-mL Erlenmeyer flasks
funnel, glass tubing, and rubber tubing (or fume hood)
100-mL graduated cylinder
heating mantle and rheostat
ice bath
250-mL round-bottom flask
50-mL round-bottom flask with glass stopper
125-mL separatory funnel
thermometer

CHEMICALS:

anhydrous calcium chloride, about 2 g
10% aqueous sodium hydroxide, 25 mL
1-butanol, 18.5 g
conc. sulfuric acid, 50 mL
sodium bromide, 30 g

TIME REQUIRED: approximately 4 hours

STOPPING POINTS: after the reflux period; while the 1-bromobutane is being dried with $CaCl_2$

>>>> ***SAFETY NOTE 1*** During the reflux period, this reaction releases gaseous HBr, which is both corrosive and toxic (10–20 times more toxic than carbon monoxide and about as toxic as chlorine). The reaction must be carried out in a fume hood, or else the reflux condenser must be fitted with an HBr trap (see Experimental Note 1). If you use a trap, be sure the funnel is not submerged; otherwise, the trap liquid may be pulled into the reaction flask!

>>>> ***SAFETY NOTE 2*** Take *extreme care* in shaking a separatory funnel containing concentrated H_2SO_4. Vent frequently. Any accident, even a leaky stopcock, might result in acid being sprayed on yourself, your neighbors, or your work area. Any splashes on your skin or clothing should be flushed immediately with copious amounts of water (see Sections 1.1D and 1.1E).

PROCEDURE

Place 30 g of sodium bromide and 30 mL of water in a 250-mL round-bottom flask. Swirl the flask until most of the salt has dissolved. Add 18.5 g of 1-butanol, and cool the flask to 5–10° in an ice bath. Slowly add 25 mL of concentrated sulfuric acid. (CAUTION: strong acid!) Fit the flask with a reflux condenser in the fume hood. If a hood is not available, fit the condenser with a gas trap (see Experimental Note 1 and Figure 10.2, page 179). Add a few acid-resistant carborundum boiling chips, and heat the mixture at reflux for 30 minutes, using a heating mantle. During the reflux period, the reaction mixture will form two layers.

Allow the reaction flask to cool (or use an ice bath) to a temperature at which it can be handled. Add 2–3 fresh carborundum boiling chips and equip the flask for simple distillation. Distil until the temperature of the distilling mixture reaches 110–115°. The distillate consists of two phases (1-bromobutane and water), which are most apparent at the start of the distillation. At the end of the distillation, the 1-bromobutane should no longer be visible in the drops of distillate (see Experimental Note 2). The residue of the distillation (strong acid!) should be discarded by pouring it onto ice, diluting it with water, then flushing it down the drain with generous amounts of additional water.

Transfer the distillate to a 125-mL separatory funnel, and add about 25 mL of water. Shake the funnel and allow the phases to separate. Drain the lower layer of 1-bromobutane into an Erlenmeyer flask (see Experimental Note 3). Discard the upper layer.

Add 25 mL of ice-cold, concentrated sulfuric acid to the 1-bromobutane. (CAUTION: See Safety Note 2!) Swirl the flask to mix the contents. If the mixture becomes warm, chill the flask in an ice bath. Then transfer the mixture to the separatory funnel. Concentrated sulfuric acid ($d = 1.84$) is more dense than 1-bromobutane ($d = 1.28$). Therefore, 1-bromobutane now forms the upper layer. Shake the funnel gently or swirl it to avoid an emulsion, then allow it to stand for 5–10 minutes (see Experimental Note 4). Drain off the lower layer (CAUTION: strong acid!) and discard by pouring it onto ice, diluting the ice mixture, and flushing the solution down the drain.

Extract the bromobutane remaining in the separatory funnel with 25 mL of water to remove the bulk of the residual sulfuric acid. 1-Bromobutane is more dense than this aqueous solution; therefore, the bromobutane now forms the lower layer. Shake the funnel, then drain this lower layer into a clean, dry flask (or a second separatory funnel). Discard the aqueous layer remaining in the separatory funnel, then return the bromobutane to the funnel. Extract the bromobutane with 25 mL of 10% NaOH solution. (CAUTION: caustic! See Section 1.1D.) In this extraction, as in the previous one, the bromobutane forms the lower layer in the separatory funnel. Drain the bromobutane into another clean flask, add 2 g of anhydrous calcium chloride, stopper the flask tightly, and allow the mixture to stand until the liquid is clear. (Overnight is best.) Because 1-bromobutane is quite volatile, a glass-stoppered flask is the preferred drying vessel.

Decant the clear bromobutane into a dry, 50-mL, round-bottom flask using a disposable pipet or dropper to transfer the residual liquid. Take care not to transfer any solid calcium chloride. Add 2–3 boiling chips, distil the dried 1-bromobutane with a dry distillation apparatus, and collect the fraction boiling at 98–103°. (If the distillate is cloudy—that is, wet—it should be redried and redistilled.) A typical yield is 17 g (50%), n_D^{20} 1.4392–1.4400.

EXPERIMENTAL NOTES

1) The arrangement of an HBr trap is shown in Figure 10.2. The liquid in the beaker is approximately 5% aqueous NaOH. (CAUTION: caustic!)

2) To verify that no oil is codistilling with the water toward the end of the distillation, collect a few milliliters of the distillate in a test tube. Shake or flick the test tube with your finger. If any oil droplets are present, they will become visible when you hold the test tube up to the light.

3) Pay careful attention to the identification of the layers in the separatory funnel during this experiment. During one extraction, the product is in the lower layer; during another, it is in the upper layer. It is prudent to save all layers in labeled flasks until the completion of the experiment to avoid inadvertently throwing away the wrong layer.

4) Emulsions are common in the extraction, and standing time will be necessary to allow the phases to separate. Even so, the interface between the layers may be indistinct, and judgment may be necessary when the layers are separated.

Problems

10.9 (a) What are the purposes of using sulfuric acid in the reaction of 1-butanol with sodium bromide?

(b) Could concentrated HI be substituted for sulfuric acid in this experiment? Explain.

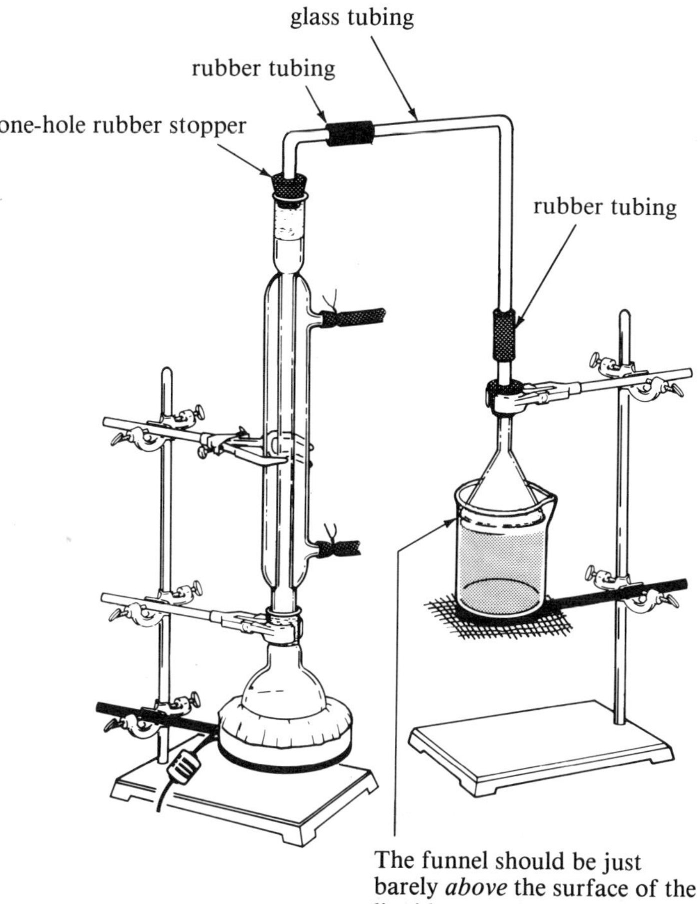

Figure 10.2 A gas trap for HBr, which must be used if Experiment 10.1 is not carried out in a fume hood.

10.10 Which of the following types of boiling chips would be suitable for the reflux of reactants in Experiment 10.1? Explain.

(a) marble ($CaCO_3$)

(b) alumina (Al_2O_3)

(c) silicon carbide (SiC)

10.11 To prevent gaseous HBr from contaminating the laboratory, one student corks the top of his reflux condenser. Why is this a bad idea?

10.12 (a) If you poured conc. sulfuric acid into 1-butanol, would you expect a temperature rise?

(b) If you poured conc. hydrochloric acid into 1-butanol, would you expect a temperature rise? Explain your answers.

10.13 Write the equation that shows how bromine (Br_2) is formed in the reaction mixture in Experiment 10.1.

10.14 Suggest a reason why the product 1-bromobutane does not react with water in the reaction mixture and revert to 1-butanol.

10.15 Suppose that you have only 15 g of 1-butanol to use in Experiment 10.1. What is your theoretical yield of 1-bromobutane?

10.16 If you want to synthesize 100 g of 1-bromobutane, how much 1-butanol should you start with?

10.17 (a) Write a flow diagram tracing 1-bromobutane through the extraction process in Experiment 10.1. In each step, state whether 1-bromobutane forms the upper layer or the lower layer (if two phases are present).
(b) Would this work-up procedure remove a nongaseous alkene by-product from the 1-bromobutane? Explain.

10.18 If Experiment 10.1 were carried out using the following alcohols in place of 1-butanol, what organic products and by-products would you expect in each case?
(a) 2-hexanol
(b) benzyl alcohol ($C_6H_5CH_2OH$)
(c) methanol

10.19 If 1-hexadecanol (n-$C_{16}H_{33}OH$) were subjected to this reaction and work-up procedure, what problems would be encountered? (*Hint*: Look up the physical constants of the reactant and product in the *Handbook of Chemistry and Physics*.)

»»» EXPERIMENT 10.2 «««

Synthesis of t-Butyl Chloride from t-Butyl Alcohol

	$(CH_3)_3COH$	+	HCl	→	$(CH_3)_3CCl$	+ H_2O
	2-methyl-2-propanol (*t*-butyl alcohol)		conc. hydrochloric acid (12*N*)		2-chloro-2-methylpropane (*t*-butyl chloride)	
MW:	74.12		—		92.57	
weight:	18.5 g		82.6 g (70 mL)		23.1 (theory)	
moles:	0.25		0.84		0.25 (theory)	

Table 10.2 Physical properties of reactants and products

Name	Bp (°C)	n_D^{20}	Density (g/mL)	Solubility
t-butyl alcohol[a]	82.2	1.3878	0.79	sol. H_2O
t-butyl chloride	52	1.3857	0.84	insol. H_2O
methylpropene (isobutylene)	−6.9	—	—	insol. H_2O

[a] Mp 25.5°

Discussion

Tertiary alcohols can be converted to alkyl chlorides very readily. In this experiment, you will convert *t*-butyl alcohol to *t*-butyl chloride by shaking the alcohol in a separatory funnel with concentrated HCl, then purifying the product by a simple distillation.

When you mix the reactants, you will observe that the mixture is initially a homogeneous solution because *t*-butyl alcohol is miscible with water. (Concentrated HCl is a 12*N* aqueous solution; therefore, it is mostly water.) *t*-Butyl alcohol is more soluble than 1-butanol (solubility = 8.3 g/100 mL) because its branched hydrocarbon chain is more compact and thus less hydrophobic than a four-carbon continuous chain. Immediately after the reactants are mixed, a second phase is formed. This second phase is the water-insoluble *t*-butyl chloride. The difference in solubilities between the alcohol and the halide is attributed to the hydrogen bonding between *t*-butyl alcohol and water and the lack of hydrogen bonding between *t*-butyl chloride and water.

The reaction of a tertiary alcohol and a hydrohalic acid proceeds by way of a carbocation (S_N1 reaction), as was shown on page 168. The carbocation combines with a halide ion (the nucleophile) to yield the product alkyl halide. However, the carbocation can also lose a proton to a molecule of water or alcohol to yield an alkene. Therefore, when *t*-butyl alcohol is treated with conc. HCl, some methylpropene (isobutylene) is also formed. Because elimination reactions are favored by high temperatures, chilling the HCl before mixing it with the alcohol helps suppress the formation of the alkene.

Methylpropene is a gas under the reaction conditions; therefore, it evaporates from the reaction mixture as it is formed. No special work-up procedure is necessary to remove this by-product from the desired product. However, when the reaction mixture is being shaken in the separatory funnel, the funnel must be vented frequently to release the gas. Furthermore, when the separatory funnel is sitting in an iron ring, its stopper must be removed to allow the methylpropene to escape.

After the reaction is complete, it is necessary to separate the *t*-butyl chloride from the excess acid. *t*-Butyl chloride is less dense than water; consequently, it forms the upper layer in the separatory funnel. The bulk of excess acid and water is removed simply by draining the lower layer. The residual acid in the separatory funnel is removed by a water wash, followed by a wash with aqueous $NaHCO_3$. The *t*-butyl chloride is then washed with water to remove any residual salts. Finally, the wet *t*-butyl chloride is dried with anhydrous $CaCl_2$ and distilled.

The drying should be done promptly because *t*-butyl chloride reacts slowly with water at room temperature to yield *t*-butyl alcohol and HCl in a solvolysis reaction (the same reaction that was performed in Experiment 8.2). If the *t*-butyl chloride is wet when it is distilled, noticeable amounts of HCl will be evolved.

t-Butyl chloride is extremely volatile. Because evaporation decreases the yield, any container of *t*-butyl chloride should be kept tightly stoppered between laboratory periods or during drying. To minimize evaporation during the distillation, the receiving flask should be chilled in an ice bath, as shown in Figure 10.3.

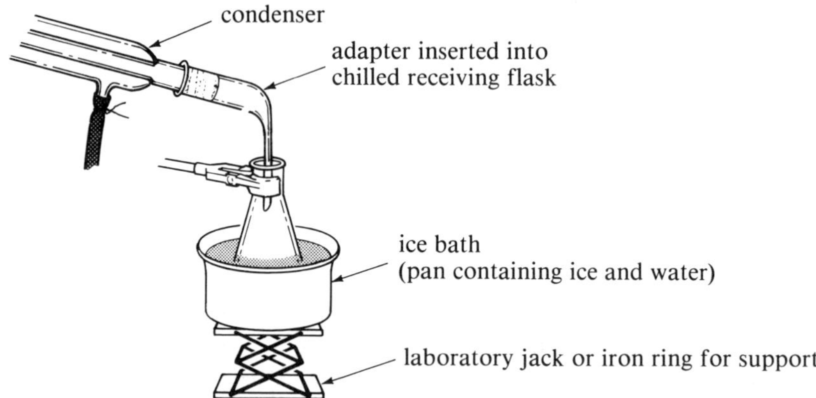

Figure 10.3 An ice-chilled distillation receiver.

Experimental

EQUIPMENT:

distillation assembly with adapter
dropper or disposable pipet
50-mL and two 125-mL Erlenmeyer flasks
glass stopper
10-mL and 100-mL graduated cylinders
ice bath
litmus paper
tared receiving flask (at least 20-mL capacity)
50-mL and 100-mL round-bottom flasks
250-mL separatory funnel
thermometer

CHEMICALS:

anhydrous calcium chloride, about 3 g
5% aqueous sodium bicarbonate, 10 mL
t-butyl alcohol, 18.5 g
conc. hydrochloric acid, 70 mL

TIME REQUIRED: about $1\frac{1}{2}$ hours

STOPPING POINT: while the *t*-butyl chloride is being dried with $CaCl_2$

>>>> **SAFETY NOTE 1** This experiment calls for the use of concentrated HCl. Any splashes on your skin or clothing should be flushed with copious amounts of water (see Sections 1.1D and 1.1E). Concentrated HCl emits gaseous HCl fumes, so you should use a fume hood while measuring and transferring this solution. Never sniff a flask containing conc. HCl.

››› ***SAFETY NOTE 2*** Gaseous methylpropene (flammable) is formed as a by-product in this experiment. Be sure that neither you nor your neighbors have burners on.

PROCEDURE

Measure 70 mL (82.6 g) of concentrated hydrochloric acid (12*N*) into a graduated cylinder. (CAUTION: See Safety Note 1.) Transfer the acid to a 125-mL Erlenmeyer flask and cool the flask in an ice bath. While the HCl is cooling, weigh 18.5 g of *t*-butyl alcohol into a 50-mL Erlenmeyer flask (see Experimental Note). When the HCl has been chilled to 5–8°, transfer it to a 250-mL separatory funnel; then add the *t*-butyl alcohol in one portion. The alcohol will initially dissolve in the acid; then, within seconds, the solution will turn cloudy.

Occasionally swirl the separatory funnel with its top off during the next five minutes. Thereafter, shake the funnel occasionally, with frequent venting to release the pressure of the gaseous methylpropene formed (see Safety Note 2). When the funnel is sitting in the iron ring, the stopper should be removed.

The total reaction period should be about 20 minutes. During this time, a clear layer of *t*-butyl chloride forms on top of the acidic solution. Carefully drain off the faintly cloudy lower layer, leaving the upper layer of *t*-butyl chloride in the separatory funnel. (CAUTION: The lower layer is still fairly concentrated HCl, which was used in excess. Pour it into a large amount of water before flushing it down the drain.)

Carefully wash the *t*-butyl chloride in the separatory funnel with 10 mL of water, followed by 10 mL of 5% sodium bicarbonate solution. (CAUTION: CO_2 is given off. If the earlier draining of the aqueous acid has not been done carefully, copious amounts of CO_2 may be generated, with potentially disastrous results because of pressure build-up in the separatory funnel.) The aqueous layer from this wash must be alkaline to litmus paper. If it is not, wash the *t*-butyl chloride with a second 10-mL portion of 5% sodium bicarbonate solution. Finally, wash the *t*-butyl chloride with 10 mL of water. If this water wash is not neutral to litmus paper, wash with another 10-mL portion of water.

Drain the cloudy *t*-butyl chloride into a 100-mL round-bottom flask. Add 3 g of anhydrous calcium chloride, stopper the flask tightly, and let it sit overnight. The cloudy liquid should turn clear during this period.

Decant the dry *t*-butyl chloride from the calcium chloride into a 50-mL round-bottom flask. Add 2–3 boiling chips, distil the product (flammable) by simple distillation, and collect the fraction boiling at 48–52° in a tared container chilled in an ice bath (see Figure 10.3). A typical yield is 17 g (73%), n_D^{20} 1.3857.

EXPERIMENTAL NOTE

t-Butyl alcohol (mp 25.5°) solidifies near room temperature. If the material in the reagent jar is solid, place the jar in a warm water bath until the *t*-butyl alcohol liquefies.

Problems

10.20 A student reasons, "I need 10 g of HCl for a reaction; therefore, I should weigh out 10 g of conc. hydrochloric acid."

(a) What is wrong with the student's reasoning?

(b) If the student did weigh 10 g of conc. hydrochloric acid, how many actual grams of HCl would the student have?

10.21 If an open flask of conc. hydrochloric acid is placed near an open bottle of 6*N* ammonium hydroxide in the laboratory, the general area becomes covered with a white coating. Explain.

10.22 Gaseous methylpropene is formed as a by-product in Experiment 10.2. Is this gas heavier or lighter than air? What are the safety implications of this physical property?

10.23 It is stated in the experimental procedure that *t*-butyl chloride forms the top layer in the separatory funnel. If you were not convinced that this is the case, how would you check the identity of each layer?

10.24 A chemist uses the procedure in Experiment 10.2 to convert 3-ethyl-3-pentanol to 3-chloro-3-ethylpentane.

(a) Would you expect the starting alcohol to dissolve in the conc. hydrochloric acid? Explain.

(b) If the alcohol is insoluble, how would this affect the procedure?

(c) What would be the principal by-product in this reaction? Would this by-product necessitate any special safety precautions? If so, list them.

(d) Write the equations for the steps in the reaction or reactions that would occur if the final 3-chloro-3-ethylpentane were not dried completely.

»»» EXPERIMENT 10.3 «««

Synthesis of Cyclohexene by the Dehydration of Cyclohexanol

	[cyclohexane ring]–OH	+ H_2SO_4/H_3PO_4	→ [cyclohexene ring]	+ H_2O
	cyclohexanol	mixed sulfuric and phosphoric acids	cyclohexene	
MW:	100.16	—	82.15	
weight:	20.0 g	7 mL, combined	16.4 g (theory)	
moles:	0.20	—	0.20 (theory)	

Table 10.3 Physical properties of reactants and products

Name	*Bp (°C)*	n_D^{20}	*Density (g/mL)*	*Solubility*	*Bp of azeotrope with water (°C)*
cyclohexanol[a]	161.1	1.4641	0.96	sl. sol. H_2O	98
cyclohexene	83.0	1.4465	0.81	insol. H_2O	71

[a] Mp 25.1°

Discussion

A secondary alcohol, such as cyclohexanol, undergoes dehydration by an E1 mechanism. The steps in this mechanism are shown on page 169. The key intermediate in the mechanism is a cyclohexyl cation, which can undergo substitution as well as elimination. To prepare an alkene in good yield, it is necessary to suppress the substitution reaction. In this experiment, the substitution reaction is suppressed by: (1) the use of strong acids with anions that are relatively poor nucleophiles; (2) a high reaction temperature, which favors elimination; and (3) distillation of cyclohexene from the reaction mixture as it is formed.

Any strong acid can be used as the dehydrating agent. Sulfuric, phosphoric, and oxalic acids, and even potassium hydrogen sulfate, have all been employed successfully. The anions of these acids (sulfate, hydrogen phosphate, and oxalate ions) are all poor nucleophiles, and thus substitution reactions are not favored. Other strong acids, such as HBr, have nucleophilic anions, and thus yield more substitution than elimination products. Concentrated sulfuric acid alone causes both oxidation and polymerization of the product alkene. Fewer side reactions occur when concentrated phosphoric acid is used as the dehydrating agent, but the rate of the alkene formation is slow. Consequently, a mixture of sulfuric and phosphoric acids is used as the dehydrating agent in this experiment.

Removal of the alkene by distillation as it is being formed in the reaction mixture is an excellent technique for preventing side reactions. Unfortunately, this technique is applicable only for dehydration reactions that produce low-boiling alkenes, such as cyclohexene. Removal of the alkene reduces tar (polymer) formation by minimizing the contact time between the acid and the alkene. Water is also removed from the acidic reaction mixture in this distillation, which prevents the reverse reaction (reconversion of the cyclohexene to cyclohexanol) from occurring.

The distillate is an azeotrope of cyclohexene and water (90% cyclohexene–10% water, bp 71°). Although cyclohexanol is high-boiling (bp 161°), it also forms an azeotrope (bp 98°) with water. Even with careful distillation, the cyclohexene–water distillate will be contaminated with some cyclohexanol, which must be removed in the subsequent work-up.

The crude distillate is transferred to a separatory funnel and the aqueous layer is drawn off. Since cyclohexanol is slightly water-soluble, it is removed from the crude cyclohexene by a water extraction. Next, the cyclohexene is extracted with saturated NaCl solution as a preliminary drying step. Because cyclohexene forms an azeotrope with water, it must be scrupulously dried at this point or the final product will be wet. Anhydrous $CaCl_2$ is the drying agent of choice because it forms molecular complexes with alcohols, as well as with water, and thus removes the last traces of cyclohexanol. After drying, the cyclohexene is purified by distillation.

Experimental

EQUIPMENT:

distillation assembly
125-mL Erlenmeyer flask
50-mL flask with ground-glass stopper
10-mL graduated cylinder
ice bath
50-mL and 100-mL round-bottom flasks
125-mL separatory funnel

CHEMICALS:

anhydrous calcium chloride, about 2 g
cyclohexanol, 20 g
85% phosphoric acid, 5 mL
saturated NaCl solution, 10 mL
conc. sulfuric acid, 2 mL

TIME REQUIRED: about $2\frac{1}{2}$ hours

STOPPING POINT: while the cyclohexene is being dried with $CaCl_2$

›››› ***SAFETY NOTE 1*** Both sulfuric and phosphoric acids are strong, corrosive acids. If any acid is splashed on your skin or clothing, wash immediately with copious amounts of water (see Sections 1.1D and 1.1E).

›››› ***SAFETY NOTE 2*** Cyclohexene is very volatile and very flammable. It should be stored in a glass-stoppered bottle with a lightly greased stopper. Its distillations should be carried out slowly, with an efficient condenser, and into a flask that is well-chilled in an ice bath. There should be no open flames in the vicinity.

Because cyclohexene vapors are heavier than air, they will accumulate in the sink and drain. As an added precaution against fire, water washes containing traces of cyclohexene should be disposed of in the fume-hood sink.

PROCEDURE

Place 20.0 g of cyclohexanol in a 100-mL round-bottom flask. Add 5 mL of 85% phosphoric acid and 2 mL of concentrated sulfuric acid (CAUTION: strong acids!). Mix the acidic solution by swirling, add 2–3 carborundum boiling chips, and equip the flask for simple distillation with a receiver adapter on the condenser. Slowly distil the contents of the distillation flask into a 125-mL Erlenmeyer flask chilled in an ice bath (Figure 10.3, page 182). (CAUTION: flammable. See Safety Note 2.) Adjust the rate of the distillation so that it takes about 45 minutes. Stop the distillation when about 8 mL of residue remains in the distillation flask (see Experimental Note). Approximately 17 g of water and crude cyclohexene will be collected in the receiver. The residue (CAUTION: strong acid!) should be poured onto ice, diluted with water, and flushed down the drain with a generous amount of additional water.

Transfer the distillate to a 125-mL separatory funnel. Drain the lower aqueous layer, then wash the cyclohexene remaining in the separatory funnel with 10 mL of water followed by 10 mL of saturated NaCl solution. Discard the aqueous layers in the hood sink when the extraction is completed. Because of the volatility of cyclohexene, conduct these washings as quickly as possible.

Drain the cyclohexene from the separatory funnel into a 50-mL, standard-taper, round-bottom flask. Add 2 g of anhydrous calcium chloride, and stopper the flask snugly with a lightly greased, ground-glass stopper. Allow the material to dry for at least 20 minutes, with occasional swirling to hasten the drying. Overnight drying is better.

After drying, remove the grease from the joint of the flask with a tissue, then carefully, but quickly, decant the bulk of the cyclohexene into a dry 50-mL round-bottom flask. (Because of the volatility of cyclohexene, filtration of the $CaCl_2$ is not advised.) Add 2–3 boiling chips to the cyclohexene, and distil it into a tared, 125-mL flask chilled in an ice bath (CAUTION: flammable).

There should be no distillation forerun. The bulk of the cyclohexene distils at 78–83°. A typical yield is 12.0 g (73%), n_D^{20} 1.4468. The product should be stored in a bottle with a lightly greased, ground-glass stopper.

EXPERIMENTAL NOTE

Toward the end of the first distillation, the residue turns yellow to dark brown, and the temperature of the distillate may approach 83°.

Problems

10.25 Without referring to the discussion, list the techniques used in Experiment 10.3 to maximize the yield of cyclohexene and minimize the yield of by-products.

10.26 A chemist desires to dehydrate 1-octanol by the procedure in Experiment 10.3.

(a) What is the expected product?

(b) Would a higher or lower temperature be needed in the first distillation? Explain.

10.27 The procedure in Experiment 10.3 states that you should leave about 8 mL of residue when distilling the crude cyclohexene. What are the reasons for this?

10.28 Write equations that show the mechanism of the dehydration of 3-pentanol.

10.29 A student's reaction mixture in Experiment 10.3 becomes black and tarry. Using equations, describe what has occurred.

»»› EXPERIMENT 10.4 ‹««

Synthesis of Cyclohexanone by the Oxidation of Cyclohexanol

In this experiment, you will oxidize a secondary alcohol (cyclohexanol) to a ketone (cyclohexanone) using one of two procedures: *oxidation with hypochlorous acid* or *oxidation with chromic acid.* Your instructor will tell you which procedure to use.

Each procedure has advantages and disadvantages. The classical procedure is the chromic acid oxidation. Its principal advantage is the ready availability of sodium dichromate. The principal disadvantages are that chromium salts are suspected carcinogens and disposal of the spent chromium salts down the drain is environmentally unwise.

The newer procedure is the hypochlorous acid oxidation. The principal advantage of this technique is that the reagents, sodium hypochlorite and acetic acid, are relatively safe. The principal disadvantage is that sodium hypochlorite is not readily available in the concentration required ($2M$) and must be prepared just prior to use. It is assumed that the storeroom, not the student, will prepare this reagent.

Oxidation of alcohols can be carried out under acidic or alkaline conditions. For the oxidation of secondary alcohols, acidic conditions are preferred. Under alkaline conditions, the product ketone can undergo further oxidation by way of its "enol" form. Thus, in base, the ketone would be obtained in lower yield and would be contaminated by other organic by-products. In this experiment, both procedures utilize acidic conditions.

alkaline oxidation of ketones:

$$\gt CH-\overset{O}{\overset{\|}{C}}- \rightleftarrows \gt C=\overset{OH}{\overset{|}{C}}- \xrightarrow{[O]} \gt C=O + HO\overset{O}{\overset{\|}{C}}-$$

a ketone — an enol, which can be oxidized at the carbon–carbon double bond — cleavage products

A. Hypochlorous Acid Oxidation

$$C_6H_{11}-OH + NaOCl \xrightarrow{CH_3CO_2H} C_6H_{10}{=}O + H_2O + NaCl$$

cyclohexanol — sodium hypochlorite — cyclohexanone

	cyclohexanol	sodium hypochlorite	cyclohexanone
MW:	100.16	74.46	98.15
weight:	10.0 g	78 mL ($2M$)	9.8 g (theory)
moles:	0.10	0.16	0.10 (theory)

Table 10.4 Physical properties of reactants and products

Name	Bp (°C)	n_D^{20}	Density (g/mL)	Solubility
cyclohexanol[a]	161.1	1.4641	0.96	sl. sol. H_2O
cyclohexanone	155.6	1.4507	0.95	sol. H_2O

[a] Mp 25.1°

Discussion

Secondary alcohols can be oxidized to ketones using sodium hypochlorite and acetic acid.* The active oxidizing agent is hypochlorous acid, HOCl, a thermally unstable weak acid ($pK_a = 7.4$) that is generated in the reaction mixture by a neutralization reaction.

$$NaOCl + CH_3CO_2H \rightarrow HOCl + CH_3CO_2Na$$

Sodium hypochlorite is the active ingredient in liquid laundry bleaches. However, the concentration of the hypochlorite in these bleaches (around 3%) is too low to be useful for alcohol oxidation. Liquid "swimming pool chlorine" (around 12% chlorine) is also a solution of sodium hypochlorite and, when fresh, has a sufficient concentration of hypochlorite to be effective. However, the availability and concentrations of these solutions are seasonally dependent. The 2*M* sodium hypochlorite used in this experiment is generated prior to use by passing chlorine gas over a solution of sodium hydroxide (see Experimental Note 1, page 192).

$$Cl_2 + 2\,NaOH \rightarrow NaOCl + NaCl$$

In the experiment, sodium hypochlorite is added to a mixture of cyclohexanol and acetic acid. A large excess of acetic acid is used to ensure that the reaction mixture remains acidic during the reaction period. The reaction is quite exothermic and the temperature must be carefully controlled. This is accomplished by two techniques: the addition of only small portions of the sodium hypochlorite solution at a time, and cooling the reaction mixture when necessary in an ice bath.

An excess of sodium hypochlorite must be added for complete reaction. Because the actual concentration of the hypochlorite will vary from run to run, a starch–iodide test is used to determine when the excess is present. Under acidic conditions, hypochlorite ions oxidize iodide ions to iodine. The presence of the iodine is detected by its formation of a blue to blue-black complex with starch.

$$^{-}OCl + 2\,H^+ + 2\,I^- \rightarrow I_2 + Cl^- + H_2O$$

After the oxidation of the cyclohexanol is complete, the excess hypochlorite is destroyed by addition of sodium bisulfite solution.

$$^{-}OCl + HSO_3^- \rightarrow Cl^- + H^+ + SO_4^{2-}$$

* N. M. Zuczek and P. S. Furth, *J. Chem. Ed.* **1981**, *58*, 824; R. V. Stevens, K. T. Chapman, and H. N. Weller, *J. Org. Chem.* **1981**, *45*, 2023.

The main steps in the work-up procedure are a steam distillation, an extraction, and a final distillation. The steam distillation removes the cyclohexanone from the bulk of the reaction mixture. (Cyclohexanone, bp 155.6°, and water form an azeotrope that boils at 95° and contains 61.6% water.) The distillate consists of water, cyclohexanone, and acetic acid. The inorganic salts and a large portion of the acetic acid remain in the distillation residue.

The acetic acid in the distillate is neutralized before the extraction step; otherwise, it would be carried into the final distillation and contaminate the product. The sodium acetate formed by neutralization is insoluble in the ether used for the extraction.

An ether extraction is used in procedure A (but not in procedure B) to minimize the mechanical loss of product during the transfer steps. Cyclohexanone is slightly water-soluble; therefore, NaCl is added to the steam distillate just before the extraction step to salt out the ketone.

The bulk of the ether used for extraction is removed by a preliminary distillation. Because the theoretical yield of cyclohexanone is only 9.8 g, the ketone is distilled from a small flask to minimize vapor loss.

Experimental

EQUIPMENT:

2–4 disposable pipets
three 250-mL, two 125-mL, and one 50-mL Erlenmeyer flasks
glass stopper
100-mL graduated cylinder
ice bath
litmus paper
two 100-mL and one 50-mL round-bottom flasks
separatory funnel
starch–iodide test paper
steam distillation assembly (see Figure 10.4, page 193)
thermometer

magnetic stirrer (optional)

CHEMICALS:

anhydrous magnesium sulfate, 2 g
anhydrous sodium carbonate, 10 g
cyclohexanol, 10 g
diethyl ether, 65 mL
glacial acetic acid, 25 mL
sodium bisulfite, 5 g
sodium chloride, 20 g
2*M* sodium hypochlorite solution, about 85 mL (see Experimental Note 1)

TIME REQUIRED: about 4 hours

STOPPING POINTS: after the addition of the sodium bisulfite; at the end of the steam distillation when the sodium chloride has been added; when the ether solution of cyclohexanone is being dried with magnesium sulfate

>>>> ***SAFETY NOTE*** Glacial acetic acid and sodium hypochlorite solution are corrosive to the skin. Glacial acetic acid also emits acetic acid vapors. Carry out this reaction in a well-ventilated room or in a fume hood.

PROCEDURE

Place 10 g of cyclohexanol in a 250-mL Erlenmeyer flask and add 25 mL of glacial acetic acid. Chill the solution in an ice bath. Transfer 75 mL of approximately 2*M* sodium hypochlorite solution (see Experimental Note 1) to a 125-mL Erlenmeyer flask or separatory funnel. Using a disposable pipet or the separatory funnel, add 1–2 mL of the sodium hypochlorite solution to the cyclohexanol–acetic acid solution, swirl the flask (or stir with a magnetic stirrer), and measure the temperature. The cyclohexanol solution initially turns a pale lemon color, which rapidly fades. A rise in the temperature of the solution will also be observed. Continue adding small volumes of sodium hypochlorite solution with swirling or stirring until the temperature reaches 35°, then cool the solution to 30° in an ice bath. Continue the addition and the cooling until the addition is complete. The addition requires 30–40 minutes. At the end of the addition, an upper organic layer in the reaction mixture will be evident.

Test the aqueous layer of the reaction mixture with starch–iodide test paper (see Experimental Note 2). If the test is negative, add an additional 2 mL of sodium hypochlorite solution. Cool the reaction mixture if necessary. Repeat this procedure until a positive test is obtained. At this point, the reaction mixture should be pale yellow. Add a final 3 mL of sodium hypochlorite solution and let the reaction mixture stand for 15 minutes.

To destroy the excess hypochlorite, add saturated sodium bisulfite solution (about 5 g of bisulfite to 5 mL of water), 1 mL at a time, until a negative starch–iodide test is obtained.

Transfer the reaction mixture to a 250-mL round-bottom flask, add 2–3 carborundum boiling chips, and equip the flask for steam distillation (see Experimental Note 3). Use about 25 mL of water to wash the residual reaction mixture out of the Erlenmeyer flask. Add about 100 mL of water to the dropping funnel of the steam distillation assembly. Steam distil the reaction mixture into a 250-mL Erlenmeyer flask, collecting about 75 mL of distillate. Add water to the distillation flask as necessary to maintain a constant volume. The boiling range is 94°–100°. (CAUTION: Cyclohexanone is flammable!)

The steam distillate contains cyclohexanone, acetic acid, and water. To neutralize the acetic acid, add very small portions of anhydrous sodium carbonate to the distillate (CAUTION: considerable effervescence) until the aqueous layer is alkaline to litmus paper (see Experimental Note 2). (About 10 g of anhydrous sodium carbonate is necessary.) Next, add 20 g of solid sodium chloride and swirl the flask to saturate the aqueous layer. (If all the sodium chloride dissolves, add an additional 10 g.)

After the solution is saturated with NaCl, add 25 mL of diethyl ether and decant the liquid into a 250-mL separatory funnel. Use about 10 mL of additional ether to wash the Erlenmeyer flask, transfer this ether wash to the separatory funnel, extract the cyclohexanone, and drain off the lower aqueous layer. Pour the ether layer into a 100-mL round-bottom flask. Re-extract the aqueous layer with an additional 25 mL of ether, combine the ether extracts, and dry them with about 2 g of anhydrous magnesium sulfate. Cap the round-bottom flask with a glass stopper and let the mixture stand overnight.

Decant the ether solution into another 100-mL round-bottom flask. Remove the bulk of the ether (flammable!) by distillation, and discard the distillate in the waste-ether jug. Cool the distillation flask to room temperature, and transfer the residual material to a 50-mL round-bottom flask, using a few milliliters of ether to wash the residual cyclohexanone from the 100-mL flask. Distil the product (flammable!), collecting the fraction boiling near 150° in a 50-mL Erlenmeyer flask (see Experimental Note 4). You should obtain about 7 g (71%) of the product, n_D^{20} 1.4500.

EXPERIMENTAL NOTES

1) Relatively fresh sodium hypochlorite solution should be used. It is assumed that this solution will be prepared by the storeroom. If the instructor wishes the student to gain experience in the technique of handling a toxic gas, use the following procedure for the preparation of sodium hypochlorite.

Sodium hypochlorite, 2M. Before proceeding, receive individual instruction on the valve system of the chlorine tank and on the safety procedures required in your laboratory. Gaseous chlorine is extremely toxic and it is imperative that you know these procedures before starting.

Place 40 g of solid sodium hydroxide (caustic!) in a 500-mL Erlenmeyer flask and add 200 mL of water. Swirl the mixture until the hydroxide dissolves, then cool the solution to 20° or colder in an ice bath. Place the Erlenmeyer flask on a triple-beam balance in a fume hood. Place a thermometer in the hydroxide solution and position the gas inlet tube approximately 1 inch *above the surface of the liquid.* Pass chlorine gas above the liquid surface and follow the course of the reaction by both the gain in the weight of the flask and by the rise in the temperature. When the temperature of the solution reaches 40°, stop the chlorine gas flow and rechill the solution to 20°. Return the flask to the balance and continue passing chlorine gas over the solution until it has gained 28 g. The procedure takes about 30 minutes.

The sodium hypochlorite solution will slowly de-gas in an uncapped vessel, losing about 50% of its gained weight in a week. If the vessel is properly capped, the solution will maintain its weight for weeks.

2) *Starch–iodide test paper* turns blue-black in a positive test and remains white in a negative test. If the upper organic layer impregnates starch–iodide paper (or litmus paper), the test is invalid. Therefore, do not dip the test paper in the solution. The best procedure is to remove a few drops of the aqueous layer with a disposable pipet and place the material on a moistened test strip.

3) A *steam distillation apparatus* that is adequate for this experiment is shown in Figure 10.4. The only purpose of the dropping funnel is to replenish water in the distillation flask as it boils away. A simple distillation apparatus without the dropping funnel may be used if a 500-mL distillation flask is available. This size flask is large enough to contain all the water required for the steam distillation.

4) If frothing occurs, decrease the amount of heat being applied to the distillation flask so that the froth does not enter the distillation head. To maintain a reasonable rate of distillation, insulate the upper part of the flask and the distillation head (see Experiment 5.2). Your instructor may be able to supply you with a commercial antifoaming agent, which will also help control the frothing.

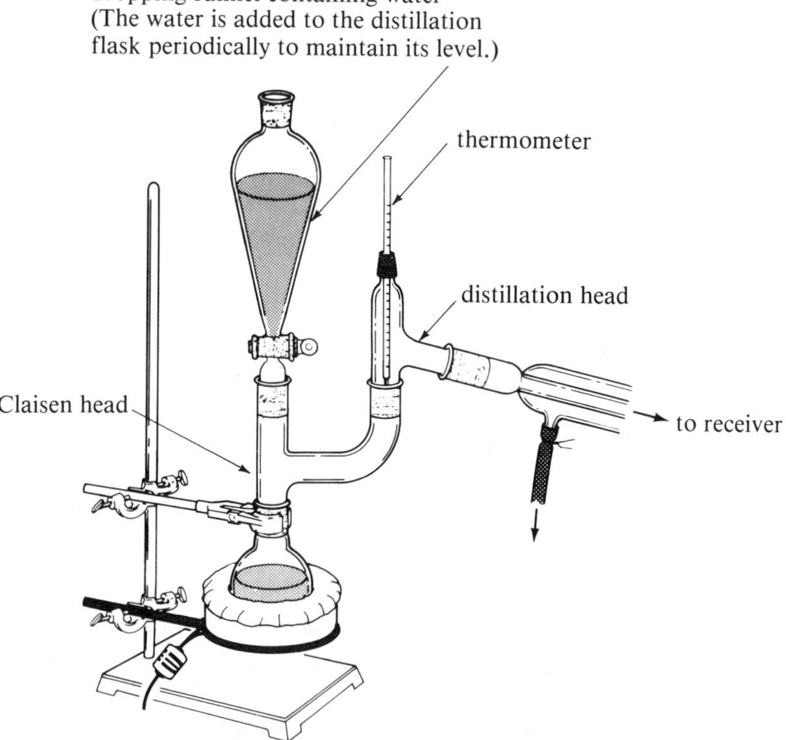

Figure 10.4 *A steam distillation apparatus that allows water to be added to the distillation flask during the distillation.*

B. Chromic Acid Oxidation*

$$3\ \text{C}_6\text{H}_{11}\text{–OH} + Na_2Cr_2O_7 \cdot 2H_2O + 4H_2SO_4 \longrightarrow 3\ \text{C}_6\text{H}_{10}\text{=O} + Cr_2(SO_4)_3 + Na_2SO_4 + H_2O$$

	cyclohexanol	sodium dichromate dihydrate	sulfuric acid	cyclohexanone
MW:	100.16	298.00	98.08	98.15
weight:	20.0 g	23.8 g	26.5 g (15 mL)	19.6 g (theory)
moles:	0.20	0.080	0.27	0.20 (theory)

Discussion

In chromate oxidations, the chromium is reduced from the +6 to the +3 oxidation state. Chromium(VI) is found in chromium trioxide (CrO_3), chromate ions (CrO_4^{2-}), dichromate ions ($Cr_2O_7^{2-}$) and trichromate ions ($Cr_3O_{10}^{2-}$). These chromate ions are interconvertible; which ions predominate depends on the pH of the solution. In acidic solution (as you will be using), dichromate ions predominate.

$$2\,[CrO_4]^{2-} + 2H^+ \rightleftarrows [Cr_2O_7]^{2-} + H_2O$$

chromate ion *in base* — dichromate ion *in acid*

For convenience, you will be using water as the solvent, with sodium dichromate as the oxidizing agent because this dichromate is soluble in water. The acid that has been found most suitable for this oxidation is sulfuric acid, which accelerates the oxidation and yields a water-soluble inorganic product, $Cr_2(SO_4)_3$ (a property that simplifies the work-up procedure). Empirically, it has been determined that a 20% excess of dichromate salt over the alcohol (0.40 mol dichromate to 1.0 mol alcohol instead of 0.33 mol to 1.0 mol) and a stoichiometric amount of H_2SO_4 (1.33 mol H_2SO_4 to 1.0 mol alcohol) result in the highest yield.†

The oxidation reaction is spontaneous and exothermic. Therefore, the dichromate solution is added slowly to a chilled mixture of the alcohol, acid, and water. The slow addition, the dilution, and the chilling all help keep the reaction within bounds.

At the start of the oxidation reaction, a definite color change can be noted. The starting dichromate solution is orange, while reduced chromium(III) ions are green. When the first aliquot of dichromate is added to the alcohol solution, it colors the reaction mixture yellow. As the reaction takes

* The physical constants of reactants and products are shown in Table 10.4, page 189.

† A. S. Hussey and R. H. Baker, *J. Org. Chem.* **1960**, *25*, 1434.

place, the mixture becomes green. After the first color change to green, additional color changes are hard to detect. For this reason, the *temperature* is used to monitor the reaction. A very definite temperature rise can be detected in the reaction mixture after each small addition of dichromate. It is important to keep the temperature of the reaction mixture below 35° until the end of the oxidation. At higher temperatures, the hydrocarbon portion of the molecules can be oxidized, resulting in a lower yield of the ketone and the formation of undesirable organic by-products.

After all the dichromate solution has been added and the reaction is complete, oxalic acid is added to the reaction mixture to destroy any unreacted dichromate ions.

$$3\,HO\overset{\overset{O}{\|}}{C}-\overset{\overset{O}{\|}}{C}OH + Cr_2O_7^{2-} + 8\,H^+ \rightarrow 6\,CO_2 + 2\,Cr^{3+} + 7\,H_2O$$

oxalic acid

The cyclohexanone is separated from the inorganic salts by steam distillation. (Cyclohexanone, bp 155.6°, and water form an azeotrope that boils at 95° and contains 61.6% water.) After the steam distillation, sodium chloride is added directly to the distillate to convert this water to a saturated NaCl solution and to "salt out" the cyclohexanone.

After it is separated from the NaCl solution, the cyclohexanone is dried and distilled. Because of the azeotrope, the cyclohexanone must be dried thoroughly or the final distilled product will be wet.

Experimental

EQUIPMENT:

250-mL beaker
disposable pipet or dropper
distillation apparatus
50-mL Erlenmeyer flask
25-mL graduated cylinder
ice bath
250-mL and 50-mL round-bottom flasks
250-mL separatory funnel
steam distillation apparatus (see Figure 10.4, page 193)
thermometer

CHEMICALS:

anhydrous magnesium sulfate, 1 g
cyclohexanol, 20 g
oxalic acid, 4 g
sodium dichromate dihydrate, 23.8 g
conc. sulfuric acid, 15 mL

TIME REQUIRED: about $3\frac{1}{2}$ hours

STOPPING POINTS: after the addition of the oxalic acid; at the end of the steam distillation, when the salt has been added to the distillate; when the cyclohexanone is being dried

»»> ***SAFETY NOTE*** This experiment calls for conc. H_2SO_4, which is a strong and corrosive acid, and a dichromate salt, which is a powerful oxidizing agent. Keep in mind that these compounds will affect your skin in a manner similar to the way they affect organic compounds used as reagents. In addition, chromium salts (oxidized or reduced) are toxic, are potential carcinogens, and cause skin ulcerations and allergic reactions. Do not inhale the dust of the chromium salts. Any splashes on skin or clothing must be washed off immediately with copious amounts of water (see Section 1.1E). During portions of this experiment, you may wish to use disposable plastic gloves.

PROCEDURE

Place about 60 g of crushed ice in a 250-mL beaker and add 15 mL of concentrated sulfuric acid. (CAUTION: strong acid!) To the ice slurry, add 20.0 g of cyclohexanol. After mixing, the temperature of the mixture should be about 20°. In a separate flask, dissolve 23.8 g of sodium dichromate dihydrate (CAUTION: toxic! see Safety Note) in 15 mL of water.

Place the beaker containing the cyclohexanol–acid mixture in an ice bath, and insert a thermometer to monitor the temperature. With a dropper, add approximately 1 mL of the dichromate solution to the cyclohexanol–acid mixture. The mixture should turn yellow after the addition, then turn green within a minute. Do not add more dichromate solution until the reaction mixture is green. The temperature should rise to about 30°. Re-cool the mixture to 20° in an ice bath. Continue the addition and cooling steps, swirling after each addition and making sure the temperature does not exceed 35°, until about 4 mL of the dichromate solution remains.

Remove the beaker from the ice bath and add the remaining 4 mL of dichromate solution in one portion. Stir the mixture and allow the temperature to rise to almost 50°. When the temperature of the mixture returns to about 35°, add 4 g of oxalic acid (CAUTION: toxic and fairly strong acid) with stirring to destroy the excess dichromate. The amount of time from the start of the dichromate addition to the addition of the oxalic acid should be about 45 minutes.

Transfer the reaction mixture and 2–3 carborundum boiling chips to a 250-mL round-bottom flask, and equip the flask for steam distillation (see Experimental Note 3, page 193). Use about 70 mL of water to wash the reaction beaker and add this wash to the round-bottom flask. Place 125 mL of water in the dropping funnel of the steam distillation assembly.

Collect about 100 mL of distillate (a mixture of water and cyclohexanone), boiling range 94°–100°. During the distillation, add small portions of water from the separatory funnel to the distillation flask to maintain the volume at a relatively constant level. At the end of the distillation, cool the distillation flask to room temperature and pour the contents into a jug marked "Waste chromium salts."

Add 15 g of sodium chloride to the distillate and stir. (Not all of the sodium chloride will dissolve; if it does, add an additional 5 g.) Decant the cyclohexanone–water mixture into a 250-mL separatory funnel, taking care not to transfer any residual sodium chloride. Drain and discard the lower aqueous salt layer. Pour the cyclohexanone layer into a 50-mL flask, add about 1 g of anhydrous magnesium sulfate, and stopper the flask tightly; then allow the flask to sit overnight, if possible.

Filter or decant the dried cyclohexanone into a dry 50-mL round-bottom flask for distillation, add 2–3 boiling chips, and collect the liquid (flammable!) boiling near 150°. (If frothing occurs, see Experimental Note 4, page 193.) A typical yield is 11 g (56%), n_D^{20} 1.4501.

Problems

10.30 If a procedure requires 10.0 g of HOCl, how many mL of 2.2*M* NaOCl are needed?

10.31 List the techniques used to prevent the cyclohexanol oxidation from becoming overly vigorous in Experiments 10.4A and B.

10.32 Why is it necessary to add extra water to the distillation flask in the steam distillation of cyclohexanone when this was not necessary for the steam distillation of cyclohexene in Experiment 10.3?

10.33 In the separation of cyclohexene from water (Experiment 10.3), a saturated NaCl solution was used. Yet in this experiment, NaCl was added to the steam distillate. Explain the reason for the difference in procedure.

10.34 Draw the structure of the enol form of cyclohexanone, and predict its oxidation product(s).

10.35 A student accidentally adds 10 g of oxalic acid to the reaction mixture to destroy excess dichromate instead of the 4 g stated in the experiment. Is the student's experiment ruined? Explain.

10.36 Could an aldehyde, such as CH_3CHO, be used to destroy the excess dichromate? Using an equation, explain the advantages or disadvantages of using this reagent instead of oxalic acid.

10.37 Predict the products of the chromate oxidation of 1,2-ethanediol under the conditions of Experiment 10.4B. (Use a flow equation; do not bother to balance the equations.)

CHAPTER 11

Infrared Spectroscopy

Infrared radiation is a portion of the electromagnetic spectrum that has longer wavelengths than those of visible light. Infrared radiation is of interest to organic chemists because organic compounds absorb radiation in this region of the electromagnetic spectrum. The radiation increases the amplitudes of vibration of the bonds in the molecules; consequently, different types of bonds, and thus different functional groups, absorb infrared radiation of different wavelengths. By inspecting the **infrared spectrum**, a plot of wavelength or frequency versus absorption, obtained from a compound of unknown structure, a chemist can often identify the bond types and functional groups of the compound. Unfortunately, it is rarely possible to determine the complete structure of a compound by inspection of its infrared spectrum alone.

The theory of infrared spectroscopy is covered in your lecture text or in the references listed on page 210. Here, we will briefly describe infrared spectra and then, in more detail, interpretation of infrared spectra and the laboratory procedures used to obtain an infrared spectrum.

11.1 The Infrared Spectrum

Figure 11.1 shows three different spectra of 1-propanol, obtained on three different instruments; these spectra exemplify the different ways in which infrared spectra can be recorded.

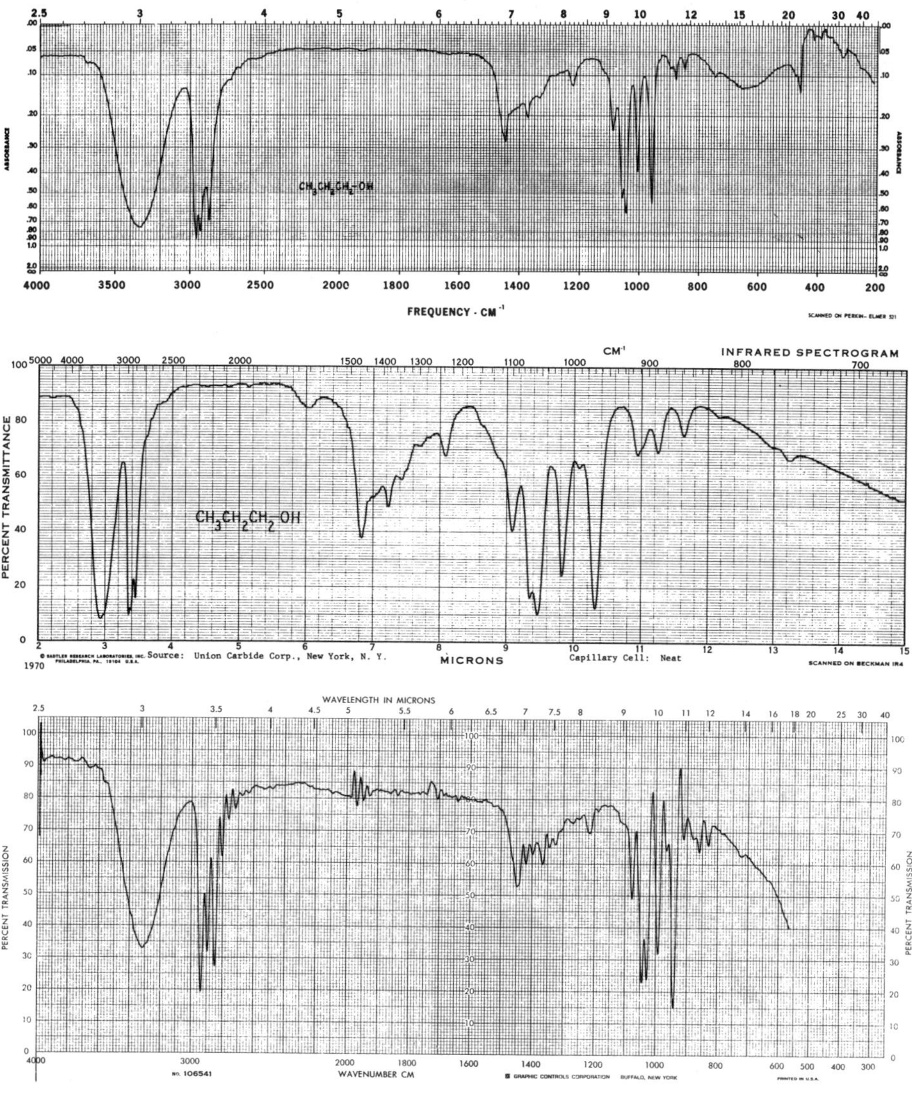

Figure 11.1 *Three infrared spectra of 1-propanol, $CH_3CH_2CH_2OH$, recorded on three different instruments. Upper two spectra © Sadtler Research Laboratories, Division of Bio-Rad Laboratories, Inc. Lower spectrum by authors.*

The lower horizontal scale of an infrared spectrum may be in **wavelengths** (λ), the distance from the crest of one wave to the crest of the next wave. In infrared spectroscopy, wavelengths are usually expressed as *micrometers* (μm), where $1\ \mu m = 10^{-4}$ cm. The older term *micron* (μ) is synonymous with micrometer.

Alternatively, the lower scale of an infrared spectrum may be in **wavenumbers**, units of frequency defined as the number of waves, or cycles, per centimeter. Wavenumbers have the units *reciprocal centimeters* (1/cm, or cm^{-1}). Wavenumbers are inversely proportional to wavelengths; that is, a shorter wavelength means a greater number of waves per centimeter. The wavenumber may be calculated from the wavelength by the following equation:

$$\text{wavenumber in cm}^{-1} = \frac{1}{\lambda \text{ in cm}} = \frac{1}{\lambda \text{ in } \mu\text{m}} \times 10^4$$

The vertical scale of an infrared spectrum shows either **per cent transmission ($\%T$) or absorbance** (A) of the radiation passing through the sample.

$$\%T = \frac{\text{intensity}}{\text{original intensity}} \times 100 \qquad A = \log\frac{\text{original intensity}}{\text{intensity}}$$

A typical infrared spectrum exhibits a "base line" at the top of the paper that represents essentially no absorption of energy by the sample. At a wavelength at which the sample absorbs radiation, the tracing shows a decrease in $\%T$ (or an increase in A). The absorption is thus recorded as a dip, called an *absorption peak* or *absorption band*; the wavelength or frequency of the minimum point of an absorption band is used to identify that band.

11.2 Absorption of Infrared Radiation

Bonds within a molecule undergo a variety of **fundamental modes of vibration**, described by such terms as stretching, bending, scissoring, rocking, and wagging. Each type of vibration absorbs infrared radiation of its own characteristic wavelength, giving rise to a large number of peaks in an infrared spectrum. Each fundamental mode is also associated with *overtones*, or *harmonics*, similar to those produced by a guitar or other stringed instrument. These overtones add still more peaks to a spectrum. Other small peaks arise from electronic "noise," caused by random electronic transitions in the circuitry of the instrument.

The net result of all these phenomena is that an infrared spectrum contains a multitude of peaks, only a few of which are important in the correlation of the spectrum with an organic structure. Do not attempt to identify every small peak in an infrared spectrum! The key to interpreting an infrared spectrum is to inspect it for the presence or absence of only a few significant absorption bands.

Besides its position, the *relative intensity* of an absorption band is also useful to the organic chemist. The relative intensities depend partly on the relative numbers of particular groups within a molecule. For example, three CH_2 groups in a molecule absorb more radiation than one CH_2 group. The relative intensity of absorption by a group of atoms also depends on the *change in dipole* when energy is absorbed. A nonpolar grouping, such as C=C in the symmetrical alkene $(CH_3)_2C{=}C(CH_3)_2$, gives rise to weak absorption compared to that of the polar carbonyl group (C=O).

11.3 Interpreting Infrared Spectra

Since the 1940's, when the infrared spectrophotometer became widely used, thousands of compounds have been subjected to this analytical technique. The result of these studies has been an empirical correlation of the positions of absorption of the various types of bonds and functional groups. This information is often summarized in charts called **correlation charts**. A typical correlation chart is shown inside the back cover of this book.

For purposes of organic structure identification, an infrared spectrum can be conveniently divided into two portions. The region from 1500–4000 cm^{-1} (2.5 μm to about 6.5 μm), to the left in the infrared spectrum, is especially useful for identification of the various functional groups. This region shows relatively strong absorption arising from stretching modes. The region to the right of about 1500 cm^{-1}, called the **fingerprint region**, is usually quite complex and often difficult to interpret; however, each organic compound has its own unique absorption pattern here. Figure 11.2 shows spectra of two alkenes; examine these spectra to see the differences in the fingerprint regions.

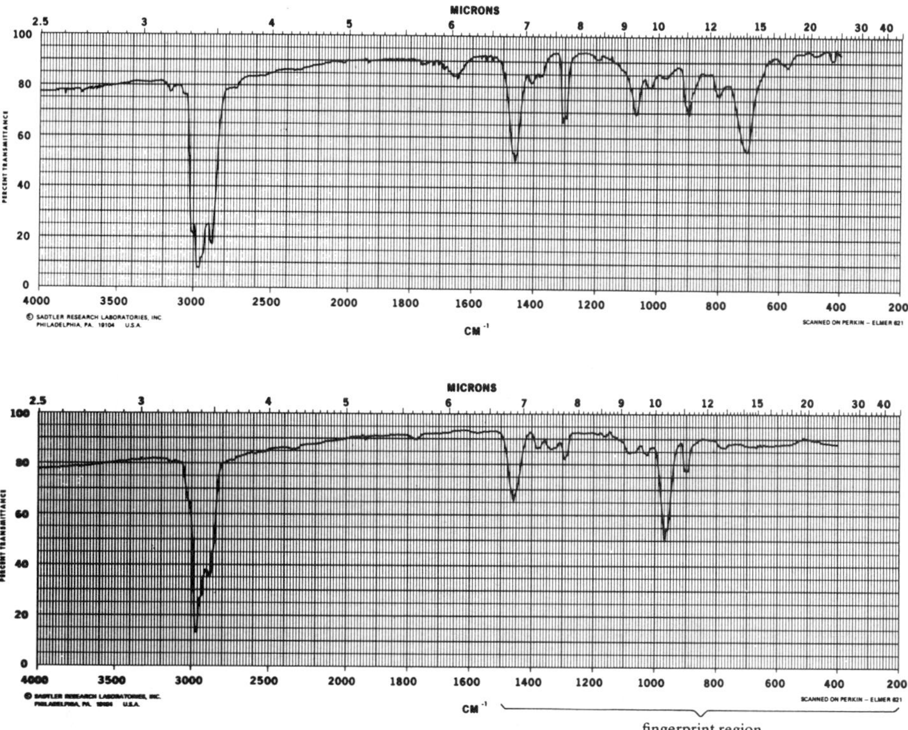

Figure 11.2 Infrared spectra of cis-3-hexene (upper spectrum) and trans-3-hexene (lower spectrum), showing the differences in their fingerprint regions. © Sadtler Research Laboratories, Division of Bio-Rad Laboratories, Inc.

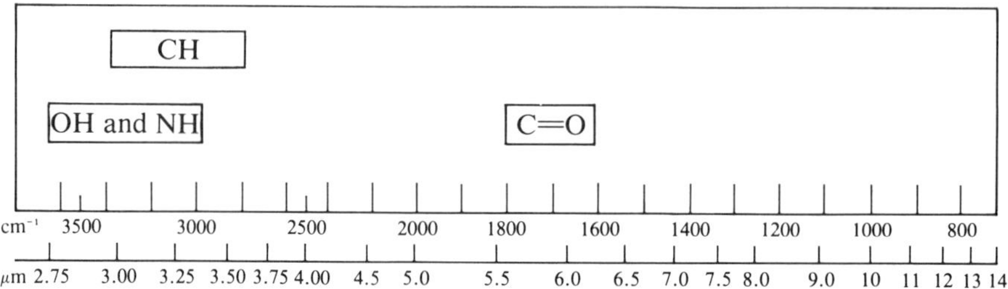

Figure 11.3 The most important parts of an infrared spectrum: the regions of OH, NH, CH, and C=O stretching vibrations.

In beginning your study of infrared spectra, do not attempt to memorize all the frequencies of absorption of the various functional groups. Instead, learn to recognize the most important regions in the spectrum: the OH–NH region, the CH region, and the C=O region. In Figure 11.3, these general regions are indicated by blocks. When you are presented with the spectrum of an unknown compound, examine these portions of the spectrum first. After you have identified the important features, then check the spectrum against the tables and discussions that follow.

11.4 Identifying Types of Compounds

A. Hydrocarbons

A distinguishing feature of the infrared spectra of hydrocarbons and other organic compounds containing CH bonds is the *CH stretching absorption.* **Alkanes** and **alkyl groups** (sp^3 CH) exhibit this absorption at 2800–3000 cm^{-1} (3.3–3.6 μm). **Alkenes, alkynes,** and **aromatic compounds** exhibit CH stretching slightly to the left of this value (at shorter wavelengths, or higher frequencies), as is shown in Figure 11.4 and also in Table 11.1.

The *C—C stretching absorption* varies significantly, both in position and in intensity, for the various types of carbon–carbon bonds. The C—C stretching of alkanes and alkyl groups is generally too weak to be of value. **Alkenes** show generally weak, but often visible, C=C absorption at about 1600–1700 cm^{-1} (5.8–6.2 μm). A highly unsymmetrical alkene shows stronger absorption than a more symmetrical alkene because a greater change in dipole results from its vibrations. Figure 11.5, the spectrum of 1-heptene, shows absorption arising from sp^3 CH, sp^2 CH, and C=C. **Alkynes** show a weak, but distinctive, C≡C absorption at 2100–2250 cm^{-1} (4.4–4.8 μm). This absorption is distinctive because C≡N

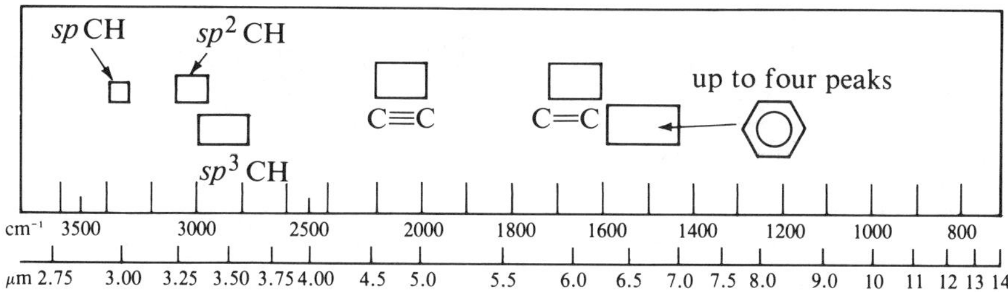

Figure 11.4 The locations of the stretching absorption of the different types of CH and C—C bonds.

Table 11.1 C—C and C—H stretching absorption

Type of bond	Position of absorption cm^{-1}	μm
C—H:[a]		
alkynyl, ≡C—H	3300	3.0
aryl, C_6H_5—H	3000–3300	3.0–3.3
alkenyl, =C(—)—H	3000–3100	3.2–3.3
alkyl, —C—H	2800–3000	3.3–3.6
C—C:		
alkynyl, C≡C	2100–2250	4.4–4.8
alkenyl, C=C[b]	1600–1700	5.8–6.2
aryl, (benzene ring)[c]	1450–1600	6.25–6.9

[a] OH and NH absorption is also observed at 3300–3500 cm^{-1} (2.8–3.0 μm). The CH of an aldehyde group appears slightly to the right, at 2700–2900 cm^{-1} (3.45–3.7 μm). The CH bending (not stretching) of a *gem*-dimethyl group, $>C(CH_3)_2$, often shows a double peak at 1375–1385 cm^{-1} (7.2–7.3 μm) and 1362–1372 cm^{-1} (7.3–7.35 μm).

[b] C=C stretching absorption may be weak or even absent for a fairly symmetrical alkene.

[c] Four peaks may be observed.

and Si—H are the only other bonds that absorb infrared radiation in this usually clear portion of the spectrum.

The spectra of **substituted benzenes** show a series of peaks (up to four) in the 1450–1600 cm^{-1} (6.25–6.9 μm) region. Substituted benzenes also exhibit absorption from CH bending vibrations in the region of 680–900 cm^{-1} (11–15 μm), at the far right of a spectrum. This absorption often reveals the positions

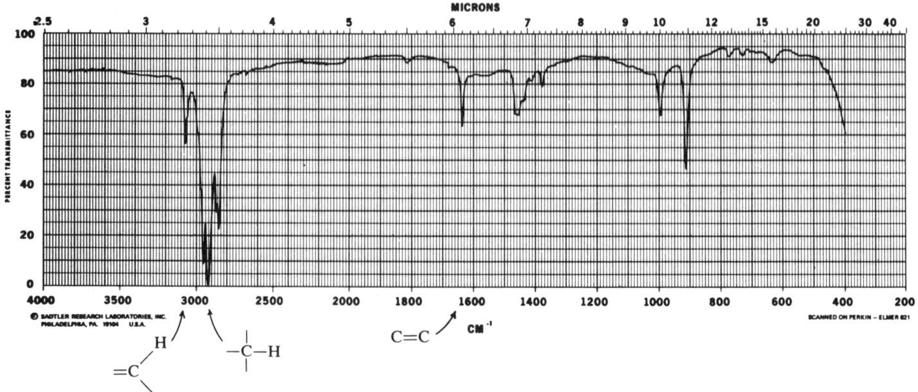

Figure 11.5 *Infrared spectrum of 1-heptene,* $CH_3(CH_2)_4CH{=}CH_2$, *showing some types of C—C and C—H absorption.*

of substitution on the ring. Table 11.2 lists the characteristic absorption patterns in this region for various types of substituted benzenes. Figure 11.6 shows the spectra of *o*-, *m*-, and *p*-chlorotoluene.

Another technique used to determine the extent and positions of substitution on the benzene ring is to look at the *overtone pattern* at 1666–2000 cm^{-1} (5–6 μm). Figure 11.7 shows the characteristic patterns. Unfortunately, these overtone bands are very weak and may not be visible unless the spectrum of a concentrated sample is taken.

B. Ethers, Alcohols, Phenols, and Amines

Ethers have a C—O stretching band that falls in the fingerprint region at 1050–1260 cm^{-1} (7.9–9.5 μm). Because oxygen is electronegative, the stretching causes

Table 11.2 *The C—H bending absorption of substituted benzenes*

Substitution	*Appearance*	*Position of absorption* cm^{-1}	μm
monosubstituted,	two peaks	730–770 690–710	12.9–13.7 14.0–14.4
o-disubstituted,	one peak	735–770	12.9–13.6
m-disubstituted,	three peaks	860–900 750–810 680–725	11.1–11.6 12.3–13.3 13.7–14.7
p-disubstituted,	one peak	800–860	11.6–12.5

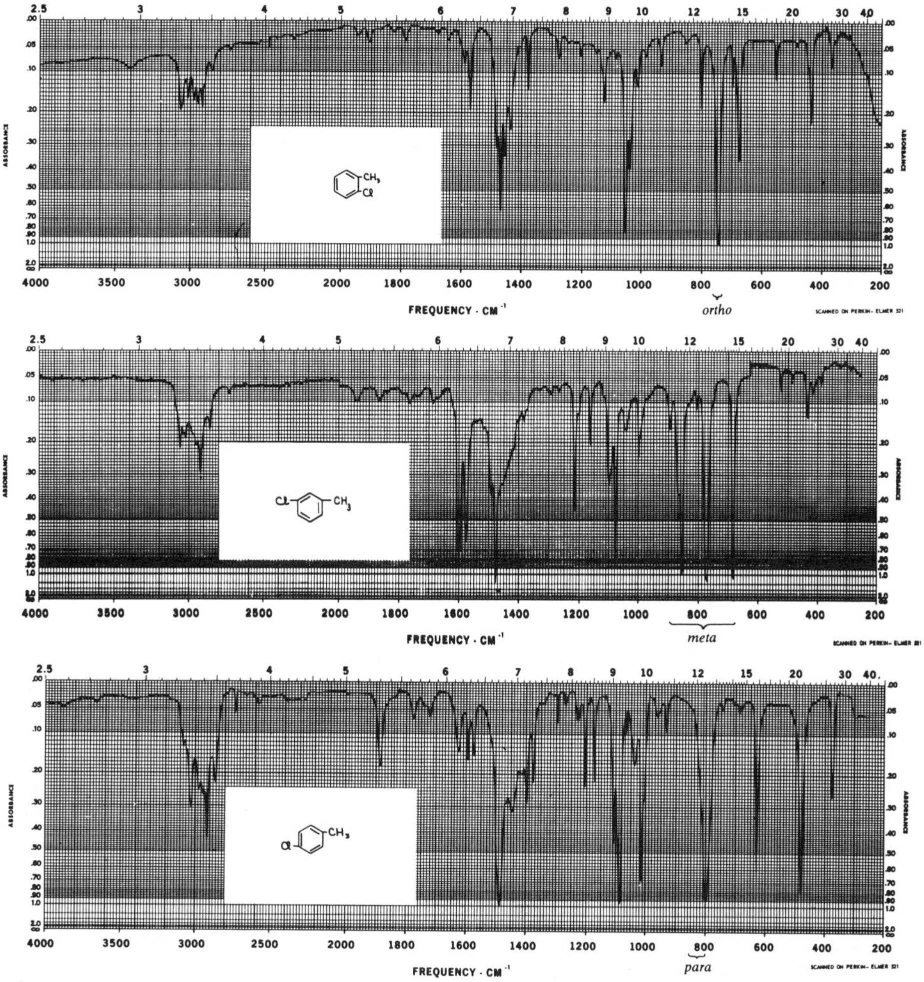

Figure 11.6 *Infrared spectra for o-, m-, and p-chlorotoluene,* (structure: Cl and CH_3 on a benzene ring) *showing the differences in absorption arising from C—H bending vibrations.*

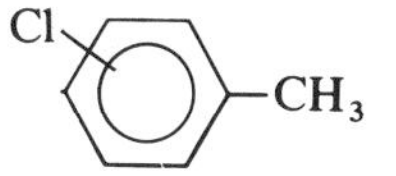

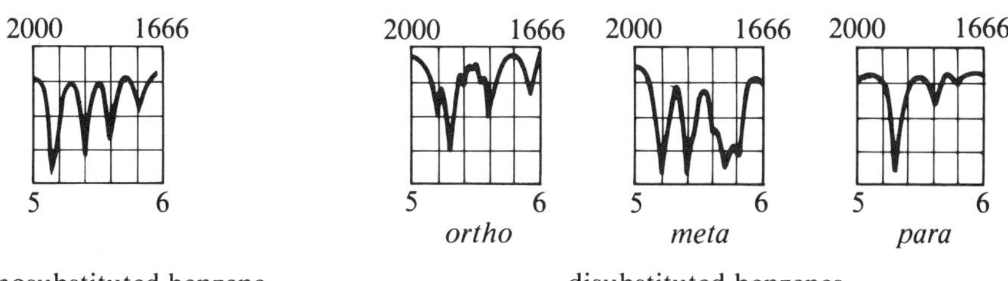

Figure 11.7 *Partial infrared spectra showing the weak overtone patterns for mono- and disubstituted benzenes at 1666–2000 cm^{-1} (5–6 μm).*

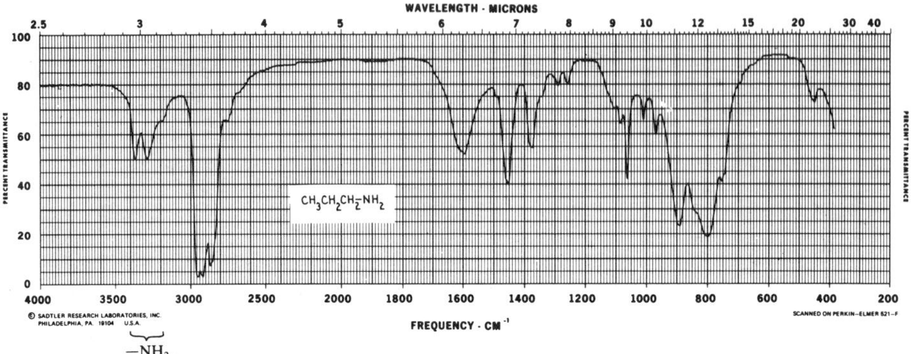

Figure 11.8 Infrared spectrum of n-propylamine, $CH_3CH_2CH_2NH_2$, showing a double NH peak. © Sadtler Research Laboratories, Division of Bio-Rad Laboratories, Inc.

a large change in bond moment; therefore, the C—O absorption is often strong. Alcohols and other compounds containing C—O bonds also show absorption in this region. For example, 1-propanol shows C—O absorption at about 1075 cm^{-1} (see Figure 11.1, page 199).

The most conspicuous feature in the infrared spectrum of an **alcohol** or **phenol** is the OH absorption, to the *left* of CH absorption, at 3000–3700 cm^{-1} (2.7–3.3 μm). (Again, refer to Figure 11.1.) The OH absorption in Figure 11.1 is of a *hydrogen-bonded OH group.* The spectrum of an alcohol in the vapor phase or in a dilute solution with a nonhydrogen-bonding solvent exhibits a sharper, weaker peak at about 3600 cm^{-1} (2.75 μm), to the left of the broad, strong OH band usually observed for hydrogen-bonded alcohols. In some cases *two* OH peaks can be observed: one for the hydrogen-bonded OH and one for the nonhydrogen-bonded OH.

Amines containing NH bonds also show absorption to the left of CH absorption. If there are two hydrogens on an amine nitrogen (—NH_2), the NH absorption appears as a double peak (see Figure 11.8). If there is only one H on the N, then one peak is observed. Of course, if there is no NH bond (as in a tertiary amine, R_3N), then there is no absorption in the NH stretching region.

C. Carbonyl Compounds

The carbonyl stretching mode gives rise to a strong, sharp peak somewhere between 1640 and 1820 cm^{-1} (5.5–6.1 μm). The positions of C=O absorption for a variety of carbonyl compounds are listed in Table 11.3.

Aldehydes and **ketones** both exhibit strong carbonyl absorption. The important difference is that an aldehyde has an H bonded to the carbonyl carbon. This particular C—H bond shows two characteristic stretching bands (just to the right of the aliphatic C—H bond) at 2820–2900 cm^{-1} (3.45–3.55 μm) and 2700–2780 cm^{-1} (3.60–3.70 μm). Both these C—H peaks are sharp, but weak, and the

Table 11.3 Carbonyl stretching vibrations[a]

Type of compound	*Position of absorption* cm^{-1}	μm	*Comments*
aldehydes, RCHO	1720–1740	5.75–5.82	The aldehyde CH is distinctive (p. 206)
conjugated	1695–1705	5.83–5.94	
ketones, $R_2C{=}O$	1705–1750	5.70–5.87	
conjugated	1660–1700	5.8–6.0	
β-diketones	1540–1640	6.1–6.5	OH absorption also present (p. 209)
carboxylic acids, RCO_2H	1700–1725	5.80–5.88	OH absorption is distinctive (p. 207)
conjugated	1680–1700	5.88–5.95	
esters, RCO_2R'	1740	5.75	Look for C—O absorption (p. 208)
conjugated	1715–1730	5.78–5.83	
amides, $R\overset{O}{\overset{\parallel}{C}}NR'_2$	1630–1700	5.9–6.0	1° and 2° amides also exhibit NH absorption (p. 209)
acid chlorides, $R\overset{O}{\overset{\parallel}{C}}Cl$	1785–1815	5.5–5.6	
conjugated	1770–1800	5.56–5.65	
acid anhydrides, $(R\overset{O}{\overset{\parallel}{C}})_2O$	1740–1840	5.44–5.75	Two carbonyl peaks
conjugated	1720–1820	5.5–5.8	
cyclic	1782–1865	5.4–5.6	

[a] Carbonyl compounds sometimes exhibit weak overtone absorption in the OH portion of the spectrum; therefore, weak absorption here does not necessarily mean that the compound is impure.

peak at 2900 cm^{-1} (3.45 μm) may be obscured by overlapping C—H absorption (see Figure 11.9). The aldehyde C—H also has a very characteristic nmr absorption. If the infrared spectrum of a compound suggests that the structure is an aldehyde, the nmr spectrum should be checked.

Carboxylic acids, either as pure liquids or in solution at concentrations in excess of about 0.01*M*, exist primarily as hydrogen-bonded dimers rather than as discrete monomers. The infrared spectrum of a carboxylic acid is therefore the spectrum of the dimer.

```
        Ö:---H—O
       //        \
   R—C            C—R
       \        //
        O—H---:O
              ..
```

Because of the hydrogen bonding, the O—H stretching absorption of carboxylic acids is broad and intense and slopes into the region of aliphatic carbon–hydrogen absorption (see Figure 11.10). The broadness of the carboxylic acid O—H band can often obscure both aliphatic and aromatic C—H absorption, as well as any

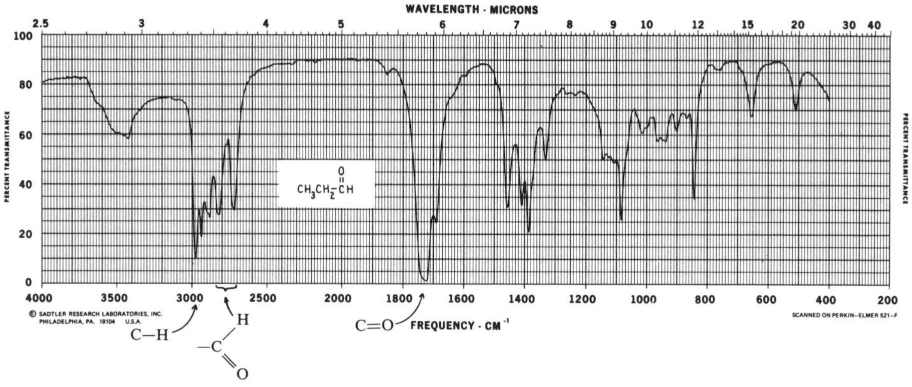

Figure 11.9 Infrared spectrum of propanal, CH_3CH_2CHO, showing the absorption by the aldehyde group. © Sadtler Research Laboratories, Division of Bio-Rad Laboratories, Inc.

other OH or NH absorption in the spectrum. The carbonyl absorption is of moderately strong intensity.

Esters exhibit both carbonyl absorption and C—O stretching absorption at 1100–1300 cm^{-1} (7.70–9.10 μm). The C—O absorption may be used to distinguish esters from some other carbonyl compounds. (Be careful—carboxylic acids, anhydrides, and some other compounds also show C—O absorption.)

$$\underbrace{\overset{O}{\overset{\|}{R C}}Cl \quad \overset{O}{\overset{\|}{R C}}CH_3}_{\text{show } C{=}O,\ \text{but not } C{-}O} \qquad \underbrace{\overset{O}{\overset{\|}{R C}}OR' \quad \overset{O}{\overset{\|}{R C}}O\overset{O}{\overset{\|}{C}}R}_{\text{show both } C{=}O \text{ and } C{-}O}$$

show C=O, but not C—O *show both C=O and C—O*

Amides may show *two* peaks in the carbonyl region. The *amide I band* arises from C=O. The *amide II band*, which appears between 1515–1670 cm^{-1} (6.0–6.6 μm), just to the right of the C=O peak, arises from NH bending. Therefore, a disubstituted, or tertiary, amide does not show an amide II band.

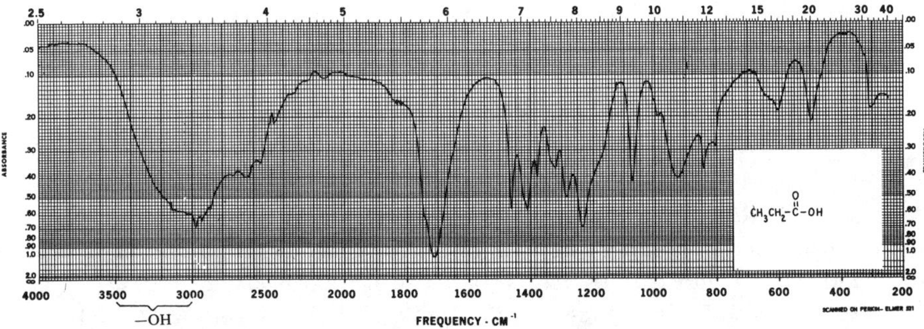

Figure 11.10. Infrared spectrum of propanoic acid, $CH_3CH_2CO_2H$, showing the broad OH absorption. © Sadtler Research Laboratories, Division of Bio-Rad Laboratories, Inc.

NH — no NH

$\overset{O}{\underset{\text{1°}}{RCNH_2}}$ $\overset{O}{\underset{\text{2°}}{RCNHR}}$ $\overset{O}{\underset{\text{3°}}{RCNR_2}}$

The NH stretching vibrations give rise to the absorption to the left of aliphatic CH absorption at 3125–3570 cm^{-1} (2.8–3.2 μm). (This is about the same region where the NH of amines and OH absorb.) *Primary amides* ($RCONH_2$) show a double peak in this region. *Secondary amides* (RCONHR), with only one NH bond, show a single peak. *Tertiary amides* ($RCONR_2$), with no NH, show no absorption in this region.

The carbonyl infrared absorption of **acid chlorides** is observed at slightly higher frequencies than that of other acid derivatives. There is no other distinguishing feature in the infrared spectrum that signifies "This is an acid chloride."

A **carboxylic acid anhydride**, which has two C=O groups, generally exhibits a *double carbonyl peak* in the infrared spectrum. Anhydrides also exhibit C—O stretching around 1100 cm^{-1} (9 μm).

Effects of structure and hydrogen bonding on carbonyl absorption. The exact position of absorption by a carbonyl group in the infrared spectrum depends on a number of factors, such as ring strain, conjugation, and hydrogen bonding.

Ring strain shifts carbonyl absorption to higher frequencies (to the *left*). For example, the relatively unstrained cyclohexanones show carbonyl absorption at 1705–1725 cm^{-1}, while the strained cyclobutanone absorbs at 1775 cm^{-1}.

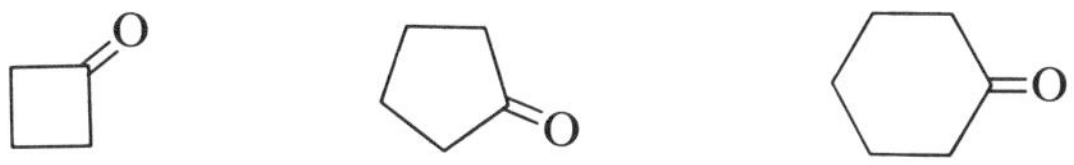

cyclobutanone	cyclopentanone	cyclohexanone
1775 cm^{-1} (5.64 μm)	1740–1750 cm^{-1} (5.72–5.75 μm)	1705–1725 cm^{-1} (5.8–5.87 μm)

increasing ring strain shifts C=O absorption to the left in the spectrum

If a carbonyl group is in **conjugation** with either C=C or an aromatic ring, the position of absorption is usually shifted to the *right* in the spectrum. Table 11.3 (page 207) lists the shifts observed for some of these carbonyl compounds.

Hydrogen bonding of a carbonyl group with an NH or OH group results in a shift of the C=O absorption slightly to the *right.* This shift is observed in the spectra of carboxylic acids. Hydrogen bonding can also arise from a solvent such as chloroform (which can form weak hydrogen bonds). The effects of hydrogen bonding are also evident in the infrared spectrum of a β-diketone. A β-diketone exists partially in the *enol form* and shows OH absorption at 2500–2700 cm^{-1} (3.7–4 μm). The carbonyl absorption of a β-diketone is broad and intense because of the intramolecular hydrogen bonding.

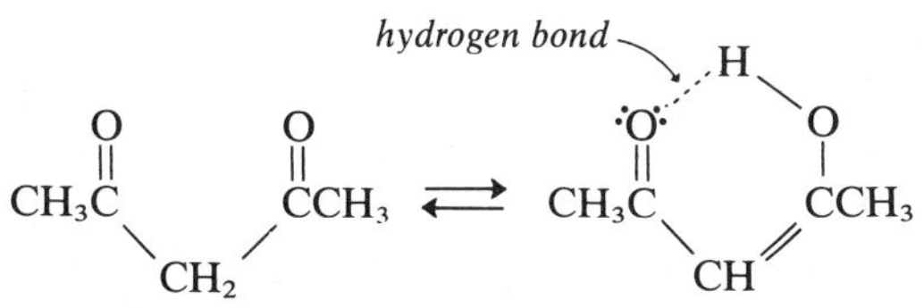

$$CH_3C(=O)CH_2C(=O)CH_3 \rightleftarrows CH_3C(=O\cdots H{-}O)CH{=}CCH_3$$

a β-diketone *keto form* — *enol form* (*hydrogen bond*)

D. Other Types of Compounds

The stretching absorption of the C—X bond of an **alkyl halide** falls in the fingerprint region of the infrared spectrum, from 500 cm^{-1} to 1430 cm^{-1} (7–20 μm). For example, alkyl chlorides absorb at about 700–800 cm^{-1} (12.5–14.3 μm). Without additional information, the presence or absence of a band in this region cannot be used for verifying the presence of a halogen in an organic compound.

Nitro compounds show two strong absorption bands in their infrared spectra: at 1500–1650 cm^{-1} (6.0–6.7 μm) and at 1250–1350 cm^{-1} (7.4–8.0 μm). In addition, aromatic nitro compounds may show absorption at 750–850 cm^{-1} (11.7–13.3 μm).

Nitriles (RC≡N) exhibit the same characteristic triple-bond absorption as do alkynes (2000–2300 cm^{-1}, 4.35–5 μm). The infrared spectrum alone cannot be used to distinguish a nitrile from an alkyne. However, a peak at 3300 cm^{-1} (3.0 μm) in the CH region would suggest the presence of a —C≡CH grouping.

Suggested Readings

Bellamy, L. J. *The Infrared Spectra of Complex Organic Molecules*, 3rd ed. New York: John Wiley and Sons, 1975.

Conley, R. T. *Infrared Spectroscopy*, 2nd ed. Boston: Allyn and Bacon, 1972.

Dyer, J. R. *Application of Absorption Spectroscopy of Organic Compounds*. Englewood Cliffs, N.J.: Prentice-Hall, 1965.

Nakanishi, K. and Solomon, P. H. *Infrared Absorption Spectroscopy*. San Francisco: Holden-Day, 1977.

Silverstein, R. M., Bassler, C. G., and Morrill, T. C. *Spectrometric Identification of Organic Compounds*, 3rd ed. New York: John Wiley and Sons, 1974.

Problems

11.1 Convert the following wavelengths to wavenumbers:
(a) 5.00×10^{-4} cm (b) 7.40 μm (c) 3.33 μ

11.2 Convert the following wavenumbers to wavelengths in μm:
(a) 3000 cm^{-1} (b) 831 cm^{-1} (c) 1650 cm^{-1}

11.3 Arrange the following list of compounds in order of increasing intensity of the C=C stretching absorption (least intense first), assuming identical molar concentrations in the sample cell:
(a) $CH_3CH{=}CHCH_2CH_3$ (b) $CH_2{=}CHCH_2CH_3$
(c) $CH_2{=}CCl_2$

11.4 The infrared spectrum of a student's sample shows a weak absorption band at 3710 cm^{-1} (2.7 μm), yet the student is positive that the compound is not an alcohol or amine. Explain.

11.5 True or false? A double peak at around 3300 cm^{-1} (3.0 μm) always indicates the presence of $—NH_2$. Explain your answer.

11.6 A student runs the infrared spectrum of cyclopentanone using chloroform as the solvent. The infrared spectrum shows a *double* carbonyl peak. Explain.

11.7 Tell how you would distinguish between each of the following pairs of compounds by their infrared spectra alone.

(a) $CH_3CH_2CO_2H$ and $CH_3CH_2CO_2CH_3$

(b) $CH_3C(=O)CH_2CH_3$ and $CH_3CO_2CH_2CH_3$

(c) cyclohexyl–$C(=O)NHCH_3$ and cyclohexyl–$C(=O)N(CH_3)_2$

(d) CH_3CH_2CHO and $CH_3C(=O)CH_3$

(e) $CH_3CH_2CH_2CO_2CH_2CH_3$ and $CH_3C(=O)CH_2CO_2CH_2CH_3$

(f) HO–C_6H_4–OH (para) and HO–C_6H_4–CO_2H (ortho)

(g) Br–C_6H_4–NH_2 (para) and Br–C_6H_4–NH_2 (ortho)

(h) $CH_3CH{=}CHCO_2CH_3$ and $CH_2{=}CHCH_2CO_2CH_3$

⟫⟫ EXPERIMENT 11.1 ⟪⟪

Infrared Spectra in Structure Determination

Discussion

A **sample cell** is a sample container that fits into the spectrophotometer in the path of the infrared radiation. Different types of cells are available for solids, liquids, solutions, and even gases. The "windows" of most common infrared cells, which allow the infrared radiation to pass through the sample, are polished sodium chloride plates. Sodium chloride is used because it is transparent to infrared radiation in the region of interest. (Sodium chloride cells are sensitive to moisture; see the Experimental Note on page 218 for their care.)

Liquid Samples

Liquid samples are usually analyzed *neat* (meaning "pure" or "without solvent") as **thin films**. A drop or two of the liquid is sandwiched between two

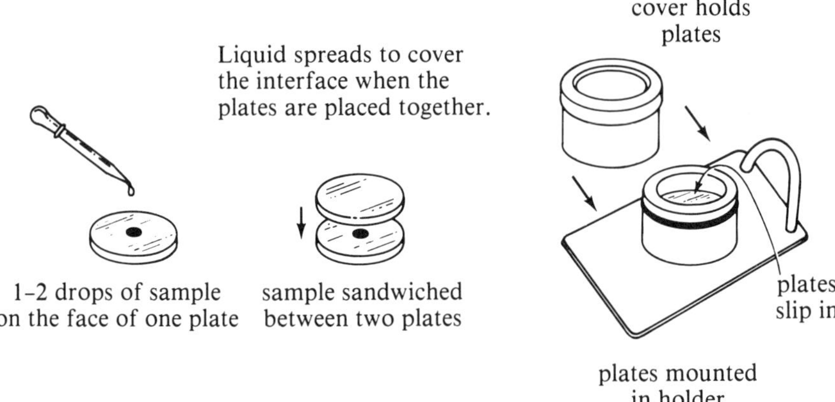

Figure 11.11 Preparing a liquid sample for infrared spectroscopy by the thin-film technique.

NaCl plates, then the plates are mounted in a holder and placed in the spectrophotometer. Figure 11.11 illustrates this technique.

Few problems are encountered in thin-film sampling, and these are easily solved. *Too much liquid* between the plates gives rise to a spectrum in which many of the peaks are too strong—in the 0–10% transmission range, or even "off the paper." Also, leakage around the edges of the plates can contaminate the plate holder. To remove the excess liquid, wipe part of the sample off the plates (gently) with a dry tissue. Then, rerun the spectrum.

Too little sample between the plates results in a spectrum with weak peaks. More sample may be added to the plates and the spectrum rerun. If a spectrum shows acceptably strong peaks at the start, followed by progressively weaker peaks, the sample may be evaporating. One solution to this problem is to stop the scanning during the run, put more sample between the plates, then continue the scan. However, a preferred technique for sampling a volatile liquid is as a solution.

Solid Samples

A solid sample must be made transparent to infrared radiation before its spectrum is run. Three ways to do this are to include the sample in (1) a mull; (2) a solution; or (3) a KBr pellet.

1) A **mull** is prepared by grinding the solid sample with an inert carrier (petroleum jelly or mineral oil) to the consistency of toothpase or thick gravy. The mull is then sandwiched between NaCl plates and run as a liquid thin film.

The older (and less satisfactory) technique for preparing a mull is to grind the sample with mineral oil in an agate mortar. It is important to grind the sample very thoroughly. Even moderately large solid particles in the thin film cause reflection and scattering of the infrared radiation.

A proper balance of mineral oil to sample is surprisingly difficult to achieve and is best determined experimentally. Begin by grinding together a

1 : 1 mixture of sample and mineral oil (10–20 mg of solid with 2–3 drops of mineral oil). Use the spectrum to show you if more sample or more mineral oil is needed. If the ratio of sample to mineral oil is correct, the strongest sample (not mineral oil) peaks will dip to about 40–50% transmission.

A better way to prepare a mull is to use the ground-glass joint of a flask and a stopper for the grinding. Place a little petroleum jelly (such as Vaseline) on the inside of the joint and sprinkle 0.2–0.3 g of the solid on the surface of the jelly. Grind the sample by twisting the ground-glass stopper and pressing the stopper and joint together. Occasionally, remove the stopper, scrape up the material that has oozed out with a metal spatula, return it to the center of the joint, then continue grinding. The entire procedure should take no more than three minutes. (Mineral oil is not viscous enough to be used in this way.)

Mineral oil and petroleum jelly are mixtures of hydrocarbons, and consequently show hydrocarbon absorption bands in the infrared spectrum. Before running a mull spectrum, first run the spectrum of a thin film of mineral oil or jelly alone for reference. By doing this, you will be aware which absorption bands in your sample spectrum arise from the carrier. Some workers place a thin film of mineral oil or jelly in the reference beam of the spectrophotometer to block out its absorption from the sample's spectrum. (The instrument automatically subtracts absorption of the reference beam from absorption of the sample beam.) This procedure produces a cleaner-looking spectrum, but decreases the sensitivity in the CH and C—C regions of the spectrum.

After running a mull spectrum, clean the NaCl plates gently with dry solvent and a tissue before returning them to the dessicator.

2) A **solution spectrum** is obtained by dissolving the sample in a suitable solvent and placing the solution in an infrared **solution cell** (Figure 11.12). A matched cell filled with solvent alone is placed in the reference beam to block solvent absorption bands. Because the solvent bands are not always entirely

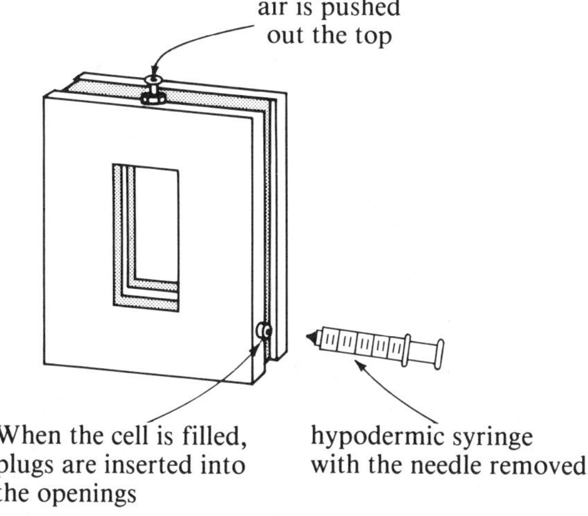

Figure 11.12 Filling an infrared solution cell.

blocked out, you should be aware of the solvent's absorption frequencies. As with mineral oil or petroleum jelly, placing solvent in the reference beam decreases the sensitivity of the spectrum in the regions where the solvent absorbs radiation.

The solvent chosen must dissolve the sample and must not contain any functional groups that would interfere with the spectrum of the sample. Typical infrared solvents are CCl_4, $CHCl_3$, and CS_2. Of these, CCl_4 is the most useful because its only absorption is at 700–800 cm^{-1} (12–14 μm), at the extreme right of the spectrum. The other solvents show some absorption in the more useful regions of the infrared spectrum, but one of these solvents may be necessary if the sample is insoluble in CCl_4. Commercial CCl_4 and $CHCl_3$ contain ethanol as a stabilizer and must be freshly distilled. In addition, CCl_4, $CHCl_3$, and CS_2 are all toxic. Carbon disulfide presents the added hazard of an extremely low ignition temperature. A hot plate, a steam bath, or even a radiator pipe can ignite CS_2 vapors. Safe handling procedures, such as transferring these solvents only in a fume hood, should be observed.

After carrying out a solution spectrum, flush the cell thoroughly with the solvent. Dry the cell by placing a syringe barrel filled with dessicant in one port of the cell and drawing air through the cell with a second syringe attached to the other port. Then return the cell to the dessicator.

3) A **KBr pellet**, or **wafer**, is obtained by putting great pressure on a mixture of finely ground sample (dry!) and specially dried potassium bromide until a clear or translucent pellet is formed. Either a hydraulic press or a special die, tightened by wrenches, must be used to prepare the pellet. Although the KBr pellet can produce the best solid-sample spectrum, the equipment is not available in all undergraduate laboratories. If your laboratory uses KBr pellets, your instructor will show you how to prepare them.

Instrument Adjustment and Calibration

Almost all infrared spectrophotometers require adjustments of the wavelength selector, recorder paper, and pen position before a spectrum can be run. How these adjustments are made depends upon the individual instrument.

For careful work, a **background spectrum** and a **calibration spectrum** should be carried out. A background spectrum—that is, a spectrum of the empty plates or cells—shows the level of electronic noise and alerts you to contamination of the plates or cells. A calibration spectrum is the spectrum of a polystyrene film provided by the instrument manufacturer. The film is mounted, ready for use, in a cardboard holder. The spectrum of polystyrene, with marked absorption positions, is printed on the cardboard. To calibrate the instrument, simply run the spectrum of the polystyrene film and note deviations, if any, of absorption positions from those on the card. If the deviations are large, notify your instructor that the instrument needs adjusting.

Experiment 11.1 consists of running infrared spectra of two knowns and three unknowns. Your instructor will tell you which parts of this experiment to do and will demonstrate the use of the instrument.

Experimental

EQUIPMENT:

dropper or disposable pipet
infrared spectrophotometer
NaCl plates and holder
pair of solvent cells, if solution spectra will be run
round-bottom flask with ground-glass stopper (or mortar and pestle, preferably agate)

CHEMICALS:

benzoic acid, 0.5 g
1-butanol, a few drops
petroleum jelly (or mineral oil) for mulls; KBr for pellets; or $CHCl_3$ and CCl_4 for solutions

TIME REQUIRED: 2–3 hours, depending on the spectrophotometer

›››› ***SAFETY NOTE*** Chloroform and carbon tetrachloride are volatile and toxic. Use the fume hood.

PROCEDURE

A. Infrared Spectra of 1-Butanol and Benzoic Acid

Obtain samples of 1-butanol and benzoic acid. Run the infrared spectrum of 1-butanol as a thin film (see Experimental Note). Then, run the infrared spectrum of benzoic acid in chloroform solution, unless your instructor requests that you use another technique. The optimum concentration of benzoic acid solution depends upon the path length of the solvent cell. Check with your instructor; otherwise, start with a 10% solution and run a trial spectrum. Depending on the quality of the trial spectrum, add either benzoic acid or chloroform to the solution to bring it to optimum concentration.

The spectra of 1-butanol and benzoic acid are shown in Figure 11.13. The spectra you obtain should appear very similar to these. Label the spectra with your name and laboratory notebook page; the date; the name or formula of the compound (or the unknown number); and the sampling technique (thin film, 10% $CHCl_3$, KBr pellet, etc.). If more than one spectrophotometer is being run by your class, identify which one you used.

B. Infrared Spectrum of an Unknown Liquid Sample

Obtain a liquid unknown from your instructor and run a thin-film infrared spectrum (see Experimental Note). The compound you receive will be one of the following four compounds. Using the infrared spectrum as your only source of information, identify the compound.

1) cyclohexanol *2*) methyl benzoate
3) cyclohexylamine *4*) cyclohexanone

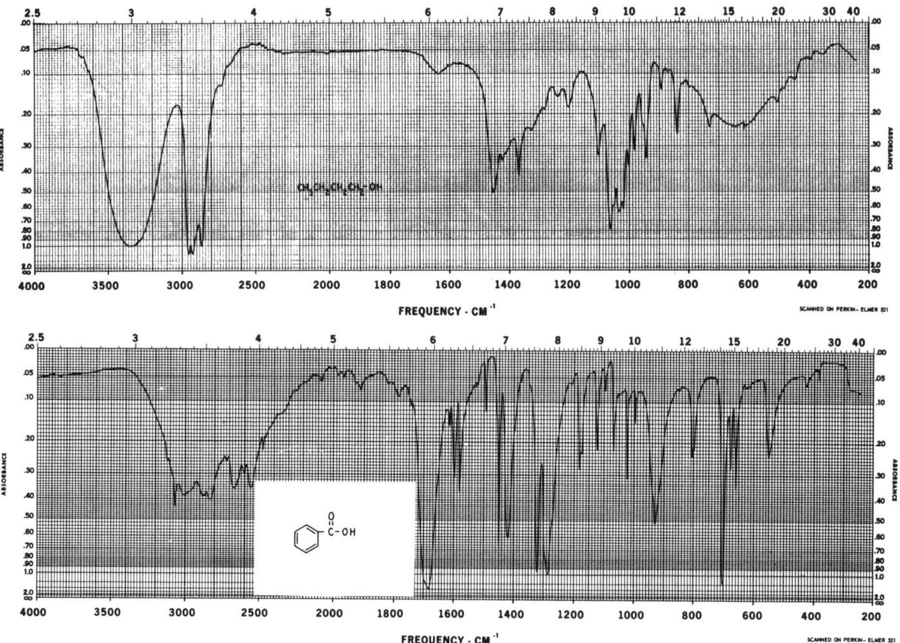

Figure 11.13 Infrared spectra of 1-butanol, $CH_3(CH_2)_3OH$ (upper spectrum), and benzoic acid, $C_6H_5CO_2H$ (lower spectrum). © *Sadtler Research Laboratories, Division of Bio-Rad Laboratories, Inc.*

C. Infrared Spectrum of an Unknown Solid Sample

Obtain a solid sample from your instructor and run the spectrum as a mull, a KBr pellet, or a CCl_4 solution. The compound you receive will be one of the following four compounds. Identify the compound from its infrared spectrum.

1) benzophenone, $C_6H_5\overset{O}{\overset{\|}{C}}C_6H_5$
2) biphenyl, $C_6H_5{-}C_6H_5$
3) diphenylmethanol, $(C_6H_5)_2CHOH$
4) tetradecylamine, $CH_3(CH_2)_{12}CH_2NH_2$

D. Identification of an Unknown Sample

Obtain an unknown compound from your instructor and run its infrared spectrum as a thin film, mull, KBr pellet, or solution. Your compound will be one of the 40 compounds in the following list:

1) octane
2) 3,3-dimethylpentane
3) 1-hexene
4) 2-octene

5) 1-pentyne
6) ethylbenzene
7) 5-phenyl-1-pentyne
8) phenylethyne
9) *t*-butyl chloride
10) 2-methyl-1-propylamine
11) *N*-methyl-1-butylamine
12) 1,2,3,6-tetrahydropyridine, 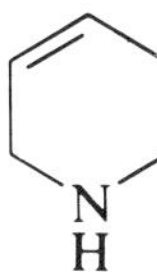

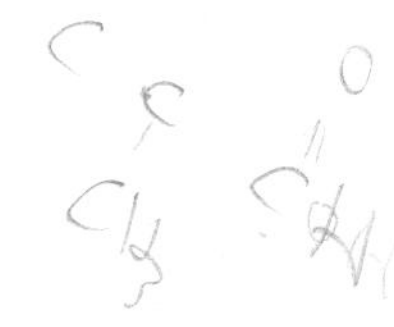

13) acetonitrile, CH_3CN
14) valeronitrile, $CH_3(CH_2)_3CN$
15) phenylacetonitrile, $C_6H_5CH_2CN$
16) di-*n*-propyl ether
17) phenyl *n*-propyl ether
18) 1-heptanol
19) hydracrylonitrile, $HOCH_2CH_2CN$
20) 2-buten-1-ol
21) 2-propyn-1-ol
22) *p*-cresol, $CH_3O-C_6H_4-OH$
23) 2-hexanone
24) 3-hydroxy-3-methyl-2-butanone
25) 1-phenyl-2-butanone
26) benzoylacetonitrile, $C_6H_5\overset{O}{\overset{\|}{C}}CH_2CN$
27) 4-methylbenzophenone, $CH_3-C_6H_4-\overset{O}{\overset{\|}{C}}C_6H_5$
28) 2-methylpropanal
29) 2-hexenal
30) 3,3-dimethylbutanoyl chloride, $(CH_3)_3CCH_2\overset{O}{\overset{\|}{C}}Cl$
31) 3-cyclohexen-1-carbonyl chloride, $C_6H_9-\overset{O}{\overset{\|}{C}}Cl$
32) hexanoic anhydride, $\left[CH_3(CH_2)_4\overset{O}{\overset{\|}{C}}\right]_2O$

33) naphthalic anhydride,

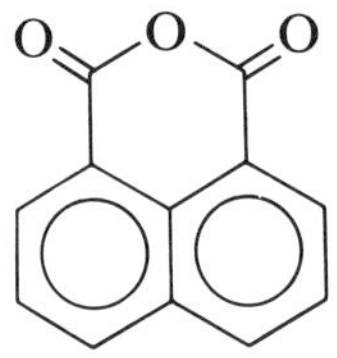

34) octanamide, $CH_3(CH_2)_6\overset{O}{\overset{\|}{C}}NH_2$

35) benzamide, $C_6H_5\overset{O}{\overset{\|}{C}}NH_2$

36) pivalic acid (2,2-dimethylpropanoic acid)

37) 4-pentenoic acid

38) *o*-phthalic acid, $C_6H_4(CO_2H)_2$

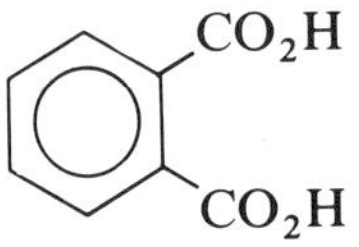

39) methyl lactate (methyl 2-hydroxypropanoate)

40) *n*-dodecyl acetate, $CH_3\overset{O}{\overset{\|}{C}}O(CH_2)_{11}CH_3$

EXPERIMENTAL NOTE

Sodium chloride plates and cells are relatively expensive because each plate is cut from a single giant crystal and then polished. The plates and cells are also fragile and sensitive to moisture. Atmospheric moisture, as well as moisture in a sample, causes the polished plate surfaces to become pitted and fogged. Fogged plates scatter and reflect infrared radiation instead of transmitting it efficiently; poor-quality spectra are the result. Fogged plates may also retain traces of compounds from previous runs, giving rise to false peaks. If it is necessary to use fogged plates, be sure to clean them carefully, then run a blank spectrum of the plates without sample before running the spectrum of your sample. By doing this, you can determine if your spectrum will contain extraneous peaks.

To prevent fogging, use only scrupulously dried samples for spectral work. Handle the plates *only by their edges*, never by their flat surfaces, because moisture from your fingers will leave fingerprints. NaCl plates and solution cells must be cleaned with a dry solvent such as CH_2Cl_2 after use and stored in a dessicator.

Fogged thin-film plates can be polished by rubbing them with a circular motion on an ethanol–water-saturated paper towel laid on a hard surface. Check with your instructor before you attempt to polish a plate!

Problems

11.8 A student attempts to run a spectrum of a mineral oil mull. The student adjusts the base line as high as it will go, but cannot get it above 40% *T*. Suggest at least two possiible reasons for the student's problem.

11.9 A student carries out an infrared spectrum of a solid compound in CCl_4 solution. The student is surprised when the pen leaves the base line at 85% *T* and rises to the *top* of the paper between 725–800 cm^{-1} (12–14 μm), then returns to the base line. What caused this "reverse" peak?

11.10 Unknowns (36), (37), and (38) in Experiment 11.1D are all carboxylic acids. How would you differentiate them?

11.11 The infrared spectrum (thin-film) of a compound with the molecular formula C_7H_5N shows weak absorption at 3100 cm^{-1} (3.2 μm), moderately strong absorption at 2230 cm^{-1} (4.5 μm), and three peaks between 1400–1500 cm^{-1} (6.7–7.1 μm) as the principal absorption. Suggest a structure for this compound.

CHAPTER 12

Nuclear Magnetic Resonance Spectroscopy

Nuclear magnetic resonance (nmr) spectroscopy is based on the absorption of radio waves by certain nuclei in molecules exposed to a strong magnetic field. Here we will briefly summarize the theory of nmr and the interpretation of proton nmr spectra. We suggest that you supplement your study of nmr spectroscopy by consulting your lecture text or one of the references on page 228.

12.1 The Nmr Spectrum

Protons, or hydrogen nuclei, in an organic molecule have spin and thus have small magnetic moments. When a sample is placed in an applied magnetic field (H_0), the magnetic moments of the protons become aligned either parallel or antiparallel to the field.

parallel antiparallel

When energy of the proper frequency, for a particular applied magnetic field, is supplied to the sample, some of the protons in the parallel spin state absorb energy and "flip" to the higher-energy antiparallel spin state. In an nmr spectrometer, this absorption of energy is detected, amplified, and recorded.

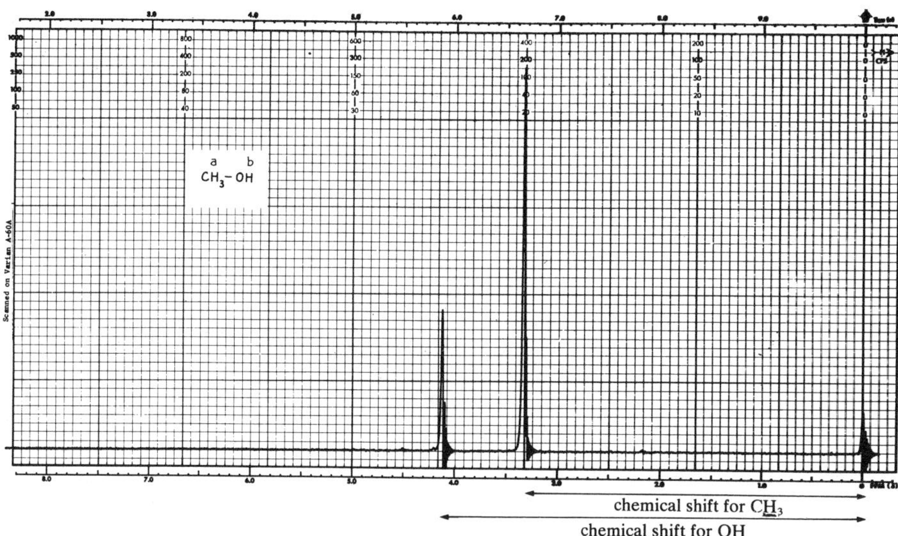

Figure 12.1 Nmr spectrum of methanol, showing the chemical shifts (solvent, CCl_4). The fluctuations just to the right of each signal are called ***ringing*** and are always observed if the instrument is properly adjusted, or "tuned." Ringing arises from distortion of the magnetic field when the sample is scanned. © Sadtler Research Laboratories, Division of Bio-Rad Laboratories, Inc.

The application of a magnetic field to molecules induces small magnetic fields within the molecules themselves. These induced fields vary throughout the different parts of a molecule because of differences in structure. Because of these induced molecular magnetic fields, the magnetic environments of protons within the molecules are not all the same; some protons absorb energy of slightly higher or lower frequency than other protons at a particular applied H_0. It is this property that gives rise to the nmr spectrum.

Depending on the type of instrument, either the magnetic field or the applied radiofrequency is held constant, while the other is varied; both types of instruments give identical spectra. A typical nmr spectrometer provides a field strength of 14,092 gauss. At this field strength, the frequency of radio waves that causes flips is about 60 MHz, where 1 Hertz (Hz) = 1 cycle per second and 1 megaHertz (MHz) = 1×10^6 Hz.

Figure 12.1 is the nmr spectrum of methanol. The vertical scale is **intensity of absorption**; therefore, unlike an infrared spectrum, the base line is at the bottom of the paper and positions of absorption are observed as peaks, not dips. The horizontal scale is in **δ values,*** which are *parts per million (ppm) of the applied radiofrequency*. At 60 MHz, a δ value of 1.0 ppm equals 60 Hz.

The spectrum of methanol shows three absorption peaks. The small peak at the far right arises from *tetramethylsilane* (TMS), a compound added directly to an nmr sample and used as an internal standard. The TMS peak is always set

* The *tau* (τ) *scale*, where $\tau = 10 - \delta$, is sometimes used for expressing ppm.

at $\delta = 0$ ppm by the operator of the instrument just before the spectrum is run. The other two peaks in the spectrum arise from methanol.

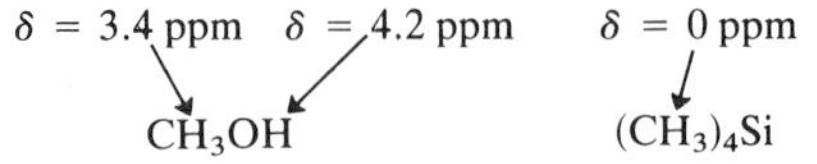

$\delta = 3.4$ ppm $\delta = 4.2$ ppm $\delta = 0$ ppm

CH_3OH $(CH_3)_4Si$

The difference (in ppm) between the TMS peak and the signal for a particular proton is called the **chemical shift** for that proton. Because TMS is set at zero, the chemical shift may be read directly from the scale of δ values at the bottom of the recorder paper. The two types of protons in methanol exhibit chemical shifts of $\delta = 3.4$ and 4.2 ppm. Chemical shifts for a variety of types of protons are listed inside the back cover of this book.

12.2 Interpreting Nmr Spectra

A. Chemical Shifts

The inductive effect plays a role in the positions of absorption in nmr spectroscopy. A *more shielded proton* is one surrounded by a relatively greater electron density than another proton. A shielded proton absorbs *upfield* (to the right, closer to TMS) in the nmr spectrum. A less-shielded, or *deshielded,* proton is one with a relatively lesser electron density and absorbs *downfield* (to the left) in the nmr spectrum.

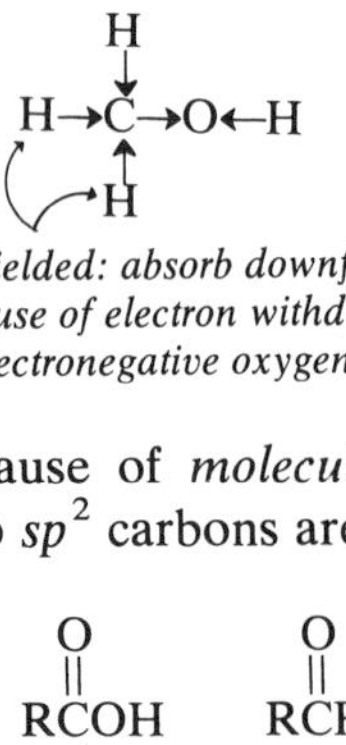

H
H→C→O←H
H

deshielded: absorb downfield because of electron withdrawal by electronegative oxygen atom

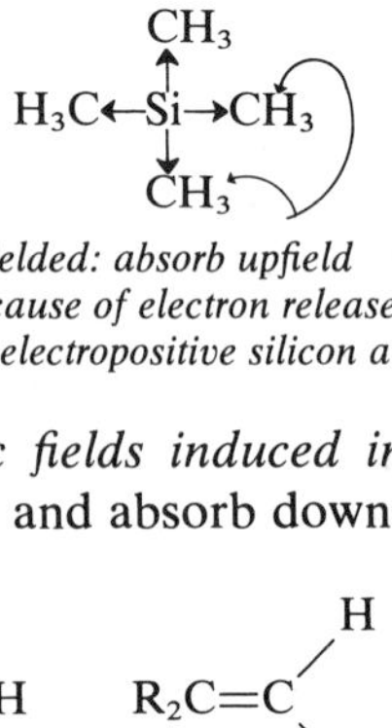

CH_3
H_3C←Si→CH_3
CH_3

shielded: absorb upfield because of electron release by electropositive silicon atom

Because of *molecular magnetic fields induced in the pi bonds,* protons attached to sp^2 carbons are deshielded and absorb downfield from alkyl protons.

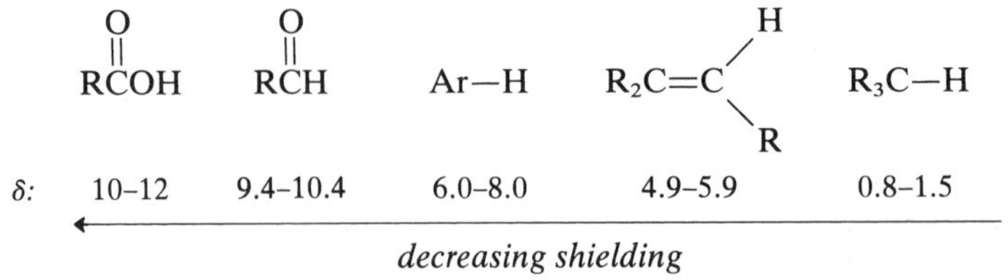

	RC(=O)OH	RC(=O)H	Ar—H	R_2C=C(H)(R)	R_3C—H
δ:	10–12	9.4–10.4	6.0–8.0	4.9–5.9	0.8–1.5

← *decreasing shielding*

A proton in an aldehyde or carboxyl group absorbs far downfield, outside the usual instrument scan, because of the combination of the effects of the nearby pi bond and the electron-withdrawing oxygen. The signals for these protons are observed in an **offset scan**, an extension of the instrument scan placed above the rest of the spectrum. Figure 12.2 shows the nmr spectrum of an aldehyde with the offset scan.

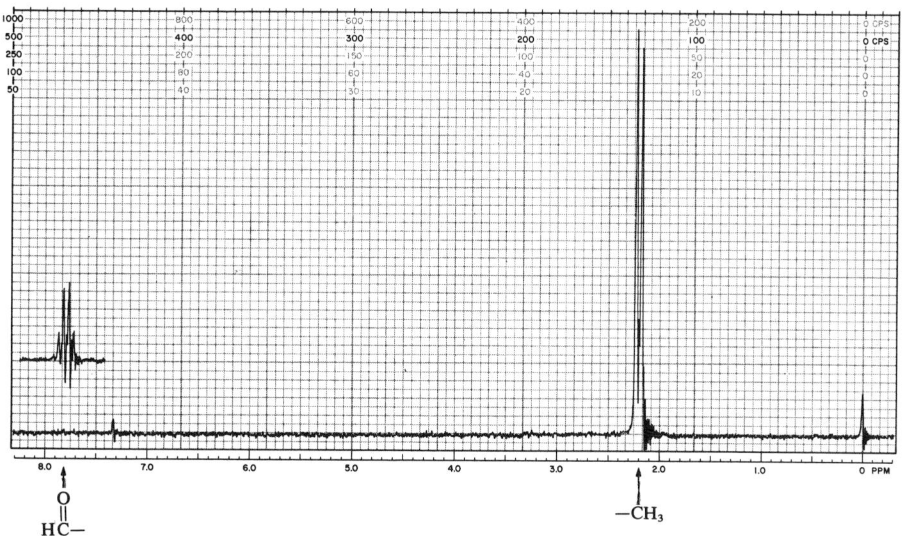

Figure 12.2 *Nmr spectrum of acetaldehyde, CH_3CHO, showing the offset scan for the aldehyde proton. From NMR Spectra Catalog, compiled by N. S. Bhacca, L. F. Johnson, and J. N. Shoolery, Varian Associates, 1962.*

B. Areas Under the Peaks

Protons that are in the same magnetic environment in a molecule have the same chemical shift in an nmr spectrum. Such protons are said to be *magnetically equivalent protons*. Protons that are in different magnetic environments have different chemical shifts and are said to be *nonequivalent protons*. Equivalent protons in nmr spectroscopy are generally the same as chemically equivalent protons.

three equivalent protons (but nonequivalent to CH_2) → CH_3CH_2Cl ← *two equivalent protons (but nonequivalent to CH_3)*

```
Cl        H ← cis to Cl
  \      /
   C = C                three nonequivalent protons
  /      \
H         H ← trans to Cl
```

The **areas under the signals** in an nmr spectrum are proportional to the number of equivalent protons giving rise to that signal.

	CH_3OH	CH_3CH_2Cl	$ClHC{=}CH_2$	$CH_3CH_2CH_3$
area ratio:	3:1	3:2	1:1:1	6:2, or 3:1

A typical nmr spectrometer is equipped with an **integrator**, which gives a signal that shows the relative areas under the peaks in a spectrum. The integration tracing appears as a series of steps superimposed upon the nmr spectrum; the height of the step over each absorption peak is proportional to the area under that peak. From the relative heights of the steps of the integration curve, the relative areas under the peaks may be determined. For example, assume that the heights of the steps of an integration curve were measured with a ruler and found to be 33 mm, 100 mm, and 50 mm. For determining the relative numbers of equivalent protons, these values are converted to ratios of small whole numbers as follows:

1) Divide each height by the smallest of the heights:

$$\frac{33}{33} = 1.00 \qquad \frac{100}{33} = 3.03 \qquad \frac{50}{33} = 1.51$$

2) Multiply the quotients by the integer that will convert them to the smallest whole numbers possible (rounding off, if necessary). In this case, multiplying by 2 gives the ratio of 2.00 : 6.06 : 3.02 or (rounded) 2 : 6 : 3.

3) The numbers of protons giving rise to the three nmr signals are in the ratio of 2 : 6 : 3. The actual numbers of protons of each type in the molecule are therefore 2, 6, and 3; 4, 12, and 6; or some other multiple of 2, 6, and 3.

C. Spin–Spin Splitting

The nmr signal for a proton (or group of equivalent protons) is split if the proton "sees" neighboring protons on adjacent carbons that are nonequivalent to it.

equivalent, not split CH_3CH_3

these protons split the signal for the CH_2 protons CH_3CH_2Cl *these protons split the signal for the CH_3 protons*

The splitting of signals is called **spin–spin splitting**, and protons splitting the signals of one another are called **coupled protons**. (Spin–spin splitting arises from the parallel and antiparallel spin states of the neighboring protons.) Protons that have the same chemical shift do not split the signals of each other. Only neighboring protons that have *different chemical shifts* cause splitting.

For many compounds, we can predict the number of spin–spin splitting peaks in the nmr absorption of a particular proton (or a group of equivalent protons) by counting *the number* (n) *of neighboring protons nonequivalent to the proton in question and adding 1*. This is called the $\boldsymbol{n + 1}$ **rule**. Some different types of splitting patterns that follow the $n + 1$ rule are listed in Table 12.1.

two neighboring H's: split into 2 + 1, or 3, peaks

CH_3CH_2Cl

three neighboring H's: split into 3 + 1, or 4, peaks

split into two H_b H_a *split into two*

H_3C–(benzene ring)–OCH_3

not split (H₃C) *not split* (OCH_3)

H_b H_a

In some cases, the magnetic environments of chemically nonequivalent protons are so similar that the protons exhibit the same chemical shift. In this

Table 12.1 *Some simple spin–spin splitting patterns in nmr spectra*

Partial structure[a]	Number of neighboring nonequivalent H's	Signal appearance	Signal name
—C$\underline{H}$	0		singlet
>CHC$\underline{H}$	1		doublet
—CH_2C$\underline{H}$	2		triplet
CH_3C$\underline{H}$	3		quartet
$(CH_3)_2$C$\underline{H}$	6		septet

[a] In each example, only the signal for the underlined proton is shown.

case, no splitting is observed. For example, toluene ($C_6H_5CH_3$) has four groups of chemically nonequivalent protons, yet the nmr spectrum shows only two absorption peaks (one for the CH_3 protons and one for the ring protons).

D. Coupling Constants

The separation between any two peaks of a split signal is called the **coupling constant *J*** and varies with the environment of the protons and their geometric relationship to each other. The symbol J_{ab} means the coupling constant for H_a split by H_b or for H_b split by H_a. For any pair of coupled protons showing doublets, the J value is the *same* in each of the doublets.

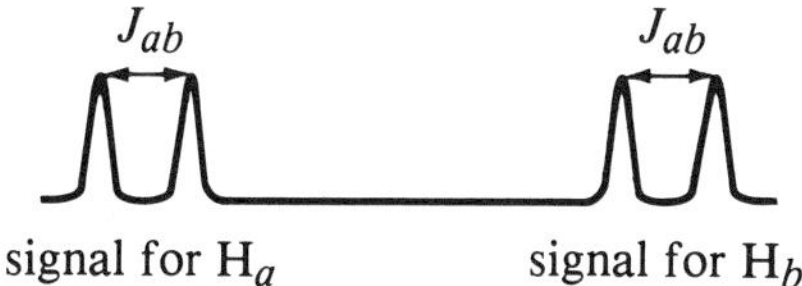

If the signal for a proton is split into a triplet, the distance between two adjacent peaks in the triplet is J_{ab}; thus, the width of the triplet is $2J_{ab}$. Similarly, the width of a quartet is $3J_{ab}$, and so forth.

J_{ab} J_{ab} J_{ab}

signal for H_a signal for H_b

Table 12.2 Selected coupling constants

Partial structure	Coupling constant, J_{ab} (Hz)
$>CH_a-CH_b<$	6–8
$>C=C(H_a)(H_b)$	0–3.5
H_a C=C H_b (trans)	11–18
H_a C=C H_b (cis)	6–14
$>CH_a-C(=O)H_b$	1–3
H_a, H_b (benzene ring)	7–10

Because J values arise from internal phenomena, and are thus independent of the strength of H_0, they are reported in Hz and not in δ values. The top of the nmr spectrum shows a scale in Hz for easy computation of J values. Table 12.2 lists some typical coupling constants.

Coupling constants allow us to determine which protons are coupled with one another in a complex spectrum. If the J values are the same, the protons may be coupled. If the J values are *different*, the protons in question are not splitting the signals of one another, but are being split by some other proton.

E. Complexities in Nmr Spectra

Leaning. If the chemical shifts of coupled protons are fairly close, unsymmetrical splitting patterns (such as doublets or triplets) are observed. The signals seem to "lean" toward each other, with the inner peaks larger than the outer peaks.

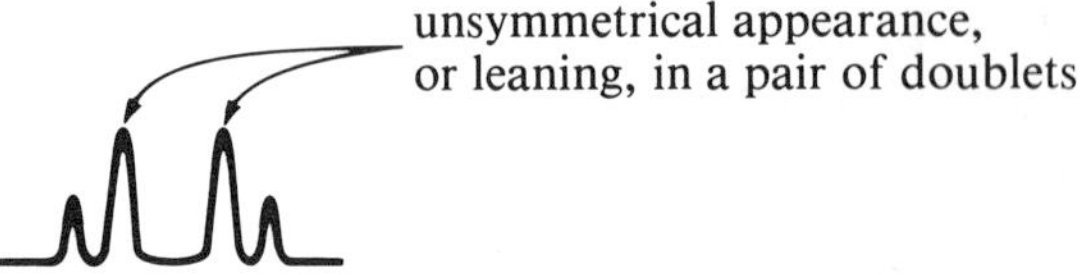

Protons split by more than one type of neighboring proton usually do not obey the $n + 1$ rule.

split by two types of neighboring H's

$CH_3CH_2CH_2Cl$

$$\begin{array}{ccc} H & & H \\ & C{=}C & \\ H & & C_6H_5 \end{array}$$

The nmr spectrum of styrene is shown in Figure 12.3. The three alkenyl (vinyl) protons give rise to twelve peaks, not six as would be predicted by the $n + 1$ rule. Eight to twelve peaks are typical of $-CH{=}CH_2$ groups.

Chemical exchange. On the basis of the preceding discussions, one would expect the nmr spectrum of methanol to show a doublet (for the CH_3 protons) and a quartet (for the OH proton). If the spectrum of methanol is determined at a very low temperature (−40°) or in specially prepared solvents such as CCl_4 stored over Na_2CO_3, this splitting is observed. However, if the spectrum is run at room temperature, only two singlets are observed, as can be seen in Figure 12.1, page 221.

The reason for this behavior of methanol is that alcohol molecules (and also amines) undergo rapid reaction with each other in the presence of a trace of acid, exchanging OH (or NH) protons in a process called **chemical exchange**. This exchange is so rapid that neighboring protons cannot distinguish any differences in spin states.

$$CH_3OH + H'^{+} \rightleftarrows CH_3\overset{+}{O}H(H') \rightleftarrows CH_3OH' + H^{+}$$

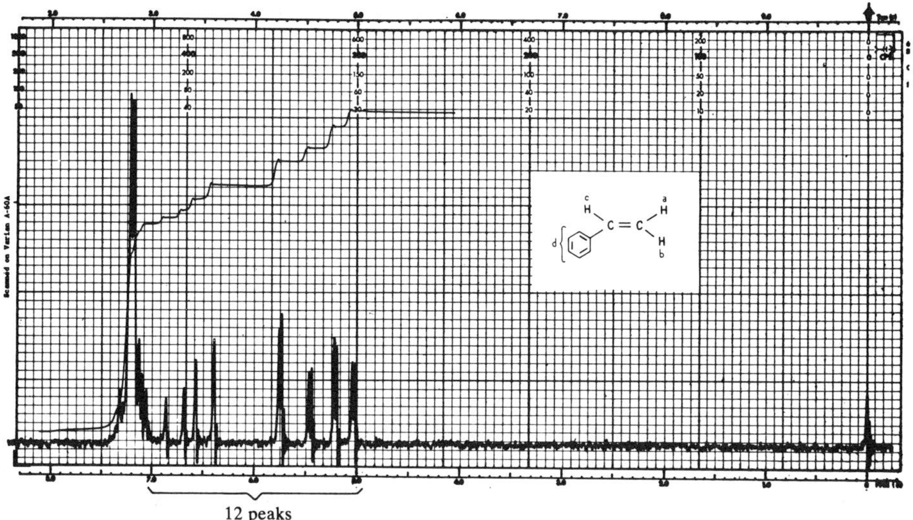

Figure 12.3 Nmr spectrum of styrene, $C_6H_5CH{=}CH_2$. © Sadtler Research Laboratories, Division of Bio-Rad Laboratories, Inc.

Hydrogen bonding. Unlike most chemical shifts for CH protons, the chemical shifts for OH and NH protons are dependent on the physical environment. For example, these chemical shifts are temperature-dependent. Of more practical importance to most organic chemists is that the chemical shifts of OH and NH protons are *concentration-dependent* because of hydrogen bonding. In a nonhydrogen-bonding solvent (such as CCl_4) and at low concentrations (1% or less), OH proton absorption is observed at a δ value of around 0.5 ppm. At the more usual, higher concentrations, the absorption is observed in the 4–5.5 ppm region. The OH proton absorption can be shifted even farther downfield by hydrogen-bonding solvents.

12.3 Use of Deuterium in Proton Nmr Spectroscopy

Any nucleus with spin can absorb radio waves when it is subjected to a magnetic field. However, each type of nucleus has its own unique combination of magnetic field strength and radiofrequency required for resonance. The average resonance frequency of a proton in a magnetic field of 14,092 gauss is 60 MHz. By contrast, the resonance frequency for deuterium ($^{2}_{1}H$) at 14,092 gauss is only 2.3 MHz. Therefore, in an nmr spectrum obtained at 60 MHz, normal protons ($^{1}_{1}H$) absorb energy, but deuterium nuclei do not.

The fact that deuterium does not absorb energy in proton nmr spectra has several important consequences. Nmr solvents are often deuteriated so that they do not interfere with proton nmr spectra. Chloroform-d (deuteriochloroform, $CDCl_3$) is a common solvent for nmr work.

Another use of deuterium in nmr spectra is the substitution of a deuterium atom for a hydrogen atom in a compound to simplify an otherwise complex spectrum or to identify a particular proton. For example, the substitution of deuterium can be used to identify a peak arising from OH or NH protons. The sample is shaken with a few drops of D_2O, the NH or OH protons undergo chemical exchange, and the peak in the nmr spectrum that changes is thus identified as the OH or NH peak. In this case, a new peak will appear at a δ value of about 5 ppm because of the formation of DOH.

$$\underset{\text{old absorption}}{ROH} + D_2O \rightleftarrows ROD + \underset{\text{new absorption}}{DOH}$$

Suggested Readings

Dyer, J. R. *Applications of Absorption Spectroscopy of Organic Compounds.* Englewood Cliffs, N.J.: Prentice-Hall, 1965.

Jackman, L. M. and Sternhell, S. *Applications of Nuclear Magnetic Resonance in Organic Chemistry*, 2nd ed. New York: Pergamon Press, 1969.

Mathieson, D. W. *Nuclear Magnetic Resonance for Organic Chemists.* New York: Academic Press, 1967.

Paudler, W. W. *Nuclear Magnetic Resonance.* Boston: Allyn and Bacon, 1971.

Silverstein, R. M., Bassler, G. C. and Morrill, T. C. *Spectrometric Identification of Organic Compounds*, 3rd ed. New York: John Wiley and Sons, 1974.

Problems

12.1 Calculate the following conversions. [In (b)–(d), assume a standard 60-MHz nmr spectrometer.]

(a) 60 MHz to Hz

(b) 7.0 Hz to a δ value

(c) $\delta = 3.0$ ppm to Hz

(d) $\delta = 0.50$ ppm to Hz

12.2 (a) Give two reasons why TMS is a good internal standard.

(b) Would tetraethylsilane be a good internal standard? Explain.

12.3 List the following protons in order of increasing chemical shift (smallest shift first).

$Cl_2CHCH_2CH_2Cl$
(a) (b) (c)

12.4 Pyrrole shows two principal CH signals in its nmr spectrum: at $\delta = 6.5$ and 6.7 ppm. Is pyrrole an aromatic compound or a conjugated diene?

pyrrole

12.5 How many types of nonequivalent protons does each of the following compounds contain? (*Example:* CH_3CH_2Cl has two types, CH_3 and CH_2.)

(a) $(CH_3)_2CHCl$ (b) $CH_3CH(CH_2Cl)_2$

(c) C_6H_5–Cl (d) $(CH_3)_3COH$ (e) $CH_3\overset{O}{\overset{\|}{C}}CH_3$

12.6 In the preceding problem, determine the relative areas under the principal signals for each compound.

12.7 What would be the splitting pattern (singlet, doublet, etc.) observed for each of the following indicated protons?

$H\overset{O}{\overset{\|}{C}}OCH_2CH_3$ — (a) H, (b) CH_2, (c) CH_3

$(CH_3)_2CHOCH_3$ — (d) $(CH_3)_2$, (e) CH, (f) OCH_3

Cl–C_6H_4–CH_3 — (g), (h) ring protons; (i) CH_3

Cl–C_6H_4–Cl — (j), (k) ring protons

12.8 An nmr spectrum shows a singlet at $\delta = 7.3$ ppm, a triplet at 4.3, a triplet at 2.9, and a singlet at 2.0. The relative areas are 5, 2, 2, and 3, respectively. Which of the following compounds is compatible with this spectrum?

(a) H_3C–C_6H_4–CH_3

(b) $CH_3\overset{O}{\overset{\|}{C}}O$–$C_6H_4$–$CH_2CH_3$

(c) C_6H_5–$CH_2CH_2O\overset{O}{\overset{\|}{C}}CH_3$

(d) C_6H_5–$\overset{O}{\overset{\|}{C}}OCH_2CH_2CH_3$

EXPERIMENT 12.1

Nmr Spectra in Structure Determination

Discussion

In some laboratories, students are expected to operate the nmr spectrometer; in other laboratories, students submit prepared samples to a spectrometer operator. Your instructor will specify the procedure used in your laboratory and will instruct you in the use of the instrument if you are to obtain your own spectra.

An nmr spectrum is usually carried out with a solution of the sample in a suitable solvent. The solution, along with a small amount of TMS, is placed in the sample container, called an **nmr tube**, which is capped and placed in the instrument. The sample tube is rotated rapidly, spun by an air stream in the instrument. This spinning results in an averaged magnetic field throughout the entire sample.

A typical nmr tube holds 0.4–0.5 mL of solution. The concentration of sample is usually 10–30%. The concentration of TMS should be 1–2%; however, it is time-consuming to measure the TMS this accurately. Also, TMS is volatile; see the Experimental Note on page 235. For routine spectra, add just enough TMS to the solution so that its signal can be clearly discerned. (Add one drop to start with.) Your instructor may provide solvent that already contains the proper amount of TMS.

All solid material must be removed from the solution before it is placed in the nmr tube. Suspended or solid particles (particularly magnetic particles like iron dust from a syringe needle) cause broadening of the absorption peaks in the spectrum.

The best solvents for nmr spectra are those that contain no protons, such as CCl_4, CS_2, or $CDCl_3$. For water-soluble compounds, D_2O may be used; however, TMS is water-insoluble and a different internal standard must be used. Almost all deuteriated solvents contain traces of nondeuteriated solvent. Therefore, a background spectrum of the solvent should be run before the sample is dissolved in it. Deuteriated solvents are very expensive and must not be wasted. Solubility of a sample should be tested in the nondeuteriated analog, such as $CHCl_3$ or H_2O, before a deuteriated solvent is used.

All nmr solvents must be extremely pure. Reagent grade CCl_4, for example, is not suitable for nmr spectra because it contains ethanol added as a stabilizer. If there is any doubt about the suitability of a solvent for nmr work, run a background spectrum of the solvent plus a drop or two of TMS.

Information from an Nmr Spectrum

If an nmr spectrum shows well-spaced multiplets, you may be able to arrive at the structure of an unknown compound from its spectrum and a minimum amount of other information. On the other hand, if the spectrum shows extensive fine structure, your chances of deducing the structure from the nmr spectrum alone are probably slim.

In a spectrum with well-spaced multiplets:

1) Identify the different proton groups in the spectrum (alkyl, aryl, etc.), using their chemical shifts as a guide.

2) Determine which protons are coupled. (Measure the J values, if necessary, to be sure they match.)

3) From the relative areas under the peaks (determined from the integration line), calculate the relative ratios of the different types of protons.

4) Draw up a list of partial structures (such as CH_3CH_2-, $>CHCH<$) consistent with each of the singlets or multiplets, their chemical shifts, and their relative areas. In drawing up this list, consider *all* the information available (infrared spectrum, elemental analysis, etc.).

5) Consider the list of partial structures as a jigsaw puzzle. Fit the pieces together, trying various combinations. Your final solution must be consistent with *all* the data.

If the spectrum is too complex for complete interpretation, then simply glean as much information as you can from the spectrum. In general, you should attempt to identify the types of protons (alkyl, aryl, etc.). Calculate the ratios of the various groups of protons from the relative areas of the multiplets. Determine which types of protons are *absent* in the spectrum as well as those present. For example, if the spectrum shows no absorption for aryl protons, you can eliminate the phenyl group as part of the structure. Take what clues you can, and then turn your attention to other sources of information, such as the infrared spectrum, physical constants, and chemical information.

Experimental

EQUIPMENT:

nmr sample tube
nmr spectrometer
syringe

CHEMICALS:

carbon tetrachloride, 1–5 mL
isopropylbenzene (cumene), 0.1 g
tetramethylsilane (TMS), about 0.1 g
unknown samples, about 0.1 g each*

TIME REQUIRED: approximately 2–3 hours, depending on the instrument, its state of tune, and the experience of the operator

›››› ***SAFETY NOTE*** Carbon tetrachloride (CCl_4) is volatile and toxic. Always use this solvent in the fume hood.

* Suggested unknowns for part D are listed in the instructor's guide.

PROCEDURE

A. Nmr Spectrum of a Known Compound

Obtain a small sample of isopropylbenzene, and prepare about 2 mL of a 10% solution in carbon tetrachloride. Fill an nmr tube about one-third full with this solution. Using a syringe, add 1–2 drops of TMS to the tube (see Experimental Note). Invert the tube several times to mix the TMS with the solution. Because the caps provided for nmr tubes often leak, hold your finger (protected by a small piece of polyethylene sheeting or kitchen wrap) over the end of the tube while inverting it.

After the operation of your nmr spectrometer has been demonstrated, run the spectrum of the solution, making all instrument adjustments necessary to obtain a good spectrum. After the spectrum is completed, increase the amplitude and retrace the portion of the spectrum from 3–4 ppm to bring out the septet (7 peaks). Finally, integrate the spectrum. The spectrum you obtain should look similar to that shown in Figure 12.4.

B. Nmr Spectrum of an Unknown Compound

Obtain a small sample of an unknown from your instructor and run the nmr spectrum as described in part A. The unknown will be one of the following five compounds. Identify the structure, using the nmr spectrum as your only source of information.

1) 1-chloropropane
2) 1-chlorobutane
3) 2-chlorobutane
4) 1-chloro-2-methylpropane
5) 2-chloro-2-methylbutane

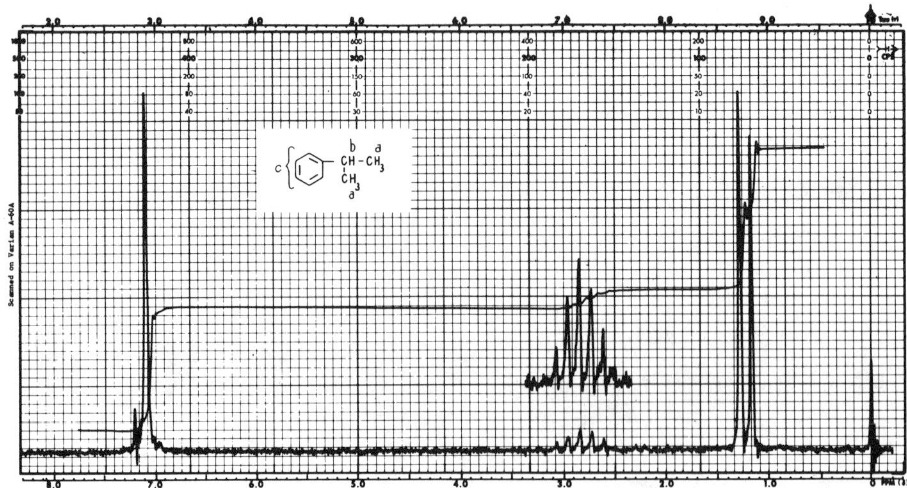

Figure 12.4 *The nmr spectrum of isopropylbenzene (cumene), $C_6H_5CH(CH_3)_2$ (solvent, CCl_4). © Sadtler Research Laboratories, Division of Bio-Rad Laboratories, Inc.*

C. Nmr Spectrum of an Unknown Compound

Obtain a small sample of an unknown from your instructor, and proceed as in part B. Your unknown will be one of the compounds in the following list:

1) 1,2-ethanediamine (ethylenediamine), $NH_2CH_2CH_2NH_2$

2) 2,2-dimethoxypropane, $(CH_3)_2C(OCH_3)_2$

3) trimethyl orthoformate, $(CH_3O)_3CH$

4) 1,4-dimethylbenzene (*p*-xylene)

5) 2-methyl-3-buten-2-ol

6) 3-methylphenol (*m*-cresol), H_3C —OH

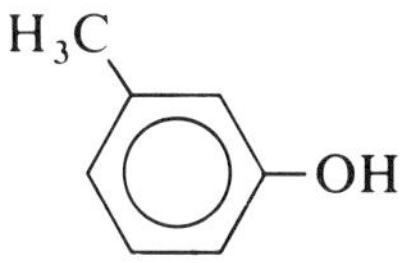

7) *o*-cresol, CH_3 —OH

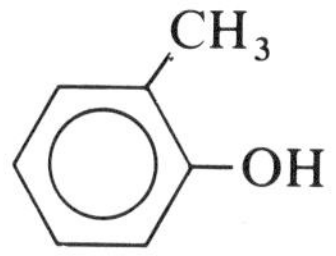

8) *p*-cresol, H_3C— —OH

9) 3-phenyl-2-propen-1-ol (cinnamyl alcohol) (probably *trans*)

10) diphenylmethanol (benzhydrol)

11) 1,2-diphenylethanol

12) 2,2-dichloroethanol

13) 2-chloroethanol

14) 1-chloro-4-methoxybenzene (*p*-chloroanisole)

15) tetrahydrofuran, O

16) diisopropyl ether

17) di-*n*-propyl ether

18) *m*-ethylaniline, CH_3CH_2 —NH_2

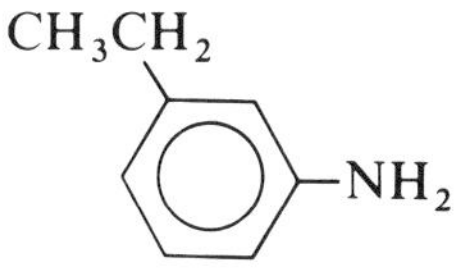

19) *o*-ethylaniline, CH_2CH_3 —NH_2

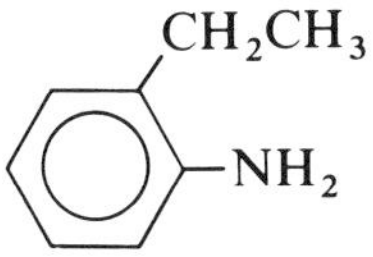

20) diisobutylamine, $(CH_3)_2CHCH_2NHCH_2CH(CH_3)_2$

21) 1-phenyl-1-ethylamine (α-phenylethylamine), $C_6H_5\overset{\overset{\displaystyle CH_3}{|}}{C}HNH_2$

22) 1-methyl-1-propylamine (*sec*-butylamine), $CH_3CH_2\underset{\underset{\displaystyle CH_3}{|}}{C}HNH_2$

23) *n*-butylamine, $CH_3CH_2CH_2CH_2NH_2$

24) 1-bromo-2-phenylethene (*β*-bromostyrene)

25) 1,4-dichloro-2-butene

26) 1,4-dichlorobutane

27) methylcyclopentane

28) 1-octene

29) naphthalene,

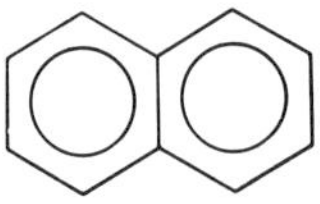

30) phenylethyne (phenylacetylene)

31) 1-phenyl-1-butanone (butyrophenone), $C_6H_5\overset{\overset{O}{\|}}{C}CH_2CH_2CH_3$

32) 1-phenyl-1-propanone (propiophenone), $C_6H_5\overset{\overset{O}{\|}}{C}CH_2CH_3$

33) 2-butanone, $CH_3\overset{\overset{O}{\|}}{C}CH_2CH_3$

34) 3-chloro-2-butanone

35) 4-heptanone, $CH_3CH_2CH_2\overset{\overset{O}{\|}}{C}CH_2CH_2CH_3$

36) 3-hexanone, $CH_3CH_2\overset{\overset{O}{\|}}{C}CH_2CH_2CH_3$

37) 2-hexanone, $CH_3\overset{\overset{O}{\|}}{C}CH_2CH_2CH_2CH_3$

38) *trans*-2-hexenal, $\underset{CH_3CH_2CH_2}{\overset{H}{}}\!\!\diagdown\!C{=}C\!\diagup\!\underset{H}{\overset{CHO}{}}$

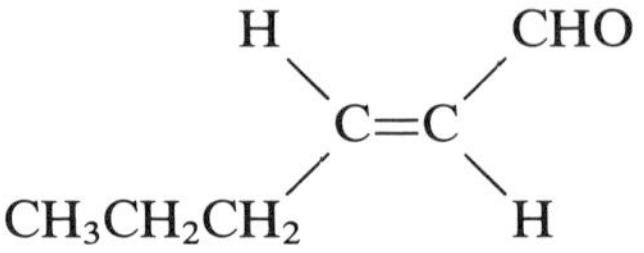

39) 2-methyl-1-phenyl-1-propanone (isobutyrophenone),
$C_6H_5\overset{\overset{O}{\|}}{C}CH(CH_3)_2$

40) 3-buten-2-one, $CH_3\overset{\overset{O}{\|}}{C}CH{=}CH_2$

D. Structure Determination Using Infrared and Nmr Spectra

Obtain a sample of an unknown from your instructor. Run the infrared and nmr spectra. Identify the compound using these two spectra as your only sources of information.

EXPERIMENTAL NOTE

TMS boils at 28° and is therefore quite volatile. If the signal for TMS in your spectrum is not clear, add one more drop to the nmr tube, mix, and rerun the spectrum. If you are using premixed solvent–TMS, keep the bottle tightly capped.

Too much TMS in the sample solution results in a massive signal that saturates the instrument. If the nmr tube is not spinning at the proper speed, a large TMS signal will also be accompanied by **spinning side bands**, which can be mistaken for sample signals. If your sample contains an excessive amount of TMS, prepare a fresh solution, add a smaller amount of TMS than before, and rerun the spectrum. Alternatively, set the nmr tube (uncapped) in the fume hood and allow the TMS to evaporate.

Problems

12.9 Sodium 3-(trimethylsilyl)propanesulfonate, $(CH_3)_3SiCH_2CH_2CH_2SO_3^-\ Na^+$, is a commonly used internal standard for nmr spectra of D_2O solutions. What would be the advantages and disadvantages of using this compound as an internal standard?

12.10 A student is running the nmr spectrum of a D_2O solution of a compound using acetone as an internal standard. How can the student relate the δ values of his spectrum to TMS?

12.11 The following types of protons (underlined) all show nmr absorption between δ = 10–16 ppm (offset scan). How would you use the infrared spectrum to identify the type of compound?

$RC(=O)\underline{H}$ $C_6H_5O\underline{H}$ $RCO_2\underline{H}$ $RC(=O)CH{=}C(O\underline{H})R$ (intramolecularly hydrogen-bonded enol)

12.12 How would you distinguish between the following pairs of compounds using nmr spectroscopy? (Include in your answer the expected splitting patterns and relative areas of the nmr absorption.)

(a) p-H_3C–C_6H_4–$C(=O)CH_3$ and m-H_3C–C_6H_4–$C(=O)CH_3$

(b) $(CH_3)_3CI$ and $(CH_3)_3CCl$

(c) H_3C–C_6H_4–C_6H_4–CH_3 and H_3C–C_6H_4–CH_3

(d) $CH_3CO_2CH_2CH_3$ and $CH_3CH_2CO_2CH_3$

»»› CHAPTER 13 ‹««

The Chemistry of Alkenes

Ninety percent of the organic chemicals produced in the United States today come from petroleum. Most of these chemicals are synthesized from *alkenes*, which are obtained as products of cracking and reforming reactions carried out at the refinery. The principal alkenes obtained in this way are *ethylene* (ethene), *propylene* (propene), and the *butylenes* (butenes).

Ethylene is the precursor of a number of industrial and consumer chemicals. Virtually all of its products, as well as the products from other alkenes, arise from the *addition of reagents to the double bond*. These addition products may then be subjected to further reaction, such as elimination.

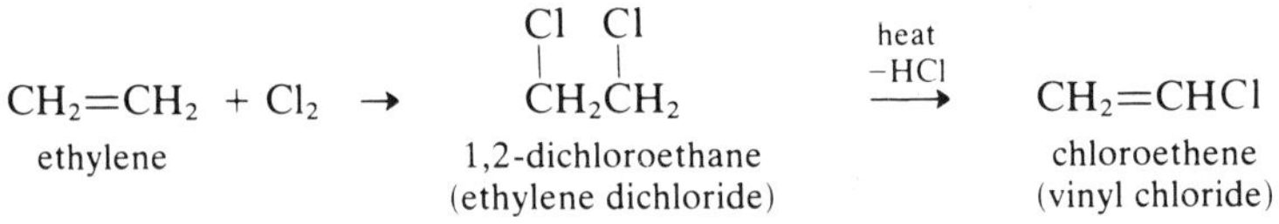

$$\underset{\text{ethylene}}{CH_2{=}CH_2} + Cl_2 \rightarrow \underset{\substack{\text{1,2-dichloroethane}\\\text{(ethylene dichloride)}}}{\overset{Cl\ \ Cl}{CH_2CH_2}} \xrightarrow[-HCl]{\text{heat}} \underset{\substack{\text{chloroethene}\\\text{(vinyl chloride)}}}{CH_2{=}CHCl}$$

Alkenes can be *oxidized* readily to a variety of products. For example, ethylene can be converted to any of the following compounds, depending on the oxidizing agent and reaction conditions.

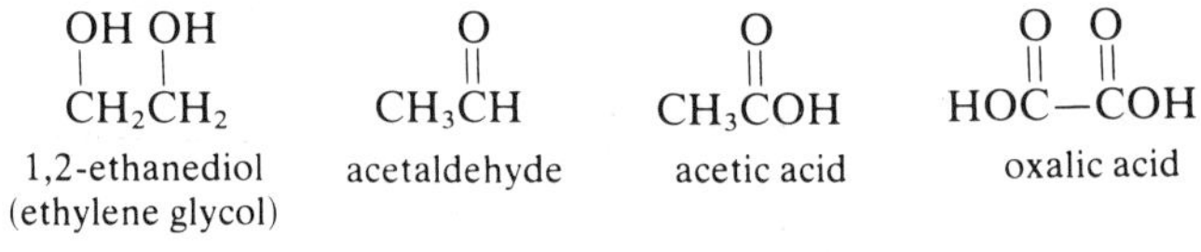

$$\underset{\substack{\text{1,2-ethanediol}\\\text{(ethylene glycol)}}}{\overset{OH\ OH}{CH_2CH_2}} \qquad \underset{\text{acetaldehyde}}{CH_3\overset{O}{\overset{\|}{C}}H} \qquad \underset{\text{acetic acid}}{CH_3\overset{O}{\overset{\|}{C}}OH} \qquad \underset{\text{oxalic acid}}{HO\overset{O}{\overset{\|}{C}}-\overset{O}{\overset{\|}{C}}OH}$$

Alkenes can also undergo addition reactions with other alkenes to yield higher-molecular-weight products. Two important reactions of this type are *Diels–Alder reactions*, which yield cyclohexenes and are used industrially to synthesize a number of useful products (insecticides, for example), and *polymerization reactions*, which yield polymers. Half the ethylene produced in the United States is converted to polyethylene, which is used for kitchen wrap, cleaners' bags, jars and bottles, and other inexpensive plastic items.

a Diels–Alder reaction:

$$CH_2{=}CH{-}CH{=}CH_2 + HC(CHO){=}CH_2 \xrightarrow{\text{heat}} \text{(cyclohexene-CHO)}$$

a polymerization reaction:

$$x\ CH_2{=}CH_2 \xrightarrow{\text{catalyst}} \text{⌇}{-}CH_2CH_2CH_2CH_2CH_2CH_2{-}\text{⌇} \quad \text{or} \quad {+}CH_2CH_2{+}_x$$

ethylene — polyethylene

13.1 Electrophilic Addition to Alkenes

Most addition reactions of alkenes are initiated by an **electrophile** (E^+), hence the term **electrophilic addition.** Experiment 13.1 is the acid-catalyzed reaction of cyclohexene with water to yield cyclohexanol.

$$\text{cyclohexene} + H_2O \xrightarrow{H_2SO_4} \text{cyclohexyl}{-}OH$$

cyclohexanol

An electrophilic addition reaction of an alkene usually occurs by a stepwise mechanism. The initial step is the addition of the electrophile (often H^+) to the pi bond. The intermediate formed by this addition is either a carbocation or some other electron-deficient structure that resembles a carbocation. The acid-catalyzed reaction of an alkene with water is a reaction that proceeds by way of a carbocation.

Step 1, formation of a carbocation:

$$\text{cyclohexene} + H_3O^+ \rightleftharpoons \text{cyclohexyl}^+ + H_2O$$

cyclohexyl cation

The second step is the combination of the carbocation with a nucleophile. When water is the nucleophile, the product of this step is a protonated alcohol, which is in equilibrium with the alcohol itself.

Step 2, reaction of the carbocation and water:

$$\text{cyclohexyl}^+ + :\ddot{O}H_2 \rightleftharpoons \text{cyclohexyl}{-}\overset{+}{\ddot{O}}H_2$$

a protonated alcohol

Step 3, an acid–base equilibrium:

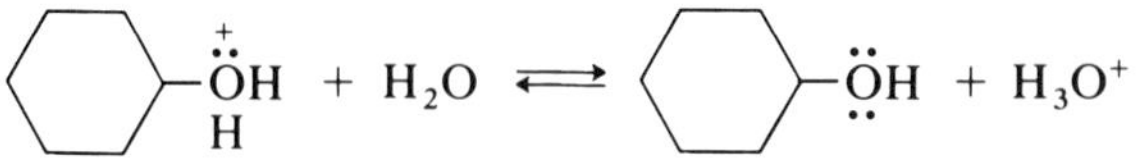

$$C_6H_{11}\text{–}\overset{+}{\ddot{O}}H_2 + H_2O \rightleftharpoons C_6H_{11}\text{–}\ddot{O}H + H_3O^+$$

cyclohexanol

The synthesis of cyclohexanol from cyclohexene is the reverse reaction of the dehydration of cyclohexanol (Experiment 10.3). In that experiment, the product cyclohexene was distilled from the reaction mixture as it was formed to prevent its reconversion to cyclohexanol. In Experiment 13.1, the cyclohexene is allowed to remain in contact with the H_2O–H_2SO_4 mixture to maximize the yield of cyclohexanol, and the product is removed from the reaction mixture by steam distillation.

13.2 Oxidation of Alkenes

A. Oxidation Products

Considering only the organic products, the common alkene oxidation reactions may be grouped into two major classifications: *those oxidations that result in double-bond cleavage* and *those that do not.* The actual products of an oxidation reaction are determined in part by the structure of the alkene and in part by the

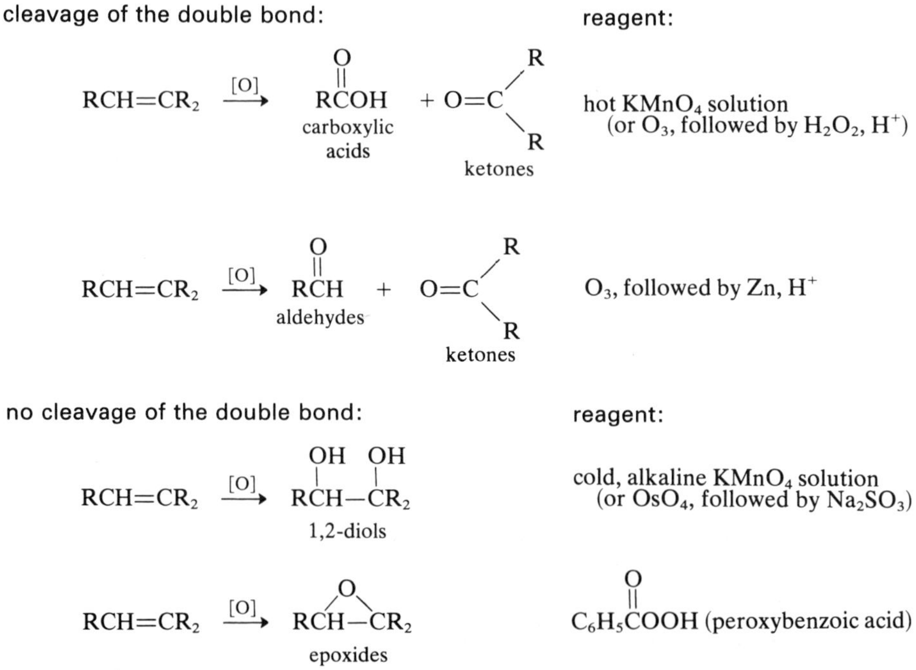

cleavage of the double bond: reagent:

$$RCH{=}CR_2 \xrightarrow{[O]} \underset{\text{carboxylic acids}}{RCOH} + \underset{\text{ketones}}{O{=}CR_2}$$ (RCOH with C=O)

hot $KMnO_4$ solution (or O_3, followed by H_2O_2, H^+)

$$RCH{=}CR_2 \xrightarrow{[O]} \underset{\text{aldehydes}}{RCH} + \underset{\text{ketones}}{O{=}CR_2}$$ (RCH with C=O)

O_3, followed by Zn, H^+

no cleavage of the double bond: reagent:

$$RCH{=}CR_2 \xrightarrow{[O]} \underset{\text{1,2-diols}}{RCH(OH){-}C(OH)R_2}$$

cold, alkaline $KMnO_4$ solution (or OsO_4, followed by Na_2SO_3)

$$RCH{=}CR_2 \xrightarrow{[O]} \underset{\text{epoxides}}{RCH{-}CR_2}$$ (with O bridging the two carbons)

C_6H_5COOH (C=O; peroxybenzoic acid)

Figure 13.1 Typical products from the laboratory oxidations of alkenes. In the cleavage reactions, note that an RCH= portion of an alkene is oxidized to RCHO or RCO_2H, while R_2C= is oxidized to $R_2C{=}O$.

oxidizing agent and reaction conditions. Figure 13.1 shows the types of products that can be obtained and the usual reagents used to obtain them.

In Experiment 13.2, cyclohexene is oxidized to *cis*-1,2-cyclohexanediol with cold, alkaline potassium permanganate ($KMnO_4$) solution. Because a permanganate oxidation can also result in cleavage of the carbon–carbon double bond, a by-product of this reaction is the anion of a diacid.

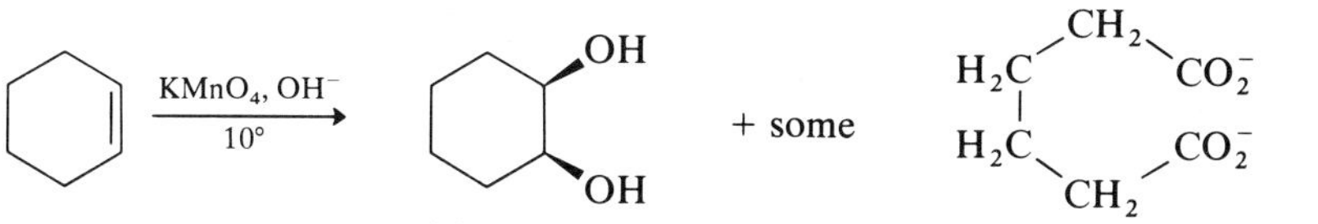

$KMnO_4$, OH^-, 10° ; + some

cis-1,2-cyclohexanediol

anion of 1,6-hexanedioic acid (adipic acid)

Because the result is the addition of two hydroxyl groups to the alkene, this reaction is sometimes referred to as a **hydroxylation reaction**. Because the hydroxylation of cyclohexene with cold, alkaline $KMnO_4$ solution yields only *cis*-1,2-cyclohexanediol (and not the *trans* isomer), this hydroxylation is called a ***cis*-addition**.

B. Mechanism of the Hydroxylation

The hydroxylation of cyclohexene with $KMnO_4$ to yield the *cis*-1,2-diol was first carried out by the Russian chemist Vladimir Markovnikov in 1878. Although this reaction has been known for over 100 years, its mechanism is still not completely understood. The reaction is believed to involve the initial addition of the permanganate ion to the pi bond to yield an unstable, cyclic manganate ester, which in turn is rapidly hydrolyzed to the *cis*-diol.

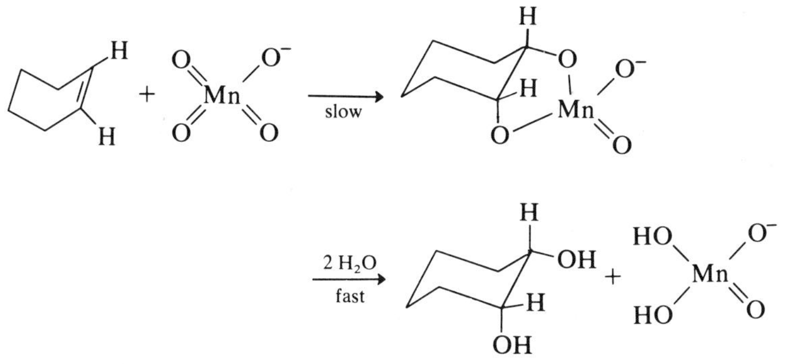

slow

$2\ H_2O$, fast

The fact that a *cis*-diol (and not *trans*) results from the permanganate hydroxylation supports the proposed mechanism. More support arises from experiments using oxygen-18. When $KMnO_4$ labeled with ^{18}O is used to hydroxylate an alkene, both oxygen atoms in the product diol are ^{18}O. This observation means that both oxygens of the diol must have come from the MnO_4^- ion and not from the water or hydroxide ions in the reaction mixture.

Although we can trace the steps of the alkene oxidation quite well, the reactions that the permanganate ion undergoes are not as well understood. From the balanced equation (page 248) we can see that two MnO_4^- ions are needed to oxidize three cyclohexene molecules. Therefore, the reaction must be more complex than a simple cyclic ester would suggest. In addition, Mn is at a +7 oxidation state in MnO_4^-; +5 in $H_2MnO_4^-$ (from the hydrolysis of the manganate ester); but +4 in the observed product, MnO_2.

$$\overset{+7}{Mn}O_4^- \xrightarrow{\text{alkene}} H_2\overset{+5}{Mn}O_4^- \xrightarrow{?} \overset{+4}{Mn}O_2$$

It has been suggested that the $H_2MnO_4^-$ ion reacts with MnO_4^- to yield MnO_4^{2-} (+6 oxidation state), which also can oxidize an alkene to a diol and thus be converted to MnO_2.

$$H_2MnO_4^- + MnO_4^- + 2\,OH^- \rightarrow 2\,MnO_4^{2-} + 2\,H_2O$$

$$2\,MnO_4^{2-} \xrightarrow{\text{alkene}} MnO_2$$

Another hypothesis is the $H_2MnO_4^-$ undergoes *disproportionation* (an oxidation–reduction reaction) to yield MnO_4^- and MnO_2.

$$3\,H_2MnO_4^- \rightarrow \underset{\text{oxidized}}{MnO_4^-} + \underset{\text{reduced}}{2\,MnO_2} + 2\,OH^- + 2\,H_2O$$

13.3 Diels–Alder Reactions

A **Diels–Alder reaction** is a **cycloaddition reaction**, a reaction in which two molecules undergo addition to yield a cyclic product. Various types of cycloaddition reactions are known. Because a Diels–Alder reaction involves two double bonds (*four* pi electrons) of a conjugated diene and one double bond (*two* pi electrons) of the other reactant, this type of reaction is called a [4 + 2] cycloaddition. In a Diels–Alder reaction, the conjugated diene is referred to as the *diene*; the other compound is called the *dienophile*.

a [4 + 2] cycloaddition:

$$HC(=CH_2)-HC=CH_2 + CH_2=CH_2 \xrightarrow{\text{heat}} \text{cyclohexene}$$

the diene (*four π electrons*) — *the dienophile* (*two π electrons*)

Diels–Alder reactions are concerted *cis*-addition reactions in which the pi molecular orbitals are converted to the molecular orbitals of the product in a single step. No catalyst is needed; these reactions proceed by heating the reactants, often in a high-boiling solvent. Best results are usually obtained when the dienophile has an electron-withdrawing unsaturated group (such as a carbonyl group) in conjugation with its reacting pi bond.

$$CH_2=CH-\overset{\delta+}{C}(=O^{\delta-})OC_2H_5$$

a carbonyl carbon is electron-withdrawing

Because they are one-step *cis*-additions, Diels–Alder reactions are stereospecific: *cis* dienophiles yield *cis* products and *trans* dienophiles yield *trans* products. In bridged-ring products, the nonreacting electron-withdrawing group (such as a carbonyl group) usually is *endo* (*trans* to the bridge), not *exo* (*cis* to the bridge), probably because of favorable pi-orbital interactions in the transition state.

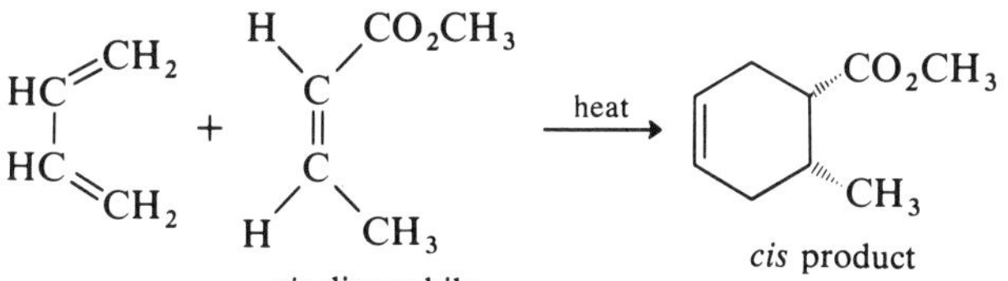

heat

cis dienophile

cis product

heat

trans to bridge

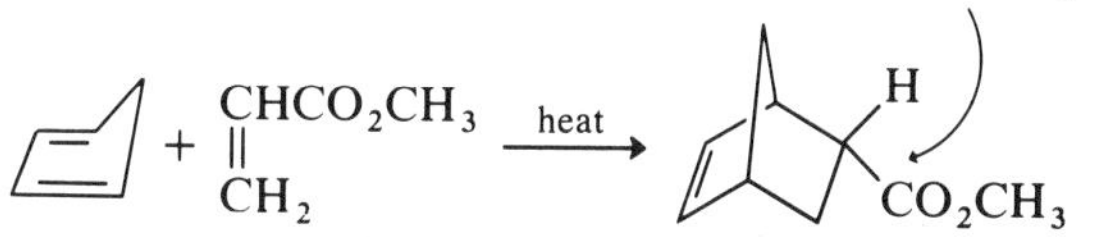

endo product

In Experiment 13.3, you will carry out a Diels–Alder reaction between anthracene (the "diene") and maleic anhydride (the dienophile).

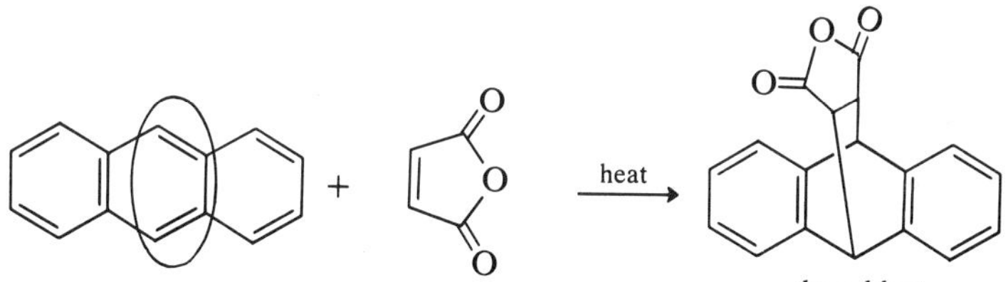

heat

anthracene
"diene" portion circled

maleic anhydride
the dienophile

the adduct

In this reaction, a portion of an aromatic ring acts as the diene. Because the product still contains two aromatic rings, the energy requirements for the addition reaction are not unfavorable. (Because the product is a three-dimensional bridged-ring system, we suggest you make models for a better visualization.)

13.4 Polymerization of Alkenes

The polymerization of alkenes or other compounds containing carbon–carbon double bonds are *addition reactions* in which no small molecules, such as H_2O, are lost. The polymer products of this type of reaction are thus called **addition polymers.** Experiment 13.4 is the preparation of an addition polymer. (Experiment 16.5 is the preparation of a nylon, a *condensation polymer*; see page 308.)

In the presence of an acidic catalyst, alkenes undergo **cationic polymerization.** An alkene that yields a relatively stable carbocation readily undergoes polymerization by this route. Polymerization is almost always an undesired side reaction in any reaction calling for the addition of an electrophile to an alkene. For example, when cyclohexene is hydrated (Experiment 13.1), you will observe that the reaction mixture becomes dark. The dark color arises from the formation of polymers, along with carbonaceous material of unknown structure.

When a polymer is the desired product, a **free-radical initiator** often gives superior laboratory results compared to a cationic catalyst. In Experiment 13.4, you will polymerize styrene using a free-radical initiator.

$$x\ C_6H_5\text{—}CH{=}CH_2 \xrightarrow{\text{free-radical initiator}} \text{—}[CH(C_6H_5)\text{—}CH_2]_x\text{—}$$

styrene

polystyrene
used for molded plastic items and for making styrofoam

A free-radical initiator is a compound that readily decomposes (usually thermally) to yield free radicals. Two common free-radical initiators are dibenzoyl peroxide and 2,2′-azobis(2-methylpropanenitrile), usually called azobisisobutyronitrile and abbreviated **AIBN**.

$$C_6H_5COO\text{—}OCC_6H_5\ (\text{with } C{=}O) \xrightarrow{\text{heat}} 2\,C_6H_5C(=O)\dot{O} \xrightarrow{\text{heat}} 2\,C_6H_5\cdot + 2CO_2$$

dibenzoyl peroxide

$$(CH_3)_2C(CN)\text{—}N{=}N\text{—}C(CN)(CH_3)_2 \xrightarrow{\text{heat}} 2(CH_3)_2\dot{C}(CN) + N{\equiv}N$$

AIBN

The polymerization reaction is initiated by this decomposition of the initiator to yield free radicals. These free radicals react with the alkene (styrene in the following example) to yield new free radicals in a propagation step of the polymerization. In this case, the resonance-stabilized *benzylic* free radical is formed. The alternative radical shown in the following equation is not resonance-stabilized and is therefore formed at a much lower rate.

$$(CH_3)_2\dot{C}(CN) + CH_2{=}CH(C_6H_5) \rightarrow (CH_3)_2C(CN)\text{—}CH_2\text{—}\dot{C}H(C_6H_5) \quad \text{and not} \quad (CH_3)_2C(CN)H\text{—}CH(C_6H_5)\text{—}CH_2\cdot$$

benzylic position

The dimeric free-radical intermediate is formed by a similar propagation step:

$$(CH_3)_2C(CN)CH_2\dot{C}H(C_6H_5) + CH_2{=}CH(C_6H_5) \rightarrow (CH_3)_2C(CN)CH_2CH(C_6H_5)\text{—}CH_2\dot{C}H(C_6H_5) \xrightarrow{CH_2=CH(C_6H_5)} \text{and so forth}$$

monomeric radical

dimeric radical

As the propagation steps continue, long-chain polystyrene intermediates are formed. Termination of the reaction can occur by the combination of two long-chain intermediate free radicals or by their disproportionation.

termination by combination:

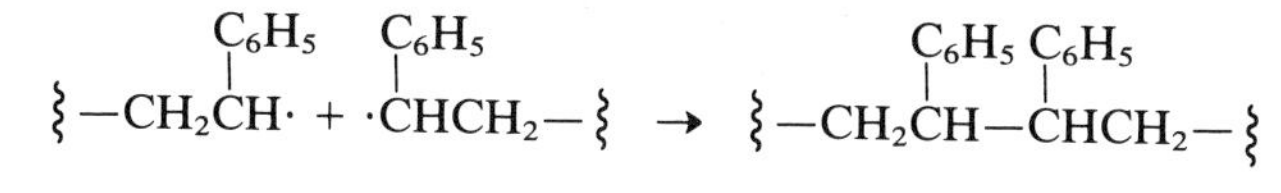

$$\overset{\displaystyle C_6H_5}{\underset{}{\sim}-CH_2\overset{|}{C}H\cdot} + \cdot\overset{\displaystyle C_6H_5}{\overset{|}{C}}HCH_2-\sim \rightarrow \sim-CH_2\overset{C_6H_5}{\overset{|}{C}}H-\overset{C_6H_5}{\overset{|}{C}}HCH_2-\sim$$

termination by disproportionation:

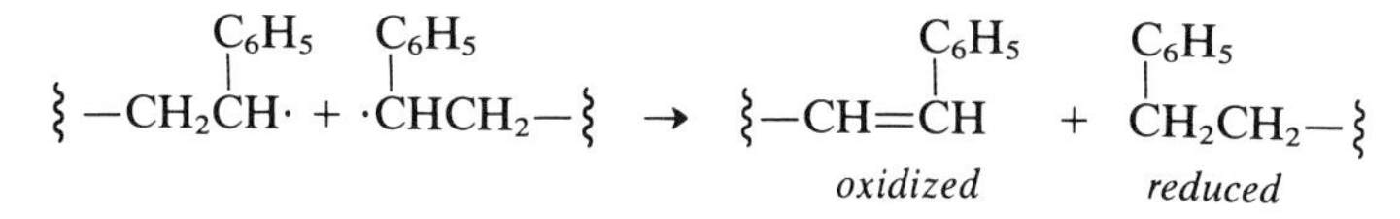

$$\sim-CH_2\overset{C_6H_5}{\overset{|}{C}}H\cdot + \cdot\overset{C_6H_5}{\overset{|}{C}}HCH_2-\sim \rightarrow \underset{\textit{oxidized}}{\sim-CH=\overset{C_6H_5}{\overset{|}{C}}H} + \underset{\textit{reduced}}{\overset{C_6H_5}{\overset{|}{C}}H_2CH_2-\sim}$$

Problems

13.1 Suggest a reason why the addition of HCl gas to propene yields mainly 2-chloropropane and not 1-chloropropane. (Use equations in your answer.)

13.2 Predict the principal products of the addition of conc. HCl to 1-methylcyclohexene. (Use equations, showing all the carbon and hydrogen atoms in the reactants, intermediate, and products—not simply hexagons for the rings.)

13.3 *trans*-2-Pentene is treated with gaseous HBr. Predict the products, including stereochemistry.

13.4 Predict the products of the reaction of each of the following compounds with *hot*, alkaline, aqueous $KMnO_4$.

(a) *trans*-2-pentene

(b) 1-methyl-1-cyclohexene

13.5 Predict the principal products (including stereochemistry) of the reactions of the compounds in Problem 13.4 with *cold*, alkaline, aqueous $KMnO_4$.

13.6 Predict the products of the following Diels–Alder reactions. (Include stereochemistry where appropriate.)

(a) $CH_2=CHCH=CH_2 + CH\equiv CCO_2CH_3 \xrightarrow{\text{heat}}$

(b) 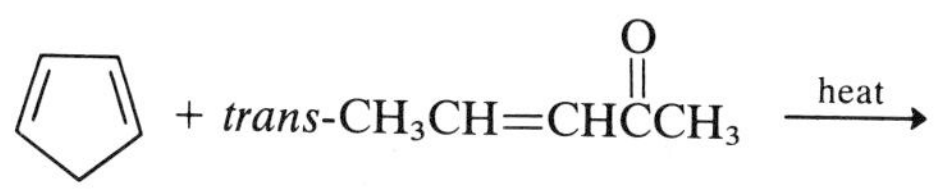 cyclopentadiene $+$ *trans*-$CH_3CH=CH\overset{O}{\overset{||}{C}}CH_3 \xrightarrow{\text{heat}}$

(c) furan $+$ (bicyclic enone) $\xrightarrow{\text{heat}}$

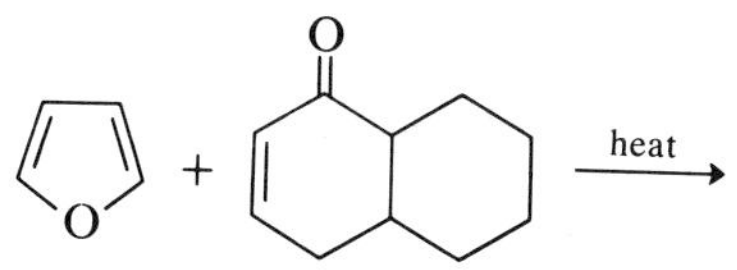

13.7 Suggest a reason why the cationic polymerization of propene yields a head-to-tail polymer instead of head-to-head or tail-to-tail.

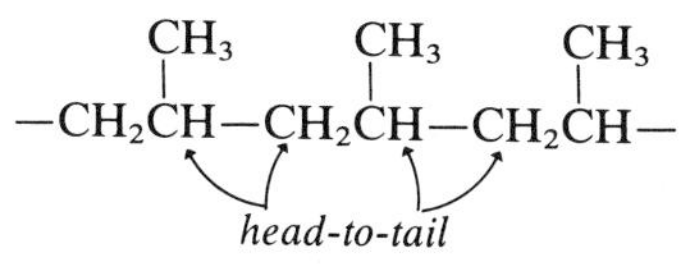

$$-CH_2\overset{CH_3}{\overset{|}{C}}H-CH_2\overset{CH_3}{\overset{|}{C}}H-CH_2\overset{CH_3}{\overset{|}{C}}H-$$

head-to-tail

13.8 Most organic compounds do not dissociate into free radicals when they are heated. Suggest reasons (showing structures) why AIBN dissociates so readily.

13.9 Suggest synthetic routes to the following compounds, starting with an alkene or cycloalkene (or dienes) and other appropriate reagents.

(a) cyclopentanone
(b) *sec*-butyl methyl ether
(c) 2-methyl-2-butanol
(d)

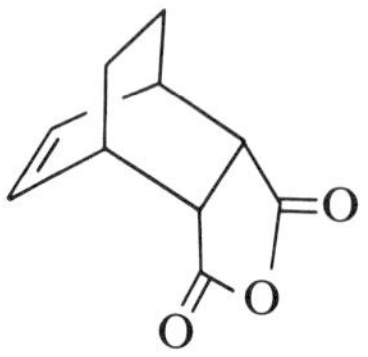

»»» EXPERIMENT 13.1 «««

Synthesis of Cyclohexanol from Cyclohexene

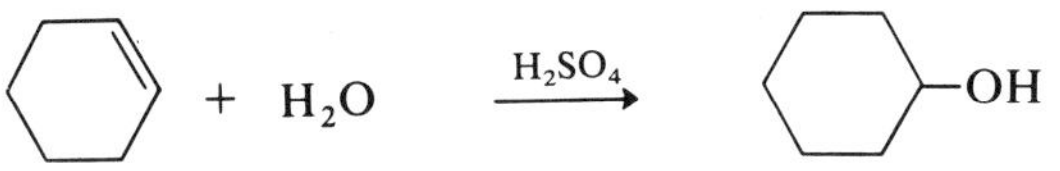

$$\text{cyclohexene} + H_2O \xrightarrow{H_2SO_4} \text{cyclohexanol (-OH)}$$

	cyclohexene	water	sulfuric acid	cyclohexanol
MW:	82.15	18.02	98.08	100.16
weight:	8.2 g	3.4 g	12.9 g (7 mL)	10.0 g (theory)
moles:	0.10	0.19	0.13	0.10 (theory)

Table 13.1 Physical properties of reactants and products

Name	*Bp (°C)*	n_D^{20}	*Density (g/mL)*	*Solubility*	*Bp of azeotrope with water (°C)*
cyclohexene	83.0	1.4465	0.81	insol. H_2O	71
cyclohexanol[a]	161.1	1.4641	0.96	sl. sol. H_2O	98
sulfuric acid	338	—	1.84	sol. H_2O	—

[a] Mp 25.1°

Discussion

In this experiment, cyclohexanol is prepared by shaking a mixture of cyclohexene, water, and conc. sulfuric acid. When the sulfuric acid is first mixed with the water, the solution becomes hot; therefore, the solution must be cooled prior to the addition of cyclohexene. (Otherwise, the volatile cyclohexene would simply evaporate.)

In the reaction, cyclohexene is protonated to yield a cyclohexyl cation. This carbocation can undergo reaction with the hydrogen sulfate ion present in solution, as well as with water. Thus, cyclohexene is converted to protonated cyclohexanol plus some cyclohexyl hydrogen sulfate.

$$\text{cyclohexyl}^{+} \underset{}{\overset{H_2O}{\rightleftharpoons}} \text{—}\overset{+}{O}H_2$$

protonated cyclohexanol

$$\text{cyclohexyl}^{+} \overset{HSO_4^-}{\rightleftharpoons} \text{—OSO}_2\text{OH}$$

cyclohexyl hydrogen sulfate

Cyclohexene is insoluble in the water–acid solution, while protonated cyclohexanol and cyclohexyl hydrogen sulfate are soluble; therefore, when the original two-phase mixture becomes homogeneous, the reaction is complete. In practice, the degree of homogeneity is difficult to distinguish because of darkening of the reaction mixture (from cationic polymerization of the cyclohexene and associated reactions).

Cyclohexanol is isolated from the reaction mixture by dilution with water followed by steam distillation. When the water is added, a two-phase system results: an upper layer of cyclohexanol and a lower layer of aqueous acid. When this mixture is heated, the cyclohexyl hydrogen sulfate by-product is converted to additional cyclohexanol.

$$\text{—OSO}_3\text{H} + H_2O \xrightarrow[S_N1]{\text{heat}} \text{—}\overset{+}{O}H_2 + HSO_4^-$$

$$\overset{H_2O}{\rightleftharpoons} \text{—OH} + H_3O^+$$

Steam distillation separates the organic compounds from the sulfuric acid solution. The distillate consists primarily of cyclohexanol and water (plus a small amount of cyclohexene). Because cyclohexanol and water form an azeotrope boiling at 98°, the bulk of the material distils at this temperature.

The theoretical yield of cyclohexanol in this experiment is only 10.0 g, and cyclohexanol is soluble in water to the extent of 3.6 g/100 mL. Therefore, NaCl is added to the steam distillate to salt out the alcohol. As a further precatuion against mechanical loss (droplets clinging to the separatory funnel, etc.), diethyl ether is used to extract the cyclohexanol from the salt-water mixture.

After separation, the organic layer is dried prior to distillation. Anhydrous magnesium sulfate is a suitable drying agent; however, anhydrous potassium carbonate is the agent of choice because it will neutralize any trace of acid that may have been carried through the work-up procedure. In the final distillation, a trace of acid would cause dehydration of the cyclohexanol.

*Experimental**

EQUIPMENT:

aluminum foil
distillation apparatus
two 50-mL and one 125-mL Erlenmeyer flasks
glass cloth
10-mL graduated cylinder
long-stem funnel
disposable pipet or dropper
two 50-mL round-bottom flasks with ground-glass stoppers
250-mL round-bottom flask
125-mL separatory funnel

CHEMICALS:

anhydrous potassium carbonate, 2 g
cyclohexene (reagent grade), 8.2 g
diethyl ether, 20 mL
sodium chloride, 20 g
conc. sulfuric acid, 7 mL

TIME REQUIRED: about 3 hours plus an overnight drying prior to the final distillation

STOPPING POINTS: after shaking the cyclohexene with H_2SO_4; when NaCl has been added to the steam distillate; when the ether extract is being dried

››››› ***SAFETY NOTE 1*** A stoppered flask containing conc. sulfuric acid will be shaken by hand; it is imperative that this flask not leak!

››››› ***SAFETY NOTE 2*** Cyclohexene is volatile and flammable; weigh it in the fume hood if possible. This compound is also a good solvent. Perform the experiment only in a flask with a ground-glass stopper. (A cork or rubber stopper is not suitable.)

PROCEDURE

Place 3.4 mL of water in a 50-mL standard-taper round-bottom flask and carefully add 7 mL of conc. sulfuric acid (see Safety Note 1). Cool the flask in an ice bath until it is cold to the hand. Weigh 8.2 g of cyclohexene into a 50-mL Erlenmeyer flask, then add it to the aqueous acid in one portion (see Experimental Note). Stopper the round-bottom flask with a lightly greased,

* Beginning with this experiment, the Safety Notes and precautionary phrases will emphasize only new or unique hazards. Notes concerning previously encountered hazards will not be emphasized to the same extent. We assume that the student has gained enough experience to be aware of routine dangers in the laboratory.

ground-glass stopper. Holding the stopper *firmly* in place, shake the flask vigorously for 10 minutes. Then let the stoppered flask stand at room temperature for a few minutes. Repeat the shaking and standing until the mixture no longer forms two phases when standing (about one hour, but higher yields can sometimes be obtained if the reaction mixture is allowed to stand for a longer time after the shaking period). Transfer the contents of the flask to a 250-mL distillation flask, using 120 mL of water for washing the mixture into the larger flask. (Use a long-stem funnel for the transfer so that the acidic mixture does not contact the joint of the flask.)

Add 3–4 boiling chips, fit the flask onto a distillation apparatus, and distil 75–100 mL. The initial distillate will appear oily (a cyclohexanol–water mixture). The residue of the distillation (still fairly concentrated acid) should be diluted with water and flushed down the drain with a generous amount of water.

Add 20 g of solid sodium chloride to the distillate and stir. Most, but not all, of the salt will dissolve. Add 10 mL of diethyl ether (*flammable*!), transfer the mixture to a separatory funnel, and use an additional 10 mL of diethyl ether to wash the distillation receiver. After extraction and separation, transfer the ether layer to a 50-mL round-bottom flask with a loose-fitting glass stopper (to prevent pressure build-up); add 2 g of anhydrous potassium carbonate; and allow the mixture to stand overnight.

After drying, decant the liquid into a dry 50-mL round-bottom flask for distillation. The forerun consists of diethyl ether (bp 34.6°) and cyclohexene (bp 83.0°) and should be discarded into a waste bottle in the fume hood (not in the sink). After the forerun has distilled, the residue may froth. To minimize the frothing, wrap the distillation flask and head with glass cloth and aluminum foil. Collect the cyclohexanol, bp 155–162°, in a tared receiver. Determine your yield and the refractive index. A typical yield is 5.6 g (56%), n_D^{20} 1.4641.

EXPERIMENTAL NOTE

Reagent grade cyclohexene must be used in the experiment. Impure cyclohexene darkens when shaken with H_2SO_4 to the extent that two layers cannot be detected.

Problems

13.10 Explain why cyclohexanol is more soluble in 65% H_2SO_4 than in a more dilute solution of acid.

13.11 Write the equations for the mechanisms of the conversion of cyclohexyl hydrogen sulfate to (a) cyclohexanol; (b) cyclohexene.

13.12 In Experiment 13.1, anhydrous calcium chloride would be a poor choice as a drying agent. Why?

13.13 Write an equation for the cationic polymerization of cyclohexene.

»»» EXPERIMENT 13.2 «««

Synthesis of cis-1,2-Cyclohexanediol from Cyclohexene

$$3\,\text{(cyclohexene)} + 2KMnO_4 + 4H_2O \xrightarrow[(CH_3)_3COH]{NaOH} 3\,\text{(cis-1,2-cyclohexanediol)} + 2MnO_2 + 2KOH$$

	cyclohexene	potassium permanganate	water (excess)	*cis*-1,2-cyclohexanediol
MW:	82.15	158.03		116.16
weight:	2.5 g	3.2 g		3.5 g (theory)
moles:	0.030	0.020		0.030 (theory)

Table 13.2 *Physical properties of reactants, solvents, and products*

Name	*Bp (°C)*	*Mp (°C)*	*Solubility*
cyclohexene	83.0	—	insol. H_2O; sol. $(CH_3)_3COH$
t-butyl alcohol[a]	82.2	25.5	sol. H_2O
ethyl acetate	77.06	—	sl. sol. H_2O
cis-1,2-cyclohexanediol	—	99–101	sl. sol. H_2O and $CH_3CO_2CH_2CH_3$

[a] Bp of azeotrope with water, 79.9° (11.8% water, 88.2% alcohol).

Discussion

In this experiment, a chilled, aqueous solution of potassium permanganate is added slowly to a cold solution of cyclohexene and sodium hydroxide in a mixed solvent of *t*-butyl alcohol and water. The alcohol (miscible with water) allows the water-insoluble cyclohexene to be dissolved in the otherwise aqueous solution. The aqueous–organic solution is prepared by dissolving the cyclohexene in the alcohol and the NaOH in water, then mixing these two solutions and chilling before adding the permanganate.

1) Mix (cyclohexene) and $(CH_3)_3COH$

2) Dissolve NaOH in H_2O

3) Combine and chill before adding $KMnO_4$ solution slowly

The $KMnO_4$ is dissolved in warm water to hasten its dissolving; then this solution is also chilled before being added to the cyclohexene solution. Because the reaction is exothermic, the reaction mixture is chilled during the addition. Both the chilling and the slow addition of the $KMnO_4$ solution to the cyclohexene solution help prevent unwanted further oxidation to adipate ions.

A solution of $KMnO_4$ is deep purple. The inorganic product, MnO_2, is a gelatinous, muddy-brown precipitate. A color change can be observed at the start of the addition, but the initial MnO_2 precipitate masks further color changes.

After the oxidation, the MnO_2 is removed by filtration. Because it is gelatinous, this precipitate will clog filter paper. To prevent this, a **filter aid** (Filter-Cel, Celite) is used to adsorb the MnO_2. Filter aids are *diatomaceous earths* (silica), which have large surface areas and are thus good adsorbents. In this experiment, the filter aid is added directly to the reaction mixture, which is then brought to a near-boil to coagulate the solids. The mixture is vacuum-filtered while still hot. The filter aid adsorbs and retains a substantial amount of the product, as well as MnO_2. Therefore, the filter cake is thoroughly washed with a mixture of *t*-butyl alcohol and water after the filtration.

The volume of the filtrate is reduced by simple distillation. *t*-Butyl alcohol and water form an azeotrope boiling at 79.9° and composed of almost 90% alcohol. Thus, most of the *t*-butyl alcohol is removed from the crude product in this step.

The distillation residue contains the desired diol, water, and some residual *t*-butyl alcohol. Cyclohexanediol is water-soluble; therefore, it must be salted out of the aqueous residue. Solid potassium carbonate is used rather than NaCl because K_2CO_3 converts any adipic acid to dipotassium adipate. Most of the dipotassium adipate remains in the carbonate–water layer, but some contaminates the organic layer. This impurity is removed later in the work-up procedure.

The diol initially separates from the water–K_2CO_3 mixture as an oil. After this oil solidifies (see Experimental Note 4, page 251), the crude diol is filtered with vacuum, then is dissolved in an excess of ethyl acetate. The solid that does *not* dissolve is dipotassium adipate, which is removed by filtration before the excess ethyl acetate is boiled away. Crystallization of the remaining crude diol from the ethyl acetate affords a fairly pure product.

Experimental

EQUIPMENT:

400-mL beaker
distillation assembly
one 50-mL, two 125-mL, and one 250-mL Erlenmeyer flasks
filter paper
funnel
25-mL graduated cylinder
hot plate
ice bath
500-mL round-bottom flask (see Experimental Note 3)
125-mL separatory funnel
stirring rod
thermometer
vacuum filtration assemblies (one large and one small)

CHEMICALS:

t-butyl alcohol, 120 mL
Celite or other filter aid, 1–2 g
cyclohexene, 2.5 g
ethyl acetate, 25–40 mL
potassium carbonate, 25–100 g
potassium permanganate, 3.2 g
sodium hydroxide, 0.5 g

TIME REQUIRED: about 6 hours, plus time to dry the crystals, determine the melting point, and run two infrared spectra

STOPPING POINTS: after filtering the MnO_2; after concentration of the filtrate; while the oil is solidifying during the salting out; after filtration of the ethyl acetate solution; while the product is crystallizing or drying

>>>> ***SAFETY NOTE 1*** Potassium permanganate is toxic and a strong oxidant. If you get any on your skin, wash immediately with soap and water. (*Cosmetic note*: A solution of $NaHSO_3$ or slightly acidified Na_2SO_3 will remove MnO_2 stains from skin and glassware.)

>>>> ***SAFETY NOTE 2*** Sodium hydroxide is caustic. If a pellet (or the solution) touches your skin, wash immediately with copious amounts of water. Pick up spilled pellets with tweezers and discard in the sink with generous amounts of running water.

PROCEDURE

Dissolve 3.2 g of potassium permanganate (see Safety Note 1) in 100 mL of warm water with swirling, then cool the solution in an ice bath to about 15° (see Experimental Note 1). Transfer the cool solution to a separatory funnel.

Measure 100 mL of *t*-butyl alcohol into a 400-mL beaker (see Experimental Note 2) and add 2.5 g of cyclohexene. In a separate flask, dissolve 0.5 g of solid sodium hydroxide in 2–3 mL of water (see Safety Note 2), then add this solution to the cyclohexene–alcohol solution. Cool the combined solution to −5°–0° by adding 40–50 g of crushed ice to the beaker.

Place the beaker in an ice bath, and add the permanganate solution dropwise from the separatory funnel at a rate of about two drops per second. (About 20 minutes is required for the addition.) During the addition, stir the mixture in the beaker vigorously with a glass stirring rod or a magnetic stirrer. If the temperature of the reaction mixture reaches 15°, stop the addition and allow the mixture to cool to about 10° before resuming the addition.

After the addition, add 1–2 g of a filter aid and heat the mixture almost to boiling with frequent stirring. Filter the hot mixture with vacuum. Use a mixture of 20 mL water and 20 mL *t*-butyl alcohol to rinse the filter cake by pouring a small portion of the rinse solution onto the filter cake with the suction off, allowing it to soak into the solid, then turning on the vacuum to suck the liquid into the filtrate. Repeat the procedure several times until all the rinse solution has been used.

Distil the filtrate until the volume of the residue is about 50 mL (see Experimental Note 3). Transfer the hot residue to a tared 250-mL Erlenmeyer flask, and determine the weight of the contents. In a separate container, weigh out 1 g of potassium carbonate for each gram of residual solution. Add the carbonate to the solution about 1 g at a time, swirling until it dissolves. When an oil begins to separate from the solution, stop the addition and allow the oil to solidify (see Experimental Note 4). Then add the rest of the carbonate in one portion and stir or swirl until it dissolves. Separate the solid, impure diol from the mixture by vacuum filtration. Transfer the solid to a 125-mL Erlenmeyer flask, add 20 mL of ethyl acetate (*flammable*!) and bring the mixture to a boil on a hot plate to dissolve the diol. Filter the mixture by gravity with fluted filter paper to remove the solid dipotassium adipate, then wash the filter paper with a small amount of hot ethyl acetate. If the filtrate is cloudy, repeat the warming and filtration.

Add 2–3 boiling chips to the filtrate, and remove most of the ethyl acetate by boiling it on a hot plate in the fume hood. Be careful not to overheat the flask. The diol may oil out, but should solidify when cool. Crystallize this product from ethyl acetate and dry the crystals at least overnight.

A typical yield of diol is 1.7 g (48%), mp 85–89°. A second crystallization from ethyl acetate yields purer product: 1.4 g (40%), mp 93–96°; however, repeated crystallizations are required to raise the melting point to the literature value of 99–101°. Run infrared spectra on cyclohexene and its *cis*-1,2-diol. Identify the key absorption bands in each spectrum.

EXPERIMENTAL NOTES

1) Because of the dark color of the $KMnO_4$ solution, it is difficult to determine when it has all dissolved. Decant the solution into another flask, then inspect the original flask for solid $KMnO_4$. If any solid remains, return the solution to the flask and continue swirling. If necessary, rewarm the solution.

2) *t*-Butyl alcohol melts at 25.5° and often solidifies in its bottle. If this should occur, place the bottle in a pan of warm water until the alcohol melts.

3) If a 500-mL flask is not available, use a smaller flask and stop the distillation periodically to add more liquid.

4) Allowing the first portion of the oil to solidify before adding more K_2CO_3 results in larger crystals, which are easier to filter. If the initial oily diol does not solidify, dilute the mixture with a small amount of water to allow some of the oil to go back into solution. When chilled, the remaining oil should solidify. If this modification is necessary, add an additional gram of K_2CO_3 for each mL of water added.

Problems

13.14 When the water–alcohol solution is cooled by the addition of crushed ice, the temperature drops below 0°. Why does this happen?

13.15 Would you expect oxidation products from *t*-butyl alcohol to contaminate the crude product in Experiment 13.2? Explain.

13.16 In Experiment 13.2, a student did not add the $KMnO_4$ solution dropwise, but added it in one portion. Predict and explain the consequences of this act.

13.17 In Experiment 13.2, a student observes a small carbonyl peak in the infrared spectrum of *cis*-1,2-cyclohexanediol. Why?

»»» EXPERIMENT 13.3 «««

(SUPPLEMENTAL)

Synthesis of 9,10-*Dihydroanthracene-9,10-endo-α,β-succinic Anhydride: A Diels–Alder Reaction*

anthracene + maleic anhydride —xylene, heat→ 9,10-dihydroanthracene-9,10-*endo*-α,β-succinic anhydride

	anthracene	maleic anhydride	9,10-dihydroanthracene-9,10-*endo*-α,β-succinic anhydride
MW:	178.24	98.06	276.30
weight:	2.0 g	1.0 g	2.8 g (theory)
moles:	0.011	0.010	0.010 (theory)

Table 13.3 Physical properties of reactants and products

Name	*Mp (°C)*	*Solubility*
anthracene	216	insol. H_2O, sol. xylene
maleic anhydride	60	decomposes in H_2O, sol. xylene
9,10-dihydroanthracene-9,10-*endo*-α,β-succinic anhydride	262–264	slow decomposition in H_2O, sl. sol. xylene

Discussion

This Diels–Alder reaction is carried out by boiling the reactants in xylene (a mixture of dimethylbenzenes, bp 140°). Both reactants are soluble in xylene, and the reaction is rapid because of the high boiling temperature of this solvent. As the reaction mixture cools, the product crystallizes. The product is isolated by filtration; if the starting anthracene was pure, it requires no further purification.

It is difficult to remove the high-boiling xylene by "air-drying" the crystals. In addition, exposure to the air results in the hydrolysis of anhydrides by atmospheric moisture.

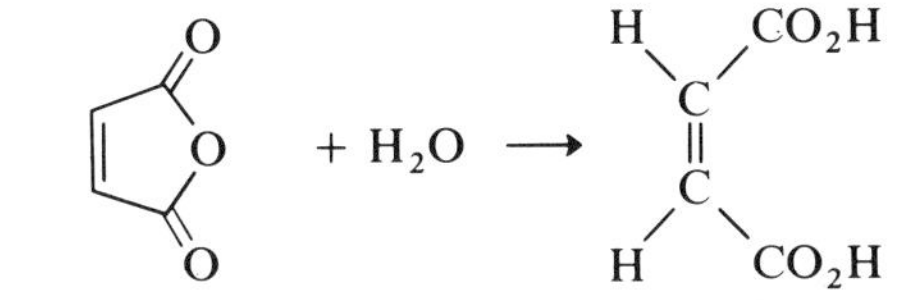

maleic anhydride + H_2O → maleic acid

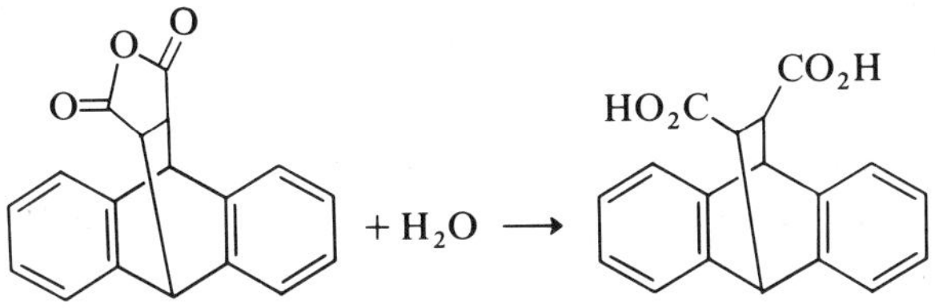

the product anhydride + H_2O → a diacid

For this reason, the product is dried under an inverted beaker along with some paraffin wax, which dissolves xylene vapors and thus acts as a "dessicant."

Experimental

EQUIPMENT:

25-mL and 1000-mL beakers
25-mL graduated cylinder
ice bath
melting-point assembly with a 300° thermometer
mortar and pestle (to grind the maleic anhydride, if necessary)
reflux condenser
50-mL round-bottom flask
spatula
vacuum filtration assembly with a small Büchner funnel
watch glass

CHEMICALS:

anthracene, 2.0 g
maleic anhydride (fine powder), 1 g
xylene (mixed isomers), 35 mL

TIME REQUIRED: about 2 hours, plus time for drying the product and taking melting points

STOPPING POINTS: while the reaction mixture is cooling to room temperature; while the product is being dried

>>>> ***SAFETY NOTE*** *Maleic anhydride* is a powerful and volatile irritant! Avoid skin contact and inhalation of its vapors. *Anthracene* is carcinogenic to laboratory animals. You may wish to use disposable gloves when working with this compound. Clean your work area and glassware thoroughly after using this compound.

PROCEDURE

Place 2.0 g of anthracene (see Experimental Note 1) in a 50-mL round-bottom flask and add 25 mL of xylene. Add 1.0 g of well-ground maleic anhydride (see Experimental Note 2), fit the flask with a reflux condenser, and heat the mixture at reflux for 30 minutes. Allow the flask to cool to room temperature, then chill it in an ice bath (see Experimental Note 3). Filter the solid with vacuum and wash the product with 10 mL of ice-cold xylene. Dry the solid on a watch glass next to a small beaker of paraffin wax shavings, both covered by a beaker (see Experimental Note 4). A typical yield is 2.5 g of colorless product, mp 262–264°. If the sample is tinged with yellow and has a depressed melting point (the result of starting with impure anthracene), recrystallize the product from xylene (20 mL per gram).

EXPERIMENTAL NOTES

1) A good grade of anthracene should be used; otherwise, the product will be tinged yellow and exhibit a depressed melting point with decomposition. If only crude (off-white or yellowish) anthracene is available, crystallize it from absolute (anhydrous) ethanol (65 mL per gram; more if 95% ethanol is used).

2) Maleic anhydride is commercially available as pressed rolls. In order for this reaction to proceed in a reasonable period of time, the maleic anhydride must be finely powdered by grinding in a mortar (CAUTION: see Safety Note).

3) A mixture of *o*-, *m*-, and *p*-xylenes is the best solvent. If *p*-xylene (mp 13°) is used, the whole reaction mixture may solidify when chilled.

The product is fairly insoluble in xylene. Once crystallization begins, it cannot be redissolved by heating. Therefore, the use of activated carbon, followed by a hot filtration to remove a trace of color, is not recommended.

4) If the relative humidity is low, the product may be air-dried; however, the rate of evaporation of xylene is slow. If the product is exposed to air at 45% relative humidity for 4 days, it will show about a 4° melting-point depression.

Problems

13.18 Predict the products of hydrolysis of the following anhydrides:

(a) $CH_3CH_2\overset{O}{\overset{\|}{C}}O\overset{O}{\overset{\|}{C}}CH_3$ (b) 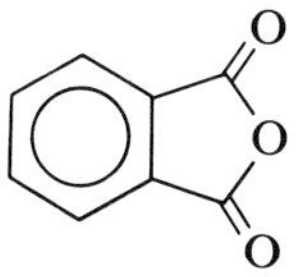 [phthalic anhydride]

13.19 What products would be obtained from a Diels–Alder reaction of anthracene with the following dienophiles?

(a) *trans*-$CH_3CH{=}CHCO_2C_2H_5$

(b) $CH{\equiv}CCO_2CH_3$

(c) 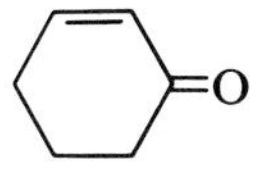 [2-cyclohexen-1-one]

»»» EXPERIMENT 13.4 «««

(SUPPLEMENTAL)

Synthesis of Polystyrene

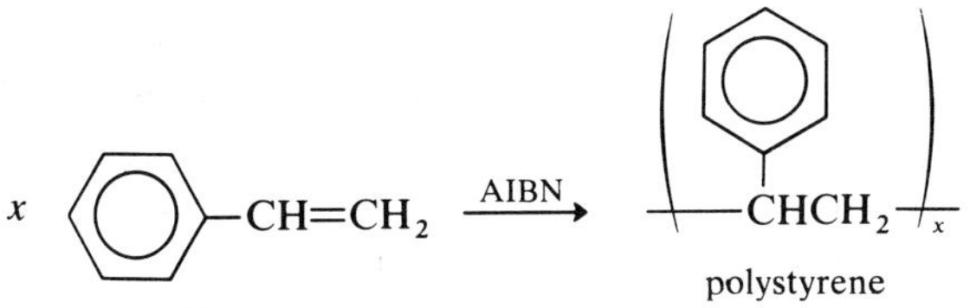

$$x\ C_6H_5{-}CH{=}CH_2 \xrightarrow{\text{AIBN}} {-}\!\!\left[CH(C_6H_5)CH_2\right]\!_x{-}$$

	styrene	polystyrene
MW:	104.16	indeterminant
weight:	5.0 g	5.0 g (theory)
moles:	0.048	—

Discussion

Polystyrene is prepared from styrene with a free-radical initiator, AIBN (see Section 13.4). Commercial styrene polymerizes slowly, but spontaneously, when exposed to air because O_2 (a diradical) can also initiate the free-radical polymerization.

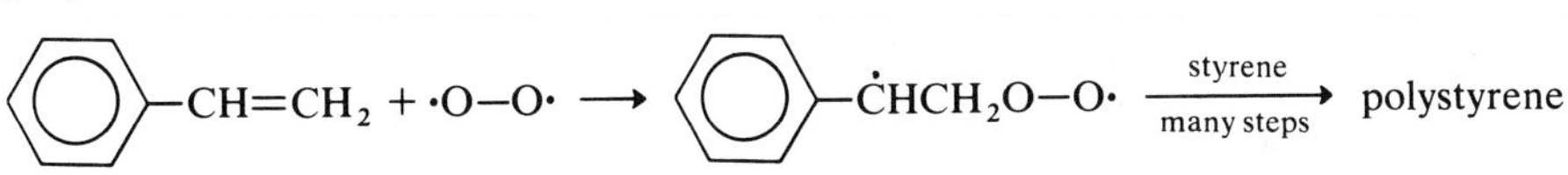

$$C_6H_5{-}CH{=}CH_2 + \cdot O{-}O\cdot \longrightarrow C_6H_5{-}\dot{C}HCH_2O{-}O\cdot \xrightarrow[\text{many steps}]{\text{styrene}} \text{polystyrene}$$

Because of this spontaneous polymerization, commercial styrene contains 10–50 ppm of 4-*t*-butylpyrocatechol, which acts as a free-radical inhibitor and retards the formation of the polymer. The pyrocatechol breaks the free-radical reaction cycle by reacting with any free radicals present and forming a resonance-stabilized free radical that is not reactive enough to add to the double bond of styrene.

$$(CH_3)_3C\text{-}C_6H_3(OH)\text{-}OH + R\cdot \longrightarrow (CH_3)_3C\text{-}C_6H_3(OH)\text{-}O\cdot + RH$$

4-*t*-butylpyrocatechol — growing polymer — a resonance-stabilized free radical

After all the *t*-butylpyrocatechol has been consumed, the styrene in a bottle begins to polymerize and becomes progressively more viscous. Partially polymerized styrene, as well as fresh styrene, may be used in this experiment.

Of course, the presence of a free-radical inhibitor in styrene inhibits a laboratory free-radical polymerization as well as the spontaneous, oxygen-initiated polymerization. This problem is alleviated by using a relatively large amount of AIBN in the laboratory polymerization. When the mixture of styrene plus *t*-butylpyrocatechol and AIBN is heated, the AIBN decomposes to free radicals and nitrogen gas. Enough free radicals are generated both to deactivate the *t*-butylpyrocatechol and to initiate the polymerization reaction.

Even under carefully controlled conditions, polymerization leads to a rather heterogeneous mixture. While it is convenient to refer to polystyrene as a single product, a synthetic polymer is actually a mixture of molecules of varying chain lengths and thus varying molecular weights. One may refer to the *average molecular weight* of a polymer, but not to a specific molecular weight. Because the chain lengths of polymeric molecules depend on reaction conditions, average molecular weights must be measured experimentally.

Polymers with a high average molecular weight are said to have a *high degree of polymerization* and are insoluble in most solvents. The polystyrene in this experiment has a *low* degree of polymerization. The solubilities of low-molecular-weight polymers in various solvents follow the same rules as for monomeric compounds. Polystyrene is a relatively nonpolar hydrocarbon; consequently, we would predict that polystyrene is soluble in nonpolar, nonhydrogen-bonding solvents and insoluble in polar, hydrogen-bonding solvents. In this experiment, you will test these predictions.

Experimental

EQUIPMENT:

hot water bath
mortar and pestle
small test tube, flask, or beaker
stirring rod or spatula
test tube

CHEMICALS:

AIBN, 0.1 g
styrene, 5.0 g
95% ethanol, ethyl acetate, dichloromethane, and toluene, a few mL of each

TIME REQUIRED: 1 hour plus overnight polymerization

STOPPING POINT: while the polymerizing styrene is sitting overnight

>>>> ***SAFETY NOTE*** Although the chemicals in this experiment are used in small amounts, keep in mind the toxicity and flammability of each.

PROCEDURE

Weight 5.0 g of styrene into a test tube (see Experimental Note 1). Add 0.1 g of AIBN and mix as thoroughly as possible with a stirring rod or spatula. Cork the test tube loosely and heat the mixture in a hot water bath for about 30 minutes. (The amount of heating time is not critical.) Allow the mixture to stand in your locker at least overnight (see Experimental Note 2).

Remove the solid polymer from its container by breaking the tube. Weigh the solid and calculate the yield (see Experimental Note 3). Observe the brittleness of the polymer. Break the polymer into small pieces and test its solubility in water, 95% ethanol, ethyl acetate, dichloromethane, and toluene. In these tests, allow the polymer to stand in the solvent with occasional mixing for at least $\frac{1}{2}$ hour before making a decision about solubility.

EXPERIMENTAL NOTES

1) The normally liquid styrene (mp −31°) may have partially polymerized in its bottle. Unless it is completely solidified, it is adequate for this experiment.

2) If the polystyrene is not solid after standing overnight, add another 0.10 g of AIBN, reheat the mixture, and allow it to sit another night in your locker.

3) Because the molecular weight of polystyrene is indeterminate, the usual yield calculations cannot be used. Therefore, assume the theoretical yield to be 5.0 g. You will not obtain this yield because some styrene (bp 145.2°) evaporates during the polymerization.

Problems

13.20 Write resonance structures for the following free radical: $C_6H_5CHCH_3$

13.21 Write resonance structures for the free radical of *t*-butylpyrocatechol, shown on page 256.

13.22 Assume that two trimeric free radicals (each containing three units of styrene) undergo combination to yield a hexameric molecule of polystyrene in this experiment. Write the complete formula for the hexameric molecule, showing the end groups.

13.23 If a bottle of styrene contains 25 ppm of 4-*t*-butylpyrocatechol, how many moles of 4-*t*-butylpyrocatechol are present in 5.0 g of styrene?

CHAPTER 14

Aromatic Substitution Reactions

In the 1800's and early 1900's, most industrial organic chemicals were obtained from **coal tar**, the condensable distillate from the destructive distillation (heating in the absence of air) of coal. Coal tar is rich in aromatic compounds: benzene, toluene, the xylenes (dimethylbenzenes), pyridine, and naphthalene, along with larger aromatic ring systems.

When petroleum and its products (**petrochemicals**) became plentiful and cheap in the 1940's, the use of coal tar as a major source of organic compounds declined. In 1950, benzene became industrially available from petroleum by **catalytic reforming**, a process in which alkanes are heated in the presence of a catalyst and undergo cyclization and dehydrogenation.

Benzene is used as a solvent, a gasoline additive, and a chemical intermediate in the production of such diverse materials as aspirin and polystyrene.

Although aromatic compounds are useful and some are biologically indispensable, many aromatic compounds are carcinogenic. Benzene itself is a moderate carcinogen, and many aromatic compounds with four or more fused rings are very potent carcinogens.

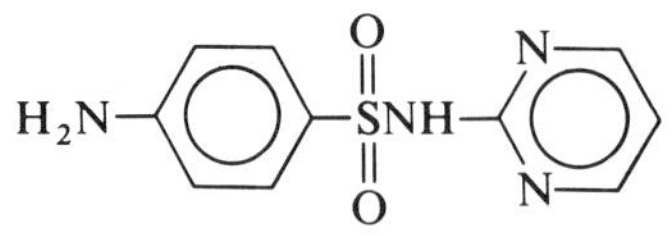

sulfadiazine
a sulfa drug

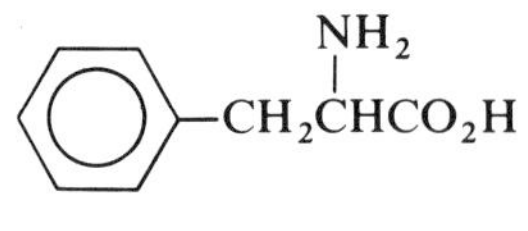

phenylalanine
an essential part of proteins

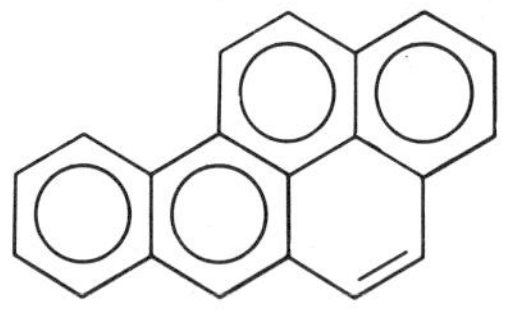

benzo[*a*]pyrene
a carcinogen

14.1 Electrophilic Aromatic Substitution Reactions of Benzene

Benzene does not undergo reactions typical of alkenes, such as oxidation with $KMnO_4$ or addition of Br_2. However, in the presence of the appropriate Lewis-acid catalyst, benzene can undergo an *electrophilic aromatic substitution reaction*. In this type of reaction, another atom or group is substituted for a hydrogen on the ring. Nitration is a typical aromatic substitution reaction:

$$C_6H_6 + HNO_3 \xrightarrow[50°]{H_2SO_4} C_6H_5{-}NO_2 + H_2O$$

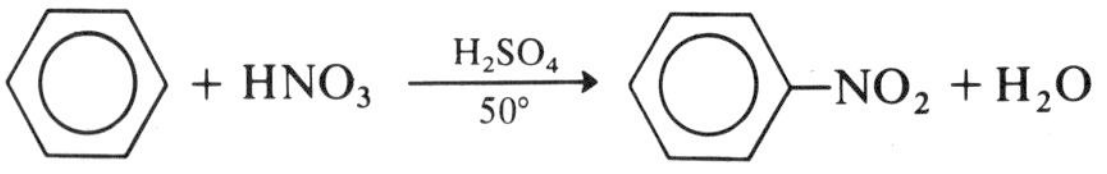

benzene

nitrobenzene
85%

The mechanism for the nitration of benzene involves: (1) formation of the electrophile ($^+NO_2$); (2) attack of this electrophile at a ring carbon; and (3) loss of H^+ to a base such as HSO_4^- in the reaction mixture to yield nitrobenzene.

formation of the electrophile:

$$H_2SO_4 + H\ddot{O}{-}NO_2 \xrightleftharpoons{\text{fast}} HSO_4^- + H_2\overset{+}{O}{-}NO_2$$

$$H_2\overset{+}{O}{-}NO_2 \xrightarrow{\text{fast}} H_2\ddot{O} + {}^+NO_2$$

electrophilic attack and loss of H^+:

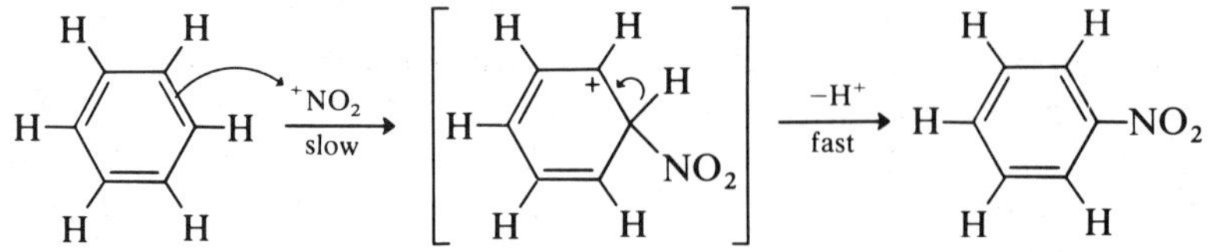

$$C_6H_6 \xrightarrow[\text{slow}]{^+NO_2} [C_6H_6NO_2]^+ \xrightarrow[\text{fast}]{-H^+} C_6H_5{-}NO_2$$

intermediate
benzenonium ion

14.2 Substitution of a Second Group

In Experiments 14.1 and 14.2, two monosubstituted benzene compounds are each subjected to a *second* substitution reaction.

$$C_6H_5COCH_3 \text{ (with C=O)} + HNO_3 \xrightarrow[10^\circ]{H_2SO_4} m\text{-}O_2NC_6H_4COCH_3 + H_2O$$

methyl benzoate → methyl *m*-nitrobenzoate 74%

$$C_6H_5NHCCH_3 \text{ (with C=O)} + HNO_3 \xrightarrow[10^\circ]{H_2SO_4} O_2N\text{-}C_6H_4\text{-}NHCCH_3 + H_2O$$

acetanilide* → *p*-nitroacetanilide 68%

In theory, each of these reactions could produce three isomeric products. For example, methyl benzoate could yield *ortho* (1,2); *meta* (1,3); and *para* (1,4) disubstituted products. If the substitution were random, the product ratio would be 40% *ortho*, 40% *meta*, and 20% *para*. (The differences in percentages arise from the fact that methyl benzoate contains two *ortho* positions and two *meta* positions, but only one *para* position.) Yet, when an ester like methyl benzoate is subjected to nitration, the yield is about 70% of the *meta* product. By contrast, when acetanilide is nitrated, the yield is about 70% of the *para* product. Table 14.1 shows the actual product ratios as per cents of the mononitrated products (not per cent yields) isolated in one particular study of these reactions.

There are several reasons that the second substitution is not random. These reasons include *steric hindrance*, in which a bulky first substituent hinders attack at the *ortho* positions; *inductive effects*; and *resonance effects*. Let us consider the mechanisms of the two substitution reactions that you will perform in light of these effects.

A. *Ortho,para*-Directors

The mechanism for a second substitution on a benzene ring is similar to that of the first substitution. Let us first examine the mechanism for the nitration of acetanilide by considering the intermediates for *meta* and *para* substitution.

Table 14.1 Product ratios in aromatic nitration

	Product ratio (%)		
Starting material	***ortho***	***meta***	***para***
acetanilide	19	2	79
methyl benzoate	28	68	4

* An amide with a phenyl substituent on the nitrogen is called an **anilide**. Acetanilide is therefore the *N*-phenyl amide of acetic acid.

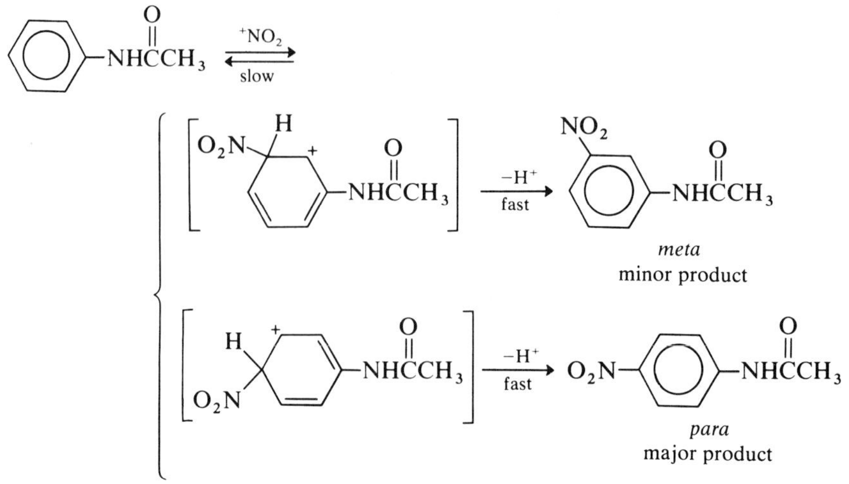

If we write the resonance structures for the *para* intermediate, we see that the positive charge of this intermediate can be resonance-stabilized by the pi electrons of the benzene ring and *also by the unshared pair of electrons of the nitrogen atom of the amide group.*

resonance structures of the *para* intermediate:

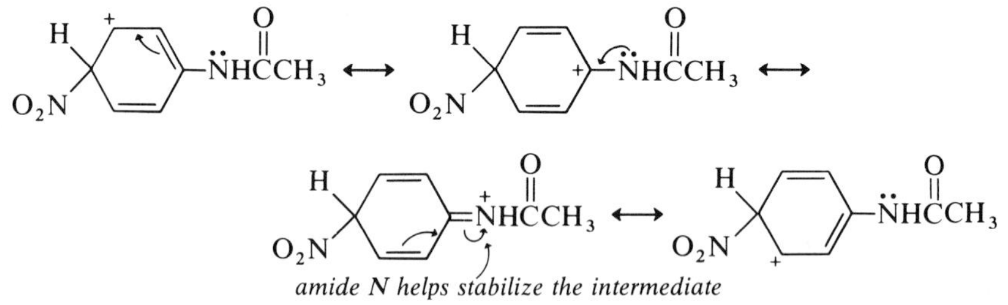

amide N helps stabilize the intermediate

The *meta* intermediate is not as stabilized as the *para* intermediate because the amide nitrogen's unshared electrons cannot help delocalize the positive charge. (Draw the resonance structures to verify this statement for yourself.) Because the *para* intermediate is more stabilized than the *meta* intermediate, its rate of formation is faster. Because the formation of these intermediates is the rate-limiting step in the reaction, *para*-substitution proceeds faster than *meta*-substitution, and the *para* product predominates in the product mixture.

The rate of *ortho*-substitution is similarly enhanced by resonance stabilization of the intermediate; however, the steric hindrance by the acetamide group prevents the formation of a high yield of the *ortho* product.

B. *Meta*-Directors

The nitration of methyl benzoate yields predominantly the *meta* product. Again, the resonance structures of the intermediates can be used to explain this observation.

resonance structures of the *meta* intermediate:

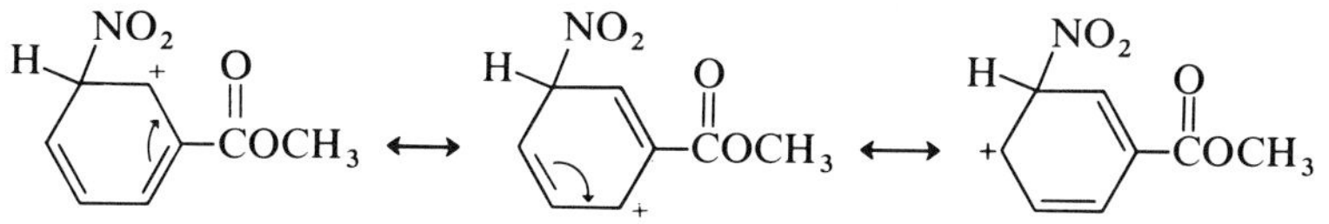

resonance structures of the *para* intermediate:

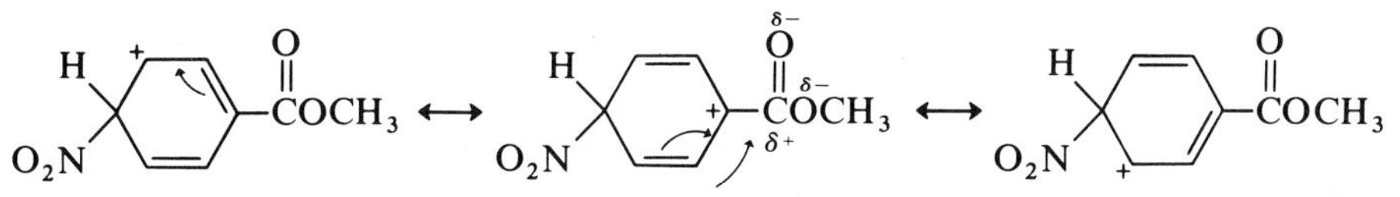

adjacent + charges destabilize the p-intermediate

The second resonance structure for the *para* intermediate shows the positive charge of the benzenonium ion adjacent to the partial positive charge of the carbonyl group. This proximity of two positive charges destabilizes the *para* intermediate relative to the *meta* intermediate. If you draw the resonance structures for the *ortho* intermediate, you will find the same destabilization. The *meta* intermediate is the most stabilized of the three and is formed the most rapidly. Thus, the *meta* product predominates.

14.3 Sulfa Drugs

The use of herbs, plants, and potions as medicines has its roots in antiquity. However, with the advent of scientific thought and systematic testing, physicians' pharmacopoeia have been expanded to include a variety of drugs that even allow successful intercession in the course of an otherwise fatal disease. Among the first chemotherapeutic agents used to treat bacterial infections in humans were the drugs related to sulfanilamide—the **sulfa drugs**.

Although sulfanilamide was first synthesized in 1908, it was not until 1932–1934 that the German scientist G. Domagk reported that a related compound, Protonsil, could cure human infections. By the end of the decade, Domagk had received the Nobel prize in medicine and the sulfa drugs were hailed as "wonder drugs."

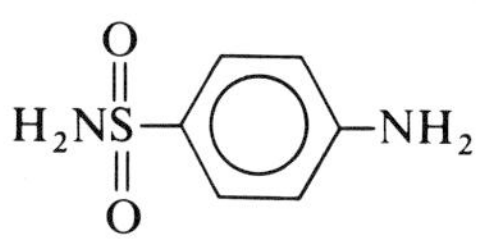

sulfanilamide

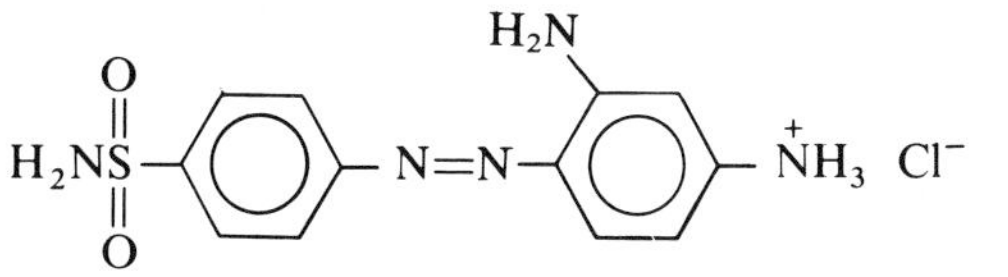

Protonsil

Shortly after Protonsil's introduction, systematic studies revealed that (1) a metabolite of Protonsil is *p*-acetamidobenzenesulfonamide, and (2) sulfanilamide itself is a bacteriostatic agent against the same bacteria as Protonsil. It was concluded that Protonsil is converted in the host to sulfanilamide, which is the actual active drug. Sulfanilamide, in turn, is converted to *p*-acetamidobenzenesulfonamide for elimination in the urine.

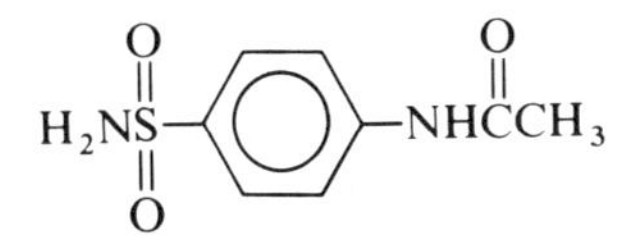

p-acetamidobenzenesulfonamide

On the basis of these conclusions, attention was then directed to the synthesis of structural variants of sulfanilamide. Since that time, over 5000 related compounds have been prepared and tested. Of these, only a few have been found sufficiently useful and safe to be marketed.

Sulfa drugs are not effective against all pathogenic bacteria, but only against those bacteria that use *p*-aminobenzoic acid (PABA) in the biosynthesis of folic acid, which is a necessary growth factor, or vitamin, in all animals. Sulfa drugs are not effective against bacteria (or animal cells) that do not use *p*-aminobenzoic acid, but instead use preformed folic acid.

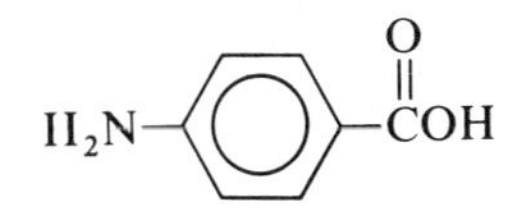

p-aminobenzoic acid (PABA)

It is currently believed that these two structurally similar compounds, sulfanilamide and *p*-aminobenzoic acid, compete for the same enzyme surface. At a sufficiently high concentration, the sulfa drug interferes with the bacterial biosynthesis of folic acid by occupying the majority of the enzyme sites, thus preventing the bacteria from growing or reproducing.

For the most part, sulfa drugs have been replaced by the broad-spectrum antibiotics such as penicillins and tetracyclines. However, a few sulfa drugs are still in medical use. For example, sulfadiazine (page 260), introduced in 1940, and sulfisoxazole (Gantrisin), introduced in 1946, are used to combat urinary infections.

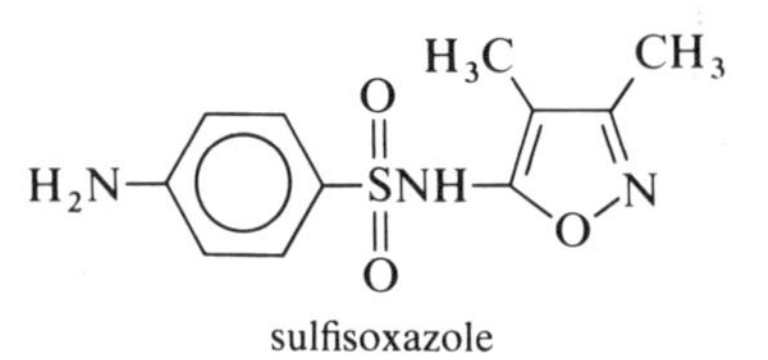

sulfisoxazole

14.4 Multistep Syntheses

Experiment 14.3 is a three-step conversion of acetanilide to sulfanilamide. Most syntheses carried out in organic research and manufacturing consist of a series of reactions in which the product of one reaction is the starting material for the next reaction. Such a set of reactions is called a **multistep synthesis**. Each reaction in a multistep synthesis is conducted as if it were an individual experiment. However, when viewed as a whole, there are two aspects of a multistep synthesis that should be considered: calculation of yield and purification of intermediate products.

There are two ways to calculate yields in a multistep synthesis: the overall yield of the entire sequence of reactions, and the yield of each intermediate step. The *theoretical overall yield* of the final product is calculated from the *limiting reagent* of the entire scheme. This is the reagent used in the smallest molar (or equivalent) quantity, as calculated from the complete set of balanced equations. A properly designed multistep synthesis always has the limiting reagent in the first step of the sequence.

$$\underset{\text{limiting reagent}}{A} + B \xrightarrow{\text{step 1}} C \xrightarrow{\text{step 2}} D \xrightarrow{\text{step 3}} E$$

The *actual overall yield* must be reported in a way that differentiates it from the yield of only the final step: for example, "per cent yield of E from A." Then, the yield of the final step alone can be reported as "per cent yield of E from D."

Purification of intermediate products in a multistep synthetic scheme is an important practical consideration. Unfortunately, how much purification is desirable is a question that usually must be answered from the experimental results of each individual synthetic scheme. Purification usually reduces the yield of an intermediate product and thus the overall yield of the sequence; yet, the lack of intermediate purification may result in a complex product mixture in the final step. In general, intermediate products should be at least partially purified.

In Experiment 14.3, the three-step conversion of acetanilide to sulfanilamide, the solid intermediate obtained from Step 1 is carried into Step 2 without drying; therefore, an accurate weight of the product cannot be obtained and an intermediate per cent yield cannot be calculated. The solid products from Steps 2 and 3 are dried; thus, an intermediate yield of the combined Steps 1 and 2 and an overall yield can be calculated. All intermediates in this scheme are solids, and partial purification is achieved by filtration. Additional purification of the intermediates is not recommended in this particular scheme.

Problems

14.1 Draw the resonance structures for the intermediate in the *meta* nitration of acetanilide and explain why the *meta* product is not formed in high yield.

14.2 Predict whether each of the following groups would be an *o, p*-director or a *m*-director. Explain your answers.

(a) $-N(CH_3)_2$ (b) $-\overset{+}{N}(CH_3)_3\ Cl^-$ (c) $-CO_2H$
(d) $-CHCl_2$ (e) $-Si(CH_3)_3$

14.3 Predict the principal product(s) of aromatic mononitration of the following compounds:

(a) 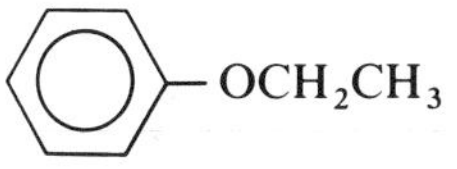 $C_6H_5-OCH_2CH_3$ (b) 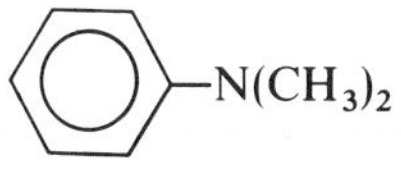 $C_6H_5-N(CH_3)_2$ (be careful!)

(c) 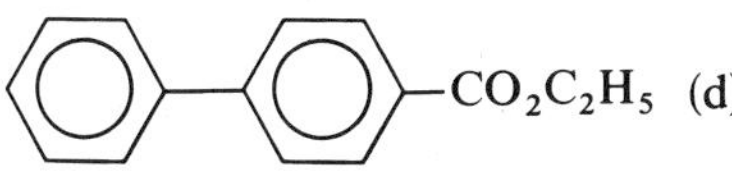 $C_6H_5-C_6H_4-CO_2C_2H_5$ (d) 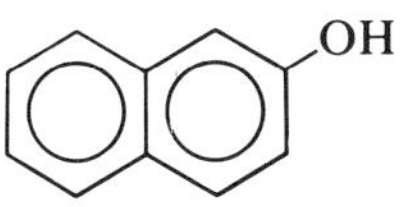 [naphthalene ring with OH]

14.4 Predict the relative rates of nitration of the following compounds. Explain your answer.

(a)  $CH_3O-C_6H_4-OCH_3$ (b) $CH_3O_2C-C_6H_4-CO_2CH_3$

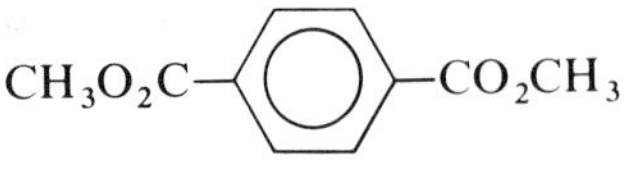

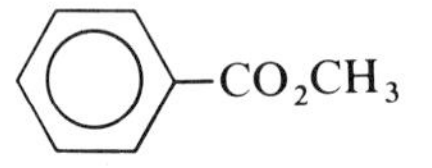

(c) $C_6H_5-CO_2CH_3$ (d) 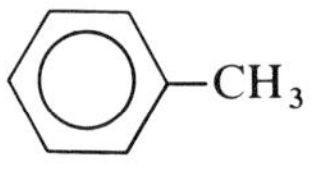 $C_6H_5-CH_3$

14.5 Phenol can be nitrated with nitric acid alone (no H_2SO_4).

(a) Write equations to show the formation of the electrophile and the reaction(s) of this electrophile with phenol.

(b) Explain why no sulfuric acid is needed in this reaction.

14.6 Refer to the equations for the steps in the conversion of acetanilide to sulfanilamide in Experiment 14.3 (pages 273–274). Write equations to show how you would convert acetanilide to sulfadiazine (page 260).

14.7 (a) Assume that you carried out Experiment 14.3 using 1.0 mol acetanilide, 3.0 mol $ClSO_3H$, and 2.0 mol of every other reagent, and that the yield of product was 85 g. What is the limiting reagent?

(b) In (a), what is the overall yield (expressed in the proper form)?

»»» EXPERIMENT 14.1 «««

Synthesis of p-Nitroacetanilide from Acetanilide

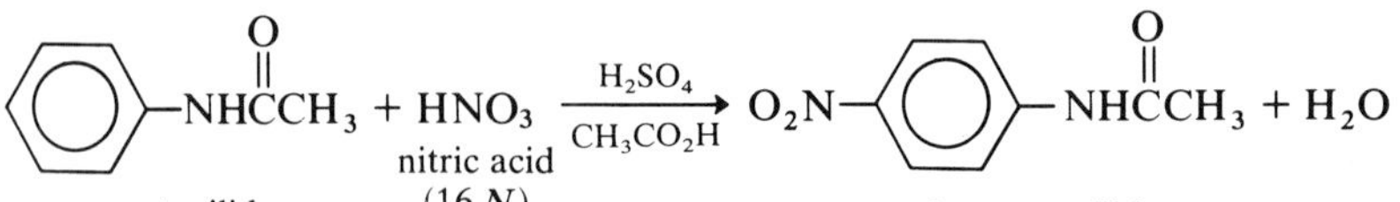

$$C_6H_5-NHCCH_3(=O) + HNO_3 \xrightarrow[CH_3CO_2H]{H_2SO_4} O_2N-C_6H_4-NHCCH_3(=O) + H_2O$$

	acetanilide	nitric acid (16 *N*)	*p*-nitroacetanilide
MW:	135.17	63.02	180.18
weight:	6.5 g	3.5 mL	8.7 g (theory)
moles:	0.048	0.056	0.048 (theory)

Table 14.2 Physical properties of reactants and products

Name	**Mp (°C)**	**Solubility**
acetanilide[a]	114	sl. sol. cold H_2O; sol. hot H_2O
p-nitroacetanilide	216	insol. cold H_2O; sol. hot H_2O; sol. hot CH_3CH_2OH
o-nitroacetanilide (a by-product)	93–94	moderately sol. cold H_2O; sol. hot H_2O; sol. CH_3CH_2OH

[a] Appreciably volatile at 95°.

Discussion

The aromatic nitration of acetanilide is an exothermic reaction; the temperature must be carefully controlled by chilling, stirring, and the slow addition of reagents. Acetanilide is first dissolved in the solvent, glacial acetic acid, by warming. The solution is cooled, then sulfuric acid is added; however, even with cooling, the temperature of the solution rises almost 40°. Both the acetanilide solution and the nitrating solution (a mixture of HNO_3 and H_2SO_4) must be chilled to about 10° before the reaction is begun.

To prevent dinitration of the acetanilide, the nitrating mixture is added in small portions to the acetanilide solution (and not vice versa) so that the concentration of HNO_3 is kept at a minimum. After all the HNO_3–H_2SO_4 solution has been added, the reaction mixture is allowed to warm slowly to room temperature. If the reaction mixture has been kept excessively cold during the addition, there will be a relatively large amount of unreacted HNO_3 present, which may cause the temperature to rise *above* room temperature. If this should happen, the mixture must be rechilled.

The work-up procedure consists of removal of the acids and crystallization of the product. Every trace of acid must be removed because hydrogen ions catalyze the hydrolysis of the amide to *p*-nitroaniline or its protonated cation. Most of the acid is removed by pouring the reaction mixture onto ice and water, then filtering the flocculent yellow precipitate of *p*-nitroacetanilide. The last traces of acetic acid are removed by neutralization. Because bases also catalyze the hydrolysis of amides, the neutralizing agent used is disodium hydrogen phosphate (Na_2HPO_4). This reagent reacts with acids to yield NaH_2PO_4. The result is a buffered solution with a pH near neutral.

$$CH_3\overset{\overset{\displaystyle O}{\|}}{C}OH + HPO_4^{2-} \rightleftarrows CH_3\overset{\overset{\displaystyle O}{\|}}{C}O^- + H_2PO_4^-$$

The crude product is air-dried before crystallization. If all of the acid was removed, the product will be light yellow. A deep yellow to yellow-orange product is indicative of the presence of *p*-nitroaniline from hydrolysis (see Experiment 16.2). Unfortunately, *p*-nitroaniline is difficult to remove from *p*-nitroacetanilide by crystallization.

Experimental

EQUIPMENT:

250-mL beaker
dropper or disposable pipet
two 50-mL and one 125-mL Erlenmeyer flasks
10-mL graduated cylinder
hot plate
ice bath
spatula
stirring rod
thermometer
vacuum filtration assembly
watch glass

CHEMICALS:

acetanilide, 6.5 g
disodium hydrogen phosphate, 15 g
95% ethanol, 60 mL
glacial acetic acid, 10 mL
conc. nitric acid, 3.5 mL
conc. sulfuric acid, 15 mL

TIME REQUIRED: about two hours to crude product; 15–20 minutes for crystallization; two overnight dryings; 15 minutes for melting-point determination

STOPPING POINTS: during either of the two drying periods or while the product is crystallizing from ethanol

>>>> ***SAFETY NOTE 1*** A mixture of concentrated nitric and sulfuric acids is used as the nitrating mixture. Use extreme caution when preparing and using this mixture.

>>>> ***SAFETY NOTE 2*** Nitro compounds are toxic and can be absorbed through the skin. You may wish to wear disposable plastic gloves during portions of this experiment.

PROCEDURE

Place 6.5 g of acetanilide in a 125-mL Erlenmeyer flask, add 10 mL of glacial acetic acid (CAUTION: *strong irritant*), and warm the flask on a hot plate in a fume hood until the acetanilide dissolves. Cool the flask in an ice bath to about 20°; then add 10 mL of cold, conc. sulfuric acid. The temperature of the mixture will rise to about 60°. Chill the solution to about 10° in an ice bath. (The solution will become very viscous.)

Mix 3.5 mL of conc. nitric acid and 5 mL of conc. sulfuric acid in a 50-mL flask, and chill the flask in an ice bath. When both solutions are cold, slowly add the HNO_3–H_2SO_4 solution, 1 mL at a time, to the acetanilide solution. Keep the reaction flask in an ice bath so that the temperature of the reaction mixture is maintained between 10–20°. Stir the reaction mixture carefully after each addition. The entire addition requires about 15 minutes.

After the addition is completed, allow the reaction flask to stand at room temperature for 30 minutes. Monitor the temperature; if it rises above 25°, chill the flask in an ice bath. Should the rechilling be necessary, allow the flask to stand for 30 minutes or more at room temperature after the rechilling.

Pour the reaction mixture into a 250-mL beaker containing 100 mL of water and 25 g of cracked ice. Using a large Büchner funnel, filter the heavy lemon-yellow precipitate with vacuum. Press out as much aqueous acid from the filter cake as possible with a spatula or clean cork while suction is being applied (CAUTION: see Safety Note 2). The precipitate is voluminous; use care in transferring it to the Büchner funnel or a substantial amount of product will be lost.

Transfer the filter cake to a clean 250-mL beaker, and add 100 mL of 15% aqueous disodium hydrogen phosphate. Stir the mixture to a paste-like consistency and refilter using vacuum. Wash the beaker with two 30-mL portions of cold water. Finally, wash the filter cake with an additional 50 mL of cold water. Press the filter cake with a spatula or clean cork to remove as much water as possible, then dry the solid overnight on a watch glass. Determine the yield and melting point.

The crude product can be purified by crystallization from 30–60 mL of 95% ethanol. (The crude product dissolves very slowly, even with heating; avoid using an excess of solvent.)

Problems

14.8 List at least two reasons for the choice of glacial acetic acid as the solvent for the nitration of acetanilide.

14.9 What would be the effects of each of the following changes of reaction conditions in this experiment, assuming that all other conditions are held constant? Explain your answers.

(a) increasing the amount of glacial acetic acid from 10 mL to 20 mL

(b) increasing the amount of nitric acid from 3.5 mL to 7.0 mL

(c) decreasing the amount of sulfuric acid from 15 mL to 5 mL

(d) allowing the temperature of the reaction mixture to stay at 40° during the addition of the HNO_3–H_2SO_4 solution

14.10 Write equations for the hydrolysis of *p*-nitroacetanilide in (a) aqueous acid; (b) aqueous hydroxide.

14.11 Predict what would happen during the crystallization of *p*-nitroacetanilide from 95% ethanol if all the acidic material had not been neutralized previously. Use an equation in your answer.

»»» EXPERIMENT 14.2 «««

Synthesis of Methyl m-Nitrobenzoate from Methyl Benzoate

$$C_6H_5COOCH_3 + HNO_3 \xrightarrow{H_2SO_4} m\text{-}NO_2C_6H_4COOCH_3 + H_2O$$

	methyl benzoate	nitric acid	methyl *m*-nitrobenzoate
MW:	136.14	63.02	181.15
weight:	6.1 g	4 mL	8.16 g (theory)
moles:	0.045	0.06	0.045 (theory)

Table 14.3 Physical properties of reactants and products

Name	*Mp (°C)*	*Bp (°C)*	*Density (g/mL)*	*Solubility*
methyl benzoate	—	198–200	1.09	insol. H_2O; sol. CH_3OH
methyl *m*-nitrobenzoate	78	—	—	insol. H_2O; sol. CH_3OH
methyl *o*-nitrobenzoate (a by-product)	−13	275	—	insol. H_2O; sol. CH_3OH

Discussion

The solvent for this reaction is concentrated sulfuric acid, which dissolves methyl benzoate by a *protonation reaction.* Because the protonation reaction is exothermic, the sulfuric acid must be chilled before it is mixed with the ester.

$$C_6H_5{-}\overset{\ddot{O}:}{\overset{\|}{C}}{-}\ddot{\underset{\cdot\cdot}{O}}CH_3 + H_2SO_4 \rightleftarrows C_6H_5{-}\overset{:\overset{+}{O}H}{\overset{\|}{C}}{-}\ddot{\underset{\cdot\cdot}{O}}CH_3 + HSO_4^-$$

In the nitration of methyl benzoate, as in the nitration of acetanilide, careful control of the reaction temperature and thorough mixing of the reactants during the reaction are essential. The nitrating mixture (HNO_3–H_2SO_4) is added in small portions to the chilled methyl benzoate–H_2SO_4 solution (not ester to nitrating mixture) to avoid an excess of HNO_3 in the reaction mixture and thus possible dinitration. After each addition, the viscous reaction mixture must be thoroughly stirred to avoid a localized high concentration of HNO_3.

The reaction is completed by allowing the chilled solution to warm to room temperature. If the reaction mixture was kept too cold during the reaction period, there may be an excess of unreacted HNO_3 present. If so, the temperature of the reaction mixture may exceed room temperature. If this occurs, the reaction mixture must be rechilled.

The work-up procedure consists of three main steps: (1) dilution with water and filtration of the product to separate it from the acids; (2) a methanol wash to remove the by-product methyl *o*-nitrobenzoate and unreacted starting ester; and (3) a final crystallization.

The first step in the work-up is pouring the reaction mixture onto cracked ice and water. The organic product and by-products separate as a solid. Any lumps of solid should be broken up with a spatula so that traces of acid will not be trapped inside. A filtration, followed by a water wash of the filter cake, removes virtually all of the acid from the organic material. (Because an *organic* acid is not used as solvent, as in Experiment 14.1, further treatment is not necessary to remove the last traces of acid.)

The organic solid consists of the desired methyl *m*-nitrobenzoate plus a substantial amount (up to 30%) of the *o*-isomer. Unreacted methyl benzoate is present as a contaminant. Both of these impurities are liquid and both are soluble in methanol, while the product methyl *m*-nitrobenzoate is a solid and only slightly soluble in methanol. Therefore, the impurities can be washed away with cold methanol. (Because the product is slightly soluble, an excess of methanol should be avoided.) After the washing with methanol, the crude product is air-dried, then crystallized from methanol.

Experimental

EQUIPMENT:

250-mL beaker
disposable pipet or dropper
two 50-mL and one 125-mL Erlenmeyer flasks
10-mL and 100-mL graduated cylinders
ice bath
spatula
stirring rod
thermometer
vacuum filtration assembly
watch glass

CHEMICALS:

methanol, 30 mL (95% ethanol may be substituted; see Experimental Note 3)
methyl benzoate, 6.1 g
conc. nitric acid, 4 mL
conc. sulfuric acid, 16 mL

TIME REQUIRED: two hours to crude product; 15–30 min. for crystallization; two overnight dryings; 15 min. for a melting-point determination

STOPPING POINTS: during either of the drying periods or when the product is crystallizing

>>>> ***SAFETY NOTE 1*** A mixture of concentrated nitric and sulfuric acids is used as the nitrating mixture. Use extreme caution when preparing and using this mixture.

>>>> ***SAFETY NOTE 2*** Nitro compounds are toxic and can be absorbed through the skin. You may wish to wear disposable plastic gloves during portions of this experiment.

PROCEDURE

Pour 12 mL of conc. sulfuric acid into an Erlenmeyer flask and chill to 0–10° in an ice bath. While the acid is cooling, weigh 6.1 g of methyl benzoate into a 10-mL Erlenmeyer flask. In another 50-mL Erlenmeyer flask, mix 4 mL of conc. nitric acid and 4 mL of conc. sulfuric acid and also cool this mixture in an ice bath.

When the original 12 mL of sulfuric acid is cold, add the methyl benzoate to it. The temperature of the solution will rise to about 25°. Rechill the acid–ester solution until its temperature is below 10°, then add 1 mL of the H_2SO_4–HNO_3 nitrating mixture dropwise from a dropper or disposable pipet. Mix thoroughly with a stirring rod after the addition. Keeping the temperature of the reaction mixture close to 10° by cooling (see Experimental Note 1), slowly add the rest of the nitrating mixture dropwise with stirring. The addition requires 20–40 minutes.

After the addition of the nitrating mixture is completed, remove the reaction flask from the ice bath and allow it to stand at room temperature for 15 minutes. If the temperature of the reaction flask exceeds room temperature, cool the flask, then allow it to stand at room temperature for an additional 15 minutes.

Pour the reaction mixture, with stirring, into a mixture of 100 mL of water and 50 g of cracked ice in a beaker. The crude product will form a white, curd-like precipitate. Break up the soft lumps with a spatula, then filter the precipitate with vacuum. Wash the filter cake thoroughly with three 10-mL portions of water to remove the acid and, using vacuum, suck the filter cake as dry as possible. Your instructor may wish you to take an infrared spectrum of this crude material before proceeding (see Experimental Note 2). Finally, wash the filter cake thoroughly with two 10-mL portions of cold methanol (see Experimental Note 3).

Air-dry the solid on a watch glass overnight. Determine the melting point and yield of the crude product (see Experimental Note 4). Crystallize the product from 10 mL of methanol, and air-dry the crystals overnight. A typical yield is 4.8 g.

EXPERIMENTAL NOTES

1) Because of the small volume, it may be necessary to tip the flask slightly so that the solution covers the thermometer bulb.

2) By comparing the infrared spectra of the crude product before and after the methanol wash, it is possible to verify that the *o*-isomer is indeed being leached out of the solid filter cake.

3) 95% Ethanol may be substituted for methanol washing and crystallization. If the acid has been carefully removed, the extent of ester exchange (leading to ethyl *m*-nitrobenzoate) will be minimal.

4) The melting point of the crude material is ill-defined. The solid begins to soften between 58° and 68°, shows liquid at about 70°, and finally melts at about 77°. The melting point of the crystallized product is much sharper.

Problems

14.12 Explain why methyl benzoate, when dissolved in sulfuric acid, is protonated on the carbonyl oxygen instead of on the methoxyl oxygen. Use formulas in your answer.

14.13 When methyl benzoate is subjected to mononitration, would you expect dinitro products to be important by-products? Explain.

14.14 What would be the effects of each of the following changes in reaction conditions in this experiment? Explain your answers, using equations where appropriate.

(a) allowing the methyl benzoate solution to warm to 50° during the addition of the HNO_3–H_2SO_4 mixture

(b) neglecting to wash the crude product with water before washing it with methanol

(c) washing the crude product with methanol at room temperature instead of using chilled methanol

»»» EXPERIMENT 14.3 «««

(SUPPLEMENTAL)

The Synthesis of Sulfanilamide: A Multistep Procedure

Step 1, Chlorosulfonation of acetanilide:

$$C_6H_5\text{–}NHCCH_3 (C{=}O) + 2ClSO_3H \longrightarrow ClS(O_2)\text{–}C_6H_4\text{–}NHCCH_3 (C{=}O) + H_2SO_4 + HCl$$

acetanilide; chlorosulfonic acid; *p*-acetamidobenzenesulfonyl chloride

	acetanilide	chlorosulfonic acid	*p*-acetamidobenzenesulfonyl chloride
MW:	135.16	116.52	—
weight:	5.0 g	13.0 mL (21.5 g)	—
moles:	0.037	0.18	—

Step 2, Preparation of p-Acetamidobenzenesulfonamide:

$$ClS(O_2)\text{–}C_6H_4\text{–}NHCCH_3 (C{=}O) + 2NH_3 \longrightarrow H_2NS(O_2)\text{–}C_6H_4\text{–}NHCCH_3 (C{=}O) + NH_4Cl$$

p-acetamidobenzenesulfonamide

Step 3, Conversion to Sulfanilamide:

$$H_2NSO_2-C_6H_4-NHCOCH_3 + H_2O + HCl \longrightarrow H_2NSO_2-C_6H_4-\overset{+}{N}H_3\ Cl^- + CH_3COOH$$

$$H_2NSO_2-C_6H_4-\overset{+}{N}H_3\ Cl^- + NaHCO_3 \longrightarrow H_2NSO_2-C_6H_4-NH_2 + NaCl + H_2O + CO_2$$

p-aminobenzenesulfonamide
(sulfanilamide)

MW:	—	172.22
weight:	—	6.4 g (theory)
moles:	—	0.037 (theory)

Discussion

The three-step synthesis of sulfanilamide is essentially the same procedure that was originally used to prepare the compound in 1908. A large number of different sulfanilamides can be synthesized by substituting amines for the NH_3 used in Step 2.

The first step in the synthesis is the chlorosulfonation of acetanilide. The reaction can be quite violent. For this reason, the acetanilide is first melted, then resolidified in the reaction vessel to decrease the surface area exposed to the chlorosulfonic acid and thus decrease the vigor of the reaction.

Chlorosulfonic acid, the half-acid chloride of sulfuric acid, undergoes dissociation to HCl and SO_3, which is the actual sulfonating agent. The reaction of SO_3 with acetanilide yields an initial arylsulfonic acid, which reacts with a second molecule of chlorosulfonic acid to yield the arylsulfonyl chloride. As in the case of the nitration of acetanilide, almost all of the substitution takes place at the *para* position.

$$HOSO_2Cl \rightleftharpoons SO_3 + HCl$$

$$CH_3CONH-C_6H_5 + SO_3 \rightleftharpoons CH_3CONH-C_6H_4-SO_2OH$$

$$CH_3CONH-C_6H_4-SO_2OH + ClSO_2OH \rightleftharpoons CH_3CONH-C_6H_4-SO_2Cl + H_2SO_4$$

After a short heating period, the principal components in the reaction mixture are unreacted chlorosulfonic acid (which was used in excess), sulfuric acid, and the product. The work-up consists of adding the reaction mixture to an ice–water mixture, which converts the excess chlorosulfonic acid to sulfuric acid, and filtering the solid product. The wet arylsulfonyl chloride should not be allowed to stand for an extended period of time because it undergoes slow hydrolysis.

$$CH_3C(=O)NH-C_6H_4-SO_2Cl + H_2O \xrightarrow{\text{slow}} CH_3C(=O)NH-C_6H_4-SO_2OH + HCl$$

Step 2 in the synthetic scheme is the reaction of the sulfonyl chloride with aqueous ammonia. The two are mixed and the mixture is swirled for a minute or so for reaction to occur (the formation of a semisolid mass), then the mixture is warmed. Premature heating will drive off NH_3 gas. The *p*-acetamidobenzenesulfonamide is isolated by vacuum filtration. This product is allowed to air-dry so that a weight and an intermediate per cent yield can be determined.

Step 3 in the synthesis of sulfanilamide is the acid hydrolysis of the *N*-acetyl group of the intermediate. This hydrolysis is carried out by heating *p*-acetamidobenzenesulfonamide and hydrochloric acid. (It is important that the hydrochloric acid be diluted prior to its addition, because concentrated acid will cause the sulfonamide to char.) As the mixture is heated, the water-insoluble sulfonamide is converted to the water-soluble sulfanilamide hydrochloride.

The subsequent conversion of this water-soluble salt to the water-insoluble sulfanilamide is complicated by the fact that the sulfanilamide is amphoteric.

$$H\bar{N}-S(=O)_2-C_6H_4-NH_2 \xleftarrow{OH^-} H_2NS(=O)_2-C_6H_4-NH_2 \xrightarrow{H^+} H_2NS(=O)_2-C_6H_4-\overset{+}{N}H_3$$

sulfanilamide

If the mixture is either too acidic or too alkaline, the sulfanilamide will be in the form of a water-soluble salt; therefore, the pH must be controlled. The use of $NaHCO_3$ allows the carbonic acid–bicarbonate equilibrium to buffer the system and thus keep the pH in the proper range.

As the sulfanilamide hydrochloride is converted to sulfanilamide, carbon dioxide is released. Toward the end of the neutralization reaction, the solid sulfanilamide precipitates as a froth of paste-like consistency. Care must be taken that enough $NaHCO_3$ is added to neutralize the hydrochloride salt completely and that solid $NaHCO_3$ is not trapped in the sulfanilamide.

The final product is isolated by a vacuum filtration followed by air-drying. If the melting point of the crude product is too low or if effervescence is observed upon melting (due to trapped solid $NaHCO_3$), the product should be recrystallized from water.

Experimental

EQUIPMENT:

two 250-mL beakers
condenser
two 125-mL Erlenmeyer flasks
25-mL and 100-mL graduated cylinders
heating mantle and rheostat
hot plate
ice bath
litmus paper
50-mL round-bottom flask
spatula
vacuum filtration assembly

CHEMICALS:

acetanilide, 5.0 g
conc. aqueous NH_3, 25 mL
chlorosulfonic acid, 13 mL
conc. HCl, 5 mL
sodium bicarbonate, about 6 g

TIME REQUIRED: Step 1, 1–1½ hours; Step 2, 1 hour; Step 3, 2 hours

STOPPING POINTS: after Step 2, while the solid *p*-acetamidobenzenesulfonamide is drying

>>>> ***SAFETY NOTE*** Chlorosulfonic acid is a strong acid and strong irritant. It reacts violently with water, releasing hydrochloric acid. Use ice to decompose any residual material.

PROCEDURE

Step 1, Chlorosulfonation of Acetanilide. Place 5.0 g of acetanilide in a 125-mL Erlenmeyer flask and, *in a fume hood*, heat the acetanilide on a hot plate until it is liquid. Allow the liquid to solidify, then chill the flask in an ice bath. Add 13 mL of chlorosulfonic acid (CAUTION: see Safety Note!). Let the flask stand at room temperature until the solid has dissolved, then heat the flask on a hot plate for 10 minutes. Allow the flask to cool to room temperature and slowly pour the contents, with vigorous stirring, into a 250-mL beaker containing 100 mL of an ice–water mixture. The hydrolysis is exothermic and a milky-white precipitate is formed. Collect the solid by vacuum filtration, using two 10-mL portions of water to aid in the transfer. Press the solid with a spatula and, using vacuum, suck off as much water as possible (see Experimental Note). Transfer the wet solid to a tared 125-mL Erlenmeyer flask. You should obtain about 8.5 g of crude wet solid. If it is necessary to stop the synthesis at this point, dry the chlorosulfonate as much as possible before storing it.

Step 2, Preparation of p-Acetamidobenzenesulfonamide. In a fume hood, add 25 mL of concentrated aqueous NH_3 to the Erlenmeyer flask containing the product prepared in Step 1. Swirl the flask until the reaction mixture forms a solid mass (about one minute). Warm the flask on a hot plate in the hood for 10–15 minutes. (Do not boil the mixture.) By the end of this period, almost all the solid will have dissolved. Let the flask cool to room temperature, then chill it in an ice bath. Vacuum-filter the solid, using 10 mL of water for washing. Air-dry the product overnight. You should obtain about 2 g, mp 211–214°.

Step 3, Conversion to Sulfanilamide. Place the crude *p*-acetamidobenzenesulfonamide in a 50-mL round-bottom flask and add a solution of 5 mL of concentrated hydrochloric acid in 10 mL of water. Equip the round-bottom flask with a condenser and heat the mixture at reflux for one hour. At the end of the reflux period, cool the flask to room temperature and transfer the contents to a 250-mL beaker, using 10 mL of water to wash out the residual material.

Chill the beaker in an ice bath. Add small portions of solid $NaHCO_3$ (about 0.2 g at a time), stirring with a spatula and letting the effervescence diminish between additions. Continue adding $NaHCO_3$ until the solution is alkaline to litmus paper. About 6 g will be required. Toward the end of the addition, the mixture will become a frothy paste. Vacuum-filter the solid, using two 10-mL portions of cold water to wash the solid into the Büchner funnel. Air-dry the solid overnight. Determine the yield and melting point of the crude product.

The crude sulfanilamide may be crystallized from water or 95% ethanol to yield pure sulfanilamide, mp 163–164°. Run an infrared spectrum of the crystallized product and identify the pertinent absorption peaks.

EXPERIMENTAL NOTE

The precipitate of *p*-acetamidobenzenesulfonyl chloride is fine and gelatinous. In the next step in the synthesis, it must be transferred to an Erlenmeyer flask. The dryer you can press the chlorosulfonate, the easier the transfer step will be. Any chlorosulfonate lost in this transfer, of course, will have a detrimental effect on the overall yield of the sulfanilamide.

Problems

14.15 Write the equation for the reaction of acetanilide with SO_3, showing the intermediate.

14.16 Explain with equations why gaseous HCl is given off in the chlorosulfonation.

14.17 Write equations for the reaction of benzenesulfonyl chloride ($C_6H_5SO_2Cl$) with the following compounds:

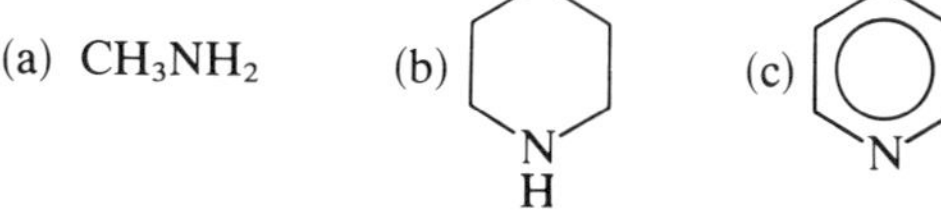

14.18 Ammonia (pK_a = 35) is one of the weakest acids known. Explain (with formulas) why sulfanilamide can lose a proton in aqueous base.

»»» CHAPTER 15 «««

The Chemistry of Aldehydes and Ketones

Many aldehydes (RCHO) and ketones ($R_2C{=}O$) are found in natural products. They are often used as flavorings and perfume, as are esters (RCO_2R). Acetophenone, the starting ketone in Experiment 15.1, has the odor of orange blossoms. Other examples follow:

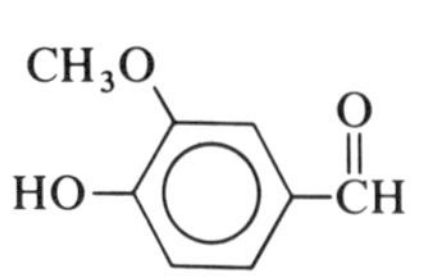

CH_3O, HO, O, CH

vanillin
vanilla flavoring

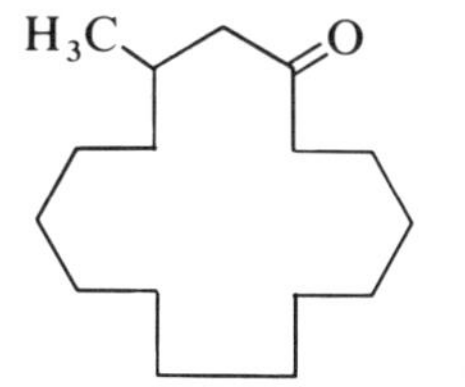

H_3C, O

muscone
in musk, used in perfumes

The chemistry of odors (including those associated with taste) is not well understood. It has been suggested that the nose contains a large number of receptor sites. Each site can accept only a molecule (or a portion of a molecule) that fits "just right" before a signal is sent to the brain. There is no direct simple correlation between structure and odor, however. Quite different compounds often have similar odors. For example, benzaldehyde, hydrogen cyanide, and mandelonitrile all have the aroma of bitter almonds, despite their disparity in structure. Mandelonitrile is the cyanohydrin addition product of benzaldehyde and HCN. These compounds are often found together in nature, a fact that may bear some relationship to their similarities in odors.

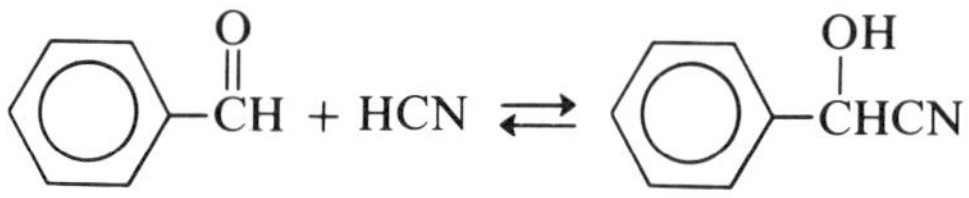

$$C_6H_5\text{-}\overset{O}{\overset{\|}{C}}H + HCN \rightleftarrows C_6H_5\text{-}\overset{OH}{\overset{|}{C}}HCN$$

benzaldehyde

mandelonitrile
a cyanohydrin

Mandelonitrile is found in the fruit pits of certain trees of the *Prunus* genus (bitter almonds, cherries, apricots, etc.). In this case, the cyanohydrin is found as a *glycoside* (that is, bonded to a sugar molecule). Amygdalin and laetrile are well-known examples of these glycosides. In the human digestive tract, these compounds are broken down to benzaldehyde, HCN, and glucose.

15.1 Addition Reactions of Aldehydes and Ketones

Most reactions of aldehydes and ketones occur because of the *polarity* of the carbonyl group:

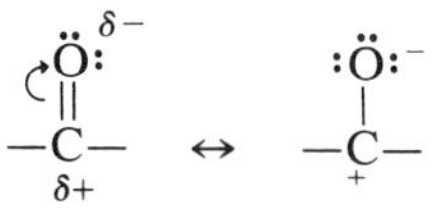

$$\overset{\ddot{O}:^{\delta-}}{\overset{\|}{-\underset{\delta+}{C}-}} \leftrightarrow \overset{:\ddot{O}:^{-}}{\overset{|}{-\underset{+}{C}-}}$$

The partially positive carbonyl carbon atom can be attacked by nucleophiles to undergo *simple addition reactions.* The Grignard reaction (Chapter 9) is an example of this type of reaction.

simple addition:

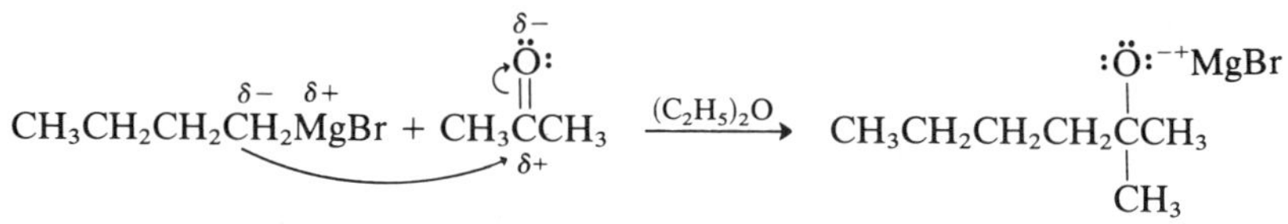

$$CH_3CH_2CH_2\overset{\delta-}{C}H_2\overset{\delta+}{MgBr} + CH_3\overset{\ddot{O}:^{\delta-}}{\overset{\|}{\underset{\delta+}{C}}}CH_3 \xrightarrow{(C_2H_5)_2O} CH_3CH_2CH_2CH_2\overset{:\ddot{O}:^{-}\,{}^{+}MgBr}{\overset{|}{\underset{\underset{CH_3}{|}}{C}}}CH_3$$

In *acidic solution*, the carbonyl oxygen is protonated. The protonation renders the carbonyl carbon even more positive, as the following resonance structures show:

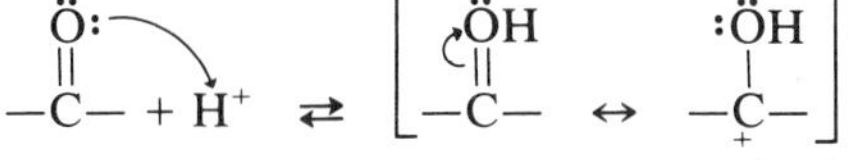

$$\overset{\ddot{O}:}{\overset{\|}{-C-}} + H^+ \rightleftarrows \left[\overset{\overset{+}{\ddot{O}}H}{\overset{\|}{-C-}} \leftrightarrow \overset{:\ddot{O}H}{\overset{|}{-\underset{+}{C}-}}\right]$$

resonance structures for the protonated carbonyl group

When the carbonyl group is protonated, it can be attacked by weak nucleophiles, such as alcohols, to yield simple addition products.

Besides simple addition reactions, aldehydes and ketones can undergo certain reactions that are the result of *simple addition followed by elimination.* In Experiment 21.1, you will identify carbonyl unknowns by means of *derivatives,* such as 2,4-dinitrophenylhydrazones. The formation of these derivatives is an example of an addition–elimination reaction.

$$(CH_3)_2C{=}O + H_2NNH\text{–}C_6H_3(NO_2)_2 \xrightarrow[\text{addition}]{H^+}$$

$$\left[(CH_3)_2\overset{OH}{\underset{|}{C}}\text{–}NHNH\text{–}C_6H_3(NO_2)_2\right] \xrightarrow[\text{elimination}]{-H_2O} (CH_3)_2C{=}NNH\text{–}C_6H_3(NO_2)_2$$

unstable a 2,4-dinitrophenylhydrazone

15.2 Reduction of Aldehydes and Ketones

Aldehydes and ketones can undergo two general types of reduction: **normal reduction** and **bimolecular reduction**, or **reductive coupling**. The normal reduction of an aldehyde or ketone yields a monomeric alcohol as the product. Bimolecular reduction, however, yields a dimeric product.

normal reduction:

$$R_2C{=}O \xrightarrow{NaBH_4} R_2CHOH$$

a bimolecular reduction, or reductive coupling:

$$2\,R_2C{=}O \xrightarrow[(2)\ H_2O,\ H^+]{(1)\ Mg} R_2\overset{OH}{C}\text{—}\overset{OH}{C}R_2$$

In general, the metal hydrides ($LiAlH_4$, $NaBH_4$) give only normal reduction products. Experiment 15.1 is the sodium borohydride reduction of a ketone, acetophenone. The product of the reaction is a secondary alcohol. Active metal reducing agents (Mg^0, Zn^0) can result in either type of reduction. Which product actually predominates depends upon both the reaction conditions and the structure of the compound undergoing reduction. Experiment 15.2 is the reductive coupling of xanthone with zinc metal.

A. Borohydride Reduction of Aldehydes and Ketones

Until the 1940's, the standard reducing agents at an organic chemist's disposal were hydrogen gas and the active metals. For the preparation of alcohols, these reducing agents have limited utility. For example, hydrogen gas usually requires high temperatures, high pressures, and an appropriate catalyst to reduce carbonyl groups. As has been mentioned, the active metals often give bimolecular products.

Today, a variety of metal hydride reducing agents are available. The discovery of the reducing powers of *sodium borohydride* (the reducing agent you will use in Experiment 15.1) occurred quite by accident during the Second World War. Herbert C. Brown (who received a Nobel Price in 1979 for his work in this area) was searching for a solvent to purify this borohydride. When Brown tried acetone, he observed that a reaction took place, and then determined that the acetone had been reduced.

$$4\,CH_3\overset{\overset{\displaystyle O}{\|}}{C}CH_3 + Na^+\ ^-BH_4 \xrightarrow{25^\circ} Na^+\ ^-B\left(\begin{array}{c} CH_3 \\ | \\ OCH \\ | \\ CH_3 \end{array}\right)_4 \xrightarrow[\text{warm}]{H_2O}$$

acetone — sodium borohydride

$$4\,CH_3\overset{\overset{\displaystyle OH}{|}}{C}HCH_3 + Na^+\ ^-B(OH)_4$$

2-propanol (isopropyl alcohol)

Shortly after Brown's discovery, the potent reducing power of *lithium aluminum hydride* ($LiAlH_4$) was reported. Both these metal hydrides are commonly used in the organic laboratory.

The reducing powers of $NaBH_4$ and $LiAlH_4$ are quite different. $LiAlH_4$ is sensitive to air and moisture and reduces practically any polar reducible group—for example, the carbonyl groups in aldehydes, ketones, carboxylic acids, carboxylic acid derivatives, and others. By contrast, $NaBH_4$ is relatively mild. It is not decomposed rapidly by water unless the two are heated. In fact, water is often used as a solvent for borohydride reductions. Although $NaBH_4$ reduces aldehydes and ketones rapidly at room temperature, it reduces esters and other carbonyl compounds slowly under the same conditions. For this reason, an aldehyde or ketone carbonyl group can be reduced by $NaBH_4$ in the presence of an ester group.

$$\underset{\text{a ketone}}{R_2C{=}O} \xrightarrow[\text{fast}]{NaBH_4,\ 25^\circ} R_2CHOH$$

$$\underset{\text{an ester}}{RCO_2R'} \xrightarrow[\text{very slow}]{NaBH_4,\ 25^\circ} RCH_2OH + HOR'$$

Neither $LiAlH_4$ nor $NaBH_4$ usually reduces isolated carbon–carbon double bonds. However, *polarized* carbon–carbon double bonds, such as those in conjugation with a carbonyl group ($C{=}C{-}C{=}O$), can be reduced by either hydride unless the reaction conditions are carefully controlled.

Mechanism of the borohydride reduction. Let us illustrate the mechanism of borohydride reduction using acetophenone, the compound you will reduce in Experiment 15.1. The reduction of the carbonyl group takes place by the nucleophilic attack of BH_4^- on the carbonyl carbon. In this reaction, the large amount of ethanol used as the solvent converts the product alkoxide to the product alcohol. If no alcohol or water solvent were present, then the product alcohol would be tied up as an alkoxyboron ion, as was shown for the reaction of acetone with $NaBH_4$ above.

reduction of acetophenone by BH_4^-:

$$C_6H_5{-}\overset{\overset{\displaystyle \ddot{O}\!:^{\delta-}}{\|}}{\underset{\delta+}{C}}{-}CH_3 \quad H{-}\overset{\overset{\displaystyle H}{|^-}}{\underset{\underset{\displaystyle H}{|}}{B}}{-}H \rightarrow \left[\ C_6H_5{-}\overset{\overset{\displaystyle :\ddot{O}:^-}{|}}{\underset{\underset{\displaystyle H}{|}}{C}}{-}CH_3 \quad \overset{\overset{\displaystyle H}{|}}{\underset{\underset{\displaystyle H}{|}}{B}}{-}H\ \right] \xrightarrow{CH_3CH_2OH}$$

$$\underset{\substack{\text{1-phenylethanol} \\ \textit{the product alcohol}}}{C_6H_5{-}\overset{\overset{\displaystyle :\ddot{O}H}{|}}{C}HCH_3} + \underset{\text{ethoxyborohydride ion}}{CH_3CH_2O{-}\bar{B}H_3}$$

The ethoxyborohydride ion that results can attack another molecule of ketone and lose a second hydrogen atom.

$$C_6H_5{-}\overset{\overset{\displaystyle O}{\|}}{C}CH_3 + CH_3CH_2O\underset{\underset{\displaystyle H}{|}}{\bar{B}}H_2 \xrightarrow{CH_3CH_2OH} C_6H_5{-}\overset{\overset{\displaystyle OH}{|}}{C}HCH_3 + (CH_3CH_2O)_2\bar{B}H_2$$

The reaction sequence continues until all four hydrogens of the original BH_4^- ion have been transferred. Therefore, one mole of $NaBH_4$ can reduce, theoretically, four moles of ketone. Because $NaBH_4$ undergoes a slow hydrolysis with atmospheric moisture, it is often impure; thus, an excess of $NaBH_4$ is generally used.

B. Reductive Coupling

Reductive coupling of ketones to 1,2-diols with an active metal such as sodium, magnesium, zinc, or aluminum proceeds by way of a *ketyl*, an anion-radical that undergoes coupling to yield the dianion of the 1,2-diol. The diol is obtained when the mixture is treated with water.

$$2R\overset{\overset{\displaystyle :\ddot{O}:}{\|}}{C}R + :Mg \longrightarrow 2R\underset{\bullet}{\overset{\overset{\displaystyle :\ddot{O}:^-}{|}}{C}}R + Mg^{2+}$$

a ketyl

$$\longrightarrow R_2\overset{\overset{\displaystyle ^-:\ddot{O}:}{|}}{C}{-}\overset{\overset{\displaystyle :\ddot{O}:^-}{|}}{C}R_2 \xrightarrow[-2\,OH^-]{2\,H_2O} R_2\overset{\overset{\displaystyle OH}{|}}{C}{-}\overset{\overset{\displaystyle OH}{|}}{C}R_2$$

a 1,2-diol

Experiment 15.2 is the reductive coupling of xanthone to yield dixanthylene, a *thermochromic compound*. (Thermochromism is described in the discussion section in the experiment.) Although this reaction does not result in a diol, the mechanism is probably similar to that for diol formation.

$$2\ \text{xanthone} \xrightarrow[CH_3CO_2H]{Zn,\ HCl} \text{dixanthylene}$$

xanthone

dixanthylene

Problems

15.1 What would be the reagents of choice for the following conversions?

(a) $C_6H_{11}{-}CO_2CH_2CH_3 \longrightarrow C_6H_{11}{-}CH_2OH$

(b) cyclohexanone $\longrightarrow$ cyclohexanol (cyclohexyl–OH)

(c) $CH_3CH{=}CHCH_2CHO \rightarrow CH_3CH_2CH_2CH_2CH_2OH$

(d) $CH_3CH{=}CHCH_2\overset{O}{\overset{\|}{C}}CH_3 \rightarrow CH_3CH_2CH_2CH_2\overset{O}{\overset{\|}{C}}CH_3$

(e) $2\ C_6H_5\overset{O}{\overset{\|}{C}}CH_3 \rightarrow C_6H_5\underset{CH_3}{\overset{OH}{C}}—\underset{CH_3}{\overset{OH}{C}}C_6H_5$

(f) cyclohexanone ($=O$) $\longrightarrow$ cyclohexanone hydrazone ($=NNH_2$)

15.2 Explain why a Grignard reagent attacks an aldehyde or ketone without the presence of a catalyst, but an alcohol requires an acidic catalyst to undergo reaction with these carbonyl compounds.

15.3 Acetone is treated with the following reagents. Write an equation for each reaction, indicating whether an acidic catalyst would be required or not.

(a) CH_3CH_2Li (b) $CH_3C{\equiv}CMgBr$
(c) $NaBH_4$ (d) $HOCH_2CH_2OH$
(e) $C_6H_5NHNH_2$

15.4 Reductive coupling reactions leading to 1,2-diols usually give better yields when aryl ketones instead of aliphatic ketones are the reactants. Suggest a reason for this.

15.5 Dixanthylene suffers from severe steric hindrance. Explain.

»»» EXPERIMENT 15.1 «««

Sodium Borohydride Reduction of Acetophenone

$$4\ C_6H_5\overset{O}{\overset{\|}{C}}CH_3 + NaBH_4 \xrightarrow{CH_3CH_2OH} 4\ C_6H_5\overset{OH}{\overset{|}{C}}HCH_3 + \text{a mixture of boron salts}$$

	acetophenone	sodium borohydride	1-phenylethanol
MW:	120.16	37.83	122.17
weight:	12.0 g	1.5 g	12.2 g (theory)
moles:	0.100	0.040	0.100 (theory)

Table 15.1 Physical properties of reactants and products

Name	*Bp (°C)*	n_D^{20}	*Density (g/mL)*	*Solubility*
acetophenone[a]	202	1.5339	1.03	insol. H_2O; sol. CH_3CH_2OH
sodium borohydride	—	—	—	sol. H_2O; sl. sol. CH_3CH_2OH
1-phenylethanol[b]	203.4	1.5275	1.03	insol. H_2O; sol. CH_3CH_2OH

[a] Mp 20.5°
[b] Mp 20°

Discussion

The reduction of an aldehyde or ketone with sodium borohydride is straightforward and usually affords a high yield of the alcohol. The usual procedure (and the one employed in this experiment) involves dissolving the borohydride in 95% ethanol and adding the carbonyl compound to this solution. To ensure complete reaction, an excess of sodium borohydride is used.

The reaction between sodium borohydride and acetophenone is exothermic. Therefore, it is important to add the acetophenone *dropwise* and to control the reaction temperature with an ice bath. After the reaction has been completed, the excess borohydride and the ethoxyborohydrides are destroyed with aqueous acid. Because hydrogen gas is evolved, this treatment with acid must be carried out in a fume hood or a very well-ventilated room.

$$\underset{\text{an ethoxyborohydride ion}}{CH_3CH_2O\bar{B}H_3} + 3\,H_2O + H^+ \rightarrow \underset{\text{boric acid}}{H_3BO_3} + CH_3CH_2OH + 3\,H_2$$

Because ethanol, the reaction solvent, is water-soluble, a clean separation of organic and inorganic products cannot be achieved by a simple extraction with water and diethyl ether at this point. (Too much product would be lost in the aqueous ethanol layer.) To circumvent this problem, the first step in the work-up is to boil off much of the ethanol. In a larger-scale reaction, the ethanol would be distilled and collected. In a small-scale reaction such as in this experiment, the ethanol can be boiled away in the fume hood. When most of the ethanol has been removed, the product 1-phenylethanol oils out.

Water and diethyl ether are then added to the residue for the extraction of the organic compounds from the inorganic salts. The ether extract is dried with either sodium sulfate or magnesium sulfate. The crude product is obtained by distilling the ether.

Because of its high boiling point, 1-phenylethanol cannot be distilled at atmospheric pressure. Although it could be vacuum-distilled, distillation could not separate it from unreacted starting material (if any) because the two compounds have boiling points only 1° apart. (An infrared spectrum can be used to determine if any ketone is present in the product.)

Experimental

EQUIPMENT:

150-mL beaker
distillation apparatus
dropper
three 50-mL Erlenmeyer flasks
10-mL and 50-mL graduated cylinders
hot plate or steam bath
ice bath
two 50-mL round-bottom flasks
125-mL separatory funnel
thermometer

magnetic stirrer (optional)

CHEMICALS:

acetophenone, 12.0 g
anhydrous magnesium sulfate (or Na_2SO_4), 1 g
diethyl ether, 40 mL
95% ethanol, 30 mL
3*M* hydrochloric acid, 10 mL
sodium borohydride, 1.5 g

TIME REQUIRED: about $4\frac{1}{2}$ hours

STOPPING POINTS: after the acetophenone has been added; after the excess ethanol has been boiled off; while the ether solution is drying

>>>> ***SAFETY NOTE*** Sodium borohydride is caustic. Do not let it come into contact with your skin. If accidental contact should occur, wash immediately and thoroughly.

PROCEDURE

Place 1.5 g of sodium borohydride (see Safety Note) in a 150-mL beaker. Add 30 mL of 95% ethanol and stir until the solid is dissolved (see Experimental Note). Weigh 12.0 g of acetophenone into a 50-mL Erlenmeyer flask, and prepare an ice bath.

Add the acetophenone dropwise to the borohydride solution while stirring the mixture continuously, preferably with a magnetic stirrer. Keep the temperature of the reaction mixture between 30–50° by controlling the rate of addition and by cooling the beaker in the ice bath as necessary. As the acetophenone is added, a white precipitate forms. The addition should take about 45 minutes. After the addition is complete, allow the reaction mixture to stand at room temperature for 15 minutes with occasional stirring (or continuous stirring if you are using a magnetic stirrer).

In a fume hood, add about 10 mL of 3*M* HCl to the reaction mixture. After the reaction has subsided, heat the mixture to boiling on a hot plate or steam bath in the fume hood until the mixture separates into two layers.

Cool the reaction mixture in an ice bath, then transfer it to a separatory funnel. Wash the residual material in the beaker into the separatory funnel with 20 mL of diethyl ether (*flammable*). If the inorganic salts precipitate, add 20–40 mL of water, as necessary, to dissolve them. Extract the aqueous layer with this ether, then with a second 20-mL portion of ether; combine the ether extracts; wash them with an equal volume of water; and dry them with anhydrous magnesium sulfate or sodium sulfate.

Decant or filter the dried solution into a tared flask, and distil the ether slowly, using a heating mantle or steam bath and an efficient condenser. Do not overheat the flask. A typical crude yield is 10 g (82%). Measure the refractive index, and run the infrared spectrum (thin film). From these two pieces of data, estimate the purity of the crude product.

EXPERIMENTAL NOTE

Fresh sodium borohydride dissolves in 95% ethanol, but partially hydrolyzed sodium borohydride will not all dissolve. This should not affect the experiment because an excess of borohydride is used. (If your borohydride is extremely poor quality, your instructor may suggest that you use a greater excess than is called for.)

Problems

15.6 Write an equation for the hydrolysis of $NaBH_4$.

15.7 Under what circumstances would you expect to find unreacted acetophenone in the product of Experiment 15.1?

15.8 What precautions would you take if you were carrying out the following borohydride reduction?

$$\overset{\overset{\large O}{\|}}{HC}CH_2CH_2\overset{\overset{\large O}{\|}}{C}OCH_3 \xrightarrow[(2)\ H_2O,\ H^+]{(1)\ NaBH_4} \overset{\overset{\large OH}{|}}{CH_2}CH_2CH_2\overset{\overset{\large O}{\|}}{C}OCH_3$$

15.9 Which compound would you expect to undergo borohydride reduction more rapidly? Explain.

(a) CH_3CH_2CHO or (b) $CH_3CH_2\overset{\overset{\large O}{\|}}{C}CH_2CH_3$

EXPERIMENT 15.2

(SUPPLEMENTAL)

Synthesis of Dixanthylene, a Thermochromic Compound

2 xanthone + 2Zn + 4HCl $\xrightarrow{CH_3CO_2H}$ dixanthylene + $2ZnCl_2$ + $2H_2O$

	xanthone	Zn	dixanthylene
MW:	196.22	65.38	360.44
weight:	2.5 g	3.2 g	2.3 g (theory)
moles:	0.013	0.049	0.0065 (theory)

Table 15.2 *Physical properties of reactants and products*

Name	*Mp (°C)*	*Solubility*
xanthone	173–174	sl. sol. CH_3CO_2H
dixanthylene	315	insol. CH_3CO_2H; sl. sol. aromatic hydrocarbons

Discussion

Thermochromism is the reversible change in the color of a substance with a change in temperature. The thermochromism of liquid crystals (Experiment 2.3) arises from the partial ordering of molecules in a *mesophase*, a state between a completely ordered solid and a random liquid. With cholesteric liquid crystals, a small change in temperature results in a change in the twist of the helical aggregate, which, in turn, causes changes in the angle of reflection of light from the surface.

The formation of a mesophase causing a change in the angle of reflection of light is not the only cause of thermochromism. The actual colors of certain compounds can undergo change with changes in temperature (just as an acid–base indicator undergoes pH-dependent color changes). A color change arises from a shift in the wavelengths of visible light being absorbed by a compound. This shift can be caused by *structural changes* (tautomerism, diradical formation, etc.) or by *physical changes*, such as a change in the conformation of the molecules. Bianthrone (Figure 15.1) and dixanthylene are two thermochromic compounds that exhibit color changes because of thermal conformational changes. The thermochromic behavior of bianthrone was first observed in 1909 and that of dixanthylene in 1928. Most of the original research has been carried out with bianthrone; therefore, we will emphasize this structure in our discussion.

Numerous suggestions as to why these compounds exhibit thermochromism have been proposed. It was recognized early that these compounds suffer from severe steric hindrance (between positions 1 and 8′ and

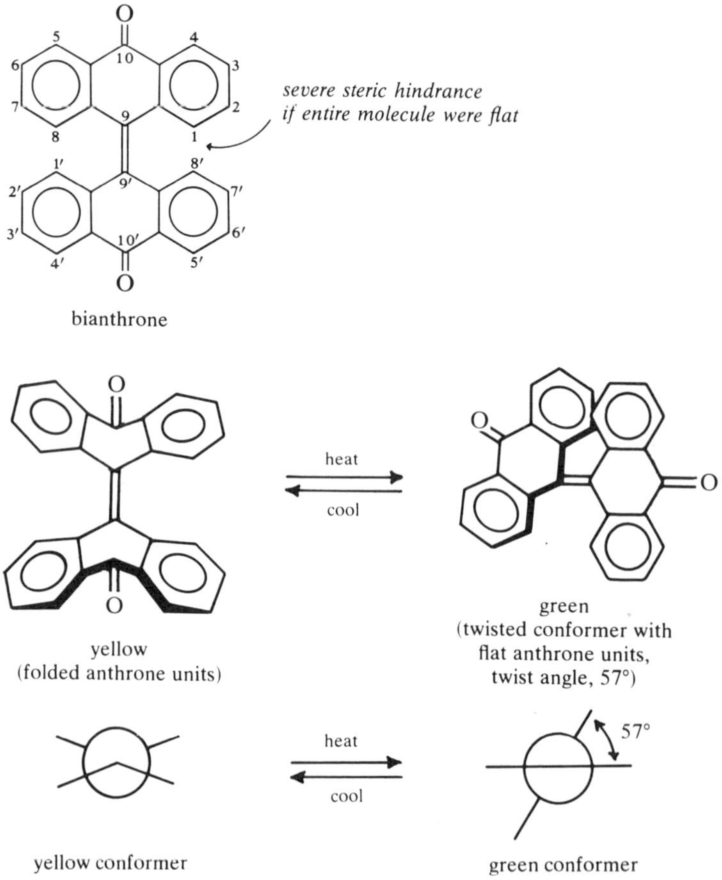

Newman projections through the 9,9′ carbons of bianthrone, showing the geometric relationship of the two anthrone units

Figure 15.1 Conformational changes of bianthrone with heat (adapted from R. Korenstein, K. A. Muszkat, and S. Sharafy-Ozeri, J. Amer. Chem. Soc. ***1973****, 95, 6177).*

positions 1′ and 8, as shown in Figure 15.1). The two ways in which one of these molecules can relieve the hindrance are by a folding of the central ring in each anthrone unit or by twisting about the central 9–9′ double bond. It was shown in 1954 by x-ray crystallography that molecules of the solid yellow bianthrone are not flat, but that each anthrone unit is folded into a "tent" form, as shown in Figure 15.1. In this conformation, there can be very little conjugation between the two aromatic rings in each anthrone unit. (There is enough conjugation, though, to give crystalline bianthrone its yellow color.)

The conformation of the molecules of a liquid cannot be determined by x-ray crystallography; therefore, the conformation of the molecules in the green liquid bianthrone could not be determined by the early investigators. A relatively recent study on the structure of one of the green forms of bianthrone depended on two techniques unavailable to earlier workers: (1) computer analysis of the molecular orbitals and strain energies of the possible conformers, and (2) nmr spectroscopic analysis of bianthrone and substituted bianthrones. The results of this study indicate that, in the green form of bianthrone, the molecules are twisted about the central double bond and that each anthrone unit is flat (and thus more conjugated), as shown in Figure 15.1.

Dixanthylene is structurally very similar to bianthrone and probably undergoes similar conformational changes. In this experiment, you will synthesize dixanthylene by the reductive coupling of xanthone and observe its thermochromic behavior.

Suggested Readings

Ault, A., Kopet, R., and Serianz, A. *J. Chem. Ed.* **1971**, *48*, 410

Day, J. H. "Thermochromism." *Chem. Rev.* **1963**, *63*, 65

Korenstein, R., Muszkat, K. A., and Sharafy-Ozeri, S. *J. Amer. Chem. Soc.* **1973**, *95*, 6177

Experimental

EQUIPMENT:

condenser
droppers or disposable pipets
two 125-mL Erlenmeyer flasks
funnel
100-mL graduated cylinder
heating mantle and rheostat
hot plate
ice bath
microburner
250-mL round-bottom flask
vacuum filtration apparatus

CHEMICALS:

p-cymene (*p*-isopropyltoluene), 50 mL
glacial acetic acid, 150 mL
conc. hydrochloric acid, 1–3 mL
xanthone, 2.5 g
zinc dust, 3.2 g

TIME REQUIRED: 4–6 hours

STOPPING POINTS: after the reflux period; while the dixanthylene is drying

PROCEDURE

Place 2.5 g of xanthone (see Experimental Note 1) in a 250-mL round-bottom flask, add 100 mL of glacial acetic acid (*strong irritant*), and swirl the flask until most of the xanthone has dissolved (about 10 minutes). Then add 0.8 g of zinc dust (see Experimental Note 2) and 2 drops of concentrated hydrochloric acid (*strong acid, irritating fumes*). Fit the flask with a reflux condenser.

From this point to the end of the reaction period (about two hours), add 2 drops of concentrated hydrochloric acid through the reflux condenser every five minutes. Since some of the acid will cling to the side of the condenser, wash it into the reaction mixture with about 1 mL of glacial acetic acid.

Heat the reaction mixture to reflux and note the time at which the mixture begins to boil. After 20 minutes of reflux, allow the mixture to cool and chill it to about 10° (not in an ice bath because the glacial acetic acid will solidify). Then add a second 0.8-g portion of zinc dust directly to the reaction mixture, not through the condenser (see Experimental Note 3). Replace the reflux condenser immediately, add a fresh boiling chip, and again heat the mixture to reflux. Repeat this addition two more times after 20-minute reflux periods. (A total of 3.2 g of zinc dust will have been added.) After the last addition, continue the reflux for 30 minutes.

Discontinue the HCl additions, and allow the reaction mixture to cool to room temperature. Vacuum-filter the mixture, using two 10-mL portions of glacial acetic acid to wash the organic solid and the residual zinc dust into the Büchner funnel. Using vacuum, suck as much acetic acid from the solid as possible.

Transfer the solid from the Büchner funnel to an Erlenmeyer flask and add 50 mL of *p*-cymene (see Experimental Note 4). Heat the mixture to boiling, then gravity-filter the blue-green solution (see Experimental Note 5). Allow the filtrate to cool to room temperature, then chill it in an ice bath. Collect the yellow solid by vacuum filtration and air-dry it in the fume hood or in a desiccator containing paraffin shavings. You should obtain about 0.5 g of product. The melting point of dixanthylene (315°) is too high to determine with the usual student melting-point apparatus.

To observe the thermochromism of dixanthylene, place 1.5–2 cm in a melting-point capillary and melt the dixanthylene by holding the capillary near (not in) a flame. The yellow dixanthylene crystals give a dark blue melt. On cooling, yellow crystals reform. (Unfortunately, dixanthylene decomposes upon melting, partially obscuring these transitions.)

EXPERIMENTAL NOTES

1) Xanthone can be obtained from the Aldrich Chemical Co. It can also be prepared by the pyrolysis of phenyl salicylate (A. F. Holleman and H. T. Clarke, *Organic Syntheses*, Collective Vol. I, 2nd ed., Henry Gilman, Ed. New York: John Wiley and Sons, Inc., 1941, page 552), a procedure suitable only for an advanced student.

2) Fresh zinc dust should be used and its exposure to the air should be minimized to prevent air oxidation.

3) If the zinc dust is added to the *hot* reaction mixture, vigorous boiling will ensue; this boiling is due in part to the small particles striking the hot solution and in part to the reaction between zinc and HCl.

Zinc dust is quite dense, and a considerable percentage of it can be lost quite easily. When adding the zinc dust to the reaction mixture, do it in such a manner that it does not come into contact with the ground-glass joint of the flask. One way to do this is to weigh the zinc dust into the crease near the edge of a creased weighing paper. Carefully center the crease over the ground-glass joint and pour the dust into the reaction mixture by tilting or tapping the paper. If you should inadvertently get dust on the joint, wash it into the flask with a little glacial acetic acid.

4) *p*-Cymene (bp 177°) has an unpleasant odor that clings to clothing. Use appropriate precautions. Other high-boiling alkylbenzenes are also suitable as crystallization solvents. The original procedure calls for 25 mL of 1,3,5-trimethylbenzene (mesitylene), bp 164°.

5) The thermochromic behavior of dixanthylene can be observed during this portion of the procedure: the hot solution is blue-green, while the cool solution is yellow.

»»» CHAPTER 16 «««

Carboxylic Acids and their Derivatives

Both in nature and in industry, carboxylic acids and their derivatives (esters, amides, acid halides, acid anhydrides, and nitriles) are among the most important organic compounds. Proteins, soaps, vinegar, synthetic textiles, and aspirin are just a few examples of the diverse chemicals that fall into this broad category.

The experiments in this chapter include only a representative few of the many reactions that carboxylic acids and their derivatives can undergo. Two esterification experiments include the synthesis of aspirin and, using a generalized procedure, the synthesis of an ester. Two hydrolysis experiments, an alkaline hydrolysis (saponification) and an acidic hydrolysis, demonstrate typical reactions of carboxylic acid derivatives. The last experiment in this chapter is the synthesis of a nylon polymer.

16.1 Hydrolysis and Saponification

All the derivatives of carboxylic acids can undergo **hydrolysis** ("cleavage by water") to yield the carboxylic acids themselves. These hydrolysis reactions can be carried out under either acidic or alkaline conditions. For example:

$$\underset{\text{an ester}}{\mathrm{R\overset{\overset{\displaystyle O}{\|}}{C}OR'}} + \underset{\text{excess}}{H_2O} \overset{H^+}{\rightleftharpoons} \underset{\text{a carboxylic acid}}{\mathrm{R\overset{\overset{\displaystyle O}{\|}}{C}OH}} + \underset{\text{an alcohol}}{\mathrm{HOR'}}$$

In the case of an ester, the acid-catalyzed hydrolysis is reversible; therefore, a large excess of water is used to drive the reaction to the carboxylic acid and alcohol side. The acidic hydrolysis of the other derivatives is not reversible.

A. Alkaline Hydrolysis of an Ester: Saponification

The alkaline hydrolysis of an ester is called **saponification.** This reaction is irreversible and thus usually gives a better yield of a carboxylic acid and an alcohol than does acidic hydrolysis of an ester.

Saponification means "the making of soap." Indeed, soaps are made just this way: the treatment of a fat (a triester of glycerol) with aqueous sodium hydroxide.

$$\underset{\text{a fat}}{\begin{array}{c} CH_2OC(=O)R \\ | \\ RC(=O)OCH \\ | \\ CH_2OC(=O)R \end{array}} + 3\,NaOH \xrightarrow{H_2O} \underset{\text{glycerol}}{\begin{array}{c} CH_2OH \\ | \\ CHOH \\ | \\ CH_2OH \end{array}} + \underset{\text{a soap}}{3\,RC(=O)O^- \; Na^+}$$

The saponification reaction itself yields a carboxylate ion and an alcohol. If a free carboxylic acid is the desired product, the reaction mixture must be acidified after the saponification is complete. The intermediate carboxylate is rarely isolated.

saponification:

$$\underset{\text{an ester}}{RCO_2R'} + OH^- \longrightarrow \underset{\text{a carboxylate}}{RCO_2^-} + \underset{\text{an alcohol}}{R'OH}$$

acidification:

$$RCO_2^- + H^+ \longrightarrow \underset{\text{a carboxylic acid}}{RCO_2H}$$

Saponification is an example of a **nucleophilic acyl substitution**, and proceeds in two steps: an **addition,** followed by an **elimination.** Step 1 is the nucleophilic attack of OH^- on the partially positive carbonyl carbon.

Step 1, addition of OH^-:

$$RC(=\ddot{O}\!:)-OR' + \,^-\!:\ddot{O}H \rightleftarrows RC(-:\ddot{O}:^-)(-:\ddot{O}H)-OR'$$

Step 2 is the elimination of the alcohol, which occurs in two stages: loss of an alkoxide ion and a concurrent acid–base reaction, resulting in the observed products, an alcohol and a carboxylate ion.

Step 2, elimination:

$$RC(-:\ddot{O}:^-)(-:\ddot{O}H)-\ddot{O}R' \longrightarrow \left[RC(=\ddot{O}:)(-:\ddot{O}H) + \,^-\!:\ddot{O}R' \right] \longrightarrow RC(=O)(-:\ddot{O}:^-) + HOR'$$

proton transfer

In Experiment 16.1, benzoic acid is prepared by the saponification of methyl benzoate, followed by the acidification of the reaction mixture.

B. Hydrolysis of Amides

Amides can be hydrolyzed by either strong acid or base: both types of reactions are irreversible. In acidic hydrolysis, the acid is not a catalyst, but a reactant, and must be present in at least stoichiometric amounts. Under acidic conditions, the products are a carboxylic acid and a protonated amine. The free amine can be generated by neutralization of the reaction mixture. The following equations represent the acidic hydrolysis of *p*-nitroacetanilide, followed by neutralization—the reaction sequence used in Experiment 16.2.

hydrolysis:

$$O_2N-C_6H_4-NHCCH_3(=O) + H_2O + H^+ \longrightarrow O_2N-C_6H_4-\overset{+}{N}H_3 + CH_3COH(=O)$$

neutralization:

$$O_2N-C_6H_4-\overset{+}{N}H_3 + \text{excess } NH_3 \rightleftharpoons O_2N-C_6H_4-NH_2 + NH_4^+$$

16.2 Esterification Reactions

Two standard procedures to prepare esters are: (1) the acid-catalyzed reaction of an alcohol with a carboxylic acid, sometimes called a **Fischer esterification**, and (2) the reaction of an alcohol or phenol with an active carboxylic acid derivative, such as an acid halide or anhydride.

A. Fischer Esterification

A Fischer esterification is the reverse reaction of the acid-catalyzed hydrolysis of an ester. Experiment 16.4 contains a generalized procedure for this type of reaction.

The mechanism of the Fischer esterification reaction is a series of protonations and deprotonations, along with an addition step and an elimination step. First, the carbonyl group of the carboxylic acid is protonated by the acidic catalyst. Next, the weakly nucleophilic alcohol attacks the carbonyl carbon.

Step 1, protonation:

$$RC(=\ddot{O}:)OH \overset{H^+}{\rightleftharpoons} \left[RC(=\overset{+}{\ddot{O}}H)OH \longleftrightarrow R\overset{+}{C}(-:\ddot{O}H)-\ddot{O}H \longleftrightarrow RC(-:\ddot{O}H)=\overset{+}{\ddot{O}}H \right]$$

resonance structures for a protonated carboxylic acid

Step 2, addition of R′OH and loss of H^+:

$$R\overset{+}{C}(OH)OH + R'\ddot{O}H \rightleftharpoons RC(OH)(OH)-R'\overset{+}{\ddot{O}}H \rightleftharpoons RC(OH)(OH)-R'\ddot{O}: + H^+$$

In the final steps, an OH group is protonated and then eliminated as water. A final deprotonation yields the ester.

Step 3, protonation, elimination, and deprotonation:

$$\underset{\text{R'O}}{\overset{\text{OH}}{\text{RC}}}-\ddot{\text{O}}\text{H} \overset{\text{H}^+}{\rightleftarrows} \underset{\text{R'O}}{\overset{:\ddot{\text{O}}\text{H}}{\text{RC}}}-\overset{+}{\ddot{\text{O}}}\text{H}_2 \overset{-\text{H}_2\text{O}}{\rightleftarrows} \underset{\underset{\text{protonated ester}}{\text{R'O}}}{\overset{:\ddot{\text{O}}-\text{H}}{\text{RC}^+}} \overset{-\text{H}^+}{\rightleftarrows} \underset{\underset{\text{the ester}}{\text{R'O}}}{\overset{\ddot{\text{O}}:}{\overset{\|}{\text{RC}}}}$$

Because the reaction mixture is an equilibrium mixture, the concentration of reactants and products may be used to determine an equilibrium constant, K:

$$K = \frac{[\text{RCO}_2\text{R}'][\text{H}_2\text{O}]}{[\text{RCO}_2\text{H}][\text{R}'\text{OH}]}$$

The equilibrium constants for Fischer esterifications are generally less than 4. From a synthetic standpoint, this means that when a 1 : 1 molar ratio of reactants is used, the equilibrium mixture contains a substantial amount of the starting carboxylic acid. For example, if $K = 4$ and a 1:1 molar ratio of alcohol to carboxylic acid is used, the highest possible yield of an ester can be 67%.

equilibrium concentrations with a 1 : 1 molar ratio of reactants when $K = 4$

$$[\text{RCO}_2\text{R}'] = 0.67 \text{ mol/liter}$$
$$[\text{H}_2\text{O}] = 0.67 \text{ mol/liter}$$
$$[\text{RCO}_2\text{H}] = 0.33 \text{ mol/liter}$$
$$[\text{R}'\text{OH}] = 0.33 \text{ mol/liter}$$

It would seem impractical to synthesize an ester by a Fischer esterification procedure except for one fact: *an equilibrium can be shifted.* To drive the reversible reaction to completion, a chemist could use one or more of the following techniques:

1) Use an excess of either RCO_2H or R'OH (the cheaper of the two). In Experiment 16.4, an excess of acetic acid is used.

2) Remove one product. In many cases, water can be distilled from the reaction mixture as it is formed, for example, as a low-boiling azeotrope with the alcohol.

3) Choose only reactants with a favorable equilibrium constant. For example, a Fischer esterification of propanoic acid with *t*-butyl alcohol should not be attempted because the K for the equilibrium is about 0.005. In fact, a Fischer esterification is impractical for any sterically hindered alcohol or hindered carboxylic acid or for a phenol because of unfavorable equilibrium constants.

B. Esterification with Active Carboxylic Acid Derivatives

Although phenyl and hindered esters cannot be prepared in good yields by Fischer esterification, these esters are easily prepared, generally in good yield, by the reaction of the phenol or alcohol with an acid anhydride or an acid chloride.

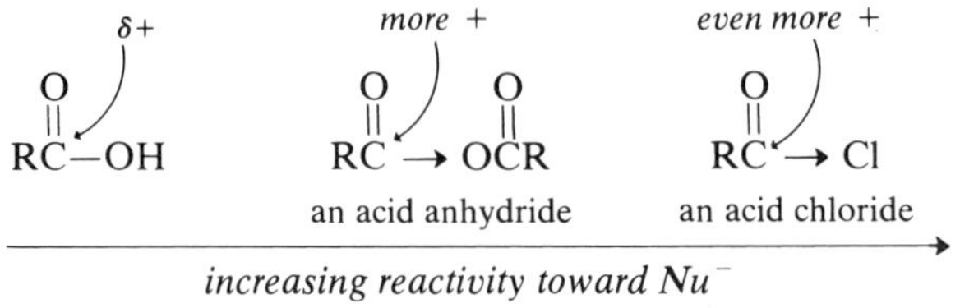

$$\overset{\delta+}{\underset{\text{}}{RC(=O)\!-\!OH}} \qquad \overset{\text{more }+}{RC(=O) \rightarrow OC(=O)R} \qquad \overset{\text{even more }+}{RC(=O) \rightarrow Cl}$$

an acid anhydride — an acid chloride

increasing reactivity toward Nu⁻ →

The reaction is irreversible because the product carboxylic acid (or chloride ion) is a weak nucleophile and cannot attack the ester.

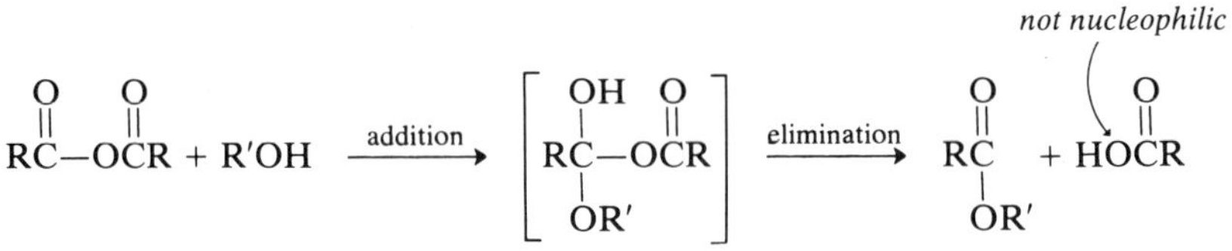

$$RC(=O)\!-\!OC(=O)R + R'OH \xrightarrow{\text{addition}} \left[RC(OH)(OR')\!-\!OC(=O)R \right] \xrightarrow{\text{elimination}} RC(=O)OR' + HOC(=O)R$$

not nucleophilic

Experiment 16.3 is the esterification of salicylic acid, which contains a phenol group, with acetic anhydride. If the esterification were attempted with acetic acid, the yield of the ester would be negligible.

C. Condensation Polymers

In Chapter 13, we briefly discussed addition polymers, which are polymers formed without the loss of a small molecule such as water. A **condensation polymer** is formed by a reaction in which small molecules, such as H_2O or ROH, are lost. The important biological polymers (proteins, nucleic acids, and polysaccharides) are condensation polymers. Two important synthetic fibers, Dacron and nylon, are also condensation polymers.

typical condensation polymerization reactions:

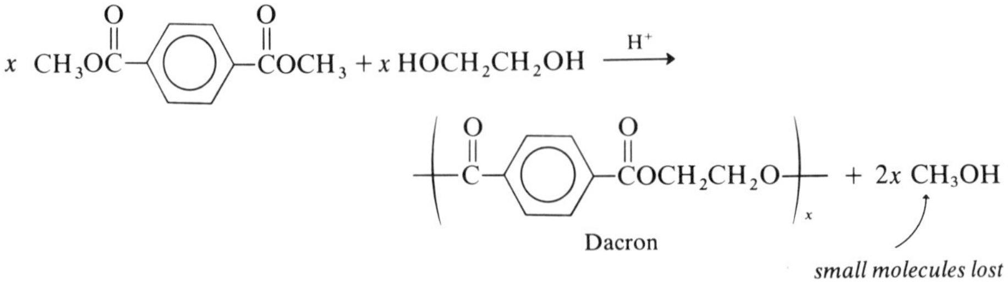

$$x\ CH_3OC(=O)\!-\!C_6H_4\!-\!C(=O)OCH_3 + x\ HOCH_2CH_2OH \xrightarrow{H^+} \left[C(=O)\!-\!C_6H_4\!-\!C(=O)OCH_2CH_2O \right]_x + 2x\ CH_3OH$$

Dacron — *small molecules lost*

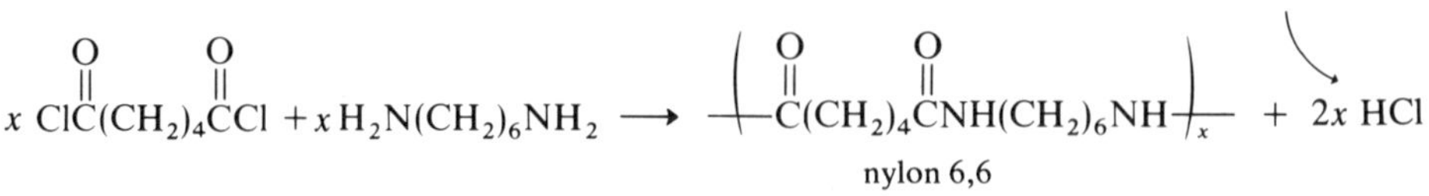

$$x\ ClC(=O)(CH_2)_4C(=O)Cl + x\ H_2N(CH_2)_6NH_2 \longrightarrow \left[C(=O)(CH_2)_4C(=O)NH(CH_2)_6NH \right]_x + 2x\ HCl$$

nylon 6,6

To undergo a condensation polymerization, each molecule must contain at least two functional groups that can undergo reaction. For example:

$$CH_3(CH_2)_3C(=O)Cl + H_2N(CH_2)_3CH_3 \longrightarrow CH_3(CH_2)_3C(=O)NH(CH_2)_3CH_3 + HCl$$

cannot undergo polymerization — an ordinary amide

$$\underset{\text{can undergo polymerization}}{ClC(=O)(CH_2)_4C(=O)Cl + H_2N(CH_2)_4NH_2} \longrightarrow \underset{\text{a polyamide}}{H_2N(CH_2)_4NHC(=O)(CH_2)_4C(=O)-\S}$$

Ring formation and polymerization are competitive reactions. Which reaction predominates depends partly upon the structure of the reactants and partly upon the experimental conditions. In general, if reactants can form a five- or six-membered ring as a condensation product, then ring formation predominates. Ring formation versus polymer formation can be controlled to an extent by the reaction conditions: a high concentration of reactants favors polymerization, while a low concentration favors ring formation. In Experiment 16.5, you will prepare a typical condensation polymer, a nylon, by the reaction of a diacid chloride with a diamine. The compounds used in this experiment do not cyclize because they cannot form a five- or six-membered ring.

Problems

16.1 Although butanoic acid (butyric acid) is a major contributor to the odor of rancid butter, *n*-butyl butyrate has the aroma of pineapple. Starting with 1-butanol and other appropriate reagents, suggest *three* routes to this ester.

16.2 Predict the products of:

(a) the acidic hydrolysis of *n*-propyl acetate (which contributes to the odor of pears)

(b) the saponification of isobutyl propanoate (which smells like rum)

16.3 Aspirin and oil of wintergreen (methyl salicylate) can both be synthesized from salicylic acid (*o*-hydroxybenzoic acid). Write an equation for the conversion of salicylic acid to oil of wintergreen.

16.4 From the mechanisms presented in this chapter, predict the products of the following reactions:

(a) $CH_3C(={}^{18}O)OH + CH_3CH_2OH \xrightarrow{H^+}$

(b) $CH_3C(=O){}^{18}OH + CH_3CH_2OH \xrightarrow{H^+}$

(c) $CH_3CO_2H + (S)\text{-}CH_3CH(OH)CH_2CH_3 \xrightarrow{H^+}$

(d) $(R)\text{-}CH_3CH(NH_2)CO_2H + (S)\text{-}CH_3CH(OH)CH_2CH_3 \xrightarrow{H^+}$

16.5 Suggest a synthesis for (a) a polyester, and (b) a polyamide from glutaric anhydride and other appropriate reagents.

[structure: O O O ring]

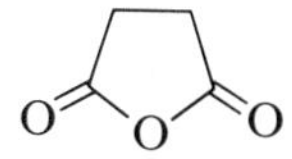

glutaric anhydride

16.6 Predict the products of the esterification of oxalic acid (HO_2CCO_2H) with 1,2-ethanediol.

»»» EXPERIMENT 16.1 «««

Saponification of Methyl Benzoate

Step 1, Saponification:

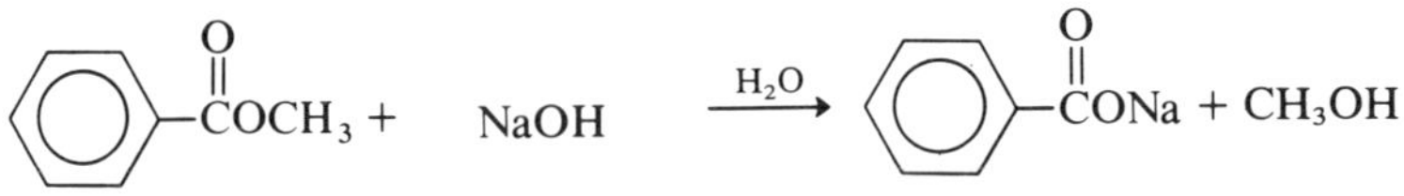

$$C_6H_5COOCH_3 + NaOH \xrightarrow{H_2O} C_6H_5COONa + CH_3OH$$

	methyl benzoate	sodium hydroxide	sodium benzoate (not isolated)
MW:	136.16	40.00	—
weight:	1.4 g	1.2 g	—
moles:	0.010	0.030	—

Step 2, Acidification:

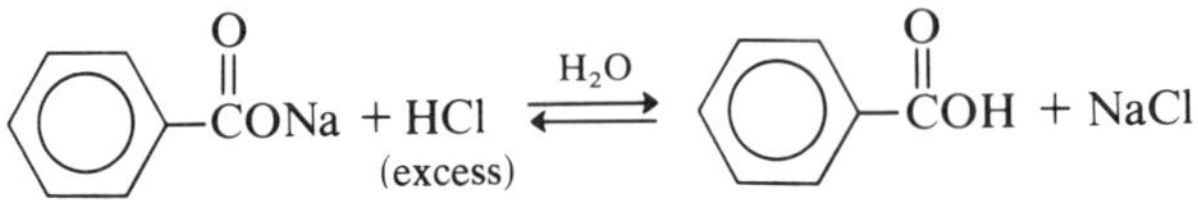

$$C_6H_5COONa + HCl \text{ (excess)} \underset{}{\overset{H_2O}{\rightleftharpoons}} C_6H_5COOH + NaCl$$

		benzoic acid
MW:	—	122.13
weight:	—	1.2 g (theory)
moles:	—	0.010 (theory)

Table 16.1 Physical properties of reactants and products

Name	*Mp(°C)*	*Bp(°C)*	*Solubility*
methyl benzoate[a]	—	215	insol. H_2O; sol. CH_3CH_2OH
benzoic acid	121.7	—	sl. sol. H_2O; sol. base

[a] Density at 20°: 1.07 g/mL.

Discussion

The conversion of an ester to a carboxylic acid by saponification is a two-step procedure. Step 1, the actual saponification, is the reaction of the ester with hydroxide ion to yield the alcohol and the carboxylate, which is not isolated. Step 2 is acidification, in which the carboxylate is converted to the carboxylic acid.

In the first step, methyl benzoate and aqueous sodium hydroxide are mixed and heated at reflux. The starting methyl benzoate is water-insoluble, but the product sodium benzoate is water-soluble. Consequently, the disappearance of the methyl benzoate layer during the reflux period can be used to monitor the progress of the saponification. When the mixture becomes homogeneous, the saponification is complete.

The reaction mixture is then cooled and acidified with concentrated HCl. (The cooling is necessary because the acidification is exothermic.) Any strong mineral acid can be used; HCl is chosen for convenience. As the reaction mixture is acidified, the water-insoluble benzoic acid precipitates. Hydrochloric acid is added until the reaction mixture is strongly acidic to ensure that all the sodium benzoate is converted to benzoic acid.

Because the acidification is exothermic, and because benzoic acid is somewhat soluble in hot water, the reaction mixture is chilled before filtration. The benzoic acid obtained from the filtration is reasonably pure, and further purification should not be necessary.

Experimental

EQUIPMENT:

condenser
dropper
125-mL Erlenmeyer flask
heating mantle and rheostat
ice bath
pH paper or litmus paper
50-mL round-bottom flask
vacuum filtration assembly

CHEMICALS:

conc. hydrochloric acid, 2–4 mL
methyl benzoate, 1.4 g
sodium hydroxide, 1.2 g

TIME REQUIRED: 1 hour plus overnight drying and time for a melting-point determination

STOPPING POINTS: when the initial reaction mixture becomes homogeneous; while the benzoic acid is air-drying or crystallizing

PROCEDURE

Step 1, Saponification. Weigh 1.4 g of methyl benzoate into a 50-mL round-bottom flask, and add a solution of 1.2 g of sodium hydroxide (*caustic*) in 15 mL of water. Add 2–3 boiling chips, fit the flask with a reflux condenser, and heat the mixture under reflux until it becomes homogeneous (20–40 min).

Step 2, Acidification. Cool the solution to room temperature in an ice bath, transfer it to an Erlenmeyer flask, and add conc. hydrochloric acid (*strong acid with irritating fumes*!) dropwise until the mixture is acidic to pH paper or litmus paper. About 2.5 mL will be required.

Cool the mixture in an ice bath and vacuum-filter the product. Air-dry the solid at least overnight, then determine the melting point and per cent yield.

Problems

16.7 Can the progress of all saponification reactions be followed by observing when two phases become one phase? Explain.

16.8 What modifications in the procedure in Experiment 16.1 would probably be necessary for the saponification of the following compounds:

(a) dimethyl terephthalate, $CH_3O_2C-C_6H_4-CO_2CH_3$

(b) trimyristin (page 342)

»»» EXPERIMENT 16.2 «««

(SUPPLEMENTAL)

Hydrolysis of p-Nitroacetanilide

Step 1, Hydrolysis:

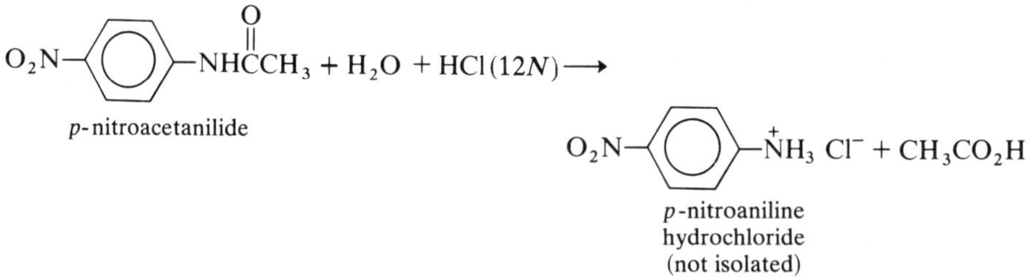

$$O_2N-C_6H_4-NHCCH_3 \ (C{=}O) + H_2O + HCl\,(12N) \longrightarrow O_2N-C_6H_4-\overset{+}{N}H_3\ Cl^- + CH_3CO_2H$$

p-nitroacetanilide → *p*-nitroaniline hydrochloride (not isolated)

	p-nitroacetanilide	HCl	*p*-nitroaniline hydrochloride
MW:	180.18	—	—
weight:	6.0 g	18 mL	—
moles:	0.033	0.22	—

Step 2, Neutralization:

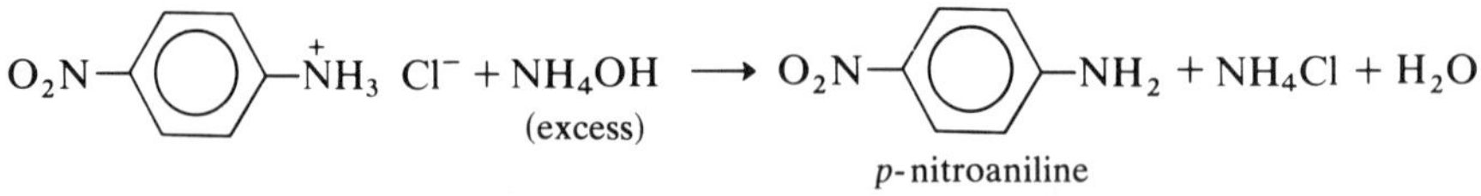

$$O_2N-C_6H_4-\overset{+}{N}H_3\ Cl^- + NH_4OH\ (\text{excess}) \longrightarrow O_2N-C_6H_4-NH_2 + NH_4Cl + H_2O$$

p-nitroaniline

	p-nitroaniline hydrochloride	*p*-nitroaniline
MW:	—	138.13
weight:	—	4.6 g (theory)
moles:	—	0.033 (theory)

Table 16.2 Physical properties of reactants and products

Name	*Mp(°C)*	*Solubility*
p-nitroacetanilide	216	sl.sol. H_2O
p-nitroaniline hydrochloride	—	sol. H_2O
p-nitroaniline	148–149	insol. cold H_2O; sl. sol. hot H_2O

Discussion

The acidic hydrolysis of *p*-nitroacetanilide is accomplished by warming it with conc. HCl. The hydrolysis must be carried out in a fume hood because the hot mixture emits gaseous HCl. Because HCl is a reactant (not a catalyst), it is important that the reaction mixture not be heated too strongly, which would drive off HCl.

p-Nitroacetanilide is insoluble in both water and aqueous acid. Both products of the hydrolysis (acetic acid and *p*-nitroaniline hydrochloride) are water-soluble. Therefore, the course of the hydrolysis can be followed by observing the amount of solid remaining in the reaction vessel. To ensure complete hydrolysis, the mixture is heated for 30 minutes after it has become homogeneous.

The hydrochloride salt is not isolated, but is converted to the amine (Step 2 on page 300). This is accomplished by neutralization with a mixture of aqueous ammonia and ice. The *p*-nitroaniline is insoluble in cold aqueous base and thus can be separated from the water-soluble inorganic compounds—NH_4Cl and NH_3.

In a large class, the neutralization reaction should be carried out in a *different fume hood* from that used for the hydrolysis, if possible. Aqueous ammonia solution emits gaseous NH_3, which reacts with gaseous HCl. The result is a fog of NH_4Cl.

Experimental

EQUIPMENT:

400-mL beaker
250-mL Erlenmeyer flask (500-mL optional)
25-mL and 100-mL graduated cylinders
hot plate
ice bath
pH paper or litmus paper
vacuum filtration assembly

CHEMICALS:

conc. aqueous ammonia (ammonium hydroxide), 25 mL
conc. hydrochloric acid, 18–28 mL
p-nitroacetanilide, 6 g

TIME REQUIRED: $1\frac{1}{2}$ hours plus overnight drying and a melting-point determination.

STOPPING POINTS: after the hydrolysis; when the product is air-drying or crystallizing

»»> ***SAFETY NOTE*** The nitroanilines are toxic and can be absorbed through the skin. Avoid contact with skin or breathing the dust. *Cosmetic Note*: *p*-Nitroaniline is a yellow dye that stains skin and clothing. If spillage should occur, clean it up immediately with hot water. If possible, use rubber or plastic gloves while working with this product.

PROCEDURE

Step 1, Hydrolysis. Place 6.0 g of *p*-nitroacetanilide, 15–20 mL of water, and 18 mL of conc. hydrochloric acid (*strong acid*) in a 250-mL Erlenmeyer flask. Warm the mixture on a hot plate to a near-boil *in a fume hood*. The solid material dissolves as the temperature of the solution approaches boiling (about 30 minutes). Maintain the temperature of the solution just undcr boiling for 30 minutes after all of the solid has dissolved. It may be necessary to add an additional 9–10 mL of acid during the heating period to maintain the volume of the mixture. At the end of the heating period, add 30 mL of cold water, then cool the yellow-to-orange solution in an ice bath until its temperature is below 30°.

Step 2, Neutralization. Place 75 mL of cold water, 25 mL of conc. aqueous ammonia (*irritating fumes*!), and about 50 g of cracked ice in a beaker. Slowly pour the orange reaction mixture into the ammoniacal mixture with vigorous stirring (CAUTION: see Safety Note). A bright yellow precipitate will form. Test the mixture with pH paper to ensure that it is alkaline. If not, add concentrated aqueous ammonia, 1 mL at a time, with stirring until the solution is alkaline.

Filter the precipitate with vacuum, using 50 mL of water to wash the last of the precipitate into the funnel. Air-dry the product and determine the yield and melting point. To obtain a more pure product, crystallize the crude *p*-nitroaniline from water, using 60–70 mL of water per gram of amine. (The amine dissolves very slowly; therefore, be careful not to use too much water.)

Problems

16.9 Draw formulas for the principal resonance structures of *p*-nitroaniline.

16.10 *p*-Nitroaniline is used in industry to synthesize other dyes. Suggest a synthesis for Acid red 176 from *p*-nitroaniline and other appropriate reagents.

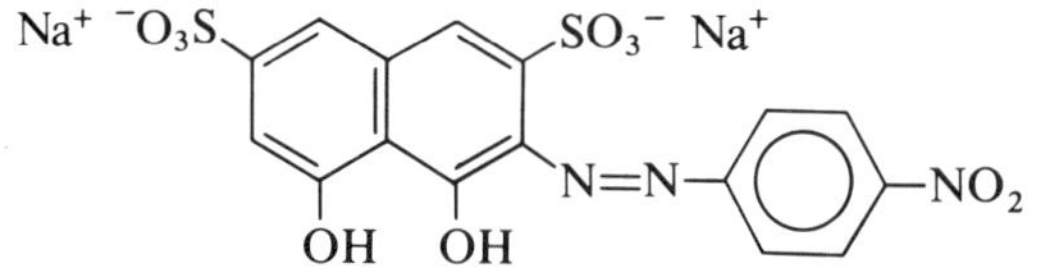

$Na^+ \ {}^-O_3S$ $SO_3^- \ Na^+$
N=N NO_2
OH OH

Acid red 176

16.11 What modifications in the procedure in Experiment 16.2 would be necessary if *p*-nitroacetanilide were hydrolyzed using H_2SO_4?

»»» EXPERIMENT 16.3 «««

Synthesis of Aspirin

$$\text{salicylic acid} + \underset{\text{acetic anhydride}}{CH_3COCCH_3} \xrightarrow{H_2SO_4} \underset{\text{acetylsalicylic acid (aspirin)}}{\text{(2-}CO_2H\text{)}C_6H_4OCCH_3} + \underset{\text{acetic acid}}{CH_3COH}$$

	salicylic acid	acetic anhydride	aspirin
MW:	138.12	102.09	180.17
weight:	2.8	5.4 g (5.0 mL)	3.6 g (theory)
moles:	0.020	0.053	0.020 (theory)

Table 16.3 Physical properties of reactants and products

Name	*Mp (°C)*	*Solubility*
salicylic acid	159[a]	sl. sol. H_2O; sol. CH_3CH_2OH
acetic anhydride[b]	—	decomposes in H_2O
aspirin	135[c]	sl. sol. H_2O; sol. CH_3CH_2OH

[a] Sublimes at its melting point.
[b] Bp 139°; density at 20°: 1.08 g/mL.
[c] Determined with rapid heating.

Discussion

The physiological properties of aspirin are discussed on page 112. Although active salicylates are found in the bark of willow trees (trees of the genus *Salix*, from whence salicylic acid derives its name), all aspirin sold today is synthesized from phenol, which, in turn, is obtained from petroleum.

$$\text{petroleum} \xrightarrow{\text{catalytic reforming}} \underset{\text{benzene}}{C_6H_6} \xrightarrow{(1)\ Cl_2\text{, catalyst};\ (2)\ NaOH\text{, heat, pressure};\ (3)\ H^+} \underset{\text{phenol}}{C_6H_5OH}$$

$$\xrightarrow{NaOH} \underset{\text{sodium phenoxide}}{C_6H_5O^- Na^+} \xrightarrow{(1)\ CO_2\text{, heat, pressure};\ (2)\ H^+} \underset{\text{salicylic acid}}{\text{(2-}CO_2H\text{)}C_6H_4OH} \xrightarrow{(CH_3C(O))_2O} \text{aspirin}$$

In this experiment, you will conduct only the last step of the commercial synthesis, the conversion of salicylic acid to aspirin.

The reaction of salicylic acid with acetic anhydride occurs rapidly. The reactants and a sulfuric acid catalyst are mixed, then warmed in a hot water bath. Solid salicylic acid is insoluble in acetic anhydride. Aspirin is soluble in a hot mixture of acetic anhydride and acetic acid, which is formed as the reaction proceeds. Thus, the course of the reaction can be followed by the disappearance of the solid salicylic acid. The reaction is essentially complete when all the salicylic acid has dissolved. Aspirin is only slightly soluble in a cold mixture of acetic anhydride and acetic acid. Therefore, as the mixture is cooled to room temperature, the aspirin precipitates.

At the end of the reaction period, the mixture contains aspirin, acetic anhydride, and acetic acid. Acetic acid is miscible with water, and acetic anhydride reacts fairly rapidly with water to yield acetic acid. By adding water to the reaction mixture and allowing the aqueous mixture to stand at room temperature for a few minutes, we can achieve a reasonable separation—the contaminants dissolve in the water, and the aspirin precipitates. Because of the presence of acetic acid and because aspirin is slightly soluble in water, about 10% of the aspirin remains in solution. Therefore, the mixture is thoroughly chilled before filtration and crystallization.

An additional small amount of aspirin can be recovered from the mother liquor if it is allowed to stand overnight. Despite this fact, it is not good practice to leave the bulk of the aspirin in acidic solution for an extended period of time because it will undergo a slow hydrolysis to yield salicylic acid and acetic acid, a typical acid-catalyzed ester hydrolysis.

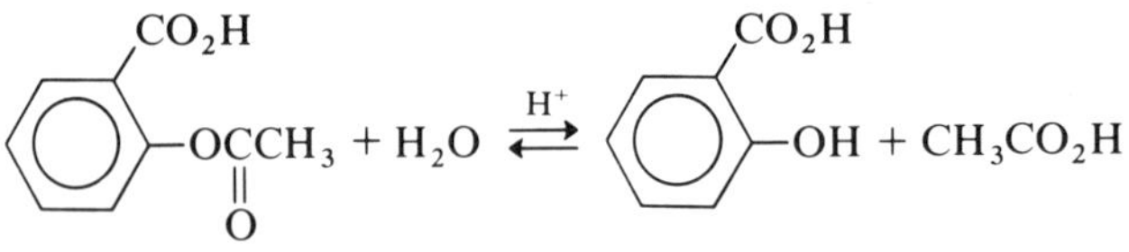

$$C_6H_4(CO_2H)\text{—}OCCH_3(=O) + H_2O \overset{H^+}{\rightleftharpoons} C_6H_4(CO_2H)\text{—}OH + CH_3CO_2H$$

Experimental

EQUIPMENT:

400-mL beaker (for hot water bath)
dropper
two 125-mL Erlenmeyer flasks
10-mL and 100-mL graduated cylinders
spatula
thermometer
vacuum filtration assembly
two watch glasses

CHEMICALS:

acetic anhydride, 5.0 mL
salicylic acid, 2.8 g
conc. sulfuric acid, 3–4 drops

TIME REQUIRED: 1 hour plus overnight drying and time for a melting-point determination

STOPPING POINTS: after the initial vacuum filtration; while the product is crystallizing from water; while the final product is being air-dried

›››› ***SAFETY NOTE*** Acetic anhydride is volatile and is a strong irritant.

PROCEDURE

Place 2.8 g of salicylic acid in a dry 125-mL Erlenmeyer flask, then add 5.0 mL acetic anhydride and 3–4 drops concentrated sulfuric acid. Mix the resultant white slurry thoroughly with a spatula, and place the flask in a warm water bath (45–50°) for 5–7 min. Swirl or stir the mixture occasionally to dissolve all the solid material. Because the reaction is slightly exothermic, a small temperature rise can be detected.

Allow the flask to cool. The aspirin begins to precipitate when the temperature of the solution is about 35–40°, and the mixture becomes semisolid. When this occurs, add 50 mL water and break up any lumps with a spatula. Allow the mixture to stand for an additional 5 minutes, then chill the flask in an ice bath and remove the crystals by vacuum filtration.

Crystallize the crude aspirin from 25 mL of warm water not exceeding 80° (see Experimental Note). Allowing the mother liquor to sit overnight may produce a second crop of crystals. Air-dry the crystals and determine the per cent yield and melting point.

EXPERIMENTAL NOTE

At temperatures exceeding 80°, aspirin forms an oil that dissolves organic impurities from the water; in this case, it may be difficult to redissolve the aspirin in water. If the solid does not dissolve in 25 mL of water, add more water from a dropper. Let the mixture warm 2–4 minutes between additions to allow the solid to dissolve.

Problems

16.12 Write an equation for the synthesis of aspirin from salicylic acid with an acid chloride instead of acetic anhydride.

16.13 The following compounds all have antipyretic and analgesic properties similar to aspirin. (Salicylic acid itself has similar physiological properties, but is too irritating to be taken internally; however, it has been used as a wart remover.) Suggest a synthesis for each from salicylic acid.

(a) sodium salicylate
(b) methyl salicylate
(c) salicylamide
(d) phenyl salicylate
(e) 2-*O*-acetyl-5-bromosalicylic acid

16.14 (a) Commercial aspirin sometimes has a distinct acetic acid odor. Why?
(b) Would ingestion of such aspirin be harmful? Explain.

»»» EXPERIMENT 16.4 «««

Synthesis of an Acetate Ester by a General Reaction Procedure

$$CH_3\overset{O}{\overset{\|}{C}}OH + ROH \xrightarrow{H^+} CH_3\overset{O}{\overset{\|}{C}}OR + H_2O$$

Discussion

Most experimental procedures in the scientific literature are written in a terse form. Nonessential details, such as flask size, reaction vessel set-up, and extraction procedures, are usually not given, although they are included in beginning laboratory texts. In addition, some literature procedures are written in a general format so that they can be used for the preparation of a variety of different, but similar, products. The weights to be used in these procedures, for example, are specified in moles instead of in grams. Physical constants of the products may or may not be given.

Experiment 16.4, the preparation of an acetate ester by a Fischer esterification, is written in a generalized form. In this experiment, considerable planning must precede the actual laboratory work. First, the actual ester to be prepared must be chosen* and its physical constants found in the literature. In this particular experiment, care should be taken that the boiling point of the ester chosen is neither too high (vacuum distillation would be required) or too low (excessive loss by evaporation). Next, the actual number of grams must be calculated. The volumes of the reagents must be estimated so that the proper-sized glassware is used. Particular attention should be given to the work-up procedure, especially the extraction steps. These should be outlined in detail. (Remember, the procedure omits the detailed steps. These must be inserted into your outline.)

In a Fischer esterification with acetic acid, the reactants are heated under reflux so that the equilibrium can be reached within a reasonable period of time. At the end of the reflux period, the reaction mixture contains the desired ester plus acetic acid, some unreacted alcohol, water, and sulfuric acid. The ester is water-insoluble, but the other materials are at least partly soluble in water. Therefore, extraction of the reaction mixture with a large volume of water removes most of the undesired material from the desired ester.

Because it is almost impossible to remove all the acetic acid from an organic material by simple water extraction, the ester is washed with a sodium bicarbonate solution to convert the acetic acid to water-soluble sodium acetate. The ester is extracted with a final portion of water to remove any traces of

* The procedure as described in this experiment is satisfactory for the preparation of *n*-butyl acetate or isoamyl acetate. Minor modifications, such as longer reflux times, might be needed for other esters.

inorganic salts introduced by the bicarbonate wash. The ester is then dried with anhydrous Na_2SO_4 or $MgSO_4$. Atmospheric distillation yields the purified product.

Experimental

EQUIPMENT:

to be determined by the student

CHEMICALS:

to be determined by the student

TIME REQUIRED: 3 hours plus overnight drying and 1 hour for the distillation

STOPPING POINTS: immediately after the reflux; while the ester is being dried

>>>> ***SAFETY NOTE*** If a volatile, "heady" ester, such as isoamyl acetate (banana oil), is to be prepared, be sure there is adequate ventilation or use a fume hood.

PROCEDURE

Heat a mixture of 0.50 mol of glacial acetic acid, 0.25 mol of a primary alcohol, and 0.5 mL of concentrated H_2SO_4 under reflux for 1.5 hours. Cool the reaction mixture, then dilute it with an amount of water 3–4 times its volume. Wash the organic layer with water, with 10–15 mL of saturated sodium bicarbonate solution, and again with water. Dry the ester with anhydrous Na_2SO_4, then purify the product by distillation. A typical yield is about 60%.

Problems

16.15 When one reactant in Experiment 16.4 is an excess of glacial acetic acid, why is it necessary to add conc. H_2SO_4?

16.16 How would you modify the procedure in this experiment to prepare 0.0025 mol (instead of 0.25 mol) of *n*-butyl acetate?

16.17 Tell whether the procedure in this experiment could be used to prepare the following esters. Indicate any modifications that would probably improve the yield.

(a) $CH_3CO_2CH_2CH_2CH_2CH_2O_2CCH_3$

(b) 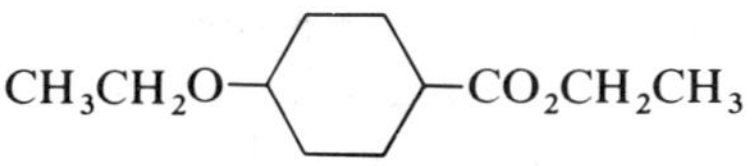 CH_3CH_2O–⟨cyclohexane ring⟩–$CO_2CH_2CH_3$

(c) $(CH_3)_3CCH_2O_2CCH(CH_3)_2$

(d) $(CH_3)_2C(OH)CH_2CH_2O_2CCH_3$

(e) C_6H_5–$CO_2CH_2CH_2CH_3$

»»» EXPERIMENT 16.5 «««

(SUPPLEMENTAL)

Synthesis of Nylon 10,6: The "Nylon Rope Trick"

$$x\ \mathrm{ClC(=O)(CH_2)_8C(=O)Cl} + x\ \mathrm{H_2N(CH_2)_6NH_2} \longrightarrow \left[\mathrm{C(=O)(CH_2)_8C(=O)\text{—}NH(CH_2)_6NH}\right]_x + 2x\ \mathrm{HCl}$$

decanedioyl chloride (sebacyl chloride); 1,6-hexanediamine (hexamethylenediamine); nylon 10,6

	decanedioyl chloride	1,6-hexanediamine	nylon 10,6
MW:	239.14	116.21	—
weight:	2.5 g (2 mL)	2.2 g	—
moles:	0.010	0.019	—

Table 16.4 Physical properties of reactants and products

Name	***Mp (°C)***	***Bp (°C)***	***Solubility***
1,6-hexanediamine	41–42	204–205	sol. H_2O; insol. $Cl_2C{=}CHCl$
decanedioyl chloride	—	220 (75 mm Hg)	decomposes in H_2O; sol. $Cl_2C{=}CHCl$
nylon 10,6	—	—	insol. H_2O and $Cl_2C{=}CHCl$

Discussion

Nylon was discovered in the 1930's by W. H. Carothers at the DuPont Company. Like the protein molecules in silk, a nylon molecule is a *polyamide*. Nylon was originally advertised as "artificial silk" and was quickly adopted by women as the ultimate textile for hosiery (sturdier than silk and more sheer than rayon, which is a cellulosic product).

The numbers following the word "nylon," as in nylon 6 or nylon 10,6, refer to the numbers of carbon atoms in the original reactant or reactants. Thus, nylon 10,6 is made from a 10-carbon diacid chloride and a 6-carbon diamine.

In this experiment, nylon is formed as a film at the interface of two immiscible liquids. One of the liquids contains the diacid chloride; the other contains the diamine. A strand, or rope (which is really a collapsed film) of nylon can be drawn continuously from the interface. As the nylon is removed, fresh reactants come into contact to yield more nylon. This preparation of nylon is an excellent lecture demonstration and has earned the name the "nylon rope trick."*

In this procedure, decanedioyl chloride and 1,6-hexanediamine are the two reactants, although another diacid chloride or diamine could be used. The mechanism of the polymerization is the same as that of simple amide formation.

Step 1, addition:

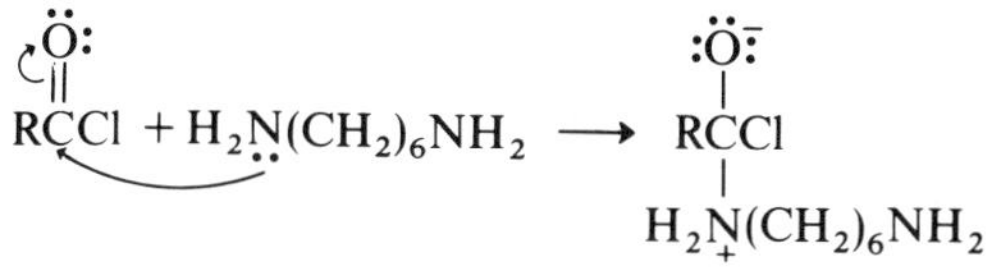

$$RC(=\ddot{O}:)Cl + H_2\ddot{N}(CH_2)_6NH_2 \longrightarrow RC(-\ddot{O}:^-)(Cl)(-\overset{+}{N}H_2(CH_2)_6NH_2)$$

Step 2, elimination of HCl:

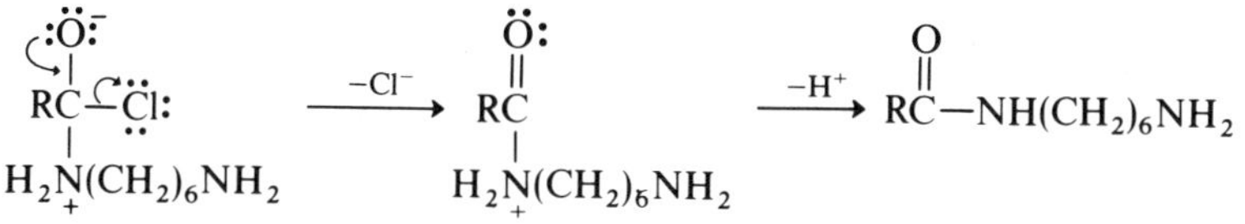

$$RC(-\ddot{O}:^-)(-\ddot{Cl}:)(-\overset{+}{N}H_2(CH_2)_6NH_2) \xrightarrow{-Cl^-} RC(=\ddot{O}:)(-\overset{+}{N}H_2(CH_2)_6NH_2) \xrightarrow{-H^+} RC(=O)-NH(CH_2)_6NH_2$$

The amine group shown at the right-hand end of the chain (Step 2) can attack another molecule of diacid chloride, which, in turn, can be attacked by another molecule of diamine, and so forth.

The fate of the eliminated HCl is quite important. If it is not removed from the site of reaction, it forms a salt with the amine group at the end of the growing polymeric chain. Salt-formation converts the nucleophilic amine group to a non-nucleophilic $-NH_3^+$ group, effectively terminating the growth of the polymer.

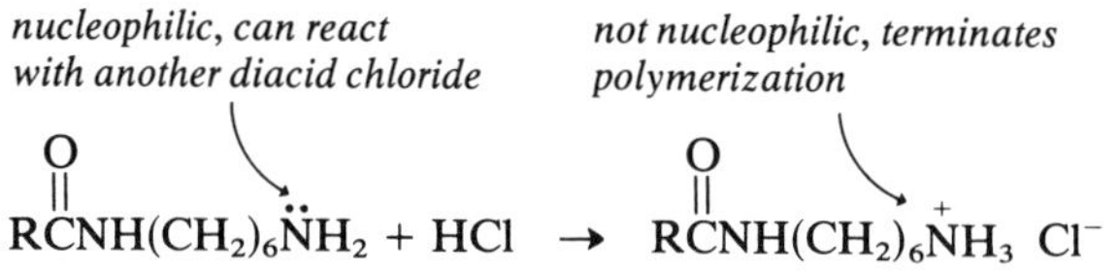

nucleophilic, can react with another diacid chloride (points to $\ddot{N}H_2$); *not nucleophilic, terminates polymerization* (points to $\overset{+}{N}H_3$)

$$RC(=O)NH(CH_2)_6\ddot{N}H_2 + HCl \rightarrow RC(=O)NH(CH_2)_6\overset{+}{N}H_3\ Cl^-$$

One reason for carrying out the reaction at an interface is to remove the HCl. The lower layer is a trichloroethene solution of the water-insoluble diacid chloride. The upper layer is an aqueous solution of the trichloroethene-insoluble diamine and Na_2CO_3. The Na_2CO_3 is a stronger base than an amine and neutralizes the HCl as it is formed.

The physical manner in which the experiment is carried out is of paramount importance to its success. Trichloroethene is heavier than water;

* T. L. Bieber, *J. Chem. Ed.* **1979**, *56*, 409; P. W. Morgan and S. L. Kwolek, *J. Chem. Ed.* **1959**, *36*, 182.

therefore, the trichloroethene solution is placed in the reaction beaker first. Then the aqueous solution is poured very carefully down the side of the beaker so that it can slide across the top of the heavier solution with minimal mixing of the two solutions. A film of nylon forms immediately at the interface. This film prevents further reaction by separating the reactants. However, when the nylon film is removed, more polymer forms at the interface.

Experimental

EQUIPMENT:

100-mL and 200-mL beakers
10-mL and 100-mL graduated cylinders
stirring rod or tongs

CHEMICALS:

decanedioyl chloride (sebacyl chloride), 2 mL
1,6-hexanediamine (hexamethylenediamine), 2.2 g
sodium carbonate monohydrate, 4.7 g
trichloroethene (trichloroethylene), 100 mL

TIME REQUIRED: $\frac{1}{2}$ hour

STOPPING POINTS: none

>>>> **SAFETY NOTE** This experiment can be messy. It is suggested that you wear disposable plastic gloves while drawing out the nylon.

PROCEDURE

Measure 100 mL of trichloroethene into a 200-mL beaker, add 2 mL of decanedioyl chloride, and stir the solution vigorously to ensure homogeneity. In a separate beaker, place 50 mL of water, 2.2 g of 1,6-hexanediamine (see Experimental Note), and 4.7 g of sodium carbonate monohydrate. Stir this mixture until the reagent dissolves.

Pour the aqueous solution carefully down the side of the first beaker to form a layer on top of the trichloroethene solution. The nylon polymer forms at the interface. Using tongs or a glass rod, reach through the aqueous layer to the interface, and slowly and constantly withdraw the film, which collapses to form a rope. If the rope breaks, insert the tongs or rod to pick up the film and start again.

EXPERIMENTAL NOTE

1,6-Hexanediamine melts around 40° and may form a solid mass in the reagent bottle. Place the bottle in a warm water bath to liquefy the diamine.

Problems

16.18 Name the following nylons:

(a) $\left[\overset{\text{O}}{\overset{\|}{\text{C}}}(\text{CH}_2)_6\overset{\text{O}}{\overset{\|}{\text{C}}}\text{NH}(\text{CH}_2)_6\text{NH} \right]_x$ (b) $\left[\text{NH}(\text{CH}_2)_5\overset{\text{O}}{\overset{\|}{\text{C}}} \right]_x$

16.19 Suggest reagents for preparing the nylons in Problem 16.18.

16.20 Write equations for the reaction that would occur if the diacid chloride used in this experiment were mixed with the aqueous Na_2CO_3.

CHAPTER 17

Stereochemistry

Stereochemistry is the study of molecular structure and chemical reactions in three dimensions. This chapter includes two experiments in which stereochemistry plays a major role: a pair of *cis,trans* isomerizations, and the isolation of one enantiomer from a racemic mixture.

17.1 Alkene Stability and Isomerization

Groups cannot rotate about a carbon–carbon double bond at room temperature. For this reason, *cis,trans* isomers of alkenes can be isolated and their relative stabilities can be compared. In general, it has been determined that a *trans* isomer is more stable than the *cis* isomer. This difference in stability is usually attributed to steric hindrance in the *cis* isomer.

steric hindrance

(R)(H)C=C(R)(H)	(R)(H)C=C(H)(R)
cis	*trans*

Isomerization of either *trans* to *cis* or *cis* to *trans* alkenes can be achieved under proper experimental conditions. A *cis-to-trans* isomerization is usually an *exothermic* process and occurs when the two isomers are allowed to equilibrate. The difference in energy between the *cis* and *trans* isomers determines the position of the equilibrium and thus the extent of isomerization.

A common experimental technique to convert a *cis* alkene to a *trans* alkene is to heat the *cis* alkene with a trace of mineral acid to protonate the double bond. The alkene becomes a carbocation in which rotation can occur because a carbocation does not contain overlapping *p* orbitals. Elimination of H^+ from the carbocation can result in either the *cis* or the *trans* alkene.

$$\underset{cis}{\text{(R)(H)C=C(R)(H)}} \underset{}{\overset{H^+}{\rightleftharpoons}} \underset{\text{intermediate carbocation}}{\text{R–CH}_2\text{–}\overset{+}{\text{C}}\text{(R)(H)}} \overset{-H^+}{\rightleftharpoons} \underset{trans}{\text{(R)(H)C=C(H)(R)}}$$

Because the protonation and elimination reactions are readily reversible in acidic solution, a thermodynamic equilibrium is established between the isomers. The lower-energy isomer predominates in the mixture, regardless of which isomer was the starting material.

A *trans-to-cis* isomerization is usually an *endothermic* process. The usual experimental technique for such a conversion is a **photoisomerization**, in which light provides the energy for the reaction.

There are two procedures by which a photoisomerization can be carried out: with a sensitizer (photocatalyst) and without. A sensitizer preferentially absorbs radiation, then transfers the absorbed energy to the alkene. The other technique involves direct irradiation of the alkene. Since the direct irradiation procedure is used in Experiment 17.1A, we will discuss the mechanism of this type of isomerization in more detail.

When an alkene is irradiated with light of the proper wavelength, it absorbs a photon. The absorbed energy causes the promotion of a pi electron from a bonding orbital to an antibonding orbital (a $\pi \rightarrow \pi^*$ transition). The electronic transition is extremely fast and the promoted electron retains the same spin as it had in the ground state.

In most compounds with molecules in the ground state, all electrons are paired and the electron spins (represented as $+1/2$ or $-1/2$) cancel. The electronic state of a molecule in which the electron spins cancel, regardless of whether the electrons are all paired or not, is called a **singlet state**. The ground state of a molecule is a singlet state referred to as S_0. The lowest-energy excited state in which the electron spins cancel is referred to as the lowest-energy excited singlet state, or S_1. In the S_1 state initially formed by photon-absorption, the geometry of a molecule is the same as it was in the gound state (*cis* or *trans*).

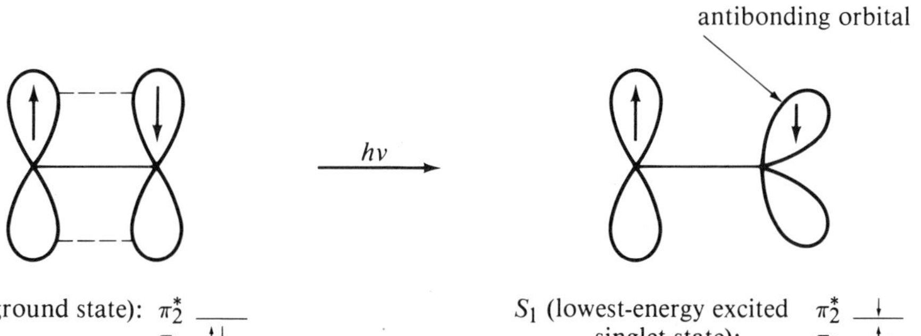

Because of orbital repulsion in the S_1 state, the excited molecule can lose energy and become more stable by changing its geometry. The central bond twists, giving a *twisted S_1 state* in which the p orbitals are at 90°. By twisting, the excited molecule loses energy by minimizing the repulsions between the unpaired electrons in the bonding and antibonding p orbitals.

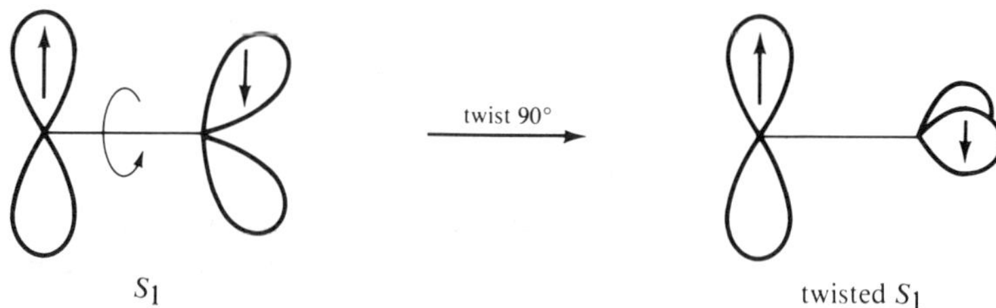

The twisted S_1 state can be reached starting with either the *cis* or the *trans* alkene, and is the common intermediate in photoisomerization.* The twisted singlet can return to the ground state of either the *cis* or *trans* alkene. The energy relationships for the interconversions are diagrammed in Figure 17.1.

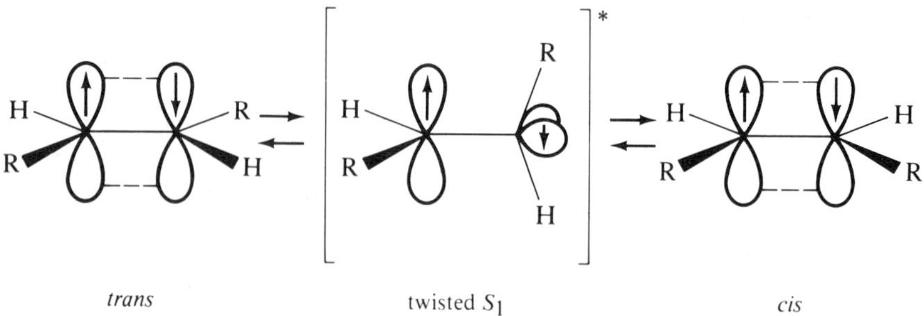

Cis,trans isomers do not have identical uv–visible spectra—that is, they absorb radiation of different wavelengths. Therefore, experimental conditions can be selected in which the *trans* isomer can be preferentially promoted to the excited state. Under these nonequilibrating conditions, the *cis* isomer will accumulate in the reaction mixture as the twisted S_1 species returns to the ground state. In Experiment 17.1, *trans*-dibenzoylethylene ($C_6H_5CO{-}CH{=}CH{-}COC_6H_5$) is converted to the *cis* isomer in just this way, then reconverted back to the *trans* isomer by an acidic equilibration.

* Direct photoisomerization of most alkenes is believed to take place through the lowest-energy singlet state, S_1. When a sensitizer is added to the reaction mixture, the isomerization may take place through a common twisted **triplet state**, T_1, an excited state in which the spin states of electrons do *not* cancel because the spin state of one pi electron has been changed.

$$T_1: \begin{array}{l} \pi_2^* \underline{\uparrow} \\ \pi_1 \underline{\uparrow} \end{array}$$

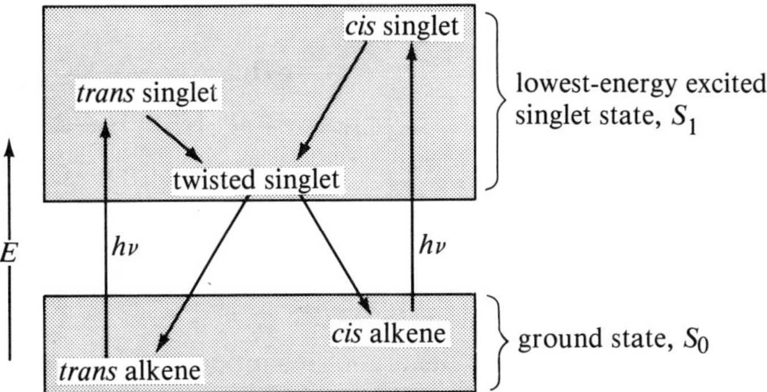

Figure 17.1 Diagram of energy changes in the photoisomerization of an alkene by direct irradiation.

17.2 Resolution of an Amine

In the laboratory, reactions of achiral compounds or of racemic mixtures almost always lead to achiral or racemic products. On the other hand, a biological system can synthesize a single enantiomer from achiral or racemic starting materials because the enzymes that catalyze these reactions are chiral.

A pair of enantiomers exhibit the same physical and chemical properties, with two exceptions: the direction in which they rotate the plane of polarization of plane-polarized light (Section 17.3) and reactions in a chiral environment. Therefore, a pair of enantiomers cannot be separated from each other by extraction, crystallization, distillation, or any of the usual physical separation procedures used in the laboratory. However, racemic mixtures can be resolved, or separated, by converting the enantiomers to diastereomers (nonenantiomeric stereoisomers), separating the diastereomers (which, unlike enantiomers, differ in their physical properties), and reconverting the diastereomers to the enantiomers.

In Experiment 17.2, (±)-α-phenylethylamine is partially resolved by the isolation of the (*S*)-(−)-amine from the racemic mixture. The reagent used to carry out this partial resolution is a naturally occurring, single stereoisomer of tartaric acid: (+)-tartaric acid,* which is found in grapes and other fruit. Commercial (+)-tartaric acid is a by-product of the wine industry and is used in soft drinks, baking powder, and other food products, as well as in organic laboratories. In this partial resolution, the racemic amine is treated with (+)-tartaric acid. The products are a pair of amine salts.

$$\underset{(\pm)\text{-}\alpha\text{-phenylethylamine}}{C_6H_5{-}\overset{\ddot{N}H_2}{\overset{|}{C}}HCH_3} \quad + \quad \underset{(+)\text{-tartaric acid}}{\begin{array}{c} O \\ \| \\ HOC \\ | \\ H{-}C{-}OH \\ | \\ HO{-}C{-}H \\ | \\ CO_2H \end{array}} \longrightarrow$$

* (+)-Tartaric acid is also referred to as L-, *d*-, dextrorotatory, and (2*R*,3*R*)-tartaric acid.

$$C_6H_5\text{—}\underset{H}{\overset{\overset{+}{N}H_3}{C}}\text{—}CH_3 \quad H\text{—}\underset{\underset{CO_2H}{HO\text{—}C\text{—}H}}{\overset{^{-}O_2C}{C}}\text{—}OH \quad + \quad CH_3\text{—}\underset{H}{\overset{\overset{+}{N}H_3}{C}}\text{—}C_6H_5 \quad H\text{—}\underset{\underset{CO_2H}{HO\text{—}C\text{—}H}}{\overset{^{-}O_2C}{C}}\text{—}OH$$

the (*S*)-amine (2*R*,3*R*)-tartrate the (*R*)-amine (2*R*,3*R*)-tartrate

a pair of diastereomers

Note that the product amine salts are *not* enantiomeric. [The enantiomer of the (*S*)-(2*R*,3*R*) salt would be (*R*)-(2*S*,3*S*).] These amine salts are diastereomers that differ in their solubilities in methanol. The less-soluble diastereomeric salt can be crystallized and separated from the other salt, which remains in the methanol solution. After separation, the enantiomeric amine is isolated from this salt by treatment with base.

$$R\overset{+}{N}H_3 \ ^{-}O_2CR' + OH^- \xrightarrow{H_2O} RNH_2 + {}^{-}O_2CR' + H_2O$$

17.3 The Polarimeter

A **polarimeter** (Figure 17.2) is an instrument used to measure the optical rotation of organic compounds. We will discuss briefly the principles behind polarimetry. The actual operation of the polarimeter will be demonstrated by your instructor.

A polarimeter uses a single wavelength of light, usually the sodium D line (589.3 nm) from a sodium lamp or from an ordinary lamp with filters. In a common

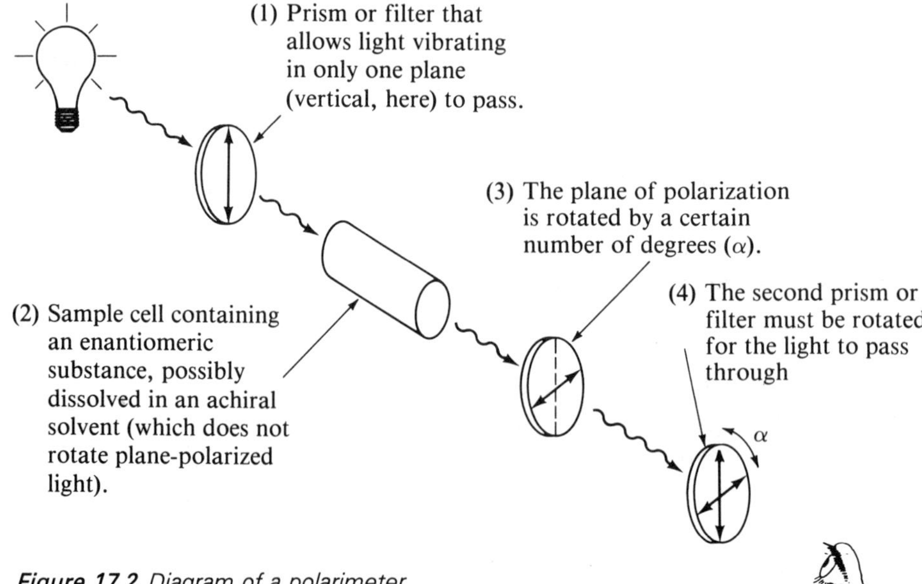

Figure 17.2 Diagram of a polarimeter.

side view of one wave showing only one plane of polarization

end-on view of a light beam vibrating in all planes perpendicular to the line of travel

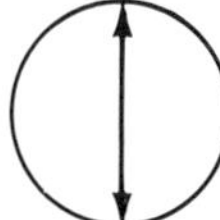

end-on view of a beam of plane-polarized light (vibrating in one plane only)

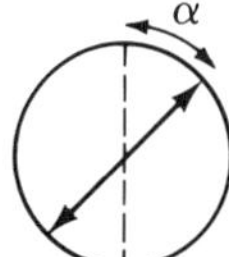

end-on view of a beam of plane-polarized light, the plane of polarization of which has been rotated from the original plane

Figure 17.3 *Planes of polarization of ordinary light and plane-polarized light.*

polarimeter, the light is passed through a *Nicol prism* ($CaCO_3$), which filters out all the light waves except for those waves vibrating in a *single plane.* Light vibrating in a single plane, rather than in all planes, is called **plane-polarized light**. Figure 17.3 depicts the vibrations of a beam of ordinary light and a beam of plane-polarized light.

When plane-polarized light is passed through a single enantiomer of an organic compound, the plane of polarization is *rotated.* This phenomenon, called **optical rotation**, is depicted in Figures 17.2 and 17.3. One enantiomer of a pair rotates the plane of polarization of the light to the right (+), while the other enantiomer rotates it to the left (−) by the same number of degrees.

The number of degrees of rotation by a sample is detected with a second Nicol prism at the eyepiece of the instrument. This prism is rotated until the light vibrating in the new plane passes through it. The angle of rotation is the angle that this second filter has been turned.

The optical rotations of a large number of compounds are known. These rotations are reproducible for a pure stereoisomer provided that the temperature, the wavelength of light, the concentration of the sample, and the pathlength of the cell are the same. For the sake of consistency, certain conventions have been adopted. Optical rotations are reported as **specific rotations,** $[\alpha]_D^{20}$ (sodium D line and a temperature of 20°). The specific rotation is the rotation observed for 1.00 g of pure sample in 1.00 mL of solution, in a cell with a pathlength of 1.00 decimeter (10.0 cm) at the specified wavelength and temperature. Rarely will a chemist use these exact conditions in determining the optical rotation of a sample. Experimental Note 4 on page 326 shows how $[\alpha]_D^{22}$ is calculated from the observed rotation α. The specific rotation of pure (S)-α-phenylethylamine, the stereoisomer isolated in Experiment 17.2, is −40.3° at 22° and is reported as $[\alpha]_D^{22} = -40\overset{\circ}{.}3$.

The **optical purity** of a sample is the per cent excess of one enantiomer compared to the amount present in the racemic mixture. A pure enantiomer is 100% optically pure, while a racemic mixture is 0% optically pure. The equation used to calculate the optical purity follows. A sample calculation of optical purity is included in Experimental Note 4, page 326.

$$\text{optical purity} = \frac{\text{calculated } [\alpha]_D{}^t \text{ of the sample}}{\text{known } [\alpha]_D{}^t \text{ of a pure sample}} \times 100$$

$$= \frac{\text{actual \% of enantiomer in mixture} - 50\%}{50\%}$$

Problems

17.1 (a) Write an equation showing the intermediate for the *cis*-to-*trans* isomerization of 2-butene. (b) Would you expect this reaction to proceed in 100% yield? Explain.

17.2 Most attempted photoisomerizations of *trans* alkenes yield mixtures of the *cis* and *trans* alkenes. Why is this so?

17.3 The photoisomerization of *cis*-2-cyclooctenone to the highly strained *trans* isomer has been accomplished. Draw the structures and suggest a reason for the success of this transformation.

17.4 Show by flow equations how you would resolve (±)-2-chloropropanoic acid using (−)-amphetamine.

$C_6H_5CH_2CH(NH_2)CH_3$

amphetamine

17.5 Suggest a route, using flow equations, for resolving (±)-2-butanol.

⟫⟫ EXPERIMENT 17.1 ⟪⟪

Cis,trans Isomerization of 1,2-Dibenzoylethylene

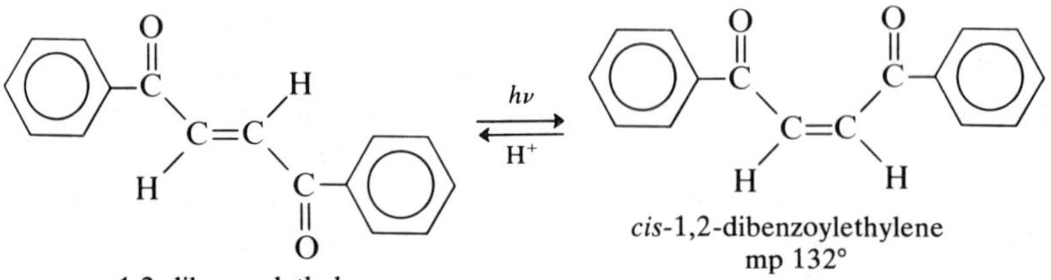

$C_6H_5C(=O)CH{=}CHC(=O)C_6H_5$ (trans) $\underset{H^+}{\overset{h\nu}{\rightleftharpoons}}$ $C_6H_5C(=O)CH{=}CHC(=O)C_6H_5$ (cis)

trans-1,2-dibenzoylethylene
mp 109°

cis-1,2-dibenzoylethylene
mp 132°

Discussion

Yellow *trans*-1,2-dibenzoylethylene can be photoisomerized to the colorless *cis* isomer without the aid of a sensitizer (photocatalyst). The procedure consists of dissolving the *trans* isomer in ethanol and exposing the solution to sunlight or bright artificial light.

The ultraviolet spectra of the *cis*- and *trans*-1,2-dibenzoylethylene isomers (Figure 17.4) reveal the physical basis of the photoisomerization experiment. The λ_{max} of each isomer is in the ultraviolet portion of the spectrum, but the absorption by the *trans* isomer tails into the visible portion of the spectrum. (This is why the *trans* isomer is yellow but the *cis* isomer is colorless.) Light consisting of the near ultraviolet and blue portions of the spectrum causes only the *trans* isomer to be converted to the excited state. By a careful selection of the wavelength, a complete conversion of the *trans* isomer to the *cis* isomer can be achieved; however, with sunlight or a light bulb, the *cis* isomer is contaminated with some *trans* isomer.

Reconversion of the *cis*-1,2-dibenzoylethylene to its *trans* isomer is very straightforward. *trans*-1,2-Dibenzoylethylene is more stable than the *cis* isomer; therefore, by heating the *cis* isomer with a trace of acid, it is almost completely isomerized to the *trans* form.

The work-up for both isomerization procedures is identical and is based upon the insolubility of both isomers in cold ethanol. The reactions are carried out in warm ethanol, the solutions are then chilled in an ice bath, and the predominant isomer in solution crystallizes.

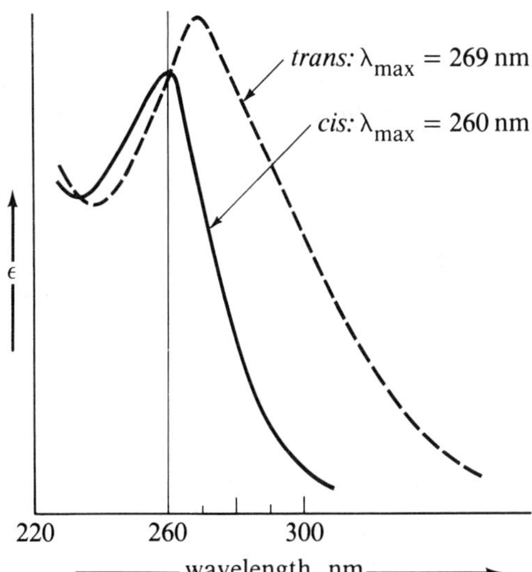

Figure 17.4 Ultraviolet spectra of cis- and trans-1,2-dibenzoylethylene.

Experimental

EQUIPMENT:

two 50-mL and two 125-mL Erlenmeyer flasks
gravity filtration assembly
hot plate
ice bath
light source (see Experimental Note 1)

black lamp (optional)
tlc assembly (optional) with Kodak Chromatogram Sheets #13181 (silica gel adsorbent with fluorescent indicator)

CHEMICALS:

trans-1,2-dibenzoylethylene, 1.0 g
95% ethanol, 100 mL
conc. hydrochloric acid, 3 drops

carbon tetrachloride (for tlc, optional), 40 mL
1,2-dichloroethane (for tlc, optional), 20 mL

TIME REQUIRED: 1–2 hours for the laboratory work and 1–7 days for the *trans*-to-*cis* isomerization (see Experimental Note 2)

STOPPING POINTS: while the *trans* isomer is being isomerized; while the products are crystallizing or drying

PROCEDURE

A. *Trans*-to-*cis* Isomerization

Dissolve 1.0 g of *trans*-1,2-dibenzoylethylene by heating it in 50 mL of 95% ethanol in a 125-mL Erlenmeyer flask. Cork the flask and place it in bright light (see Experimental Note 1). At the start of the reaction period, the mixture contains predominantly yellow crystals (*trans* isomer). As the reaction proceeds, the crystalline mass (or solution, depending on the temperature) is slowly converted to the colorless *cis* isomer (see Experimental Note 2). When the solution is colorless or only faintly yellow, chill it in an ice bath and remove the crystals by gravity filtration. Recrystallize the colorless *cis* product from ethanol, air-dry the product, measure the melting point, and calculate the per cent yield.

B. *Cis*-to-*trans* Isomerization

Dissolve 0.4 g of *cis*-1,2-dibenzoylethylene in 15 mL of 95% ethanol with heating. Add 3 drops of concentrated hydrochloric acid and warm the solution on a hot plate for 5 minutes. (Do not allow ethanol to reflux from the flask.) Chill the solution in an ice bath, filter with gravity, and wash the crystals with cold ethanol. Air-dry the solid, measure the melting point, and calculate the per cent yield.

EXPERIMENTAL NOTES

1) Almost any bright light is suitable. In bright sunlight, isomerization is complete within about one day. If artificial light is used, the flask should be positioned as close to the light as possible. With a 60-watt, white-frosted light bulb or with a "shop" fluorescent lamp, the isomerization takes about a week with an exposure of 24 hours per day. Isomerization is much more rapid (3–5 hours) when the flask is positioned 5 inches from a sunlamp (General Electric, 275-watt), but an impure product with a depressed melting point is obtained. An ultraviolet lamp induces photochemical rearrangements of dibenzoylethylene.*

2) A tlc procedure to follow the course of the isomerization has been developed by William R. Stine and David J. Sedor (Wilkes College). Spot a strip of Kodak Chromatogram Sheet #13181 (silica gel adsorbent containing fluorescent indicator) with the isomerization solution. For identification, also spot the strip with solutions of *cis*- and *trans*-1,2-dibenzoylethylene. Develop the strip in a solvent system of 80% carbon tetrachloride and 20% 1,2-dichloroethane. Allow the solvent front to move at least 7 cm up the strip (about 1 hour of development will be required). Visualize the spots with a uv black lamp.

Problems

17.6 Suggest a reason (based upon structure) why *trans*-1,2-dibenzoylethylene is colored, but the *cis* isomer is not.

17.7 (a) Would you expect the carbocation formed when 1,2-dibenzoylethylene is protonated to be more or less stable than an ordinary secondary carbocation, such as $(CH_3)_2CH^+$? Explain.

(b) Which isomer of 1,2-dibenzoylethylene would you expect to become protonated at the more rapid rate? Why?

17.8 What wavelength of light would you choose to irradiate the *trans* isomer in Experiment 17.1 to maximize the yield of the *cis* product? Explain your answer.

* H. W. Zimmerman, H. G. Durr, R. S. Givens, and R. G. Lewis, *J. Am. Chem. Soc.* **1967**, *89*, 1863.

»»» EXPERIMENT 17.2 «««

*Resolution of α-Phenylethylamine**

Step 1, Formation of the Diastereomers:

$C_6H_5CH(NH_2)CH_3$ + (+)-tartaric acid $\xrightarrow{CH_3OH}$ $C_6H_5CH(\overset{+}{N}H_3)CH_3$ $^-O_2C{-}CH(OH){-}CH(OH){-}CO_2H$

(±)-α-phenylethylamine; (+)-tartaric acid; (+)- and (−)-α-phenylethylammonium (+)-hydrogen tartrates (both diastereomers)

	(±)-α-phenylethylamine	(+)-tartaric acid	(+)- and (−)-α-phenylethylammonium (+)-hydrogen tartrates
MW:	121.20	150.09	271.29
weight:	10.0 g	12.5 g	22.2 (theory)
moles:	0.082	0.083	0.082 (theory)

Step 2, Conversion of One Diastereomer to the Amine:

$C_6H_5CH(\overset{+}{N}H_3)CH_3$ $^-O_2C{-}CH(OH){-}CH(OH){-}CO_2H$ + 2NaOH (excess) $\longrightarrow$ $C_6H_5CH(NH_2)CH_3$ + $NaO_2C{-}CH(OH){-}CH(OH){-}CO_2Na$ + $2H_2O$

(−)-α-phenylethylammonium (+)-hydrogen tartrate (one diastereomer); (−)-α-phenylethylamine

	(−)-α-phenylethylammonium (+)-hydrogen tartrate	(−)-α-phenylethylamine
MW:	271.29	121.20
weight:	11.1 (theory)	5.0 g (theory)
moles:	0.041 (theory)	0.041 (theory)

Table 17.1 *Physical properties of reactants and products*

Name	*Bp(°C)*	$[\alpha]_D$	*Solubility*
(±)-α-phenylethylamine	184–186	0°	insol. H_2O; sol. CH_3OH
(+)-tartaric acid[a]	—	+12.0° (20°C)	sol. H_2O; sol. CH_3OH
(−)-α-phenylethylamine	184–186	−40.3° (22°C)	insol. H_2O; sol. CH_3OH

[a] Mp 171–174°; $pK_1 = 3$

* The procedure for this resolution has been adapted from A. Ault, *J. Chem. Ed.* **1965**, *42*, 269; and W. Theilaker and H. G. Winkler, *Chem. Ber.* **1954**, *87*, 691.

Discussion

($\pm$)-α-Phenylethylamine is resolved by treatment with (+)-tartaric acid. The reaction is carried out by mixing methanol solutions containing the two reactants. The result is a pair of diastereomeric salts: (+)- and ($-$)-α-phenylethylammonium (+)-hydrogen tartrate. Because the reaction is exothermic, the methanol solutions are cooled to near room temperature before mixing. After mixing, the solution becomes opaque because of precipitation of the salts.

The key to success in this experiment is to obtain the proper crystalline form. The tartrate salt of the ($-$)-amine crystallizes slowly from solution as small, dense prisms. Another crystalline form containing *both* diastereomers crystallizes from a rapidly cooled solution as a voluminous mass of needlelike crystals. The needles redissolve when the solution is heated; however, the prisms do not. To obtain prismatic seed crystals, the methanol solution is heated to near-boiling, rapidly cooled in an ice bath to yield both types of crystals, then rewarmed to near-boiling to dissolve the needles. The flask is then set aside and allowed to cool slowly. If the needles also recrystallize, they will slowly redissolve and be converted to prisms, if the mixture is allowed to stand over a period of a week or more at room temperature, or the mixture can be reheated and allowed to cool. After the prismatic crystals are filtered, the salt of the (+)-amine could be isolated from the filtrate. However, this is not done in this experiment because it is time-consuming.

The remaining portion of the experiment is aimed at reconverting the diastereomeric salt to the free amine. Treatment with 10% sodium hydroxide converts the diastereomeric salt to disodium tartrate and the amine (see page 322). Disodium tartrate is water-soluble, but the amine is insoluble and forms a layer on top of the alkaline solution. If this second layer does *not* form, then not enough NaOH has been added to neutralize both the carboxyl protons and the ammonium-ion protons.

The amine is separated from the aqueous alkali by a steam distillation and then separated from the distilled water in a separatory funnel. At best, only 5 g of amine can be present; therefore, these manipulations must be carried out very carefully. The crude, wet amine is dried with anhydrous K_2CO_3. If the earlier separation was sloppy, the presence of excess water in the amine will cause the K_2CO_3 to liquefy. If this should happen, more drying agent must be added (and more product will probably be lost).

The amine should be distilled before its optical rotation is measured. Unfortunately, the amine has a high boiling point (184–188°). A small amount of high-boiling liquid is difficult to distil through the large-sized distillation apparatus typically found in a student locker. If you do not have access to a small-sized apparatus, you will have to measure the rotation of the crude, undistilled amine. In this case, if the amine is cloudy after drying, a clarification by centrifugation should be carried out.

It is necessary to know the exact weight of the amine transferred to the polarimeter tube so that the specific rotation can be calculated from the observed rotation. The weighing can be accomplished in one of two ways: (1) the polarimeter tube can be tared and then reweighed after the amine has been added, or (2) the flask containing the amine can be weighed before and after the amine is transferred.

If the quantity of amine is insufficient to fill the polarimeter tube completely, methanol is used to fill the tube before measurement. It is assumed that the methanol does not affect the measured rotation and that the $[\alpha]_D^t$ of the methanol–amine solution can be compared directly with the $[\alpha]_D^t$ of the pure amine.

Experimental

EQUIPMENT:

50-mL beaker
disposable pipet
distillation assembly
50-mL and 250-mL Erlenmeyer flasks
gravity filtration assembly
hot plate
ice bath
polarimeter
two 50-mL round-bottom flasks
125-mL separatory funnel

centrifuge (optional)
magnifying glass (optional)

CHEMICALS:

anhydrous potassium carbonate, 2–4 g
10% aqueous sodium hydroxide, 50 mL
methanol, 160–180 mL
(±)-α-phenylethylamine, 10 g
(+)-tartaric acid, 12.5 g

TIME REQUIRED: 5 hours plus an overnight crystallization and an overnight drying of the diastereomer (optional)

STOPPING POINTS: while the diastereomer is crystallizing; while the diastereomeric crystals are drying; after the steam distillation; while the amine is being dried

PROCEDURE

Dissolve 10.0 g of (±)-α-phenylethylamine* in 75 mL of methanol (CAUTION: toxic and flammable) in a 250-mL Erlenmeyer flask and mix thoroughly. In another 250-mL Erlenmeyer flask, dissolve 12.5 g of powdered (+)-tartaric acid in 75 mL of methanol. (Tartaric acid is methanol-soluble, but the solution may remain cloudy. This cloudiness will not affect your results.) Add the amine solution to the tartaric acid solution.

* α-Phenylethylamine is toxic and does not have the pronounced pharmacological activity characteristic of the β-phenylethylamines such as amphetamine (page 318).

Warm the flask on a hot plate almost to boiling, then cool it in an ice bath to crystallize the amine tartrates. Inspect the crystals carefully (with a magnifying glass if one is available). These crystals are usually voluminous and needlelike (see Experimental Note 1). Reheat the mixture almost to boiling to dissolve the needlelike crystals. The residual prismatic crystals act as seed crystals for the (−)-amine salt. Cap the Erlenmeyer flask with a 50-mL beaker to prevent the methanol from evaporating, and set the flask in your locker to cool slowly overnight. If the newly crystallized salt is still in the form of needles rather than prisms, reheat the mixture and allow it to recrystallize slowly.

Filter the prismatic crystals by gravity, then wash with 10 mL of cool methanol. If time is available, allow the crystals to air-dry. A typical yield is 8.5 g (77%), mp 190–200°.

Place the solid in a round-bottom flask and add 50 mL of 10% aqueous sodium hydroxide, or more if two phases do not form. Using a simple distillation apparatus, distil about 25 mL of liquid from the sodium hydroxide–amine mixture.

Transfer the distillate to a separatory funnel, carefully drain the lower aqueous layer, then pour the amine into a tared flask (see Experimental Note 3). (You should obtain about 5 g of wet amine, n_D^{20} 1.5222.) Add 2–4 g of anhydrous potassium carbonate. The carbonate forms a semisolid lower layer with the amine floating on top. Weigh the flask containing the amine–potassium carbonate mixture before continuing. (If the amine is cloudy, transfer it to a tared test tube and centrifuge down the suspended solid.)

Carefully transfer the amine from the flask to a polarimeter tube, using a disposable pipet (see Experimental Note 3). Reweigh the amine–potassium carbonate flask to determine the exact weight of amine transferred. Fill the rest of the polarimeter tube with methanol, taking care that no air bubbles are trapped in the tube. Measure the optical rotation of the amine solution. Calculate the specific rotation and the optical purity (see Experimental Note 4).

EXPERIMENTAL NOTES

1) The needlelike crystals (mp 168–170°) are a crystalline modification that contains both diastereomers. If the amine is isolated from these crystals, its $[\alpha]_D^{20}$ will be about −19° to −21°.

2) If the amine must stand overnight, stopper the flask with a lightly greased ground-glass stopper. Do not use a cork; amines react with cork to form a black, semiliquid material that will contaminate your product.

3) Before proceeding with the polarimetry portion of the experiment, check with your instructor on the operation of the polarimeter. The observed rotation in this experiment is quite small—about −8° to −9° for 2.1 g of (−)-amine diluted to 6.8 mL with methanol in a 10-cm cell. Therefore, you should run a blank (with the cell containing pure methanol) before measuring the optical rotation of the amine solution.

4) Sample calculation of a specific rotation:

$$[\alpha]_D^t = \frac{\alpha}{lc}$$

where $[\alpha]_D^t$ = specific rotation (sodium D line) at t°C

α = observed rotation at t°C

l = length of polarimeter cell, in dm (decimeters, where 1.0 dm = 0.10 m)

c = concentration of sample in g/mL

Assume that 2.1 g of amine is placed in a 1.0-dm cell with a total capacity of 6.8 mL. The cell is then filled with methanol. If the observed polarimeter reading is −8.75° and the temperature is 22°, then

$$[\alpha]_D^{22} = \frac{-8.75°}{(1.00\ \text{dm})\ (2.1\ \text{g}/6.8\ \text{mL})} = -28.3°$$

Sample calculation of optical purity:

for a sample with $[\alpha]_D^{22} = -28.3°$,

$$\text{optical purity} = \frac{\text{calculated } [\alpha]_D^t}{[\alpha]_D^t \text{ of pure compound}} \times 100$$

$$= \frac{-28.3°}{-40.3°} \times 100$$

$$= 70.2\%$$

Problems

17.9 In Experiment 17.2, a student obtains an amine with $[\alpha]_D^{22} = -35.3°$.

(a) What is the optical purity of the amine?

(b) Calculate the per cent of (*S*)-amine and that of (*R*)-amine in the sample.

17.10 A student wishes to isolate the (+)-amine diastereomeric salt in Experiment 17.2.

(a) How would the student do this?

(b) What would be the principal problem encountered?

17.11 When NaOH solution is added to the diastereomeric amine salt, which portion of the salt is the first to be neutralized—the carboxyl portion or the ammonium-ion portion? Explain.

17.12 Why is the amine dried with K_2CO_3 instead of with $MgSO_4$? What other drying agents could be used?

»» CHAPTER 18 ««

Enolate Condensations

In acidic or basic solution, aldehydes (and, under some conditions, ketones) can undergo **condensation reactions**, reactions in which molecules combine to yield larger molecules. With aldehydes, this particular type of condensation reaction is called an **aldol condensation**, or an **aldol addition**, because the product is both an aldehyde and an alcohol. An example of an aldol condensation is the self-condensation of acetaldehyde, a reaction discovered by the French chemist Charles Adolphe Wurtz in 1872. This condensation yields 3-hydroxybutanal, often called *acetaldol* or simply *aldol.*

$$2CH_3\overset{\overset{\large O}{\|}}{C}H \underset{\longleftarrow}{\xrightarrow{\text{aqueous NaOH}}} CH_3\overset{\overset{\large OH}{|}}{C}HCH_2\overset{\overset{\large O}{\|}}{C}H$$

ethanal (acetaldehyde) — 3-hydroxybutanal (acetaldol or aldol) 50%

18.1 Mechanism of the Aldol Condensation

Experiment 18.1 is a base-catalyzed aldol condensation; therefore, let us consider the mechanism of this condensation. In dilute base, a carbonyl compound with one or more alpha hydrogen atoms (hydrogens on the carbon adjacent to the carbonyl group) acts as a very weak acid. An alpha hydrogen can be lost to the hydroxide ion to yield water and an enolate ion.

$$CH_3\overset{O}{\overset{\|}{C}}H + OH^- \rightleftharpoons \left[{}^-CH_2-\overset{:\ddot{O}}{\overset{\|}{C}}H \longleftrightarrow CH_2{=}\overset{:\ddot{O}:^-}{\overset{|}{C}}H \right] + H_2O$$

α hydrogens

resonance structures for the enolate ion of acetaldehyde

Alpha hydrogens are slightly acidic because the product enolate ion can be stabilized by resonance. A simple carbonyl compound, such as acetaldehyde, is only weakly acidic. Even a *β*-dicarbonyl compound, the anion of which is stabilized by *two* carbonyl groups, is a weaker acid than a typical carboxylic acid by a factor of 10,000.

acidic hydrogens

	$CH_3\overset{O}{\overset{\|}{C}}CH_3$	$CH_3\overset{O}{\overset{\|}{C}}H$	$CH_3\overset{O}{\overset{\|}{C}}CH_2\overset{O}{\overset{\|}{C}}CH_3$	$CH_3\overset{O}{\overset{\|}{C}}OH$
approx. pK_a:	20	17	9	5

An aldol condensation is possible because an aldehyde (or ketone) and its enolate ion are in equilibrium in alkaline solution. An enolate anion is a nucleophile that can attack the partially positive carbon of the carbonyl group of an aldehyde or ketone. The result is an alkoxide ion, which is a strong base that reacts with water to yield the aldol product and regenerate the hydroxide ion.

an alkoxide ion

$$CH_3\overset{\delta- :\ddot{O}:}{\overset{\|}{C}}H_{\delta+} + {}^-CH_2\overset{O}{\overset{\|}{C}}H \rightleftharpoons \left[CH_3\overset{:\ddot{O}:^-}{\overset{|}{C}}H \underset{CH_2\overset{\|}{C}H}{\overset{O}{}} \right] \xrightleftharpoons{H_2O} CH_3\overset{:\ddot{O}H}{\overset{|}{C}}H-CH_2\overset{O}{\overset{\|}{C}}H + OH^-$$

An aldol condensation is a reversible reaction, and is thus subject to *thermodynamic*, rather than kinetic, control—that is, the equilibrium favors the more thermodynamically stable product. In the condensation of aldehydes, the aldol is favored. A ketone condensation product is more hindered and is not favored. Special laboratory techniques must be used to prepare the self-addition product of a ketone.

$$2\ CH_3\overset{O}{\overset{\|}{C}}CH_3 \xrightleftharpoons{\text{aqueous NaOH}} CH_3\underset{CH_3}{\underset{|}{\overset{OH}{\overset{|}{C}}}}-CH_2\overset{O}{\overset{\|}{C}}CH_3$$

steric hindrance

not favored

18.2 Crossed Aldol Condensations

An aldol condensation of two different compounds is called a **crossed aldol condensation.** To avoid mixtures of products (from self-addition as well as the crossed reaction), one reactant that has no alpha hydrogens is usually chosen. To avoid steric hindrance at the carbonyl carbon and an unfavorable equilibrium, this reactant is usually an aldehyde, such as benzaldehyde, C_6H_5CHO.

The second reactant can be any aldehyde or ketone that contains an acidic alpha hydrogen. A ketone such as acetone is ideal because it can lose an alpha hydrogen yet does not undergo self-condensation to any extent.

$$\underset{\text{acetone}}{CH_3\overset{O}{\overset{\|}{C}}CH_3} + OH^- \rightleftharpoons \underset{\textit{can attack an aldehyde C=O}}{^-CH_2\overset{O}{\overset{\|}{C}}CH_3} + H_2O$$

18.3 Dehydration of Aldols

The product of an aldol condensation is a *β-hydroxy aldehyde* or *ketone*. These compounds undergo dehydration readily because they yield products in which the double bond is in conjugation with the carbonyl group. For example, if 3–hydroxybutanal is heated in the presence of a catalytic amount of acid or base, it loses water to yield 2-butenal.

$$\underset{\text{3-hydroxybutanal}}{CH_3\overset{OH}{\overset{|}{C}}HCH_2\overset{O}{\overset{\|}{C}}H} \xrightleftharpoons[\text{heat}]{H^+ \text{ or } OH^-} \underset{\substack{\text{2-butenal} \\ \textit{favored because of} \\ \textit{conjugation}}}{CH_3CH{=}CH\overset{O}{\overset{\|}{C}}H} + H_2O$$

If benzaldehyde is used as the aldehyde in a crossed aldol condensation, dehydration occurs so readily, even in a cold mixture, that the aldol is not observed. The reason for the facile elimination of water from this type of aldol is that the double bond is in conjugation with both the carbonyl group and the benzene ring.

$$C_6H_5\overset{O}{\overset{\|}{C}}H + CH_3\overset{O}{\overset{\|}{C}}CH_3 \xrightleftharpoons[]{\substack{\text{aqueous}\\ \text{NaOH}}} \underset{\textit{not observed}}{\left[C_6H_5\overset{OH}{\overset{|}{C}}H{-}CH_2\overset{O}{\overset{\|}{C}}CH_3\right]} \xrightleftharpoons[]{-H_2O} \underset{\substack{\text{4-phenyl-3-buten-2-one} \\ \text{(benzalacetone)} \\ 90\%}}{C_6H_5CH{=}CH\overset{O}{\overset{\|}{C}}CH_3}$$

In base, this elimination is initiated by the loss of an alpha hydrogen, followed by the loss of a hydroxide ion to regenerate the catalyst. The organic product is the less hindered *trans*-alkene.

$$C_6H_5\overset{:\ddot{O}H}{\overset{|}{C}}HCH_2\overset{O}{\overset{\|}{C}}CH_3 \xrightleftharpoons[]{OH^-} C_6H_5{-}\overset{:\ddot{O}H}{\overset{|}{C}}H{-}\overset{-}{C}H\overset{O}{\overset{\|}{C}}CH_3 + H_2O$$

$$\longrightarrow \underset{\textit{trans}\text{, or } (E)}{\overset{H}{\underset{C_6H_5}{}}C{=}C\overset{\overset{O}{\|}{C}CH_3}{\underset{H}{}}} + :\ddot{O}H^-$$

The formation of the alkene is essentially irreversible, and thus provides a driving force for the entire sequence of reversible reactions starting from benzaldehyde. Overall yields in this sequence are often very good.

Crossed aldol condensations between aromatic aldehydes and ketones, followed by dehydration, are called **Claisen–Schmidt condensations**. Experiment 18.1 is the Claisen–Schmidt condensation between acetone and *two equivalents* of benzaldehyde. Step 1 in this reaction (the reaction of acetone with the *first* equivalent of benzaldehyde) is shown in the preceding two equations. Step 2 is the loss of an alpha hydrogen from this product and the subsequent reaction with the second equivalent of benzaldehyde, followed by dehydration.

acidic hydrogen

$$C_6H_5CH{=}CH\overset{O}{\overset{\|}{C}}CH_3 + H\overset{O}{\overset{\|}{C}}C_6H_5 \xrightleftharpoons{\text{aqueous NaOH}}$$

$$C_6H_5CH{=}CH\overset{O}{\overset{\|}{C}}CH_2{-}\overset{OH}{\overset{|}{C}}HC_6H_5 \xrightleftharpoons{-H_2O}$$

(*E*, *E*)-1,5-diphenyl-1,4-pentadien-3-one
(*trans,trans*-dibenzalacetone)

18.4 The Perkin Reaction

The base-catalyzed reaction of an aromatic aldehyde with a carboxylic acid anhydride to yield an α,β-unsaturated acid is called the **Perkin reaction**. The Perkin reaction is similar to an aldol condensation followed by an elimination.

a Perkin reaction:

$$\underset{\text{benzaldehyde}}{C_6H_5\overset{O}{\overset{\|}{C}}H} + \underset{\text{acetic anhydride}}{CH_3\overset{O}{\overset{\|}{C}}O\overset{O}{\overset{\|}{C}}CH_3} \xrightarrow{CH_3CO_2Na} \underset{\text{cinnamic acid}}{C_6H_5CH{=}CH\overset{O}{\overset{\|}{C}}OH} + CH_3\overset{O}{\overset{\|}{C}}OH$$

Experiment 18.2 is a Perkin reaction using benzaldehyde, phenylacetic acid, and acetic anhydride. Triethylamine is the base. The final products are the (*E*)- and the (*Z*)-α-phenylcinnamic acids.

$$C_6H_5\overset{O}{\overset{\|}{C}}H + \underset{\text{phenylacetic acid}}{C_6H_5CH_2\overset{O}{\overset{\|}{C}}OH} \xrightarrow[(CH_3CH_2)_3N]{(CH_3\overset{O}{\overset{\|}{C}})_2O}$$

(*E*)-α-phenylcinnamic acid + (*Z*)-α-phenylcinnamic acid

Although acetic anhydride is used in the reaction, it is the phenylacetic anhydride that undergoes the condensation reaction. Acetic anhydride undergoes equilibration with a carboxylic acid to yield a mixed anhydride.

$$C_6H_5CH_2\overset{O}{\overset{\|}{C}}OH + CH_3\overset{O}{\overset{\|}{C}}O\overset{O}{\overset{\|}{C}}CH_3 \rightleftharpoons \underset{\text{acetic phenylacetic anhydride}}{C_6H_5CH_2\overset{O}{\overset{\|}{C}}O\overset{O}{\overset{\|}{C}}CH_3} + CH_3\overset{O}{\overset{\|}{C}}OH$$

The mixed anhydride has the more acidic α hydrogen (adjacent to both the benzene ring and to the carbonyl group), which can be removed by the weak base triethylamine. It is this enolate-like anion that is the reactive species.

$$C_6H_5CH_2\overset{O}{\overset{\|}{C}}O\overset{O}{\overset{\|}{C}}CH_3 + (CH_3CH_2)_3N\!: \rightleftarrows C_6H_5\overset{-}{C}H\overset{O}{\overset{\|}{C}}O\overset{O}{\overset{\|}{C}}CH_3 + (CH_3CH_2)_3\overset{+}{N}H$$

most acidic α hydrogen

A. Mechanism of the Perkin Reaction

The condensation reaction proceeds by attack of the anion on the partially positive carbonyl carbon of benzaldehyde, followed by transfer of the acetyl group from the anhydride group to the alkoxide group.

Step 1, addition and acetyl transfer:

$$C_6H_5\overset{\delta+}{C}H{=}\overset{\delta-}{O} + C_6H_5\overset{-}{C}H\overset{O}{\overset{\|}{C}}O\overset{O}{\overset{\|}{C}}CH_3 \rightleftarrows \text{cyclic: } C_6H_5CH(\text{–}O^-)\text{–}CH(C_6H_5)\text{–}C({=}O)\text{–}O\text{–}C({=}O)CH_3 \rightleftarrows$$

$$H_3C\text{–}C(\text{–}O^-)(\text{–}O\text{–}CHC_6H_5)\text{–}O\text{–}C({=}O)\text{–}CH(C_6H_5) \rightleftarrows H_3C\text{–}C({=}O)\text{–}O\text{–}C_6H_5CH\text{–}CH(C_6H_5)\text{–}C({=}O)\text{–}O^-$$

intermediate

In Section 18.3 we discussed the elimination of water from an aldol product. The intermediate product in the preceding equation also undergoes spontaneous elimination. In this case, an acetate ion and a proton are lost instead of water. The proton is probably abstracted by triethylamine.

Step 2, elimination:

$$CH_3\overset{O}{\overset{\|}{C}}O\text{–}C_6H_5CH\text{–}C(H)(CO_2^-)C_6H_5 + :N(CH_2CH_3)_3 \longrightarrow CH_3\overset{O}{\overset{\|}{C}}O^- + C_6H_5CH{=}\overset{CO_2^-}{\overset{|}{C}}C_6H_5 + H\overset{+}{N}(CH_2CH_3)_3$$

Acidification during the work-up procedure converts the carboxylates to the free acids.

B. The Stereochemistry of the Elimination

In Experiment 18.2, the (*E*)-α-phenylcinnamic acid is the major product of the Perkin reaction. Only a small amount of the (*Z*) isomer is obtained. Let us consider some reasons why this is so.

The intermediate leading to the alkenyl carboxylate contains two chiral carbon atoms (carbons 2 and 3) and thus exists in 2^2, or four, stereoisomeric forms.

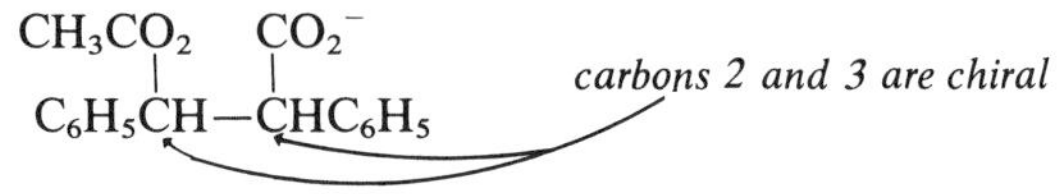

$$\begin{array}{cc} CH_3CO_2 & CO_2^- \\ | & | \\ C_6H_5CH- & CHC_6H_5 \end{array}$$

carbons 2 and 3 are chiral

The four stereoisomers of this intermediate consist of two pairs of enantiomers. One pair yields the (*E*) product in the elimination reaction and the other pair yields the (*Z*) product.

(2*R*,3*R*) (2*S*,3*S*) — enantiomers that lead to (*E*) product

(2*R*,3*S*) (2*S*,3*R*) — enantiomers that lead to (*Z*) product

Because each of the stereoisomeric intermediates contains a single β hydrogen atom, only one conformation in each case can undergo *anti*-elimination. Figure 18.1 shows Newman projections of the conformations required for the elimination. These Newman projections show why one pair of enantiomers leads to the (*E*) product and one pair leads to the (*Z*) product.

Of the two isomeric phenylcinnamic acids, the (*E*) isomer, with *cis* phenyl groups, is more stable than the (*Z*) isomer. These relative stabilities are in direct contrast to those of stilbene (1,2-diphenylethene)—*trans*-stilbene is more stable

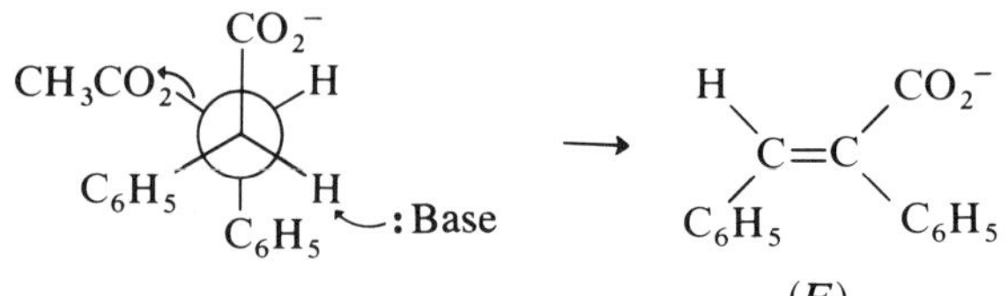

(2*S*,3*S*)
phenyl groups on the same side, as they also are in the (2*R*,3*R*) enantiomer

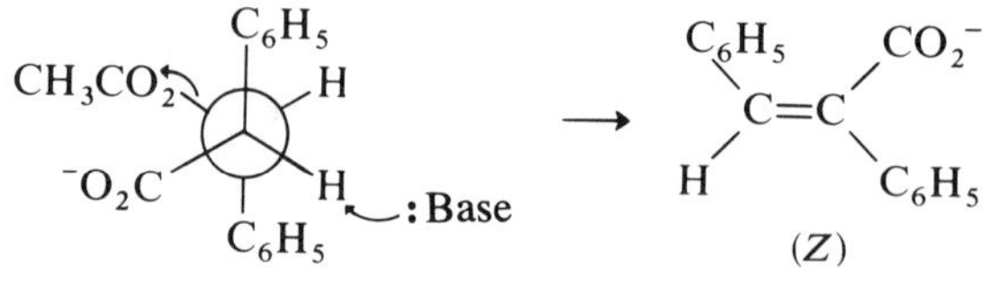

(2*S*,3*R*)
phenyl groups on opposite sides, as they also are in the (2*R*,3*S*) enantiomer

Figure 18.1. Newman projections for the conformations of the stereoisomeric intermediates required for the anti-elimination of H^+ and $CH_3CO_2^-$.

than its *cis*-isomer. The size of the carboxyl group is the structural feature that alters the relative stabilities.

$$\begin{array}{ccc} H & & CO_2H \\ & \diagdown\;\;\diagup & \\ & C{=}C & \\ & \diagup\;\;\diagdown & \\ C_6H_5 & & C_6H_5 \\ & (E) & \end{array} \quad \text{is more stable than} \quad \begin{array}{ccc} C_6H_5 & & CO_2H \\ & \diagdown\;\;\diagup & \\ & C{=}C & \\ & \diagup\;\;\diagdown & \\ H & & C_6H_5 \\ & (Z) & \end{array}$$

Because (E)-α-phenylcinnamic acid is more stable than the (Z) isomer, the *transition state* leading to the (E) isomer is of lower energy. Thus, the (E) isomer is formed at a faster rate.

The addition reaction (Step 1 on page 331) is *reversible* under the conditions of the reaction. Therefore, the four stereoisomeric intermediates are in equilibrium. As the (E) product is being formed (an essentially irreversible reaction), its "starting material"—the $(2R,3R)$ and $(2S,3S)$ enantiomeric intermediates—is being depleted. However, because all the stereoisomeric intermediates are in equilibrium, the enantiomers leading to the (E) product are replenished by the shifting equilibrium. The result is a preponderance of the (E) product at the expense of the (Z) product.

$$\left.\begin{array}{l} \text{benzaldehyde} \\ \text{phenylacetic acid} \\ \text{acetic anhydride} \\ \text{triethylamine} \end{array}\right\} \begin{array}{l} \rightleftarrows (2R,3R) + (2S,3S) \longrightarrow (E)\text{ product} \\ \\ \rightleftarrows (2R,3S) + (2S,3R) \longrightarrow (Z)\text{ product} \end{array}$$

the four stereoisomeric intermediates

Problems

18.1 Complete the following equations for acid–base reactions. (Be careful to choose the most acidic H in each case.)

(a) $CH_3CH_2\overset{\overset{O}{\|}}{C}H + OH^- \rightleftarrows$

(b) $CH_3\overset{\overset{O}{\|}}{C}CH_2\overset{\overset{O}{\|}}{C}CH_3 + OH^- \rightleftarrows$

(c) $CH_3\overset{\overset{O}{\|}}{C}CH_2\overset{\overset{O}{\|}}{C}OCH_2CH_3 + OH^- \rightleftarrows$

(d) $CH_3\overset{\overset{O}{\|}}{C}CH_2\overset{\overset{O}{\|}}{C}OH + OH^- \rightleftarrows$

18.2 Write resonance structures for the following enolate ions:

(a) $^-CH_2\overset{\overset{O}{\|}}{C}CH_3$ (b) $C_6H_5\overset{-}{C}H\overset{\overset{O}{\|}}{C}OCH_2CH_3$ (c) $CH_3O\overset{\overset{O}{\|}}{C}\overset{-}{C}H\overset{\overset{O}{\|}}{C}OCH_3$

18.3 Suggest synthetic routes, using aldol or crossed aldol condensation reactions, to the folowing compounds:

(a) $CH_3CH_2CH_2CH{=}\underset{}{\overset{\overset{CH_2CH_3}{|}}{C}}CHO$ (b) $C_6H_5CH{=}CH\overset{\overset{O}{\|}}{C}H$

(c) $(CH_3)_3CCH(OH)CH_2CO_2CH_2CH_3$

(d) 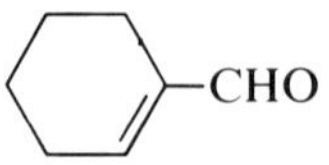 1-cyclohexenecarboxaldehyde (C_6H_9–CHO)

(e) 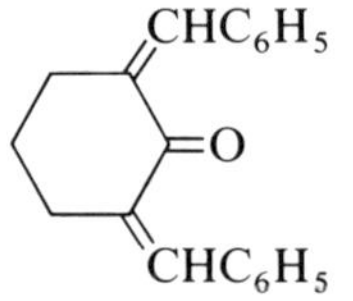 2,6-bis(CHC_6H_5)cyclohexanone

18.4 A mixture of acetic anhydride and phenylacetic acid is in equilibrium with the mixed anhydride (acetic phenylacetic anhydride), as shown on page 330.

(a) What *other* anhydride can be formed in the equilibrium mixture?

(b) How would this affect the course of the reaction?

18.5 Draw the Newman projections for the following compounds (all stereoisomers), showing all possible conformations that can undergo *anti*-elimination when treated with a strong base:

(a) 2-bromobutane

(b) 2-bromo-2,3-dimethylbutane

(c) 2-bromo-3-methyl-2-phenylpentane

18.6 Suggest routes to the following compounds by Perkin reactions:

(a) 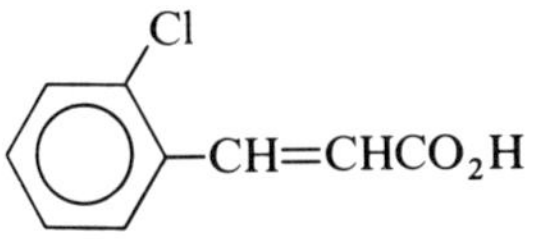 o-Cl-C_6H_4–$CH{=}CHCO_2H$ (b) 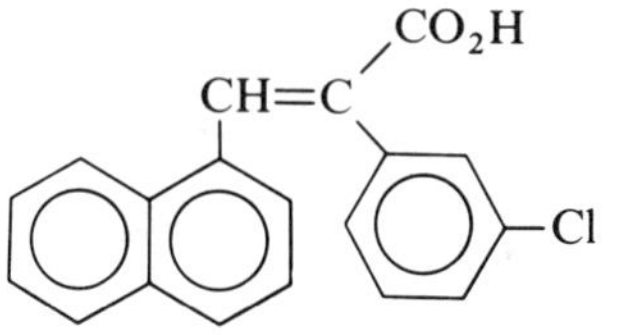 1-naphthyl–$CH{=}C(CO_2H)$–C_6H_4–Cl

18.7 Predict the major *anti*-elimination product or products when each of the following compounds is treated with sodium ethoxide. (Be sure to include the stereochemistry where appropriate; use models, if necessary.)

(a) (*R*)-3-bromohexane

(b) (1*R*,2*S*)-1-bromo-2-methyl-1-phenylbutane

»»» EXPERIMENT 18.1 «««

Synthesis of Dibenzalacetone

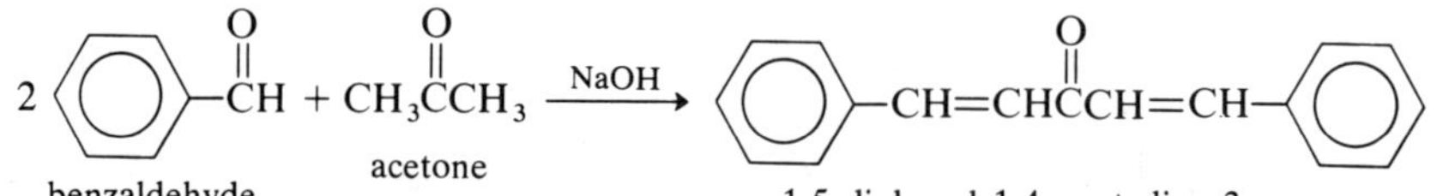

$$2\ C_6H_5CH{=}O + CH_3C({=}O)CH_3 \xrightarrow{\text{NaOH}} C_6H_5CH{=}CHC({=}O)CH{=}CHC_6H_5 + 2H_2O$$

benzaldehyde; acetone; 1,5-diphenyl-1,4-pentadien-3-one (dibenzalacetone)

	benzaldehyde	acetone	dibenzalacetone
MW:	106.13	58.08	234.28
weight:	4.2 g	1.2 g	4.7 g (theory)
moles:	0.040	0.021	0.020 (theory)

Table 18.1 Physical properties of reactants and products

Name	*Bp (°C)*	n_D^{20}	*Density (g/mL)*	*Solubility*
benzaldehyde	178.1	1.5463	1.04	sl. sol. H_2O; sol. CH_3CH_2OH and $(CH_3)_2C{=}O$
acetone	56.2	1.3588	0.79	sol. H_2O and CH_3CH_2OH
dibenzalacetone[a]	—	—	—	insol. H_2O; sl. sol. CH_3CH_2OH

[a] Mp of *trans,trans*-isomer, 110–111°.

Discussion

The reaction of acetone with benzaldehyde in the presence of base is a classical aldol condensation. Depending on the stoichiometry and reaction conditions, these reagents could be used to prepare either benzalacetone or dibenzalacetone. The series of reaction steps leading to benzalacetone and finally dibenzalacetone are shown on pages 329–330.

The reaction conditions in this experiment are chosen to favor the formation of dibenzalacetone. The reaction is carried out by adding a mixture of *two* parts benzaldehyde and *one* part acetone to an aqueous ethanol solution of NaOH. The solvent, aqueous ethanol, also favors the formation of dibenzalacetone. The reason is that the reactants and intermediates, including benzalacetone, are *soluble* in aqueous ethanol and are not removed from the reaction solution by separation or precipitation. Dibenzalacetone, however, is *insoluble* in aqueous ethanol—while the reaction mixture is being stirred, the deep yellow dibenzalacetone slowly precipitates from the solution.

The work-up consists of filtration, washing, and crystallization of the product. The critical part of the work-up is the washing. The reaction mixture, and thus the solid product, contains NaOH, which must be removed prior to crystallization. A number of washings may be required to remove it. To determine if the product has been washed sufficiently, each aqueous wash is tested with pH paper. Final purification of the dibenzalacetone is accomplished by crystallization from ethanol.

The color of dibenzalacetone is a result of the relatively large conjugated pi system, which can absorb a portion of the visible spectrum. Dibenzalacetone has been used in sun-protection preparations because it also absorbs certain wavelengths of ultraviolet light.

Experimental

EQUIPMENT:

- 150-mL beaker
- 50-mL Erlenmeyer flask
- 100-mL graduated cylinder
- hot water bath
- pH paper
- stirring rod
- vacuum filtration assembly

CHEMICALS:

acetone, 1.2 g
benzaldehyde, 4.2 g (see Experimental Note 1)
95% ethanol, about 100 mL
sodium hydroxide, 4.0 g

TIME REQUIRED: $1\frac{1}{2}$ hours plus overnight drying and time for melting-point determinations

STOPPING POINTS: when the crude product or the purified product is drying

PROCEDURE

Dissolve 4.0 g of solid sodium hydroxide (caustic!) in 40 mL of water in a beaker. Add 30 mL of 95% ethanol and cool the solution to 20°. Place 4.2 g of benzaldehyde (CAUTION: can cause dermatitis; also see Experimental Note 1) and 1.2 g of reagent-grade acetone in an Erlenmeyer flask. Swirl the flask until a homogeneous solution is obtained.

Add approximately one-half of the benzaldehyde solution to the hydroxide solution with vigorous stirring (see Experimental Note 2). Stir the mixture for 10 minutes, then add the remainder of the benzaldehyde–acetone solution. Continue stirring for another 30 minutes.

Filter the yellow solid with vacuum, press it as dry as possible, then transfer it to a clean beaker. Add 100 mL of water. Stir the mixture into a thick paste. If the pasty mixture is strongly alkaline, refilter and rewash. If the paste is near-neutral, filter with vacuum to obtain the crude product. A typical crude yield is 3.8 g. Crystallize the crude product from 95% ethanol or ethyl acetate. Determine the melting point and per cent yield.

EXPERIMENTAL NOTES

1) Benzaldehyde is readily air-oxidized to benzoic acid. Unless a freshly opened bottle of benzaldehyde is used, it is likely to contain a substantial amount of benzoic acid, which will reduce the yield. The purity of benzaldehyde can be estimated from its infrared spectrum. If necessary, benzaldehyde can be purified by vacuum distillation (Experiment 22.1, page 396).

2) A magnetic stirrer may be used; however, the precipitate may become so thick toward the end of the reaction that the stir bar will not turn.

Problems

18.8 How would benzoic acid in the benzaldehyde used in Experiment 18.1 affect the reaction?

18.9 Suggest another way (besides vacuum distillation) to purify partially oxidized benzaldehyde.

18.10 Suggest ways to modify the procedure in Experiment 18.1 so that the reaction yields benzalacetone instead of dibenzalacetone.

»»» EXPERIMENT 18.2 «««

Synthesis of (E)- and (Z)-α-Phenylcinnamic Acids

$$C_6H_5CH_2CO_2H + C_6H_5CHO + CH_3COOCOCH_3 + (CH_3CH_2)_3N \longrightarrow$$

	phenylacetic acid	benzaldehyde	acetic anhydride	triethylamine
MW:	136.16	106.13	102.09	101.19
weight:	2.7 g	3.2 g	2.2 g	2.1 g
moles:	0.020	0.030	0.021	0.021

(E)-α-phenylcinnamic acid + (Z)-α-phenylcinnamic acid

	(E)- and (Z)-α-phenylcinnamic acid
MW:	224.25
weight:	4.5 g (theory)
moles:	0.020 (theory)

Table 18.2 Physical properties of reactants and products

Name	***Mp (°C)***	***Bp (°C)***	***Solubility***
phenylacetic acid	77	—	sl. sol. H_2O; sol. $(CH_3CH_2)_2O$
benzaldehyde	—	178.1	sl. sol. H_2O; sol. $(CH_3CH_2)_2O$
triethylamine	—	89.3	sol. H_2O and $(CH_3CH_2)_2O$
acetic anhydride	—	139.5	sl. sol. H_2O; sol. $(CH_3CH_2)_2O$
(E)-α-phenylcinnamic acid[a]	173	—	sl. sol. H_2O; sol. $(CH_3CH_2)_2O$
(Z)-α-phenylcinnamic acid[b]	138	—	sl. sol. H_2O; sol. $(CH_3CH_2)_2O$

[a] $pK_a = 6.1$ [b] $pK_a = 4.8$

Discussion

Although the mechanism of the Perkin reaction of benzaldehyde and phenylacetic acid (see Section 18.4) is somewhat complex, the laboratory procedure is straightforward. The reactants are mixed with acetic anhydride and triethylamine and then heated at reflux for 30–40 minutes. After cooling, concentrated hydrochloric acid is added to convert the cinnamate anions to the carboxylic acids and the triethylamine to its hydrochloride salt. The cinnamic acids precipitate from this mixture and are extracted with diethyl ether. Some acetic acid from the hydrolysis of acetic anhydride (as well as residual acetic anhydride) also dissolves in the diethyl ether. The amine hydrochloride, a salt, remains in the aqueous layer. Washing the ether extract with water removes most of the acetic acid. (It is not necessary to remove all traces of acetic acid because acetic acid is used later in the work-up procedure.)

After the water washing, the ether solution is extracted with dilute sodium hydroxide solution. The cinnamic acids (and acetic acid) are converted to their sodium salts by the NaOH and thus dissolve in the aqueous layer. Organic impurities remain in the ether.

The (*E*)-α-phenylcinnamic acid (pK_a = 6.1) is a much weaker acid than its (*Z*) isomer (pK_a = 4.8). This fact, coupled with the insolubility of both acids in water, allows the separation of the two acids. The mixture of (*E*) and (*Z*) salts is carefully acidified to the pH at which the (*E*) carboxylate ion becomes protonated and precipitates. Because the (*Z*) acid is a stronger acid, it remains in solution as its sodium salt.

at pH = 6:

$$(E)\text{-}RCO_2^- + H_3O^+ \rightleftharpoons RCO_2H\downarrow + H_2O$$

$$(Z)\text{-}RCO_2^- + H_3O^+ \rightleftharpoons \text{no appreciable reaction}$$

The sodium acetate–acetic acid buffer system is used to control the acidity of the solution. This buffer is formed by adding acetic acid to the solution of the cinnamate salts. The first part of the acetic acid that is added reacts with NaOH to yield sodium acetate. As more acetic acid is added, the mixture becomes buffered. Even with this buffer system, the pH must be checked with pH paper, not litmus paper. Litmus paper, which turns red at pH 4.5 and blue at pH 8.3, is not sensitive at the required near-neutral pH.

At pH 6, the (*E*) acid precipitates and is removed by filtration. When the solution is acidified further with HCl (overwhelming the buffer system), the (*Z*) acid is protonated and precipitates. Only a small amount of (*Z*) acid will be obtained, and its precipitation is slow.

Experimental

EQUIPMENT:

condenser
dropper
125-mL and two 50-mL Erlenmeyer flasks
10-mL, 50-mL, and 100-mL graduated cylinders
heating mantle and rheostat
ice bath
pH paper
50-mL round-bottom flask
125-mL separatory funnel
stirring rod
vacuum filtration assembly

CHEMICALS:

acetic anhydride, 2.2 g
10% aqueous sodium hydroxide, 15 mL
benzaldehyde, 3.2 g
diethyl ether, 30 mL
glacial acetic acid, 5–6 mL
conc. hydrochloric acid, 9 mL
petroleum ether (bp 30–60°), 5–20 mL
phenylacetic acid, 2.7 g
triethylamine, 2.1 g

TIME REQUIRED: 3 hours plus overnight drying of the precipitates, crystallization, and melting-point determinations

STOPPING POINTS: after the reflux period; after the extraction with NaOH; while the (*Z*) acid is precipitating; while the (*E*) and (*Z*) acids are crystallizing or drying

>>>> ***SAFETY NOTE*** Phenylacetic acid, acetic anhydride, glacial acetic acid, and triethylamine are all strong irritants. Weigh these compounds in the fume hood and use appropriate precautions.

PROCEDURE

Place 2.7 g of phenylacetic acid, 3.2 g of benzaldehyde, 2.2 g of acetic anhydride, and 2.1 g of triethylamine in a round-bottom flask (see Safety Note). Heat the mixture under reflux in a fume hood for 30–40 minutes.

Chill the reaction flask in an ice bath, then add 4 mL of conc. hydrochloric acid followed by 25–30 mL of diethyl ether. Swirl the flask to dissolve the precipitate. It may be necessary to warm the flask in a water bath to dissolve all of the solid. (If it is necessary to warm the flask, then chill it before continuing.)

Transfer the mixture to a separatory funnel, and extract the mixture with two 20-mL portions of water. Dilute 15 mL of 10% NaOH solution with 60 mL of water. Extract the ether solution with three 25-mL portions of this diluted sodium hydroxide solution. Combine the aqueous NaOH extracts, then acidify with glacial acetic acid to pH 6 (pH paper). About 5 mL will be required. Filter the precipitate of the (*E*) acid by vacuum. (Save the filtrate.) Air-dry the solid overnight, then determine its weight and melting point.

Add 5 mL of conc. hydrochloric acid to the filtrate from the filtration of the crude (*E*) acid, then allow the mixture to stand for about 30 minutes while the (*Z*) acid precipitates. Collect this product by vacuum filtration, and allow it to air-dry overnight.

Purify both carboxylic acids by crystallization from petroleum ether (bp 30–60°). Determine the yields and melting points.

Problems

18.11 Draw a flow diagram for the separation of the (*E*)- and (*Z*)-α-phenylcinnamic acids in this experiment.

18.12 Describe what problems would arise: (a) if a student acidified the aqueous α-phenylcinnamate solution using blue litmus paper instead of pH paper as the indicator; (b) if a student acidified the same solution with conc. HCl instead of acetic acid.

18.13 When the aqueous α-phenylcinnamate is acidified with acetic acid, what species are in solution when

(a) at pH 6, the (*E*) acid precipitates?

(b) at a lower pH, the (*Z*) acid precipitates?

18.14 (a) Although you would expect to use a relatively polar solvent to crystallize a carboxylic acid, the very nonpolar petroleum ether is used in Experiment 18.2. Explain.

(b) Aside from its toxicity, would you expect benzene to be a good crystallization solvent for α-phenylcinnamic acids? Explain.

»»» CHAPTER 19 «««

Plant Products

19.1 Some Compounds Found in Plants

The flowers and seeds of plants are rich in a wide variety of organic compounds. These include triglycerides (vegetable oils), proteins, carbohydrates, alkaloids (such as caffeine, page 76), vitamins, terpenes, and other compounds. Some of the relatively low-molecular-weight compounds in flowers and seeds are highly odorous. Isolation of these compounds by steam distillation, extraction, pressing, or related techniques yields **essential oils**: concentrated mixtures of the odor-giving compounds. Essential oils are used in flavorings and perfumes. Some well-known essential oils are those from cloves, roses, mint, garlic, and vanilla beans.

The odorous compounds in essential oils are generally of two types: *terpenes* (or *terpenoids* if they contain elements other than C and H alone) and *substituted benzenes.*

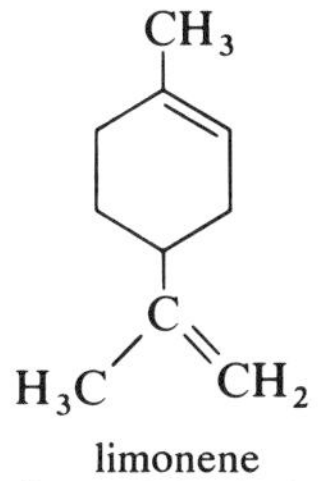

limonene
(from citrus fruits)
a terpene

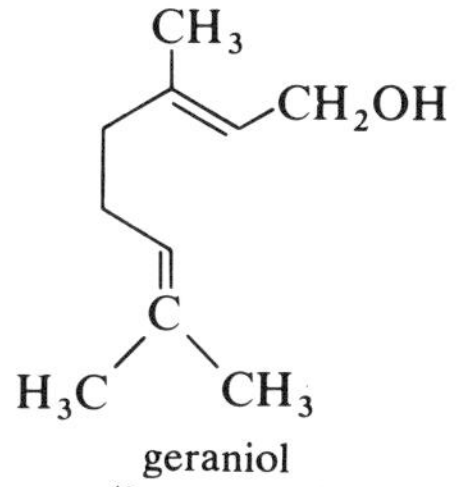

geraniol
(from roses)
a terpenoid

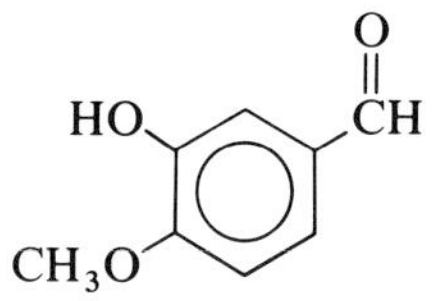

vanillin (from vanilla beans)

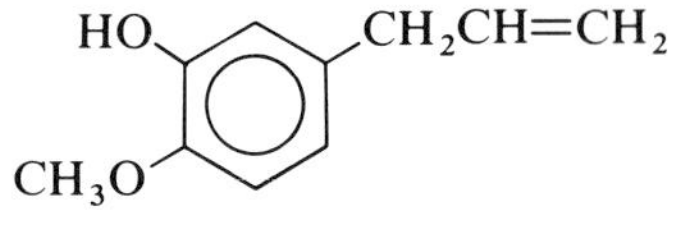

eugenol (from cloves)

two substituted benzenes

Experiment 19.1A is the isolation of *eugenol* from *cloves*, the dried flower buds of the clove tree (*Eugenia caryophyllata*). This tree grows in the West Indies, Sri Lanka, India, and Mauritius (an island in the Indian Ocean). Experiment 19.1B is the esterification of eugenol with benzoyl chloride. Preparing a solid derivative of a liquid is a common technique for purification or characterization when the quantity of the liquid is too small to be distilled.

Experiment 19.2 is the extraction of ground *nutmeg* to yield the vegetable oil *trimyristin*. Nutmeg is the powdered, dried seed kernels of the trees *Myristica moschata* and *Myristica fragrans*. These trees are native to the East Indies, but have now been introduced to Sumatra, India, Brazil, and the West Indies.

All vegetable oils and animal fats are **triglycerides**, or triesters of glycerol. The long, continuous-chain carboxylic acids that form the ester groups are called **fatty acids**. Fatty acids vary in the lengths and degree of unsaturation of their carbon chains. Generally, triglycerides found in nature are mixtures containing a variety of fatty acid units. Nutmeg is unusual in that it contains trimyristin virtually uncontaminated with other triglycerides.

$CH_2OH-CHOH-CH_2OH$
glycerol

$CH_3(CH_2)_{12}COH$ (C=O)
myristic acid
a fatty acid

$CH_3(CH_2)_{12}C(=O)-OCH(CH_2O-C(=O)(CH_2)_{12}CH_3)_2$
trimyristin
a triglyceride

19.2 Esterification with Acid Chlorides

An ester can, in some cases, be prepared by heating an alcohol with a carboxylic acid and a trace of strong acid as the catalyst.

$$\underset{\text{a carboxylic acid}}{RC(=O)OH} + \underset{\text{an alcohol}}{R'OH} \overset{H^+}{\rightleftharpoons} \underset{\text{an ester}}{RC(=O)OR'} + H_2O$$

As we discussed in Section 16.2, the direct synthesis of a phenyl ester from a phenol and a carboxylic acid is not practical because of an unfavorable equilibrium constant. Phenyl esters are prepared using *acid chlorides* (also called *acyl chlorides*) or *acid anhydrides* as acylating agents. In Experiment 16.3 (page 303), aspirin was prepared using acetic anhydride as the acylating agent; in

Experiment 19.1B, an acid chloride is used. With an acid chloride, a phenoxide ion rather than the free phenol is used as the other reactant.

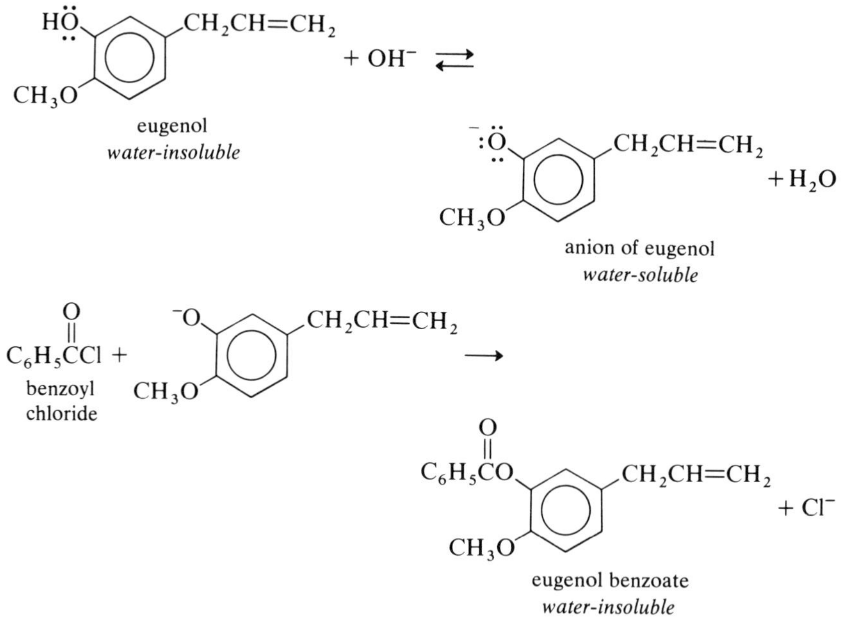

$$\text{HO-}C_6H_3(\text{OCH}_3)\text{-CH}_2\text{CH=CH}_2 + OH^- \rightleftarrows {}^-\text{O-}C_6H_3(\text{OCH}_3)\text{-CH}_2\text{CH=CH}_2 + H_2O$$

eugenol *water-insoluble* → anion of eugenol *water-soluble*

$$C_6H_5\overset{O}{\overset{\|}{C}}Cl + {}^-\text{O-}C_6H_3(\text{OCH}_3)\text{-CH}_2\text{CH=CH}_2 \longrightarrow C_6H_5\overset{O}{\overset{\|}{C}}O\text{-}C_6H_3(\text{OCH}_3)\text{-CH}_2\text{CH=CH}_2 + Cl^-$$

benzoyl chloride; eugenol benzoate *water-insoluble*

The reaction of a phenoxide ion, such as the anion of eugenol, with an acid chloride is a typical nucleophilic acyl substitution reaction: addition followed by elimination. Step 1 is the nucleophilic addition of the phenoxide ion to the carbonyl group of the acid chloride. Step 2 is the elimination of Cl^-.

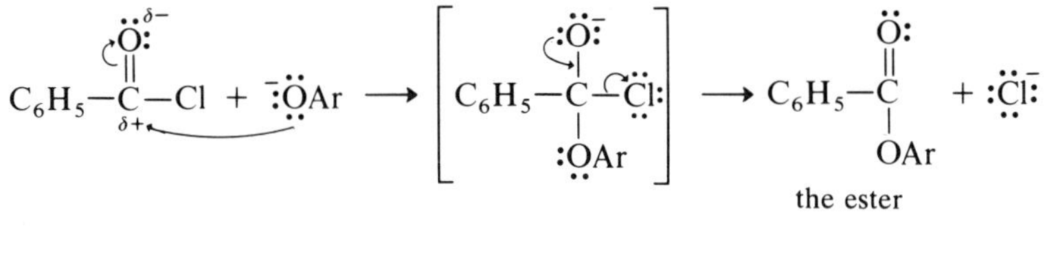

$$C_6H_5\text{—}\overset{\ddot{O}:^{\delta-}}{\overset{\|}{C}}_{\delta+}\text{—Cl} + {}^-\!:\ddot{O}Ar \longrightarrow \left[C_6H_5\text{—}\overset{:\ddot{O}:^-}{\overset{|}{C}}\text{—}\ddot{Cl}: \atop \quad\ \ |\ :\ddot{O}Ar\right] \longrightarrow C_6H_5\text{—}\overset{\ddot{O}:}{\overset{\|}{C}}\text{—OAr} + :\ddot{Cl}:^-$$

the ester

Problems

19.1 Suggest both physical and chemical techniques for showing that the positions of the carbon–carbon double bonds in limonene and geraniol are correct as they are shown in the formulas on page 341.

19.2 Suggest at least two techniques for preparing a solid derivative of vanillin (page 342.

19.3 Explain why the eugenol anion gives superior results (compared to the free eugenol) in the reaction with benzoyl chloride.

19.4 Write equations to exemplify the following reactions:

(a) Trimyristin (page 342) is refluxed in an excess of methanol to which a trace of H_2SO_4 has been added.

(b) Eugenol (page 342) is heated in a dilute solution of HCl in aqueous ethanol.

»»» EXPERIMENT 19.1 «««

Isolation of Eugenol from Cloves; Synthesis of Eugenol Benzoate

eugenol (HO, CH_3O, $CH_2CH{=}CH_2$ substituted benzene) + benzoyl chloride (C_6H_5COCl) + NaOH $\xrightarrow{H_2O}$

	eugenol	benzoyl chloride
MW:	164.20	140.57
weight:	1.2 g	1.0 mL (1.4 g)
moles:	0.007	0.010

eugenol benzoate (C_6H_5COO, CH_3O, $CH_2CH{=}CH_2$ substituted benzene) + NaCl + H_2O

	eugenol benzoate
MW:	268.30
weight:	2.0 (theory)
moles:	0.007 (theory)

Table 19.1 *Physical properties of reactants and products*

Name	***Bp (°C)***	***Density (g/mL)***	***Solubility***
eugenol	255	1.07	insol. H_2O; sol. dil. base
benzoyl chloride	197.2	1.21	decomposes in H_2O or dil. base
eugenol benzoate[a]	—	—	insol. H_2O and dil. base; sl. sol. CH_3OH

[a] Mp 69–70°

Discussion

Oil of cloves is obtained from cloves by steam distillation, which is the most common method for extracting essential oils from their sources. The clove oil is then extracted from the water layer of the distillate with dichloromethane.

Eugenol is extracted from the other components in the CH_2Cl_2 solution with 5% aqueous KOH. In this extraction, the potassium salt of eugenol is found in the upper, aqueous layer. Neutralization of the KOH extract with dilute aqueous HCl converts the eugenol anion back to the water-insoluble eugenol. The eugenol is then re-extracted from the water with CH_2Cl_2.

The eugenol–CH_2Cl_2 solution is dried with anhydrous $MgSO_4$ or Na_2SO_4 (but not with $CaCl_2$, which forms complexes with phenols as well as with alcohols and water). The crude eugenol is isolated by boiling away the CH_2Cl_2. Because of the small amount of product obtained, further purification is not practical in a student laboratory. Instead, the product is characterized by its infrared spectrum, then is carried on to part B. In part B, the liquid eugenol is converted to its solid benzoate for additional purification and characterization.

Preparation of Eugenol Benzoate

The first step in the conversion of eugenol to eugenol benzoate is dissolving the eugenol in a minimum amount of aqueous base. The esterification reaction is accomplished by adding benzoyl chloride and allowing the mixture to stand overnight. If excessive base is present in the eugenol salt solution, some of the benzoyl chloride will be converted to the benzoate ion instead of to the eugenol ester. (Benzoyl chloride also reacts with water, but this reaction is slower than the reaction with the more nucleophilic hydroxide or phenoxide ions.)

$$C_6H_5\overset{\overset{\large O}{\|}}{C}Cl + 2\,OH^- \rightarrow \underset{\text{benzoate ion}}{C_6H_5CO_2^-} + H_2O + Cl^-$$

A large excess of benzoyl chloride should also be avoided because it is difficult to remove from the product. If the weight of starting eugenol is substantially different from 1.2 g, the quantity of benzoyl chloride used should be adjusted accordingly.

Unlike eugenol, eugenol benzoate is insoluble in aqueous base. Therefore, it precipitates from the solution as it is formed. Filtration and water-washing (to remove NaOH and the sodium salts of eugenol and benzoic acid) provide the crude ester, which can be purified by crystallization.

Experimental

EQUIPMENT:

Part A
50-mL, 125-mL, and 250-mL Erlenmeyer flasks
hot plate or steam bath
100-mL round-bottom flask with glass stopper
250-mL separatory funnel
steam distillation apparatus
Part B
dropper or disposable pipet
two 50-mL Erlenmeyer flasks
funnel
10-mL and 100-mL graduated cylinders
ice bath

CHEMICALS:

Part A
ground cloves or clove buds, 10 g
dichloromethane (methylene chloride), 90 mL
5% aqueous HCl, 75 mL (or 100 mL of 1*M* HCl)
anhydrous $MgSO_4$, about 0.5 g
5% aqueous potassium hydroxide, 100 mL
Part B
benzoyl chloride, 1 mL (1.4 g)
eugenol, 1–1.5 g
methanol, 2–5 mL (95% ethanol may be substituted)
10% sodium hydroxide, 10 mL

TIME REQUIRED: about 3 hours for part A; about 1 hour (plus overnight reaction) for part B

STOPPING POINTS: after the steam distillation; while the extract is being dried; while the benzoate reaction mixture is standing overnight

PROCEDURE

A. Isolation of Eugenol from Cloves

Place 10.0 g of ground cloves or clove buds in a round-bottom flask, add 300 mL of water, and steam distil (see Experimental Note 1), collecting 100 mL of distillate. Extract the distillate with two 20-mL portions of dichloromethane (see Experimental Note 2). Extract the dichloromethane solution with two 50-mL portions of 5% KOH solution. Acidify the combined KOH extracts with 5% HCl, then extract the acidic mixture with two 25-mL portions of dichloromethane. Dry the combined dichloromethane extracts overnight in a stoppered flask with magnesium sulfate.

Decant or filter the dried dichloromethane solution into a tared Erlenmeyer flask and boil the solution almost to dryness in the fume hood (see Experimental Note 3). *Do not overheat the flask.* Allow the residual dichloromethane to evaporate before weighing the product (see Experimental Note 4). Calculate the per cent recovery of eugenol based upon the starting weight of cloves. You should obtain about 1.2 g of crude eugenol. Run an infrared spectrum and determine whether the spectrum is consistent with the structure shown on page 342.

B. Preparation of Eugenol Benzoate

To the flask containing the eugenol from part A, add 5 mL of water, then 10% aqueous sodium hydroxide dropwise with swirling until the eugenol just dissolves (see Experimental Note 5). *In a fume hood,* add 1.0 mL (1.4 g) of benzoyl chloride (CAUTION: *lachrymator and strong irritant!*) to the alkaline eugenol solution. Stir the cloudy solution vigorously, then stopper the flask and allow it to stand overnight in the fume hood.

Cool the mixture in an ice bath, separate the solid by gravity filtration, and wash it with 10–20 mL of water (see Experimental Note 6). Discard the filtrate in the *hood sink* or (better) a labeled waste jug. Air-dry the crude product overnight. Several recrystallizations from 95% ethanol or methanol are required to yield a pure product.

EXPERIMENTAL NOTES

1) The apparatus for steam distillation is described on page 193.

2) Dichloromethane has a tendency to cling to the sides of a separatory funnel and, when small quantities are used in an extraction, a large percentage of the product can be lost in the droplets. In this experiment, the dichloromethane solution should be returned to the same unrinsed separatory funnel to minimize the loss.

3) Dichloromethane is relatively expensive. Your instructor may request that you distil and collect this solvent instead of boiling it away in the fume hood.

4) One technique for determining if all the dichloromethane has evaporated is to weigh the flask periodically and plot the weight loss as a function of time.

5) The solution should be clear before you proceed and should not contain excess base. If the solution still appears cloudy after most of the material has dissolved, warm the flask to near-boiling to drive off any residual CH_2Cl_2, then cool the solution to room temperature. If excess base was used, add dilute HCl dropwise until the solution becomes cloudy. Then add 10% aqueous NaOH dropwise until the cloudiness just disappears. Cool the solution to room temperature before continuing.

6) If an oil or a semisolid separates, transfer the material to a small Erlenmeyer flask and add 10 mL of water. Then add 10% NaOH solution dropwise until the aqueous layer is strongly alkaline. Stir the mixture while cooling it in an ice bath. Eugenol and benzoic acid are soluble in aqueous base and thus can be separated from the crude eugenol benzoate (insoluble) by this procedure.

Problems

19.5 The extraction and isolation of eugenol from cloves involves several separations in which the eugenol is extracted by either CH_2Cl_2 or aqueous base. Construct a flow diagram describing the isolation. Include in your diagram: (a) which solution contains the eugenol, and (b) whether that solution is the upper or lower layer in the separatory funnel.

19.6 Why is the CH_2Cl_2 removed from the eugenol before reaction of the eugenol anion with benzoyl chloride?

19.7 Write resonance structures that show how the eugenol anion is stabilized.

19.8 Besides eugenol, oil of cloves contains *caryophyllene* and other terpenes. Explain what happens to these compounds when eugenol is isolated from cloves by the procedure in Experiment 19.1A.

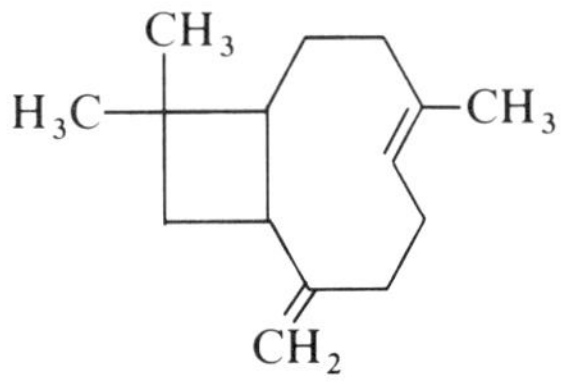

caryophyllene

19.9 List the types of organic compounds that could be extracted from their natural sources by dilute aqueous NaOH.

19.10 If an excessive amount of benzoyl chloride were used in Experiment 19.1B, what contaminants would be found in the eugenol benzoate?

19.11 Write a mechanism for the decomposition of benzoyl chloride in water.

EXPERIMENT 19.2

Isolation of Trimyristin from Nutmeg

Discussion

Trimyristin (page 342) is extracted from nutmeg by simply boiling nutmeg in diethyl ether and filtering the resultant solution. The ether is removed and the product is crystallized from acetone. The only problem usually encountered in this isolation procedure is that the trimyristin may contain some highly colored impurities. These impurities can be removed from the ether solution with decolorizing charcoal (Experiment 3.2, page 50). Trimyristin crystallizes quite slowly; therefore, rather than using an ice bath to force the crystallization, cork the crystallization flask and allow it to sit overnight.

Experimental

EQUIPMENT:

condenser for reflux
distillation assembly
dropper or disposable pipet
50-mL and 125-mL Erlenmeyer flasks
funnel
heating mantle and rheostat (or steam bath)
ice bath
two 100-mL round-bottom flasks

CHEMICALS:

acetone, 10 mL
diethyl ether, 50–100 mL
ground nutmeg, 5 g

decolorizing charcoal, 0.5 g (optional)

TIME REQUIRED: 3 hours plus overnight crystallization; time for drying the crystals and melting-point determination

STOPPING POINTS: while the trimyristin is crystallizing; while the crystals are drying

PROCEDURE

Weigh 5.0 g of ground nutmeg into a round-bottom flask and add 50 mL of diethyl ether (*flammable*!). Heat the mixture to a *gentle* reflux using an efficient condenser for one hour. The solid nutmeg in the flask may cause the mixture to bump; therefore, use the minimum heat necessary.

Cool the flask to room temperature, then filter the mixture by gravity. Remove the ether from the crude product by distillation. If the residue is dark yellow to orange, add 50 mL of ether and decolorize the solution with decolorizing charcoal. Remove the ether by distillation before proceeding. If the residue is a very pale yellow, proceed to the crystallization without the decolorizing step.

Transfer the residue from the ether distillation to a small Erlenmeyer flask with a dropper. Add 10 mL of acetone, swirl, cork the flask, then allow it to sit overnight. Separate the crystals by gravity filtration and allow them to air-dry. A typical yield is 0.7 g. Calculate the per cent recovery and determine the melting point.

Problems

19.12 Write equations that illustrate (a) the saponification, and (b) the acid hydrolysis of trimyristin.

19.13 Write the formula for a chiral triglyceride formed from two molecules of myristic acid and one molecule of stearic acid ($C_{17}H_{35}CO_2H$).

19.14 Write equations for the reaction of the following compounds with acetyl chloride (the chloride of acetic acid):

(a) eugenol (in aqueous NaOH)
(b) geraniol
(c) glycerol

19.15 *Waxes* are esters of fatty acids with long-chain alcohols. Suggest a synthesis of a wax beginning with tristearin (the tryglyceride of stearic acid) and any other reagents.

19.16 The *saponification equivalent* of an ester is the weight in grams that undergoes reaction with 1.00 mol of OH^-, or

$$\text{saponification equivalent} = \frac{\text{wt. of ester in g}}{\text{liters of } 1.00N \text{ OH}^-}$$

The saponification equivalent is determined by heating a weighed sample of ester with an excess of standardized KOH or NaOH, then titrating the excess base with standardized hydrochloric acid.

(a) What is the theoretical saponification equivalent of trimyristin?

(b) From the following data, calculate the saponification equivalent: 1.021 g of an ester is heated with 50.00 mL of 0.256*N* NaOH (aqueous ethanol solution). When reaction is complete, the mixture is titrated to the phenolphthalein endpoint with 11.21 mL of 0.100*N* HCl.

Introduction to the Chemical Literature

20.1 The Chemical Literature

The complete, written, published record of chemical knowledge is referred to as the **chemical literature**. The **primary literature**, or **original literature**, comprises the original reports of compound preparation, compound characterization, mechanistic studies, and so forth. These reports usually appear in research journals (as articles, notes, and communications) and in patent disclosures. Table 20.1 lists just a few of the journals available.

From the primary literature, information flows into the **secondary literature**. The secondary literature consists of compilations of data; articles reviewing entire areas of research; textbooks; abstracts, or summaries, of individual current research articles; and other such publications.

While the primary literature concerns itself with the reports of new findings, the goals of most secondary literature publications are to *summarize and correlate chemical knowledge*. This summarization and correlation are necessary because of the sheer quantity of chemical knowledge. There are approximately 4 million known compounds, and the number of new compounds recorded yearly by the Chemical Abstracts Service of the American Chemical Society is currently about $\frac{1}{3}$ million! Also, as new facts are discovered, chemical theories undergo modification, and previous errors in the literature are corrected. No person could possibly keep up with current chemical news and views by reading the primary literature alone. The secondary literature allows us to keep abreast of an area of study; to find the physical constants of a compound (which might have been

Table 20.1 A few chemical research periodicals

Name	Abbreviation[a]	Language
Journal of the American Chemical Society	*J. Am. Chem. Soc.*	English
Journal of Organic Chemistry	*J. Org. Chem.*	English
Canadian Journal of Chemistry	*Can. J. Chem.*	English
Journal of the Chemical Society	*J. Chem. Soc.*[b]	English
Bulletin of the Chemical Society of Japan	*Bull. Chem. Soc. Jpn.*	English, French, German

[a] These are the abbreviations used in the *Journal of the American Chemical Society*; other publications may use different abbreviations.

[b] In 1965, the *Journal of the Chemical Society* was becoming sufficiently lengthy that it was divided into three parts: *A*, *B*, and *C*. In 1972, *C* (physical organic chemistry) was renamed *Perkin Transactions I*, and *B* (organic and bio-organic) was renamed *Perkin Transactions II*. (Other section names for other areas of chemistry are also encountered.) In references to the post-1972 journals, these names are abbreviated and arabic numerals are used: for example, *J. Chem. Soc., Perkin Trans. 1.*

reported erroneously in one journal in 1947 and reported correctly in another journal in 1952); or to find a pertinent original journal article. Figure 20.1 depicts the flow of information from the research laboratory through the literature.

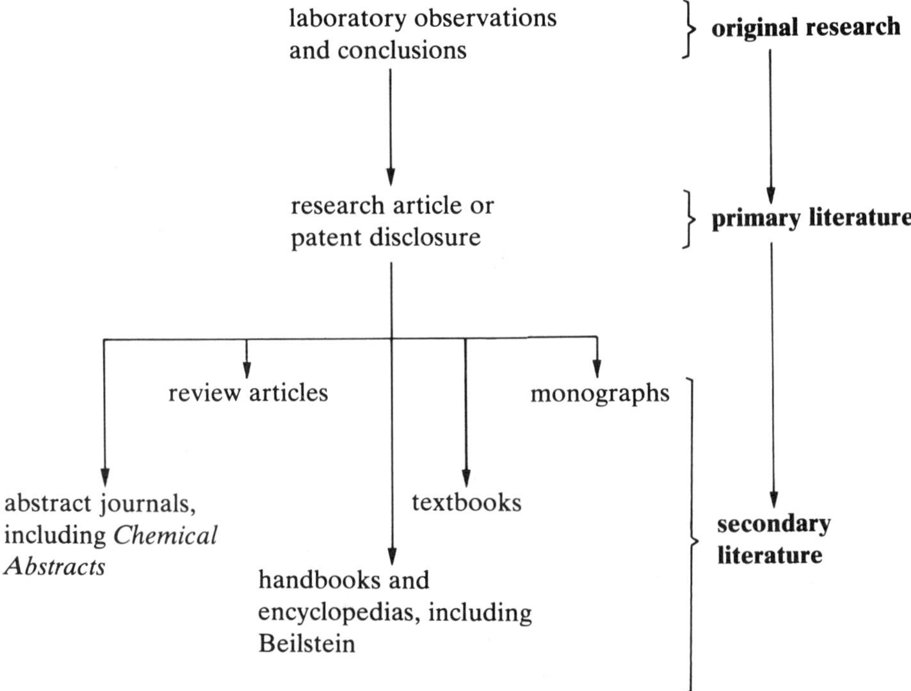

Figure 20.1 Flow of information from the laboratory into the chemical literature.

20.2 Retrieving Information from the Literature

How we retrieve information from the chemical literature depends on the type of information desired. The ultimate goal of information-retrieval would be to find the original research reports that contain the raw data from which all other reports are derived. While this is a worthy goal, it is often neither necessary nor practical. If, for example, you need only the melting point of an organic compound, you can stop your literature search once you have found the value in a reference work. Or, if you want to survey a general topic of organic chemistry (such as the Grignard reaction), you would probably look for review articles or book chapters. (It might take several years to read every original article in which Grignard reactions are mentioned!) On the other hand, if you wish to repeat a particular synthetic scheme already reported in the literature, you would want to study the original journal article to find out the exact reaction conditions used and what problems were encountered.

Another aspect of information-retrieval is that of when the information was published. Even with on-line computer retrieval systems and special publications, recent information is the most difficult to find because it has not yet been included in the secondary literature.

In the following sections, we will outline the various techniques of searching the secondary literature and then briefly introduce the use of Beilstein's *Handbuch der Organischen Chemie* and *Chemical Abstracts*. More comprehensive instructions for searching the literature are found in the following references:

Searching the Chemical Literature, R. F. Gould, Ed., Advances in Chemistry Series No. 30. Washington, D.C.: American Chemical Society, 1961.

H. M. Woodburn, *Using the Chemical Literature. A Practical Guide*. New York: Dekker, 1974.

R. E. Maizell, *How to Find Chemical Information*. New York: John Wiley and Sons, 1979.

Use of Chemical Literature, 3rd ed., R. T. Bottle, Ed. London: Butterworths, 1979.

20.3 How to Find General Information

Finding general information in the literature, such as information concerning the reduction of aldehydes, is often more challenging than finding a specific fact. The reason is that general terms such as "reduction" and "aldehydes" may take up a considerable amount of space in a comprehensive index (or they might not appear at all).

The best approach to finding information about a general topic is to start with a fairly general reference book. For example, a textbook may include a discussion with references to review articles. Some other publications that are useful for finding general information follow:

Rodd's Chemistry of Carbon Compounds, 2nd ed., S. Coffey, Ed. New York: Elsevier, 1964.

Organic Syntheses. New York: John Wiley and Sons, 1920–present: tested synthetic procedures for specific compounds.

Organic Reactions. New York: John Wiley and Sons, 1942–present: detailed discussions of important reaction types (such as Friedel–Crafts alkylations) from the laboratory point of view.

Technique of Organic Chemistry, A. Weissberger, Ed. New York: Interscience Publishers, 1949–present.

L. F. Fieser and M. Fieser, *Reagents For Organic Synthesis*, Vols. 1–9. New York: John Wiley and Sons, 1968–present.

There are many other books; monographs on specific topics; and review journals and books (such as *Chemical Reviews*, *Accounts of Chemical Research*, and *Advances in Drug Research*) that review and summarize various fields. (Check the library card catalog.)

The next step in searching for general information is to consult the subject indexes of *Chemical Abstracts*. (See the discussion on using *Chemical Abstracts* in Section 20.6.)

20.4 How to Find Information about a Specific Organic Compound

There are numerous ways to find information about a particular compound. If you are simply looking for physical constants of a compound, first check the extensive tables in the handbooks. If you cannot find the name of the compound in the tables, use the molecular formula index—the compound may be listed under a different name.

Handbooks:

CRC Handbook of Chemistry and Physics. Boca Raton, Florida: CRC Press, Inc., published annually or biennially.

Lange's Handbook of Chemistry. New York: McGraw-Hill, many editions.

Handbook of Tables for Organic Compound Identification, 3rd ed. Z. Rappoport, Ed. Boca Raton, Florida: CRC Press, Inc., 1967.

Merck Index, 9th ed. Rahway, N.J.: Merck and Co., 1976.

Heilbrons's Dictionary of Organic Compounds, 5th ed., Vols. 1–5. New York: Oxford University Press, 1982.

Spectral Compilations:

Proton Nuclear Magnetic Resonance Spectra. Philadelphia: Sadtler Research Laboratories, continuously updated.

Sadtler Standard Spectra. Philadelphia: Sadtler Research Laboratories, continuously updated: infrared spectra.

The Aldrich Library of Infrared Spectra, 2nd ed., C. J. Pouchert, Ed. Milwaukee: Aldrich Chemical Co., Inc., 1975.

If the desired information cannot be gleaned from a handbook, then you would search Beilstein, followed by *Chemical Abstracts*.

20.5 Beilstein's *Handbuch der Organischen Chemie*

A. Background

Beilstein's *Handbuch der Organischen Chemie* (Berlin: Springer-Verlag, 1918–present), commonly referred to as simply "Beilstein," is a monumental compilation of organic data. Beilstein was first published by Friedrich Karl Beilstein in 1881–1882. After the third edition, the German Chemical Society acquired the rights to this work and published the fourth edition, which is the edition used today. Although it is written in German, the organic formulas, numbers, etc. are the same as those in books written in English. For this reason, a minimum of German vocabulary enables the beginning student to find useful data. (Table 20.2 lists some abbreviations used in Beilstein.)

Table 20.2 Some abbreviations used in Beilstein[a]

Abbreviation	*German term*	*English equivalent*
Kp	Siedepunkt	boiling point
D	Dichte	density
E	Erstarrungspunkt	freezing point
F	Schmelzpunkt	melting point
n	Brechungsindex	refractive index
Zer	Zersetzung	decomposition

[a] Examples:
F: 20° (reference): a melting point of 20° followed by the original reference where the value was reported.
Kp_{760} 204° (korr): a corrected boiling point of 204° at a pressure of 760 mm Hg.

The fourth edition of Beilstein consists of a main series, called **das Hauptwerk,** of 27 volumes, plus supplemental series, or **Erganzungswerk**, each of which also has 27 volumes. The main series contains data on *all known organic compounds through 1909*, including all known syntheses, reactions, physical constants, and references to the original literature. Supplement I extends the coverage through 1919 and includes new data for compounds appearing in the main series as well as data for compounds discovered between 1910 and 1919. Later supplements have been published in whole or in part, as the volumes are completed. Table 20.3 lists the main series and first four supplements, their German names, and years of coverage. Because of the time lag between an original publication and its inclusion in Beilstein, this reference work is most useful for searching the older literature, up to about 1940.

Table 20.3 Series in Beilstein's *Handbuch der Organischen Chemie*

Series and supplements	*German name*	*Coverage*
Main Series	Hauptwerk	up to 1910
Supplement I	Erstes Erganzungswerk (E I)	1910–1919
Supplement II	Zweites Erganzungswerk (E II)	1920–1929
Supplement III	Drittes Erganzungswerk (E III)	1930–1949
Supplement IV	Viertes Erganzungswerk (E IV)	1950–1959

B. General Organization

The entries in Beilstein are arranged according to structure rather than by name. While a complete understanding of Beilstein's organizational system is not necessary for the successful use of this reference work, a general idea of the overall organization is helpful. The main series and each supplemental series have the *same organization*. Methane, for example, is the first compound listed in Volume I of the main series and is also the first compound listed in Volume I of each of the supplements. Therefore, once an entry has been located in the main series, it is an easy task to locate the same entry in the supplements.

The first four volumes in the main series (and also in the supplements) cover open-chain (acyclic) compounds; Volumes 5–16 cover compounds containing carbon rings (carbocyclic compounds); and Volumes 17–27 cover heterocyclic compounds. Within each of these three major classifications, compounds are entered in the following order: *hydrocarbons, hydroxy compounds, carbonyl compounds*, and *carboxylic acids*, followed by other compound classes that we will not discuss.

open-chain compounds (Volumes 1–4):

RH → ROH → RCHO and $R_2C{=}O$ → RCO_2H, etc.

carbocyclic compounds (Volumes 5–16):

RH → ROH and ArOH → RCHO and $R_2C{=}O$ → RCO_2H, etc.

heterocyclic compounds (Volumes 17–27):

parent heterocycle → ROH and ArOH

→ RCHO and $R_2C{=}O$ → RCO_2H, etc.

Within each of the functional-group classes, the simplest structure (lowest molecular weight and highest degree of saturation, such as methane for the open-chain hydrocarbons or methanol for the hydroxy compounds) is listed first. The "derivatives" of each main entry follow that entry; for example, cyclohexyl methyl ether is found following cyclohexanol.

While it is quite possible to locate an entry in Beilstein using only the Beilstein system, a considerable amount of time must be spent learning the system. For those who wish more information, the following references should be consulted.

How to Use Beilstein. Frankfurt/Main: The Beilstein Institute; distributed by Springer-Verlag, New York, Inc., 175 Fifth Ave., New York, N.Y. 10010.

O. Weissback, *The Beilstein Guide: A Manual for the Use of Beilstein's Handbuch der Organischen Chemie.* New York: Springer-Verlag, 1976.

J. Sunkel, E. Hoffmann, and R. Luckenbach, *J. Chem. Ed.* **1981**, *58*, 982.

Find the molecular formula in Vol. 29 (*General Formelregister*) of Supplement II or other series.

↓

Find the correct entry from (a) the German name, or (b) checking the structural formulas in the various entries.

↓

Use the page references listed in the index for the main series and each supplement and/or use the cross-references at the top of the pages to find a compound.

Figure 20.2. A summary of how to find information about a compound in Beilstein's Handbuch der Organischen Chemie.

C. How to Locate a Compound in Beilstein

The easiest way to find a compound in Beilstein is to use one of the indexes. Each volume of Beilstein contains indexes of compound names and molecular formulas. In addition, Volumes 28 and 29 of the main series and each supplemental series contain *cumulative indexes.*

Because a knowledge of German organic nomenclature is needed to locate a compound using the cumulative name index (Volume 28, *General Sachregister*), we will describe the use of the molecular formula index (Volume 29, *General Formelregister*). Volume 29 of Supplement II is contained in three books and lists all the compounds found in the main series as well as those in the first two supplements. Compounds are listed by increasing carbon content (all the C_1 compounds, followed by the C_2 compounds, etc.). The use of the molecular formula index and the key notations therein are best explained with the aid of a specific example. The general procedure used to locate a compound in Beilstein is also summarized in Figure 20.2.

Example. Locate cyclohexanol in the main series and in the first *three* supplements of Beilstein.

1) Write the structural formula for cyclohexanol and determine the molecular formula.

(cyclohexane ring)–OH, $C_6H_{12}O$

2) Find the $C_6H_{12}O$ entries in the formula index (Vol. 29) of Supplement II or other series.

These entries appear on page 221 in the first bound book of Vol. 29 of Supplement II.

3) Inspect the German listing to find the name that corresponds most closely to the English name.

Our example is simple—the German and the English names are identical: cyclohexanol. In more difficult cases, (a) use a German–English chemical dictionary, or (b) look up each entry and check the structural formulas.

4) Write down the Beilstein page references immediately following cyclohexanol.

Cyclohexanol *6*, 5, I4, II5

These references mean:

6, 5: Cyclohexanol is found in Vol. 6 (*Band* 6), page 5, of the main series.

I4: This compound is found in Vol. 6, page 4, of Supplement I (E I).

II5: The compound is found in Vol. 6, page 5, of Supplement II (E II).

5) Find cyclohexanol in the main series.

In Vol. 6, page 5, you will find:

3. *Oxy-Verbindungen* $C_6H_{12}O$.
1. Cyclohexanol, Hexahydrophenol $C_6H_{12}O$ = ...

Note that the structural formula is given immediately following the molecular formula. The information concerning cyclohexanol follows the heading. Each informational entry is referenced to the original literature. (The journal abbreviations are listed in the front of each book.)

6) Find cyclohexanol in Volume 6 of Supplements I and II.

Cyclohexanol is found on page 4 in Volume 6 of Supplement I. Note that the format is identical to that in the main series.

Cyclohexanol is found on page 5 in Volume 6 of Supplement II. Again, note that the format is identical.

7) Find cyclohexanol in the supplements published after Supplement II.

All supplements of Beilstein are cross-referenced to the main series; therefore, you can find a compound in the newer supplements without having to use another index.

At the top of page 4 in Volume 6 of Supplement I and page 5 in Volume 6 of Supplement II, you will find the following notations:

VI, 4–5 (in Supplement I)
H 6, 5–6 (in Supplement II)

These notations refer to the pages in Volume 6 of the main series (H, or *Hauptwerk*) on which the same compounds (including cyclohexanol) are found.

To find cyclohexanol in Volume 6 of Supplement III, you need only turn to the page that has "H 6, 5" centred at the top (page 10 of Volume 6 of Supplement III).

20.6 *Chemical Abstracts*

A. Background

The most comprehensive collection of modern chemical information is *Chemical Abstracts* (*Chem. Abstr.*, *CA*). *Chemical Abstracts* has been published continuously since 1907. Today, the Chemical Abstracts Service abstracts hundreds of thousands of documents per year from about 14,000 periodicals. The collective index (not the abstracts themselves) just for the five-year period 1972–1976 occupies over 90,000 pages!

The organization and goals of *Chemical Abstracts* are different from those of Beilstein. Unlike Beilstein, which covers only organic compounds (in ten-year units), *Chemical Abstracts* covers all areas of chemistry and attempts to publish concise abstracts of every article and patent as soon as possible (usually 3–4 months from the original publication data). These abstracts are organized within each printed volume into general fields (organic chemistry, biochemistry, etc.).

The format of *Chemical Abstracts* is in a state of continuous change. Originally, one volume (consisting of several bound books) was published corresponding to one year of the scientific literature. As the number of scientific publications increased, this became impractical and, beginning in 1970 (Volume 70), each volume has covered only six months of the primary literature.

The manner in which index references are given in *Chemical Abstracts* has also changed over the years. In the early years, the reference to an abstract was by volume and page number. As the pages became more crowded, columns, instead of pages, were numbered and superscript numbers (and later, letters) were added to the index references to denote the position on the page or column. Starting in 1967 (Volume 66), each abstract has been given its own number. Fortunately, these changes create no confusion for someone using *Chemical Abstracts*; the meaning of a reference becomes self-evident from the appearance of a page in a particular volume.

To a user of *Chemical Abstracts*, a far more important change is that of nomenclature, which has undergone major revisions over the decades. The older issues emphasized trivial names. The newer issues emphasize a modified IUPAC system, called the "Chemical Abstracts system." Today, very little trivial nomenclature is found in *Chemical Abstracts*. For example, toluene is now indexed as "benzene, methyl-". Fortunately, there are guides to aid the user in coping with these changes in nomenclature (see Section 20.6C).

B. Indexes

Each volume of *Chemical Abstracts* has an author index, a subject index, and a molecular-formula index covering the issues of that six-month (or one-year) period. Until 1956, *collective indexes* were published every ten years; however, from 1957 to the present, these have been published every five years. (These collective indexes are not cumulative, as are the Beilstein indexes.)

Because of the changes in the indexing system over the years (due, in large part, to computerization), the techniques for finding information before 1966 and after 1966 are somewhat different (see Section 20.6D).

C. Nomenclature

In the subject indexes of *Chemical Abstracts*, compounds are generally listed in the alphabetical order of their parents ("1-pentanol, 3-methyl" instead of "3-methyl-1-pentanol"). As we have mentioned, some of the parent index names have changed over the years. Fortunately, *Chemical Abstracts* has published guides to finding compounds in the subject indexes.

for the earlier volumes: See the nomenclature guide in the introduction to the subject index of Volume 56 (January–June, 1962).

for the later volumes: See the *Index Guides* for the 8th and 9th Collective Indexes. (These guides contain a selection of index names, an alphabetical listing of cross-references, synonyms, etc.)

D. How to Find a Compound in *Chemical Abstracts*

Pre-1966 Information. The 7th Collective Index, covering Volumes 56–65 (1962–1966), was the last of the old indexes. To find information published prior to 1966, start with this Collective Index and search backwards in time, using the 6th Collective Index next, and so forth. Both the subject indexes and the formula indexes will usually have to be consulted.

Example. Search the 7th Collective Index of *Chemical Abstracts* for abstracts dealing with the kinetics of the acetylation of cyclohexanol.

1) Write the formula and determine the molecular formula of the compound.

—OH, $C_6H_{12}O$

2) Find the $C_6H_{12}O$ listings in the formula index.

The listings start on page 511F of the first volume of the 7th Collective Formula index. You will find that there are no listings for cyclohexanol; all are contained in the subject index. (This fact is stated in the parenthetical statement immediately following the start of $C_6H_{12}O$ headings on page 511F, beginning with "See also . . . .")

3) Find the compound in the subject index.

The listings for cyclohexanol begin on page 69475 and continue to page 69495.

4) Look for the specific subject.

Acetylation of cyclohexanol is the second major entry under cyclohexanol on page 69475.

Acetylation of, *59*: 1513e
  catalysts for, *60*: 14505g
  kinetics of, *65*: 7257g, 8727h
  micro, *59*: 8632h

There are two references for our subject: Volume 65, column 7257, position g; and Volume 65, column 8727, position h.

5) Find the specific abstract. (We will consider only the second abstract here.)

At the beginning of the *65*:8727h abstract you will find:

> *Conformational analysis, XII. Acetylation rates of substituted cyclohexanols. The kinetic method of conformational analysis.* Ernest L. Eliel and Francis J. Biros (Univ. of Notre Dame, Notre Dame, Indiana). *J. Am. Chem. Soc.* **88**(14), 3334–43(1966) (Eng); . . . .

Post-1966 Information. The use of the indexes of *Chemical Abstracts* starting with the 8th Collective Index has been considerably simplified by the *Index Guide* and registry numbers. The *Index Guide* provides a cross-reference to the indexes: by first consulting the *Index Guide*, you can determine the word or compound to look up in the subject index.

The **registry number** is a unique computer-generated number that is assigned to each chemical substance to provide identification. The number for cyclohexanol is [108-93-0]. With the registry number, a chemist can consult the Chemical Abstracts Service *Registry Handbook* to find the index name and molecular formula. The first *Registry Handbook* was published in 1974 and covered the literature from 1965 to 1971. Supplements have been issued since then.

The general procedure for using the more recent indexes of *Chemical Abstracts* is as follows:

1) If you wish to find information about a structure, but have only its common (trivial) name:

(*a*) Look up the name in the *Index Guide* to find the name under which the compound is indexed.

(*b*) Look up the index name in the subject index (or the *Chemical Substance Index* if you are using the 8th or later Collective Index).

(*c*) Find the abstract(s).

2) If you have a molecular formula of the substance:

(*a*) Use the formula index to find both the indexed name and specific references.

(*b*) Use the subject index (or the *Chemical Substance Index*) to obtain additional references.

(*c*) Find the abstract(s).

3) To search for information about a general subject:

(*a*) Use the *Index Guide* to find out how the subject is indexed.

(*b*) Go to the subject index (or the *General Subject Index* starting with the 9th Collective Index).

(*c*) Find the abstract(s).

Collective Indexes 8—present:

Find the index name in the *Index Guide* or *Formula Index*.

↓

Find abstract references from both *Subject Index* (or *Chemical Substance Index*) and *Formula Index*.

Collective Indexes 1–7 (up to 1966):

Find rules for naming in the introduction to the *Subject Index*, Vol. 56.

↓

Find abstract references from both *Subject Index* and *Formula Index*.

Figure 20.3. Summary of how to find information in Chemical Abstracts.

Figure 20.3 summarizes the rules for finding information in both the earlier and later volumes of *Chemical Abstracts*.

Problems

20.1 Using the *Handbook of Chemistry and Physics*, find the melting points of the following compounds:

(a) *trans*-1,2-cyclohexanedicarboxylic acid
(b) *m*-aminophenol
(c) 2,5-dibromofuran

20.2 Using *Lange's Handbook of Chemistry*, find the refractive index of each of the following compounds:

(a) benzyl alcohol
(b) formamide
(c) 3-pentenoic acid chloride

20.3 Using the *Merck Index*, find the following information:

(a) the medical use of α-methyl-*p*-tyrosine
(b) the structure and primary literature reference to tocol
(c) what the Wenker ring closure is

20.4 Use the *Index Guide* to the 9th Collective Index to find the index name for:

(a) 1-cyclohexyl-2-propyn-1-ol
(b) Marvinol 2002

20.5 Use the *Index Guide* to the 8th Collective Index to find the registry numbers for the (*E*) and (*Z*) isomers of 13-docosenoic acid.

20.6 Use the 1965–1971 *Registry Handbook* to find the name and molecular formula of the compound with the registry number [1226-05-7].

20.7 Use the 9th Collective Index to find an abstract reference for the antimicrobial spectra of alkylbenzene sulfonic acid esters.

20.8 γ, γ, γ-Trichlorocrotonitrile ($Cl_3CCH{=}CHCN$) has been prepared by the reaction of P_2O_5 and $Cl_3CCH{=}CHCONH_2$. Use the formula index for Volumes 14–40 of *Chemical Abstracts* to find the original article.

20.9 Use Beilstein to find the following information:

(a) the journal in which the density and viscosity of a solution of 2,4-dichloroaniline in isoamyl acetate was published

(b) the boiling point at 7 mm Hg of 1-chloro-1,2-dibromocyclohexane

(c) the melting point of a mercuric chloride–hydrochloride salt of 2-methylthiazole

(d) the solubility of 2-nitro-2-methyl-1-propanol in ethanol and in water

20.10 Use Beilstein to find the journal in which a melting point of 18° for $CH_2{=}CH(CH_2)_9CO_2H$ was reported.

CHAPTER 21

Introduction to Qualitative Organic Analysis

The characterization and identification of organic compounds of unknown structure, or **qualitative organic analysis,** is an important aspect of organic chemistry. A considerable portion of research activity deals with structure identification of compounds isolated from natural products or from reactions carried out in the laboratory. Qualitative organic analysis is also useful to the student because it is an introduction to independent laboratory work and to working on a semimicro scale.

Over the years, the procedures used in structure identification have changed dramatically. Prior to the 1940's, chemists had to rely on physical constants, elemental analysis, and chemical behavior to determine the structure of a compound. Since that time, spectroscopy has come into wide usage and has had a tremendous impact on research involved with structure determinations.

Today, structure identification usually involves a combination of chemical and spectral techniques. In selected cases, a structure can be identified by its spectra alone. In most cases, however, additional information is required. Therefore, many of the characterization tests and reactions developed by the earlier chemists are still useful today, either in conjunction with, or as a substitute for, spectroscopy.

21.1 Steps in Identifying an Unknown

In Experiment 21.1, you will be issued one or more unknowns that will be limited to alcohols, phenols, aldehydes, and ketones. Then, by a combination

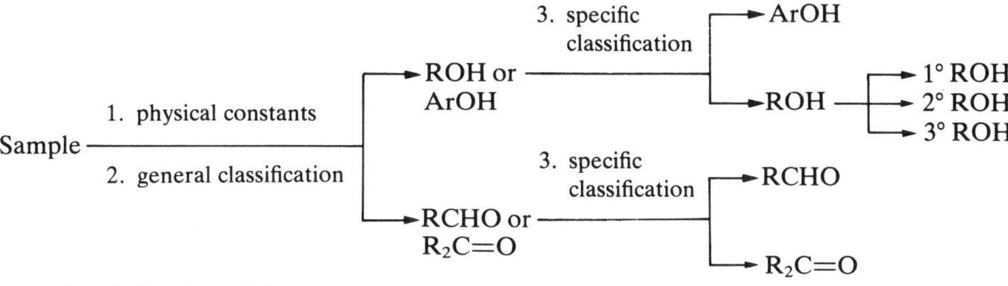

Figure 21.1 Steps in identifying an unknown.

of infrared spectroscopy and "wet chemistry," you will determine the identity of your unknown. (In this experiment, infrared spectra are not essential for the successful identification of the unknowns; the wet-chemical techniques can be used alone.)

The steps used in identifying an unknown compound follow:

1) determination of the physical constants of the unknown, including the infrared spectrum;

2) chemical classification tests to determine the *types* of functional groups present (in this experiment, OH or C=O);

3) chemical tests to identify the specific functional groups (ROH or ArOH, RCHO or $R_2C{=}O$);

4) correlation of the chemical tests with the spectral data and physical constants, which will lead to a list of possible structures for the unknown;

5) final identification of the unknown by the preparation of one or more derivatives of the unknown and comparison of the melting points of the derivatives with those reported in derivative tables.

The interrelationships of the chemical tests used in this chapter are outlined in Figure 21.1.

This general approach assumes that the unknown is listed in the derivative tables. The procedures used to identify the structure of a compound not listed in derivative tables differ principally in Step 5. We will not discuss these additional procedures in this chapter.

21.2 Physical Constants and Spectra

When you obtain an unknown, first note its appearance. Is it solid or liquid? Colored, colorless, or white? (Color may arise from a conjugated system or from a small amount of colored impurity.) If crystalline, what is the crystal type? If the unknown is a solid, determine the melting point. If the melting range is large or if the crystals appear impure, you may wish to recrystallize the unknown before proceeding. (Tlc or glc may be used to check on the purity of an unknown.)

If the unknown is a liquid, determine the refractive index. Do not attempt to obtain a boiling point by distillation without first consulting your instructor. Most organic liquids are high-boiling and decompose unless distilled under vacuum. In addition, the distillation equipment usually supplied for student use requires a large volume of organic material for a successful distillation. While there are procedures available to obtain a boiling point without distillation (micro-boiling-point determination, for example), these procedures are sensitive to small amounts of low-boiling impurities and frequently give erroneously low boiling points. In Experiment 21.1, the refractive index will be used as a physical constant *in lieu* of a boiling point.

Obtain an infrared spectrum of the compound, using the methods outlined in Chapter 11. (If nmr spectroscopy is available, carry out the chemical tests before obtaining the nmr spectrum, and use the nmr spectrum as confirmatory evidence for a structure assignment.) Inspect the infrared spectrum to determine which principal functional group(s) your unknown contains. Since the unknowns will be limited to alcohols, phenols, aldehydes, and ketones, the infrared spectrum of your unknown will exhibit OH or C=O absorption (or both).

Solubility tests. Although not necessarily a physical constant because of possible chemical reactions, the solubility of a compound in various solvents can provide a clue to its structure. Because your unknown will be limited to only four classes of compounds, only a very limited amount of solubility testing, as diagrammed in Figure 21.2, should be carried out. In testing the solubility, add 0.5–1.0 mL of the solvent to a few crushed crystals of a solid unknown or to 2–3 drops of a liquid unknown. Use the following solvents in the order given:

1) pure water (If the compound is soluble, there is no need to check the solubility in aqueous base; however, check the pH of the solution with pH paper.)
2) 5% aqueous NaOH
3) 5% aqueous $NaHCO_3$

From the solubility scheme in Figure 21.2, you can see that, in general, phenols can be differentiated from other compounds by their solubility. If the

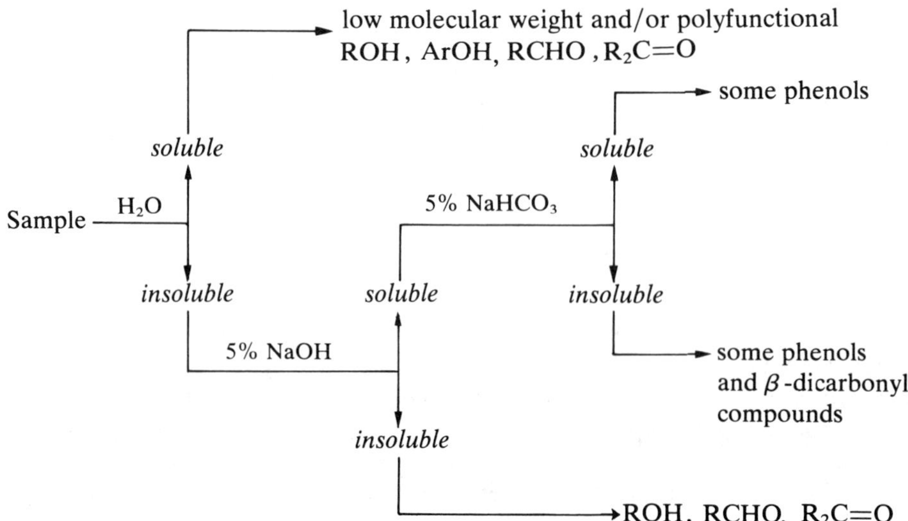

Figure 21.2 Limited solubility scheme to be used in Experiment 21.1.

phenol is soluble in water, the aqueous solution will be acidic. If a phenol is insoluble in water, then it will probably be soluble in 5% NaOH solution. A water-insoluble phenol that is sufficiently acidic to dissolve in aqueous $NaHCO_3$ must contain one or more electron-withdrawing substituents (for example, $—NO_2$, —CHO, or —X) on the aromatic ring. Such groups stabilize the conjugate base by the inductive effect and by resonance-stabilization.

For compounds other than phenols, the solubility in water is suggestive that the unknown either is of low molecular weight or has more than one hydrogen-bonding functional group.

21.3 General Classification Tests

On the basis of the infrared spectrum, you should have a good idea of the functional group(s) in your unknown. The general classification tests will confirm this identification. Of equal importance, the general tests and the specific tests described in Section 21.4 will give an indication if the functional group in your unknown behaves in a normal or an abnormal chemical manner.

Classification tests and the specific functional group tests are test-tube reactions that can be carried out quickly using a small amount of unknown. Ideally, each member of one class of compounds would give a positive test. In reality, there are interferences and exceptions that must be considered. These are discussed in Experiment 21.1.

Chemical tests are reasonably subjective. Therefore, it is advisable to run a blank (a test on a compound known to give a negative result) and also a test on a compound that will give a positive result at the same time you run the test on your unknown. The test results with your unknown can then be compared to those of the two known compounds.

The two general classification tests used in Experiment 21.1 are **the ceric nitrate test** for alcohols and phenols and **the 2,4-dinitrophenylhydrazine (DNP) test** for aldehydes and ketones. The ceric nitrate test is based upon the formation of colored complexes between Ce^{4+} ions and compounds containing OH groups. The DNP test is based upon the ability of aldehydes and ketones to form precipitates of 2,4-dinitrophenylhydrazones when treated with 2,4-dinitrophenylhydrazine.

Ideally, *one of the two tests will be positive and the other will be negative.* In such a case, you would then proceed to the specific functional group tests to identify the actual functional group. For example, if your unknown gives a negative ceric nitrate test and a positive DNP test, you would conclude that your unknown is either an aldehyde or a ketone. You would then proceed to the specific functional group tests for distinguishing aldehydes from ketones.

If both the ceric nitrate test and the DNP test are positive, you would expect that both OH and C=O groups are present. In this case, you would perform both sets of specific functional group tests described in Section 21.4 instead of just one set.

If both tests are negative, first verify that you are doing the tests correctly and that the reagents are "active." Run the tests on a few known compounds that should give positive results. After verification that the tests do indeed work with known compounds, and if the tests are still negative with your unknown, then you must conclude that you are dealing with a compound that is an exception to the general class. In such a case, you must rely heavily upon spectral data in identifying the functional group. Pay particular attention to the *limitations discussions* for clues as to the types of compounds that behave abnormally. If your unknown shows carbonyl absorption in its infrared spectrum, but fails to give a positive DNP test, you can anticipate having a considerable amount of trouble in preparing a solid derivative.

21.4 Specific Classification Tests for Functional Groups

The infrared spectrum and the general classification tests will classify your unknown as containing a hydroxyl group or a carbonyl group. The next group of tests are aimed at the identification of the specific functional group—in this case, to distinguish between an alcohol and a phenol or between an aldehyde and a ketone. Figure 21.3 outlines the scheme you will follow to determine which type of compound you have. Which test or tests you must carry out depends on the results of the general classification tests. (If your unknown does not contain an OH group, there is no point in trying to find out whether your unknown is an alcohol or a phenol!)

If the ceric nitrate test is positive, then carry out a **ferric chloride test** for phenols. If the ferric chloride test is positive, then your unknown is probably a phenol. Inspect the infrared spectrum for aromatic absorption peaks.

If the infrared spectrum shows OH absorption, the ceric nitrate test was positive, and the ferric nitrate test is negative, your unknown is presumably an alcohol. Because of the differences in reactivity of primary, secondary, and tertiary alcohols (and thus differences in procedure for the preparation of their derivatives),

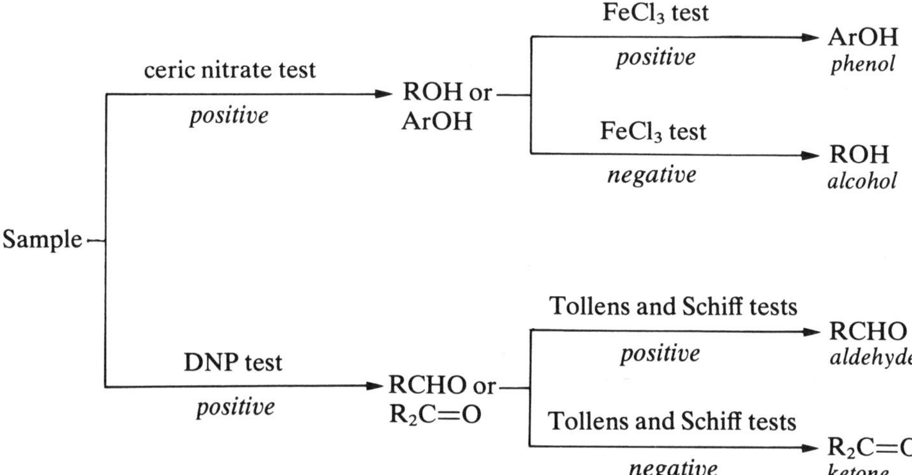

Figure 21.3 Steps in determining the specific class to which an unknown hydroxy or carbonyl compound belongs.

it is desirable to know the classification of an alcohol unknown. The **Lucas test** (reaction with HCl + $ZnCl_2$) is used to distinguish whether an alcohol is primary, secondary, or tertiary.

If the DNP test is positive, then your unknown is a carbonyl compound. The infrared spectrum should also signal this fact. Inspect the spectrum to determine whether the unknown is an aldehyde or a ketone. Run the aldehyde-identification tests; the results should be in agreement with the conclusions arrived at from the spectrum.

In Experiment 21.1, two tests are described for aldehydes: the **Tollens test** and the **Schiff test**; however, there are no specific tests for ketones. Therefore, if the aldehyde tests are negative, the unknown is tentatively identified as a ketone. Because both aldehydes and ketones are derivatized by the formation of 2,4-dinitrophenylhydrazones or semicarbazones, differentiation between the two is not imperative.

21.5 Correlation of the Chemical Tests and Spectra

Before attempting to prepare a derivative of your unknown, list the facts you have gathered. This list should include the physical constants, pertinent infrared absorption data, the results of the chemical tests, and any additional observations. Label each item as confirmed or tentative, depending upon your confidence in it.

On pages 424–432, you will find a set of derivative tables in which compounds are grouped by functional group and physical state at room temperature. Solid alcohols, for example, are listed in one table; liquid alcohols are listed in another table. Within each table, solids are listed by increasing melting point, while liquids are listed by increasing refractive index. The melting points of the derivatives of each compound are to the right of the compound's name and melting point or boiling point.

Using the physical constants of your unknown, draw up a list of possible structures for the unknown from the derivative tables. Allow a considerable amount of leeway in physical constants when listing the possibilities. For example, suppose your unknown is a solid aldehyde melting at 61–63°. In checking the derivative table of solid aldehydes, you will find that there is no compound with a melting point of 61–63° listed. However, you will find eight solid aldehydes with melting points between 50° and 85°. The chances are good that the unknown is one of these; therefore, your list would include these eight compounds. A reasonable list should contain 3–10 compounds.

21.6 Preparation of Derivatives

The goal of derivatization is to convert a liquid to a solid compound (or convert a solid to another solid compound) by a reaction that is characteristic of the particular compound class. The solid derivative is then purified by crystallization and its melting point is determined.

A little planning, and perhaps some library work, can save a considerable amount of time and effort in derivatization. For example, look up the melting points of the derivatives of the compounds on your list and determine which derivative will best distinguish between the possibilities. Using the example of the solid aldehyde cited previously, we inspect the semicarbazone column of the solid aldehyde table and discover that this derivative would be a poor choice. From the melting point of the semicarbazone, we could not distinguish between 3-nitrobenzaldehyde and 2-naphthaldehyde, nor could we distinguish between 4-(dimethylamino)benzaldehyde and 3,4,5-trimethoxybenzaldehyde, because of the proximity of the semicarbazone melting points. A 2,4-dinitrophenylhydrazone would be a much better choice for a derivative.

21.7 Final Structure Identification

The final identification of your unknown is made by (1) matching the physical constants of your unknown with those of a specific compound, and (2) matching the melting point of each derivative you prepare with the known melting point. If every structure determination proceeded smoothly, every student would determine the correct structure for his or her unknown. This is not always the case for several reasons.

Impure derivative. A very common source of student error is incomplete purification of the derivative. For example, suppose a student prepared a DNP of a liquid ketone, n_D^{20} 1.4465. The melting point of the crude DNP was determined to be 136–140°. The student concluded that these physical constants are close enough to those of 2-methylcyclohexanone (n_D^{20} 1.4483; mp of DNP, 137°) to justify an assignment of structure. Had the student recrystallized the DNP several times, he or she would have observed the melting point of the DNP approach 155°. The structure of the unknown was actually 3-methylcyclohexanone (n_D^{20} 1.4437) and not its 2-methyl isomer.

Errors. One source of error is that erroneous physical constants are sometimes reported in the literature. Another source of error is the student. (A student may spend hours recrystallizing 2,4-dinitrophenylhydrazine instead of the DNP of an unknown.)

Overlooking the facts. Seeing only what we want to see and ignoring the rest is a human frailty that can lead to, among other problems, false structure determinations. For example, a student may be convinced that a solid sample is benzaldehyde because the semicarbazone melts at 222°. Unfortunately, the student has not considered *all* the data. (Benzaldehyde is a liquid, not a solid, at room temperature.) Therefore, before drawing a conclusion, double-check *all* the facts.

>>>> ***SAFETY NOTE*** Because you will not know the name or structure of your unknown, you will not be able to assess its potential hazards. Therefore, handle each unknown as if it were a vile poison. Where appropriate, safety notes concerning the other reagents are placed with each part of the experiment.

Suggested Readings

Cheronis, N. D., Entrikin, J. B., and Hodnett, E. M., *Semimicro Qualitative Organic Analysis*, 3rd ed. New York: Interscience, 1965; reprinted, Huntington, N.Y.: R. E. Krieger Publishing Co., 1982.

Pasto, D. J., and Johnson, C. R., *Organic Structure Determination*. Englewood Cliffs, N. J.: Prentice-Hall, 1969; later revision included in Pasto and Johnson, *Laboratory Text for Organic Chemistry*. Prentice-Hall, 1979.

Shriner, R. L., Fuson, R. C., Curtin, D. Y., and Morrill, T. C., *The Systematic Identification of Organic Compounds*, 6th ed. New York: John Wiley and Sons, 1980.

CRC Handbook of Tables for Organic Compound Identification, 3rd ed., Rappoport, Z., Ed. Boca Raton, Florida: CRC Press, Inc., 1967: extensive derivative tables.

Problems

21.1 A student's unknown is soluble in water, 5% $NaHCO_3$, and 5% NaOH. The student concludes that the unknown is a phenol. Is the student correct? Explain.

21.2 State whether each of the following compounds is soluble in H_2O, 5% $NaHCO_3$, 5% NaOH, and 5% HCl:

(a) 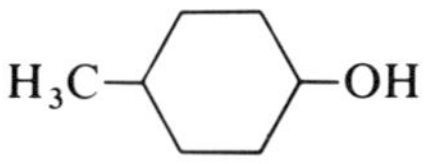 H_3C–(cyclohexane ring)–OH

(b) 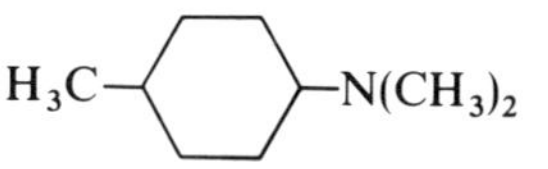 H_3C–(cyclohexane ring)–$N(CH_3)_2$

(c) $(CH_3)_3COH$

(d) 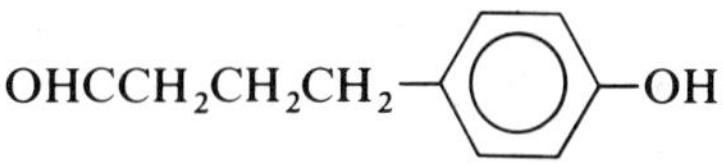 $OHCCH_2CH_2CH_2$–(benzene ring)–OH

(e) $CH_3(CH_2)_6CO_2H$

(f) $CH_3(CH_2)_4CO_2CH_3$

21.3 State whether each of the following compounds would give positive or negative results in the ceric nitrate, DNP, ferric chloride, Tollens, and Schiff tests:

(a) $ClCH_2\overset{O}{\overset{\|}{C}}CH_3$

(b) $HOCH_2CH_2CHO$

(c) 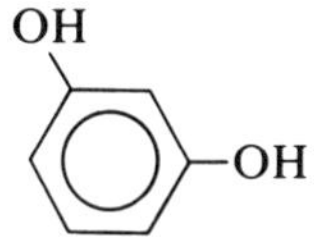 (benzene ring with OH and OH in meta positions)

»»› EXPERIMENT 21.1 ‹««

Qualitative Organic Analysis: Selected Procedures

A. DNP Test for Carbonyl Compounds

The reaction of an aldehyde or ketone with 2,4-dinitrophenylhydrazine yields a precipitate of a 2,4-dinitrophenylhydrazone, or DNP. Alcohols and phenols do not undergo reaction with this reagent.

$$\underset{\text{2,4-dinitrophenylhydrazine}}{O_2N\text{-}C_6H_3(NO_2)\text{-}NHNH_2} + \underset{\text{an aldehyde or ketone}}{R\overset{O}{\overset{\|}{C}}R} \xrightarrow{H^+} \underset{\text{a 2,4-dinitrophenylhydrazone}}{O_2N\text{-}C_6H_3(NO_2)\text{-}NHN{=}CR_2} + H_2O$$

This reaction is the basis for both the general classification test and for the preparation of a derivative of an aldehyde or ketone. In this section, we will discuss the reaction as a classification test. We will consider the use of the reaction in derivative formation later in this chapter.

In the classification test, the solid or liquid unknown is first dissolved in 95% ethanol; then a dilute, acidic solution of 2,4-dinitrophenylhydrazine is added. A positive test is signified by the formation of a yellow, orange, or red precipitate. With an aldehyde or a sterically nonhindered ketone, the precipitate usually forms immediately. The test is taken to be negative if, after standing for 15 minutes at room temperature followed by a 15-minute heating period, no precipitate has formed.

Only the formation of a true precipitate (and not an oil) is the sign of a positive test. With some carbonyl compounds, the hydrazone first oils out, then solidifies. Until the solid can be detected, the test must be considered either negative or inconclusive.

The color of the precipitate can often provide a clue concerning the structure of the unknown. The hydrazones of conjugated carbonyl compounds (aryl or alkenyl) are orange, while those of nonconjugated carbonyl compounds are yellow.

Limitations. High-molecular-weight aliphatic ketones, such as 2-octanone, 2-decanone, 6-nonanone, and 6-undecanone, may fail to react under the test conditions, or they may yield oils or low-melting solids.

False positive tests may be obtained with alcohols. Many alcohols are prepared by the reduction of aldehydes or ketones. If the alcohol is not pure, the residual aldehyde or ketone can give a false positive test. Also, some allylic and benzylic alcohols can be oxidized by hydrazines to aldehydes or ketones, which can then give a positive test. 3-Phenyl-2-propen-1-ol (cinnamyl alcohol), 4-phenyl-3-buten-2-ol, and diphenylmethanol (benzhydrol) fall into this group.

Because oxidation generally occurs when the reactants are heated, a positive test obtained only after heating the test solution should be considered inconclusive. (Check the infrared spectrum.)

Experimental

EQUIPMENT:

disposable plastic gloves
droppers
hot water bath
150-mm test tube

CHEMICALS:

2,4-dinitrophenylhydrazine test solution, 3 mL (see Experimental Note 1)
95% ethanol, 2 mL
unknown, 1 drop liquid or 25–50 mg solid

SUGGESTED COMPOUNDS FOR POSITIVE AND NEGATIVE TESTS:

liquids: 1-butanol, 2-butanol, benzaldehyde, acetophenone, cyclohexanone
solids: *o*-chlorobenzyl alcohol, diphenylmethanol (benzhydrol), *p*-chlorobenzaldehyde, benzophenone

TIME REQUIRED: 30–40 minutes

>>>> ***SAFETY NOTE*** Commercial 2,4-dinitrophenylhydrazine contains 10–20% water. Do not desiccate; removal of the water may cause the hydrazine to become explosive.

2,4-Dinitrophenylhydrazine is a potential carcinogen and can be absorbed dermally. Wear disposable plastic gloves. Wash thoroughly after using this reagent. Clean up any spills and wash used glassware carefully.

PROCEDURE

In a small test tube, dissolve 1–2 drops of liquid unknown or 25–50 mg of solid in 2 mL of 95% ethanol. (Warm the mixture if necessary.)

Add 3 mL of the 2,4-dinitrophenylhydrazine test reagent. The formation of a precipitate is a positive test. Record the color of the precipitate. If a precipitate does not form, allow the test solution to stand at room temperature for 15 minutes (see Experimental Note 2). If a precipitate has not started to form at the end of this period, warm the test solution in a water bath (60–70°) for about 15 minutes (see Experimental Note 3). Do not boil the test solution; loss of the ethanol will cause a precipitation of the reagent. If a precipitate has not begun to form at the end of the heating period, the test is considered negative.

EXPERIMENTAL NOTES

1) *Preparation of the DNP test reagent.* Place 3.0 g of 2,4-dinitrophenylhydrazine (see Safety Note) in a 125-mL Erlenmeyer flask and add 15 mL of concentrated H_2SO_4. Stir with a glass rod, breaking up any lumps of the reagent, until most of the reagent has dissolved. Cautiously add 20 mL of water, then 70 mL of 95% ethanol. Filter any undissolved reagent from the brownish-orange solution.

2) The addition of a few drops of water may aid in the formation of the precipitate. However, do not add so much water that the hydrazine reagent precipitates.

3) If the test solution is heated, regard any positive test as tentative (see *Limitations*, page 372).

B. Ceric Nitrate Test for Alcohols and Phenols

Alcohols or phenols yield colored complex ions when treated with ceric ammonium nitrate. Other oxygen functional groups (aldehyde, ketone, carboxyl, etc.) do not interfere with the complex formation except as noted in the *Limitations*.

$$\underset{\substack{\text{ceric ammonium nitrate}\\ \textit{yellow}}}{(NH_4)_2Ce(NO_3)_6} + \underset{\substack{\textit{an alcohol}\\ \textit{or phenol}}}{ROH} \xrightarrow[\text{or dioxane}]{H_2O} \underset{\textit{a colored complex}}{(NH_4)_2\overset{\overset{\displaystyle OR}{|}}{Ce}(NO_3)_5} + H^+ + NO_3^-$$

alcohols: orange-to-red complex
phenols: orange, blue, green, or brown complex (which may precipitate)

Once the presence of an alcohol or a phenol is established, an additional test on the colored complex allows us to distinguish between an alcohol and a phenol. Ceric ions, even when complexed with alcohols, are readily reduced to Ce^{3+} ions, which are colorless. If a mild reducing agent, such as dimedone, is added to the colored ceric–alcohol complex ion, the color fades. In the case of phenols, the initial color is due, in part, to phenol oxidation products (from the action of Ce^{4+} ions on the phenol). Dimedone has no effect on the color of these oxidation products; therefore, the color of a phenol test solution does *not* disappear when dimedone is added.

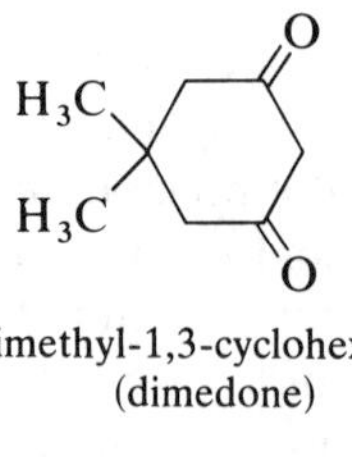

[structure: cyclohexane ring with two H_3C groups on C5 and C=O at C1 and C3]

5,5-dimethyl-1,3-cyclohexanedione
(dimedone)

$$ROH \xrightarrow{Ce^{4+}} \text{colored solution} \xrightarrow{\text{dimedone}} \text{colorless}$$

$$ArOH \xrightarrow{Ce^{4+}} \underset{\text{or precipitate}}{\text{colored solution}} \xrightarrow{\text{dimedone}} \text{little color change}$$

Limitations. The ceric nitrate test is not reliable for all alcohols and phenols. Hydroxy compounds with more than 10–12 carbons fail to give a positive test or give such a faint color change that it cannot be detected. Because the color change may be slight, it is prudent to run a blank so that the colors of the blank and unknown tests can be compared.

Easily oxidized alcohols, such as benzyl alcohol, may show an initial darkening of the original yellow color, followed by a fading of the color as the ceric ions are reduced. Some easily oxidized compounds (furfuryl alcohol, dimedone, some unsaturated ketones, etc.) may cause the initial color of the test reagent to fade with no initial darkening.

Compounds that enolize readily, such as β-diketones or β-keto esters, often give an initial dark blue color, followed by fading.

$$-\overset{\overset{\displaystyle O}{\|}}{C}CH_2-\overset{\overset{\displaystyle O}{\|}}{C}- \rightleftarrows -\overset{\overset{\displaystyle O}{\|}}{C}CH=\overset{\overset{\displaystyle OH}{|}}{C}-$$

can form colored complex with Ce^{4+}

Amino alcohols fail to give a positive test because the basic amino group causes a shift to a higher pH, resulting in a precipitation of $Ce(OH)_4$.

In summary, a *positive test for an alcohol* is an orange-to-red color that may or may not fade. (However, the color should fade when dimedone is added.) A *positive test for a phenol* is a colored precipitate or solution that does not fade even when dimedone is added. The test is *inconclusive* if a fading color other than orange-to-red is observed or if the original yellow color of the Ce^{4+} ions simply fades.

Experimental

EQUIPMENT:

dropper
stirring rod
100-mm test tubes

CHEMICALS:

ceric nitrate test solution (see Experimental Note 1)
dimedone, 10 mg
1,4-dioxane, 0.5 mL
unknown, 1–2 drops liquid or about 25 mg solid

COMPOUNDS FOR POSITIVE AND NEGATIVE TESTS:

liquids: 1-butanol, 2-butanol, *t*-butyl alcohol, phenol (liquid), benzaldehyde, acetophenone
solids: *o*-chlorobenzyl alcohol, diphenylmethanol (benzhydrol), 2-naphthol, benzophenone

TIME REQUIRED: 10–15 minutes

›››› ***SAFETY NOTE*** Dioxane is carcinogenic in laboratory animals. Use it only in the hood and wash thoroughly after using it. In some cases, acetone or water can be substituted (see Experimental Note 2).

PROCEDURE

In each of two small test tubes place 0.5 mL (about 25 drops) of 1,4-dioxane (see Experimental Note 2), 3 drops of water, and 2 drops of ceric nitrate test solution. Add 1–2 drops of liquid unknown or about 25 mg of solid unknown to one of the two test tubes. The other tube serves as the blank.

Compare the colors of the two tubes very carefully by looking down through the solutions onto white paper in good light. A faint change in the depth of the color within 10 minutes is taken as a positive test.

Add a few milligrams of dimedone to each test tube. The color in the blank test should disappear immediately upon mixing. The color in the unknown's test tube will remain if the unknown is a phenol, but will fade immediately if it is not a phenol.

EXPERIMENTAL NOTES

1) *Preparation of the ceric nitrate test reagent.* Add 100 g of ceric ammonium nitrate to 250 mL of 2*N* HNO_3. Warm the mixture to about 50° to dissolve the salt, then filter any undissolved salt. (This amount of reagent is sufficient for over 1000 tests. Adjust the quantities according to the needs of the class.)

2) 1,4-Dioxane (*p*-dioxane) often contains impurities that give a positive test. Run the test using dioxane alone before drawing any conclusions. If it is not possible to obtain dioxane that gives a negative test, use acetone in its place. (If the unknown is water-soluble, water may be used instead of dioxane.)

C. Schiff Test for Aldehydes

The Schiff test (also called the *fuchsin–aldehyde test*) is based upon the reaction of an aldehyde with a colorless form of a rosaniline dye to yield a violet or purple product. Ketones fail to undergo this reaction.

The colorless form of the dye is prepared from the pink dye *p*-rosaniline hydrochloride (fuchsin) by treatment with sulfurous acid to yield one or more leucosulfonic acids (*leuco*- means colorless or white). The equation for the formation of one of the possible leucosulfonic acids follows:

$$\left(H_2N-C_6H_4\right)_2 C{=}C_6H_4{=}\overset{+}{N}H_2\ Cl^- \xrightarrow[-2H_2O]{3H_2SO_3} \left(HO_3SNH-C_6H_4\right)_2 C(SO_3H)-C_6H_4-\overset{+}{N}H_3\ Cl^-$$

p-rosaniline hydrochloride
pink

Schiff reagent
(*a leucosulfonic acid*)
colorless

An aldehyde undergoes an addition reaction with the colorless Schiff reagent to yield a colored product. Although we show only one typical product, a mixture of products, containing varying amounts of SO_2 and RCHO, is probably formed.

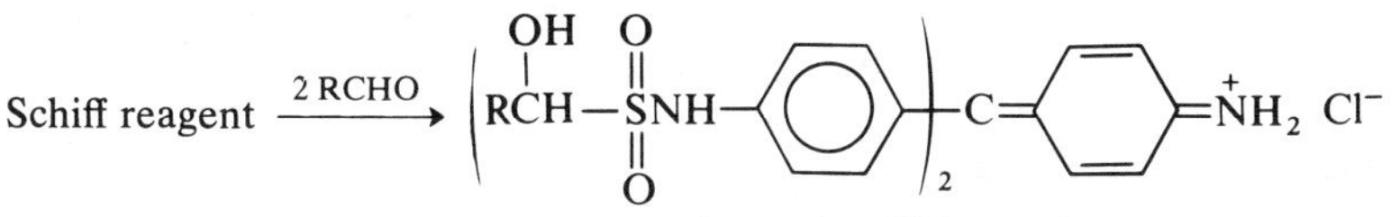

$$\text{Schiff reagent} \xrightarrow{2\,\text{RCHO}} \left(\text{RCH(OH)}-\text{S(=O)}_2\text{NH}-\text{C}_6\text{H}_4\right)_2\text{C}=\text{C}_6\text{H}_4=\overset{+}{\text{N}}\text{H}_2\ \text{Cl}^-$$

violet-purple addition product

Limitations. Although ketones do not yield the purple addition product in the Schiff test, some ketones do yield a pink color arising from the removal of SO_2 from the leucosulfonic acid and the regeneration of the original *p*-rosaniline. (This pink coloration lacks the bluish cast of the aldehyde addition products.) Because some aldehydes show only a faint purple, a blank and a known should be run simultaneously with the test on the unknown so that the colors can be compared.

Some aldehydes, such as the *o*- and *p*-hydroxybenzaldehydes, do not yield the purple addition product. Thus, if your unknown contains a phenol group, the Schiff test may give a false negative result.

Mild alkali, amines, heating, and even prolonged exposure to the air can destroy the Schiff reagent and regenerate the pink color of *p*-rosaniline (see Experimental Note 1, page 378).

Experimental

EQUIPMENT:

dropper
three 150-mm test tubes

CHEMICALS:

methanol (or 95% ethanol), 2 mL
Schiff reagent, 2 mL (see Experimental Note 1)
unknown, 1–2 drops of liquid or 25 mg of solid

COMPOUNDS FOR POSITIVE AND NEGATIVE TESTS:

liquids: 1-butanol, benzaldehyde, acetophenone
solids: 2-naphthol, *o*-chlorobenzaldehyde, benzophenone

TIME REQUIRED: 10–15 minutes

>>>> ***SAFETY NOTE*** Many aniline dyes are known carcinogens and can be dermally absorbed. Use appropriate precautions.

PROCEDURE

Place 2 mL of Schiff reagent in a test tube and add 1–2 drops of liquid unknown or about 25 mg of solid unknown (see Experimental Note 2). Shake the test tube to disperse the unknown, then allow the mixture to stand for 10 minutes. If the unknown is insoluble in the reagent, shake the mixture occasionally during the 10-minute period. Do not heat the tube.

The formation of a dark pink or purple color indicates the presence of an aldehyde group, while no color change (or a change to light pink) is considered a negative test result.

EXPERIMENTAL NOTES

1) Preparation of Schiff reagent. Dissolve 0.1 g of *p*-rosaniline hydrochloride (see Safety Note) in 100 mL of distilled water. Add 4 mL of water saturated with sodium bisulfite (about 4 g of sulfite). The color of the dye fades upon the addition of the sulfite solution. Allow the solution to stand for one hour, then add 1 mL of conc. HCl. Decolorize the resultant light pink solution with about 2 g of activated carbon. (Repeat the treatment with activated carbon, if necessary, until the reagent is colorless.)

Store Schiff reagent in a tightly stoppered brown glass bottle. Should the reagent turn pink before use, the color can be discharged by the addition of more saturated Na_2SO_3 solution and conc. HCl, followed by decolorization with activated carbon.

2) If the unknown is insoluble in the reagent solution, you may wish to dissolve the unknown in 2 mL of methanol or 95% ethanol. In this case, add 2 mL of Schiff reagent directly to the alcohol solution. (Be sure to include the alcohol in your blank.) The formation of a cloudy solution can be disregarded.

D. Tollens Test for Aldehydes (Silver Mirror Test)

Aldehydes are readily oxidized, even by such mild oxidizing agents as Ag^+. Ketones are not oxidized under these conditions. The Tollens test is the treatment of a sample with a solution containing silver–ammonia complex ions. When an aldehyde group is oxidized by this reagent, the silver ions are reduced to metallic silver, which forms a black precipitate and, if the test tube is clean, a silver mirror on the test tube.

$$\underbrace{2\,Ag(NH_3)_2^+ + 2\,OH^-}_{\textit{in Tollens reagent}} + R\overset{\overset{\displaystyle O}{\|}}{C}H \rightarrow \underset{\substack{\text{silver}\\\text{mirror}}}{2\,Ag^0} + R\overset{\overset{\displaystyle O}{\|}}{C}O^- + NH_4^+ + 3NH_3 + H_2O$$

Limitations. Some phenols, such as 1-naphthol and hydroquinone, also give a positive Tollens test. Since phenols are not to be tested with this reagent, these interferences should not be troublesome.

Experimental

EQUIPMENT:

droppers
100-mm test tube

CHEMICALS:

15% ammonium hydroxide, 3–7 drops
methanol (or ethanol), 4–5 drops
5% silver nitrate solution, 1 mL
20% sodium hydroxide solution, 1 drop
unknown, 1–2 drops of liquid or 25 mg of solid

COMPOUNDS FOR POSITIVE AND NEGATIVE TESTS:

liquids: benzaldehyde, acetophenone
solids: *p*-chlorobenzaldehyde, benzophenone

TIME REQUIRED: 10–15 minutes

>>>> ***SAFETY NOTE*** Do not add an excess of sodium hydroxide to aqueous silver nitrate. Explosions have been reported.

If Tollens reagent is allowed to become dry, explosive silver fulminate may be formed. Never store any Tollens reagent, new or used, in your locker. After the test has been completed, flush the test tubes' contents down the drain with copious amounts of water. Then wash the test tubes thoroughly and rinse them with dilute nitric acid.

PROCEDURE

In a *clean* test tube (see Experimental Note), place 1 mL of 5% $AgNO_3$ solution. Add 1 drop of 20% NaOH solution (see Safety Note). Agitate the test tube so that the dark precipitate which forms is completely dispersed throughout the liquid. Add 15% NH_4OH solution dropwise with swirling, until the precipitate almost, but not completely, dissolves. Only 3–6 drops of NH_4OH will be needed. This mixture is active Tollens reagent; it must be used within 15 minutes or be destroyed (see Safety Note).

If the unknown is a liquid, add 1–2 drops directly to the Tollens reagent. If the unknown is a solid, dissolve or disperse it in 4–5 drops of methanol or 95% ethanol before adding it to the Tollens reagent. Agitate the test solution to ensure complete mixing, then allow the tube to stand for at least 5 minutes at room temperature. Do not heat the test mixture.

The formation of a black precipitate, along with a silver mirror on the sides of the test tube, is a positive test. The formation of a black precipitate without the concurrent formation of a silver mirror is a tentatively positive test (see Experimental Note).

EXPERIMENTAL NOTE

The formation of a silver mirror on the side of the test tube is the best indication that the test is positive. If the test tube is dirty, the mirror will not form. Before performing the test, scrub the test tube with scouring powder and rinse with distilled water. (If the test tube is clean, the rinse water will drain evenly from the sides of the tube and not form droplets.)

E. Ferric Chloride Test for Phenols

When a phenol is treated with a solution of ferric chloride, colored complex ions are rapidly formed. Alcohols do not undergo this reaction; therefore, the test can be used to distinguish phenols from alcohols. The color of the complex ions may be green, blue, purple, or red. The color depends on a number of factors, including the structure of the phenol, the solvent, and the concentration. The color formed by a few phenols fades rapidly; therefore, the test solution should be inspected immediately after mixing. Because the color change may be slight, the test solution should be compared to a blank.

Limitations. Other compounds besides phenols can give positive ferric chloride tests. Any carbonyl compound, such as a β-dione or β-keto ester, that exists in at least 5% enol form may give a positive ferric chloride test. (The ceric nitrate test, the DNP test, or the infrared spectrum may be used to differentiate a carbonyl compound from a phenol.)

Phenols with a large degree of steric hindrance around the OH group, such as 2,6-di-*t*-butylphenol or 2,4,6-tribromophenol, may show no color change or a color change that is too faint to be detected. Some other phenols, such as *p*-hydroxybenzoic acid, also fail to give a positive ferric chloride test. For this reason, a negative ferric chloride test does not necessarily mean that a compound contains no phenol groups.

Experimental

***EQUIPMENT*:**

droppers
100-mm test tubes

CHEMICALS:

ferric chloride test reagent, 1–3 drops (see Experimental Note)
methanol, about 1 mL
unknown, 1–2 drops of liquid or 25 mg of solid

COMPOUNDS FOR POSITIVE AND NEGATIVE TESTS:

liquids: 1-butanol, 2-butanol, phenol (liquid), ethyl acetoacetate
solids: diphenylmethanol (benzhydrol), 2-naphthol

TIME REQUIRED: 5–10 minutes

PROCEDURE

Dissolve 1–2 drops of a liquid unknown or about 25 mg of a solid in 0.3 mL of methanol in a test tube. Add 2 drops of water, mix thoroughly, and then add 1 drop of ferric chloride test reagent. Prepare a blank using the same procedure. A positive test is a color change from light orange to green, red, or violet. The color change may be faint.

EXPERIMENTAL NOTE

Preparation of ferric chloride test reagent. Dissolve 10 g of ferric chloride hexahydrate in 100 mL of anhydrous methanol. The ferric chloride dissolves readily; no heating is necessary.

F. Alcohol Classification Test (The Lucas Test)

The Lucas test is based upon the fact that primary, secondary, and tertiary alcohols vary in their rates of reaction with a mixture of conc. HCl and $ZnCl_2$ (the *Lucas reagent*). Relative rates of reaction are determined by observing when an alcohol that is soluble in the Lucas reagent yields a product (the alkyl chloride) that is *insoluble* in the reagent.

$$\text{ROH} + \text{HCl} \xrightarrow{\text{ZnCl}_2} \text{RCl} + \text{H}_2\text{O}$$

relative rates: $3° \text{ ROH} > 2° \text{ ROH} \gg 1° \text{ ROH}$

The complete classification procedure consists of three parts: (1) an initial solubility test to determine if the test is suitable; (2) reaction with Lucas reagent to differentiate between a primary alcohol and a secondary or tertiary alcohol; and (3) reaction with conc. HCl alone to differentiate between a secondary and tertiary alcohol. Let us briefly discuss each of these steps.

Solubility. The Lucas test is generally useful only for alcohols containing six or fewer carbon atoms, because most high-molecular-weight alcohols are insoluble in the test reagent. Therefore, the first step (part 1) in the classification test is to determine whether the alcohol is soluble in the Lucas reagent. If the alcohol is soluble, then the rest of the test can be carried out. If a homogeneous solution is *not* observed, the alcohol may be insoluble, in which case the classification test cannot be used. However, a soluble tertiary alcohol may undergo such rapid reaction that its initial solubility is not observed. To differentiate between these two possibilities, the solubility of the alcohol should also be tested in conc. HCl (without $ZnCl_2$). A reactive tertiary alcohol will dissolve in the conc. HCl before it undergoes reaction, while an alcohol that is insoluble in Lucas reagent will also be insoluble in conc. HCl alone.

Lucas test. In the Lucas test (part 2), the unknown alcohol is mixed with Lucas reagent and the reaction time is noted. A *primary alcohol* shows no observable reaction after 5–10 minutes. A *secondary alcohol* shows observable reaction after 1–5 minutes and should show two distinct layers after about 10 minutes. A *tertiary alcohol* usually begins to react immediately. Because of the variables in the test, such as activity of the Lucas reagent (see Experimental Note 2), the test on the unknown alcohol is conducted simultaneously with a test on a secondary alcohol (2-butanol), and the reaction times are compared.

Test with HCl. In part 3 of the classification test, the unknown is dissolved in conc. HCl without the $ZnCl_2$. When only conc. HCl is used, secondary alcohols react very slowly and will not form an oil within the time period of this test. Tertiary alcohols, however, react rapidly and form an oil within a few minutes.

Limitations. The principal limitation of the alcohol classification test is that the unknown alcohol must be soluble in the test reagent. A second limitation is that benzylic and allylic alcohols (1°, 2°, or 3°), which can form resonance-stabilized carbocation intermediates, undergo rapid reaction with the test reagent, just as do tertiary alcohols. For example:

$$\underset{\text{1°, but allylic}}{CH_2{=}CHCH_2OH} \xrightarrow[-H_2O]{H^+} \underset{\textit{resonance-stabilized}}{\left[CH_2{=}CH{-}\overset{+}{C}H_2 \leftrightarrow \overset{+}{C}H_2CH{=}CH_2\right]} \xrightarrow{Cl^-} CH_2{=}CHCH_2Cl$$

Experimental

EQUIPMENT:

five droppers
hot water bath
six 100-mm test tubes
spatula
thermometer

CHEMICALS:

2-butanol, 5 drops
t-butyl alcohol, 5 drops
conc. hydrochloric acid, 3 mL
Lucas reagent (see Experimental Note 1), 3 mL
unknown, 5 drops of liquid or about 250 mg of solid for each test

COMPOUNDS FOR POSITIVE AND NEGATIVE TESTS:

liquids: 1-butanol, 2-butanol, *t*-butyl alcohol
solids: *o*-chlorobenzyl alcohol, diphenylmethanol (benzhydrol), benzophenone

TIME REQUIRED: 5–15 minutes for each part

PROCEDURE 1. TEST FOR SOLUBILITY

Place 1 mL of Lucas reagent (see Experimental Note 2) in a test tube, add 5 drops of liquid unknown or about 250 mg of a solid unknown, and mix with a spatula. Determine if the unknown is soluble or insoluble.

If the unknown is insoluble, carry out the same procedure substituting concentrated hydrochloric acid for the Lucas reagent, as noted in the discussion.

PROCEDURE 2. REACTION WITH LUCAS REAGENT

Place 1 mL of Lucas reagent in each of two test tubes and adjust the temperature to 26–27°, with a water bath if necessary. To one test tube, add 5 drops of 2-butanol. To the second test tube, add 5 drops of liquid unknown or about 250 mg of solid unknown and mix with a spatula. Homogeneous solutions should result. Record the times of mixing and allow the test tubes to stand undisturbed. Record the times at which the entire contents of each tube become cloudy. For 2-butanol, the elapsed time should be 1–2 minutes. Other secondary alcohols should undergo reaction within 1–5 minutes; tertiary alcohols undergo more rapid reaction; and primary alcohols give no sign of reaction within 10 minutes.

PROCEDURE 3. REACTION WITH CONCENTRATED HYDROCHLORIC ACID

Use the procedure described in part 1, substituting conc. hydrochloric acid for the Lucas reagent and *t*-butyl alcohol for 2-butanol. *t*-Butyl alcohol should form a homogeneous solution, then give a positive test in 2–3 minutes. If the unknown is a tertiary alcohol, it should have a reaction time approximately the same as that of *t*-butyl alcohol. A secondary alcohol will not undergo reaction in this period of time.

EXPERIMENTAL NOTES

1) *Preparation of the Lucas reagent.* In a fume hood, dissolve 77 g of anhydrous $ZnCl_2$ in 50 mL of chilled conc. HCl. (CAUTION: The dissolution is exothermic and a considerable amount of gaseous HCl is given off.)

It is imperative that the $ZnCl_2$ be anhydrous. Commercial anhydrous $ZnCl_2$ is sold in the form of sticks. If the sticks have not been exposed to atmospheric moisture, they are probably suitable.

2) Both zinc chloride and the Lucas reagent are extremely hygroscopic; therefore, the Lucas reagent will eventually become inactive from the absorption of atmospheric moisture. This decrease in activity is common in student laboratories where the reagent bottle is used by several students.

G. Preparation of Derivatives: 2,4-Dinitrophenylhydrazone of an Aldehyde or Ketone

The reaction of 2,4-dinitrophenylhydrazine with aldehydes and ketones was discussed in part A of this experiment. The discussion here deals with the use of this reagent to prepare a derivative of an aldehyde or a ketone. The procedure is generalized so that it can be used for a variety of aldehydes and ketones; however, keep in mind that no general procedure will work for *every* aldehyde and ketone. You must be observant and, when necessary, alter the procedure to fit your particular case.

The procedure for preparing a DNP derivative uses HCl instead of H_2SO_4 as the acidic catalyst because HCl is easier to remove from the final product. If you have trouble forming the DNP, you may wish to consider using the stronger acid sulfuric acid as the catalyst.

It is difficult to separate the starting 2,4-dinitrophenylhydrazine from the DNP product by recrystallization. Because purity is more important than yield in this experiment, the carbonyl compound should be used in molar excess.

This is difficult to do accurately because you will not know the exact molecular weight of your unknown; however, you can make a reasonable estimate of the molecular weight by comparing the physical constants of your unknown with those of known compounds. The general procedure is based on the assumption that an unknown has a molecular weight of 100 or less. If you think your unknown has a substantially higher molecular weight, adjust the proportions of reactants accordingly.

The precipitated DNP is filtered and washed thoroughly to remove the HCl. The crude DNP can be air-dried overnight and its melting point determined before crystallization. (The melting range of the crude product compared to that of the crystallized product provides an estimate of the purity of the crystallized DNP.)

An accurate melting point is necessary for identification purposes; therefore, the DNP should be crystallized to a constant melting point. Usually one recrystallization is sufficient; however, in some cases, two or three crystallizations may be necessary. While recrystallization may seem tedious, it is the only assurance of a sharp and correct melting point.

Evaluation of the melting point of a DNP is complicated by the fact that some DNP's form *polymorphs* (different crystalline forms) with different melting points. One method of checking for polymorphism is to allow the melt in the melting-point capillary to solidify, then redetermine the melting point. Assuming that no decomposition has occurred, a change in the melting point is suggestive of polymorphism. Also, some DNP's exhibit different melting points when different acidic catalysts are used in the procedure. For example, acetaldehyde yields a DNP with a melting point of 147° with H_2SO_4 as the catalyst and a DNP with a melting point of 164° with HCl as the catalyst. It is believed that these DNP's are *syn* and *anti* isomers.

Limitations. Unfortunately, not all DNP's are solid. A few open-chain ketones, such as 2-decanone or 6-undecanone (11 carbons), are oils. Ketones of this type are extremely difficult to characterize by the procedures outlined in this introductory chapter.

Difunctional carbonyl compounds, such as α-hydroxy carbonyl compounds, α,β-unsaturated carbonyl compounds, and many dicarbonyl compounds, undergo more-complex reactions than simple hydrazone formation. If you think your unknown is difunctional, check the literature.*

Experimental

EQUIPMENT:

disposable plastic gloves
two 25-mL or 50-mL Erlenmeyer flasks
10-mL graduated cylinder
hot plate or steam bath
vacuum or gravity filtration assembly

* T. D. Binns and R. Brettle, *J. Chem. Soc., Org.* **1966**, 341; F. R. Curtis, C. H. Hassally, and J. Weatherston, *J. Chem. Soc.* **1962**, 3831; L. A. Jones, C. K. Hancock, and R. B. Seligman, *J. Org. Chem.* **1961**, *26*, 228.

CHEMICALS:

2,4-dinitrophenylhydrazine, 0.5 g
95% ethanol, 10 mL
95% ethanol or ethyl acetate, 10–30 mL
conc. hydrochloric acid, 1 mL
unknown, about 0.25 g

TIME REQUIRED: About 40 minutes to crude product, 30 minutes for crystallization, plus time for air-drying the product and determining melting points

>>>> **SAFETY NOTE** 2,4-Dinitrophenylhydrazine is a potential explosive and may be carcinogenic. Read the Safety Note on page 373!

The product DNP's may be toxic and/or carcinogenic. Take appropriate precautions.

PROCEDURE

In a 50-mL Erlenmeyer flask (see Experimental Note 1), place 0.5 g of 2,4-dinitrophenylhydrazine, 10 mL of 95% ethanol, and 1 mL of conc. hydrochloric acid. Warm the mixture until the reagent dissolves, then add 5 drops or 0.25 g of the unknown (see Experimental Note 2). Heat the mixture to almost boiling, then add water dropwise until the reaction mixture is just turbid.

Let the mixture slowly cool to room temperature (see Experimental Notes 3 and 4). Filter the solid product and wash it with 50% ethanol–water. Recrystallize to a constant melting point from 95% ethanol, ethanol–water, or ethyl acetate (see Experimental Note 5).

EXPERIMENTAL NOTES

1) Do not dry your glassware with acetone. Residual acetone (liquid or vapor) will react with the 2,4-dinitrophenylhydrazine to yield a DNP, mp 126°.

2) The amount of unknown suggested in the procedure assumes a molecular weight of 100 or less. If the molecular weight of your unknown is greater than 100, use more carbonyl compound. For example, if your unknown has a molecular weight of about 200, then you would want to add at least 0.5 g.

3) Because the addition of water may cause some coprecipitation of the reagent, adding water should be avoided if possible. If you observed an immediate precipitation in the DNP classification test, simply allow the reaction mixture to cool to room temperature without adding water. If the DNP does not precipitate, then rewarm the mixture and add the water as described.

4) Low-melting hydrazones frequently form oils instead of solids. If your DNP oils out, let the reaction mixture stand overnight. Some hydrazones solidify very slowly. If, after an overnight standing, an oil is still present, decant the supernatant liquid, add water, heat the mixture, and then set the flask aside for another overnight period.

5) If there is any question in the final identification of the unknown, recrystallize the DNP from a second solvent to see if its melting point changes.

High-melting DNP's are sparingly soluble in most solvents and thus are difficult to crystallize. In such a case, one of two procedures can be used. The easiest is to boil the solid with 95% ethanol to leach out impurities and then filter the hot mixture. A better procedure is to recrystallize a small amount of the DNP from a large volume of solvent.

H. Preparation of Derivatives: Semicarbazone of an Aldehyde or Ketone

The reaction of an aldehyde or ketone with semicarbazide to yield a semicarbazone is similar to the reaction leading to a DNP.

$$\underset{\substack{\text{semicarbazide}\\\text{hydrochloride}}}{H_2N\overset{O}{\overset{\|}{C}}NHNH_2 \cdot HCl} + \underset{\substack{\text{an aldehyde}\\\text{or ketone}}}{R\overset{O}{\overset{\|}{C}}R} + CH_3CO_2Na \xrightarrow{CH_3CO_2H}$$

$$\underset{\text{a semicarbazone}}{H_2N\overset{O}{\overset{\|}{C}}NHN{=}CR_2} + H_2O + NaCl + CH_3CO_2H$$

Semicarbazones are often more difficult to isolate than DNP's; for this reason, a DNP is usually the derivative of choice.

Semicarbazide is obtained as the stable hydrochloride salt. Under the conditions of the reaction (acetic acid–sodium acetate buffer), the hydrochloride is partly converted to the free base, which is the active reagent in the mixture.

$$H_2N\overset{O}{\overset{\|}{C}}NH\overset{+}{N}H_3\ Cl^- + CH_3\overset{O}{\overset{\|}{C}}O^-\ Na^+ \xrightarrow{CH_3CO_2H} H_2N\overset{O}{\overset{\|}{C}}NHNH_2 + NaCl + CH_3\overset{O}{\overset{\|}{C}}OH$$

The work-up for a semicarbazone is similar to that for a DNP. After the reaction period, the mixture is diluted with water. If a solid precipitates, the semicarbazone has been prepared successfully. If an oil separates, attempts should be made to solidify it before assuming the reaction has failed. Chilling and scratching the side of the flask with a glass rod may cause the oil to solidify. Washing the oil and decanting the wash water may remove enough impurities for the oil to solidify. Allowing the washed oil to evaporate to dryness can be tried as a last resort.

The solid product is filtered and washed. The residual acetic acid must be scrupulously removed; semicarbazones decompose upon standing if a trace of acid is present. As in the case of the DNP, the crude semicarbazone must be recrystallized to a constant melting point.

Limitations. Most carbonyl compounds react rapidly and give crystalline products when the reaction mixture is warmed. However, aldehydes of five or fewer carbons may take days for reaction (or may undergo other reactions); therefore, preparation of semicarbazones of these compounds is not recommended.

Hindered ketones may require a longer reaction time than that suggested in the experimental procedure; however, when a semicarbazide solution is heated for a lengthy period of time, it may undergo reaction to yield biurea.

$$2\,NH_2\overset{\overset{O}{\|}}{C}NHNH_2 \xrightarrow[>1\text{ hr}]{\text{heat}} \underset{\substack{\text{biurea}\\ \text{mp } 245\text{–}250^\circ\text{d}}}{H_2N\overset{\overset{O}{\|}}{C}NHNH\overset{\overset{O}{\|}}{C}NH_2}$$

Unlike the semicarbazones of most ketones, biurea is soluble in hot water and can thus be removed from the semicarbazone.

The general procedure given in the experimental section (a hot acidic solvent system) is not satisfactory for all aldehydes and ketones. Experimental Note 1 lists a few alternative procedures.

Experimental

EQUIPMENT:

disposable plastic gloves
two 25-mL or 50-mL Erlenmeyer flasks
10-mL graduated cylinder
hot plate, steam bath, or hot water bath
ice bath
stirring rod
vacuum or gravity filtration assembly

CHEMICALS:

glacial acetic acid, 3 mL
methanol or 95% ethanol, 5–20 mL
semicarbazide hydrochloride, 0.50 g
sodium acetate, 0.4 g
unknown aldehyde or ketone, 0.5 g

TIME REQUIRED: about one hour to the crude product; $\frac{1}{2}$ hour for each crystallization; overnight drying of the crude and purified product

>>>> ***SAFETY NOTE*** Semicarbazide is a known carcinogen in laboratory animals. Use appropriate precautions.

PROCEDURE

In a 50-mL Erlenmeyer flask in the fume hood, place 0.50 g of semicarbazide hydrochloride (mp 175–185°), 0.5 mL of water, and 0.4 g of sodium acetate. Stir the mixture and crush any lumps of sodium acetate. Then add 0.5 g of the aldehyde or ketone and 3 mL of glacial acetic acid. (For alternative reaction conditions, see Experimental Note 1.)

Warm the mixture (on a steam bath, hot plate, or in a hot water bath) in the fume hood for 15–20 minutes. Do not boil the solution. Cool the solution to room temperature or chill it in an ice bath, then add 7 mL of water. Filter the precipitate by gravity or vacuum. Wash the filter cake thoroughly with three to five 5-mL portions of cold water (see Experimental Note 2). Air-dry the crude product and determine its melting point. Recrystallize to a constant melting point from water, methanol, 95% ethanol, or ethanol–water.

EXPERIMENTAL NOTES

1) (a) *For water-soluble aldehydes or ketones,* using 5 mL of water (instead of acetic acid) as the reaction solvent may give superior results. (The semicarbazones of these compounds can usually be recrystallized from water.)

(b) *For acid-sensitive compounds,* a reaction mixture consisting of 0.5 g of aldehyde or ketone, 0.5 g semicarbazide hydrochloride, 0.5 mL of pyridine (*volatile and suspected carcinogen*; *use the fume hood*!) in 5 mL of 95% ethanol may be tried.

(c) Alternatively, use 0.5 g carbonyl compound, 0.5 g semicarbazide hydrochloride, and 0.75 g sodium acetate in 5 mL of water, 95% ethanol, or methanol.

If an alcohol solvent is used and the semicarbazone does not precipitate, boil off some of the solvent in the fume hood before adding water, or repeat the experiment using only enough alcohol to dissolve the carbonyl compound.

2) If acetic acid cannot be removed from the semicarbazone, repeat the experiment using one of the preceding procedures.

I. Preparation of Derivatives: 3,5-Dinitrobenzoate of an Alcohol or a Phenol

The 3,5-dinitrobenzoate esters are excellent derivatives for identifying alcohols and phenols. The esters are often bright yellow crystalline solids that can be easily distinguished from the starting materials.

$$(O_2N)_2C_6H_3\text{–}\overset{O}{\overset{\|}{C}}Cl + ROH + C_5H_5N \longrightarrow (O_2N)_2C_6H_3\text{–}\overset{O}{\overset{\|}{C}}OR + C_5H_5NH^+\ Cl^-$$

3,5-dinitrobenzoyl chloride; *an alcohol or phenol*; pyridine; a 3,5-dinitrobenzoate; pyridine hydrochloride

In the esterification, the acid chloride of 3,5-dinitrobenzoic acid is used instead of the acid itself because the acid chloride is more reactive. Primary alcohols react with 3,5-dinitrobenzoyl chloride within a few minutes; secondary alcohols require about 20 minutes; and tertiary alcohols and phenols require about an hour.

Pyridine is added to remove HCl from the reaction mixture. Without pyridine, the HCl formed would undergo reaction with the alcohol to yield alkyl halides and alkenes.

$$\mathrm{Ar\overset{\overset{O}{\|}}{C}Cl + ROH \rightarrow Ar\overset{\overset{O}{\|}}{C}OR + \underset{\textit{can react with ROH}}{HCl}}$$

Although the reaction can be carried out in the presence of water, the work-up is simplified if the reactants are dry. If water is present, some acid chloride is converted to 3,5-dinitrobenzoic acid, which must be removed from the product by washing with base. (Small amounts of the acid are formed inevitably because the acid chloride reacts with atmospheric moisture.)

Experimental

EQUIPMENT:

disposable plastic gloves
25-mL or 50-mL Erlenmeyer flask
graduated cylinder
gravity or vacuum filtration assembly
hot plate or steam bath
stirring rod

CHEMICALS:

95% ethanol, 2–10 mL
3,5-dinitrobenzoyl chloride, 0.5 g
5% hydrochloric acid, 10–25 mL
dry pyridine, 2 mL
5% sodium bicarbonate solution, 10–25 mL
unknown, 0.5 g
(Additional chemicals may be needed; see the Experimental Note.)

TIME REQUIRED: 10–30 minutes to the crude product; 30 minutes for crystallization, and time for melting-point determinations

>>>> ***SAFETY NOTE*** Pyridine is a suspected carcinogen and has a strong odor. 3,5-Dinitrobenzoyl chloride is a lachrymator. Use a fume hood!

PROCEDURE

Place 2 mL of dry pyridine in a 50-mL Erlenmeyer flask. Add 0.5 g of the unknown and 0.5–1.0 g of 3,5-dinitrobenzoyl chloride (see Experimental Note 1). Mix the reactants, wait until the exothermic reaction subsides, then heat the flask on a hot plate or steam bath for 10 minutes or longer (see Experimental Note 2). Cool the mixture to room temperature and pour it into 20 mL of water with stirring. The insoluble ester should solidify (see Experimental Note 3).

Filter the product; wash it thoroughly with 5% HCl to remove the pyridine and then with 5% $NaHCO_3$ (or 2% NaOH for a phenyl ester) to remove the HCl and 3,5-dinitrobenzoic acid (see Experimental Note 4). Finally, wash the ester with water. Recrystallize the product to a constant melting point from 95% ethanol or ethanol–water.

EXPERIMENTAL NOTES

1) When exposed to the air, 3,5-dinitrobenzoyl chloride is hydrolyzed by atmospheric moisture to 3,5-dinitrobenzoic acid, which can react with the excess acid chloride to yield the anhydride. The anhydride is less reactive than the acid chloride toward alcohols and phenols.

$$(O_2N)_2C_6H_3\text{-}\overset{O}{\overset{\|}{C}}Cl + H_2O \xrightarrow{-2HCl} (O_2N)_2C_6H_3\text{-}\overset{O}{\overset{\|}{C}}O\overset{O}{\overset{\|}{C}}\text{-}C_6H_3(NO_2)_2$$

mp 74° mp 109°

The hydrolysis problem is particularly acute in a large student laboratory where many students use the same reagent bottle. Therefore, the melting point of the acid chloride should be checked before it is used. If the melting point is not between 68° and 76°, the acid chloride should be recrystallized from CCl_4 (*toxic and suspected carcinogen*), using a hot filtration. Alternatively, prepare fresh acid chloride by the following procedure:

Preparation of 3,5-dinitrobenzoyl chloride. In the hood, mix 1.0 g of 3,5-dinitrobenzoic acid, one drop of pyridine, and 3 mL of thionyl chloride (*strong irritant*) in a 50-mL round-bottom flask. Heat the mixture at reflux until all the carboxylic acid has dissolved. Continue heating for an additional 10 minutes (a total of 20–40 minutes).

Fit the flask with a suction apparatus attached to a water aspirator and continue heating until the bulk of the thionyl chloride has been removed. The residue is the acid chloride, which may be crystallized from CCl_4.

2) If the unknown is a secondary alcohol, use a 20-minute heating period. For a phenol or a 3° alcohol, heat for one hour.

3) The 3,5-dinitrobenzoates of some phenols separate as oils when water is added. Pour off the aqueous layer from the oil. Add 4–5 mL of diethyl ether and wash the solution with water, 2% aqueous NaOH, and another portion of water. Evaporate the solvent and crystallize the residue.

4) Prolonged treatment with aqueous base will cause saponification of the ester.

J. Preparation of Derivatives: α-Naphthylurethane of an Alcohol or a Phenol

Isocyanates (RN=C=O) are attacked by nucleophiles to yield addition products. The addition product of α-naphthylisocyanate with an alcohol or a phenol is an *alkyl* or *aryl α-naphthylurethane* (*urethan, carbamate*). These compounds are usually solids.

Nu⁻ attacks here

$$\text{N}{=}\overset{\delta+}{\text{C}}{=}\overset{\delta-}{\text{O}} \text{ (on naphthalene)} + \text{ROH} \longrightarrow \text{NH}\overset{\text{O}}{\overset{\|}{\text{C}}}\text{OR} \text{ (on naphthalene)}$$

α-naphthylisocyanate; an alcohol or phenol; an alkyl or aryl α-naphthylurethane

The reagents and glassware used in preparing a urethane derivative must be dry because water also reacts with isocyanates. In the case of α-naphthylisocyanate, the reaction with water ultimately yields *N,N′*-di-α-naphthylurea (mp 297°), which can be troublesome to remove from the desired product.

$$\underset{\text{an arylisocyanate}}{\text{ArN=C=O}} \xrightarrow{H_2O} \text{ArNHCO}_2\text{H} \xrightarrow{-CO_2} \underset{\substack{\text{an arylamine}\\ \text{(a nucleophile)}}}{\text{ArNH}_2} \xrightarrow{\text{ArN=C=O}} \underset{\text{an } N,N'\text{-diarylurea}}{\text{ArNH}\overset{\text{O}}{\overset{\|}{\text{C}}}\text{NHAr}}$$

The rate of reaction of alcohols with an isocyanate is 1° > 2° > 3°. *Primary alcohols* usually react spontaneously. *Secondary alcohols* may require a Lewis acid catalyst and/or heating. *Tertiary alcohols* may undergo elimination to yield an alkene and water (which reacts rapidly with the isocyanate to yield the diarylurea). Sometimes, using sodium acetate as a catalyst and heating the reaction mixture for a lengthy period of time allows the urethane of a tertiary alcohol to be prepared. Most *phenols* yield urethanes readily.

Limitations. Potential problems in preparing the α-naphthylurethane of a tertiary alcohol have already been mentioned.

Phenols with an electron-withdrawing group, or those that can undergo intramolecular hydrogen bonding, such as some nitrophenols, do not form urethanes readily. Using pyridine as a catalyst and heating the mixture allows the urethanes of some of these phenols to be prepared. Phenols with more than one electron-withdrawing group, such as 2,4,6-trinitrophenol (picric acid), sterically hindered phenols, and polyhydroxyphenols do not react with isocyanates. Aminophenols react at the amino group instead of at the hydroxyl group.

Experimental

EQUIPMENT:

dropper
25-mL or 50-mL Erlenmeyer flask
filtration assembly
10-mL graduated cylinder
hot plate
stirring rod

CHEMICALS:

anhydrous aluminum chloride, 50 mg (if necessary)
α-naphthylisocyanate, 1–3 g
petroleum ether (ligroin) (bp 90–110°) or heptane, 5–25 mL
pyridine, 2–20 drops (for Procedure 3)
sodium acetate, 0.1 g (for Procedure 2)
unknown, 0.5–1.0 g (2–3 g for Procedure 2)

TIME REQUIRED: 15–30 minutes for a 1° or 2° alcohol, 5–$5\frac{1}{2}$ hours for a 3° alcohol, or 30–45 minutes for a phenol; 30 minutes for crystallization; time for drying and melting-point determination

>>>> ***SAFETY NOTE*** α-Naphthylisocyanate is a lachrymator. Transfer this reagent and carry out the reaction *in the fume hood.*

PROCEDURE 1 (FOR PRIMARY AND SECONDARY ALCOHOLS)

Place 0.5–1.0 g of the unknown in a small *dry* Erlenmeyer flask, along with an equal quantity of α-naphthylisocyanate (see Experimental Note). Allow the mixture to stand a few minutes at room temperature until the spontaneous exothermic reaction subsides. If no spontaneous reaction occurs, add about 50 mg of anhydrous aluminum chloride (*strong irritant*). If reaction still does not occur, warm the mixture on a hot plate for 10–30 minutes. (Do not overheat.) (If anhydrous $AlCl_3$ is not available, the urethane may form when the mixture is heated without $AlCl_3$.) At the end of the reaction period, the mixture should be thick.

Add about 5 mL of petroleum ether (bp 90–110°) or heptane to the cooled reaction mixture. Chill and scratch the sides of the flask with a glass rod to induce crystallization. Filter, wash the solid with cold solvent, and recrystallize from petroleum ether. If the solid does not all dissolve, remove the residue (the diarylurea, insoluble in hot petroleum ether) by a hot filtration.

PROCEDURE 2 (FOR TERTIARY ALCOHOLS)

Mix 2–3 g of the tertiary alcohol, an equivalent amount of α-naphthylisocyanate, and 0.1 g of anhydrous sodium acetate. Warm the mixture 4–5 hours on a hot plate. At the end of the reaction period, proceed as described in Procedure 1. (The reaction mixture will contain a considerable amount of the diarylurea.)

PROCEDURE 3 (FOR PHENOLS)

Follow Procedure 1, except also add 2–20 drops of pyridine (*strong odor and suspected carcinogen*) and warm the reaction for 20–30 minutes. (Do *not* add $AlCl_3$.) Urethanes of phenols are sometimes crystallized from methanol, rather than from petroleum ether. If methanol is used, be sure all the α-naphthylisocyanate is removed by washing the crude product with petroleum ether; otherwise, the methanol will undergo reaction.

EXPERIMENTAL NOTE

Repeated exposure of liquid α-naphthylisocyanate to atmospheric moisture results in the formation *N,N'*-di-α-naphthylurea, which appears as a white solid in the liquid reagent and on the lip of the reagent bottle. In this case, wipe off the lip of the bottle and use only the clear supernatant liquid.

K. Preparation of Derivatives: Aryloxyacetic Acid of a Phenol

The substitution reaction of a phenoxide with chloroacetic acid yields a solid aryloxyacetic acid.

$$\underset{\textit{a phenol}}{ArOH} \xrightarrow[H_2O]{NaOH} \underset{\textit{a phenoxide}}{ArO^- \ Na^+} \xrightarrow[-Cl^-]{ClCH_2CO_2H} ArOCH_2CO_2^- \ Na^+ \xrightarrow{HCl} \underset{\textit{an aryloxyacetic acid}}{ArOCH_2CO_2H}$$

An alkoxyacetic acid of an alcohol cannot be made by this procedure.

As may be seen in the flow equation, treatment of the phenol with aqueous NaOH and chloroacetic acid (water-soluble) yields the aryloxyacetate salt. The free acid (slightly soluble in water) is obtained by acidifying the reaction mixture. The product is crystallized and characterized by melting point. If desired, the neutralization equivalent of the substituted acid may be determined by titration.

Limitations. Nitrophenols form aryloxyacetic acids with difficulty. If it is necessary to derivatize a nitrophenol, the reaction time should be increased to 3–6 hours. Sterically hindered phenols, such as 2,6-dimethylphenol, undergo reaction very slowly. In such a case, a large excess (2–5 times the recommended amount) of chloroacetic acid should be used.

Experimental

EQUIPMENT:

disposable plastic gloves
two 125-mL Erlenmeyer flasks
filtration assembly
10-mL graduated cylinder
hot plate

CHEMICALS:

40% aqueous sodium hydroxide, 10 mL
chloroacetic acid, 3 g 6*M* hydrochloric acid, 20–60 mL
unknown, 1 g

TIME REQUIRED: 15–20 minutes, plus time for crystallization and melting-point determination

»» **SAFETY NOTE** The strong sodium hydroxide solution is extremely caustic. Chloroacetic acid ($pK_a = 2.8$) is a fairly strong acid and can cause burns. Wear plastic gloves and take all appropriate precautions!

PROCEDURE

Place 10 mL of 40% aqueous sodium hydroxide solution in a 125-mL Erlenmeyer flask. Add 1.0 g of the phenol and 3.0 g of chloroacetic acid (see Experimental Note 1). Add a boiling chip, heat the solution cautiously to boiling (*it may froth!*), and continue heating with swirling for 5 minutes (see Experimental Note 2).

Cool the solution to room temperature, dilute with 10 mL of water, and acidify to about pH 3 (pH paper or Congo red paper) with 6*M* HCl. Filter the solid product and crystallize from hot water or ethanol–water (see Experimental Note 3).

EXPERIMENTAL NOTES

1) The reaction is sensitive to the concentration of base: in 30–50% aqueous NaOH, reaction generally occurs readily, but the reaction does not proceed in 10% NaOH.

2) Because of the danger of frothing, hold the flask with a clamp, not with your fingers, and swirl it on the hot plate. An alternative procedure is to heat the flask in a hot water bath with occasional swirling for one hour.

3) If the product does not precipitate from the acidified solution, it may be extracted with diethyl ether; then the ether solution may be dried, and the ether evaporated.

Problems

21.4 Using the chemical tests outlined in this chapter, how would you distinguish between the following pairs of compounds?

(a) HO–C_6H_4–C_6H_5 and C_6H_5–C_6H_4–CH_2OH

(b) $CH_3COCH_2COCH_3$ and CH_3COCH_2CHO

(c) 2-cyclohexen-1-one (=O) and 2-cyclohexen-1-ol (–OH)

(d) $CH_3CH_2CH{=}CHCH_2OH$ and $CH_3(CH_2)_4OH$

(e) 2,4-dinitrophenol (O_2N–, NO_2, –OH) and H_3C–C_6H_4–OH

21.5 What is the purpose of the pyridine in the alternative procedure for semicarbazone preparation described in Experimental Note 1(b), page 388?

21.6 Suggest a mechanism for the reaction of CH_3OH with α-naphthylisocyanate.

21.7 Suggest reasons for the following observations. (Use formulas or equations in your answers where appropriate.)

(a) An amine can destroy the Schiff reagent.

(b) Hydroquinone (1,4-dihydroxybenzene) gives a positive Tollens test.

(c) Benzyl alcohol gives a Lucas test result similar to that of a tertiary alcohol.

(d) p-Nitrophenol does not undergo reaction with α-naphthylisocyanate.

(e) p-Nitrophenol does not react readily with chloroacetic acid.

CHAPTER 22

Supplemental Techniques

The three experiments in this chapter illustrate three techniques used to separate mixtures or purify compounds. These techniques are **vacuum distillation**, **column chromatography**, and **sublimation**. They are included as supplemental techniques because they require equipment not usually found in student lockers, are time-consuming, or are of limited utility. Because you may not perform all these experiments, the description of each is included in the experimental discussion.

EXPERIMENT 22.1

Vacuum Distillation of Benzaldehyde

Discussion

A vacuum distillation is conducted in a manner similar to a simple or fractional atmospheric distillation. The theory of vacuum distillation is discussed in Section 5.2. Here, we will discuss only the practical aspects of simple vacuum distillation.

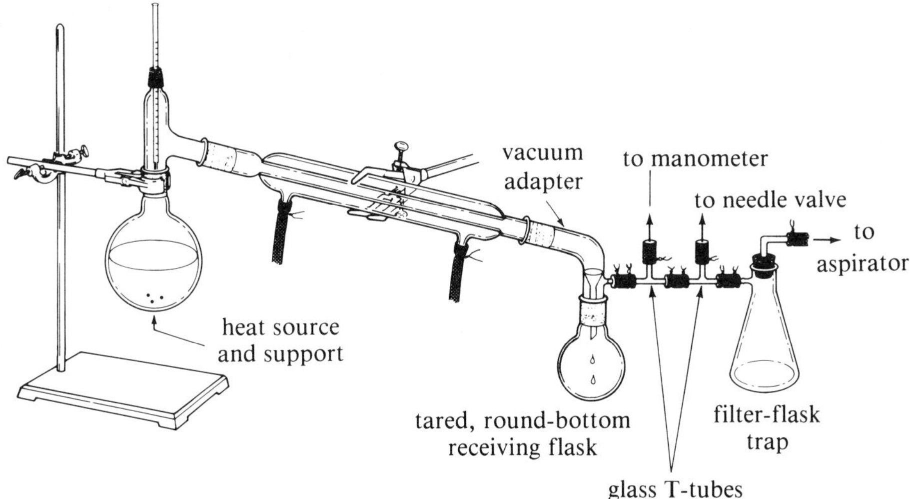

Figure 22.1 A simple vacuum distillation assembly. Each flask and the condenser are clamped in place (clamps not shown). All connecting heavy-walled rubber tubing is wired down.

Distillation Apparatus. Except for the adapter, a vacuum distillation apparatus is almost the same as that for atmospheric distillation (see Figure 22.1). The principal difference is that a vacuum distillation apparatus must be air-tight, even under vacuum. Also, any glassware used must be able to withstand being evacuated; for example, large, thin-walled Erlenmeyer flasks cannot be used. Under vacuum, most solvents boil well below room temperature; therefore, any solvent in the material to be distilled must be removed prior to a vacuum distillation.

When assembling a vacuum distillation apparatus, use a special vacuum grease to seal all glass-to-glass connections. (Stopcock grease flows too easily under heat and vacuum, and an air leak would be likely to develop during the distillation.) Use only heavy-walled vacuum tubing. Seal all rubber-tubing connections (including a rubber thermometer adapter) with a twist of wire tightened with pliers. Assemble the apparatus completely and test for air leaks by applying the vacuum before placing the compound to be distilled in the distillation flask.

Because ordinary boiling chips do not function well in a vacuum distillation, special boiling chips with smaller pores must be used to prevent bumping. Alternatively, a *very* fine capillary may be used to introduce a thin stream of air or N_2 bubbles to the boiling liquid (see Figure 22.2).

Vacuum Adapter and Fraction Collectors. The distillation in Experiment 22.1 is a simple distillation: the separation of volatile benzaldehyde from nonvolatile benzoic acid. Therefore, a simple vacuum adapter will suffice (see Figure 1.3, page 12). A simple adapter requires that the forerun and the main fraction be collected in the same receiver unless the

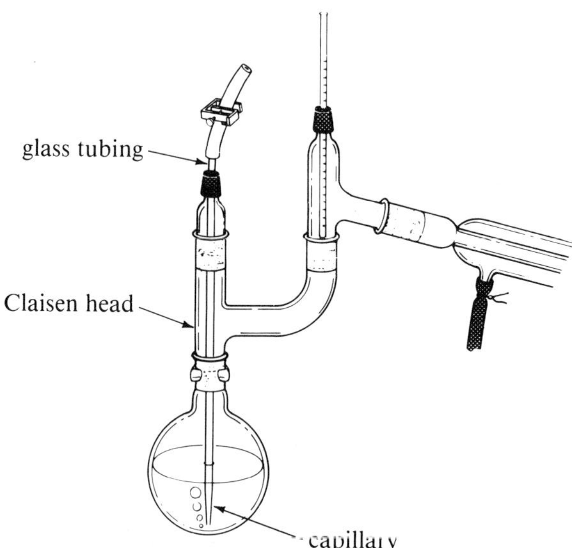

Figure 22.2 *A distillation flask equipped with a capillary tube. When the apparatus is evacuated, a thin stream of air bubbles is pulled though the capillary, which prevents bumping.*

distillation is stopped, the apparatus cooled, and the vacuum broken. To avoid this operation, if more than one fraction is to be collected, a vacuum fraction collector should be used. Three types of vacuum fraction collectors are shown in Figure 22.3.

Source of the Vacuum. A vacuum can be obtained by using either a mechanical pump or a water aspirator. Mechanical pumps must be used for distillations at pressures of less than 20 mm Hg. If a mechanical pump is to be used in this experiment, your instructor will demonstrate its use and the traps that are necessary to protect the pump. For pressures of 20 mm Hg and above (as in this experiment), a water aspirator is usually satisfactory.

One principal disadvantage of an aspirator results from fluctuating water pressure. If the water pressure drops, the vacuum in the distillation apparatus can suck water from the aspirator into the receiving flask. To ensure that such an event does not occur, position a water trap between the aspirator and the distillation apparatus (see Figure 22.1). This trap is similar to the trap for vacuum filtration except that a one-hole stopper is used.

Pressure Regulation and Measurement. Place a needle valve (or other pressure-regulation device) and a manometer (a pressure-measuring device) between the water trap and the vacuum adapter. (A typical assembly using two T-tubes is shown in Figure 22.1.) The needle valve allows you to adjust the pressure to the desired level by bleeding air into the system. If a needle valve is not available, use the base of a Fischer or Meker burner. Connect the vacuum tubing to the gas inlet and adjust the vacuum with the valve at the base of the burner normally used to adjust the height of the flame. Pressure

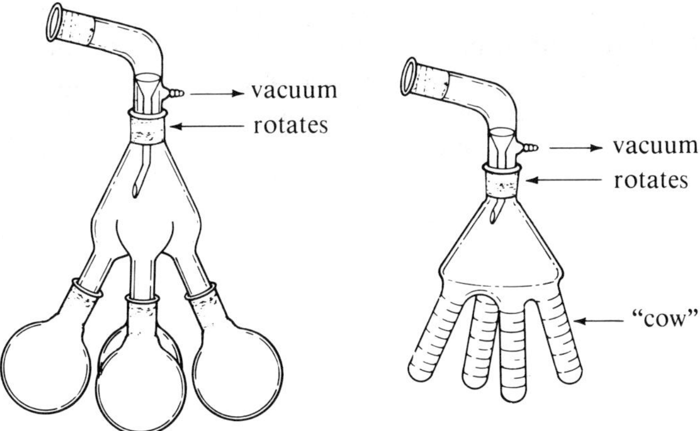

Figure 22.3 *Vacuum adapters and receivers that allow the collection of more than one fraction during the course of a distillation.*

control with a needle valve or burner is satisfactory in distillations conducted at a pressure greater than 20 mm Hg. A closed-end manometer, as shown in Figure 22.4, is suitable for measuring pressures above 20–25 mm Hg.

Vacuum Distillation of Benzaldehyde. When benzaldehyde or some other aldehyde is exposed to air, it undergoes auto-oxidation, yielding a carboxylic acid as the final product:

$$C_6H_5CHO \xrightarrow{O_2} C_6H_5COOOH \xrightarrow{C_6H_5CHO} 2\ C_6H_5COOH$$

benzaldehyde — peroxybenzoic acid — benzoic acid

For this reason, commercial aldehydes (especially those in previously opened

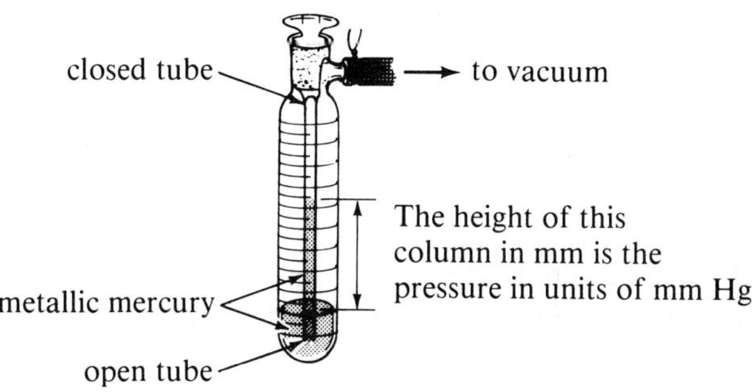

Figure 22.4 *A typical closed-tube manometer.*

bottles) are generally impure. If a pure aldehyde is desired, the material should be distilled or otherwise purified just prior to use. Benzaldehyde boils at 178° at atmospheric pressure, a temperature too high for an effective atmospheric distillation. Consequently, a vacuum distillation must be used in the distillation of this compound. A simple distillation is suitable because benzoic acid is not volatile. As an optional experiment, the distillation residue can be extracted with sodium hydroxide solution, acidified, and the solid identified as benzoic acid.

Experimental

EQUIPMENT:

A. Vacuum Distillation of Crude Benzaldehyde
carbon boiling chips (Fischer Scientific, #09-191-12); Micro-Porous boiling chips (Todd Scientific); or equivalent
heating mantle and rheostat
heavy-walled tubing with wire and pliers
manometer
needle valve
pinch clamp
simple distillation assembly (see Experiment 5.1, page 92)
trap for aspirator
two T-tubes
vacuum adapter
vacuum distillation grease

B. Isolation of Benzoic Acid (optional)
two 125-mL Erlenmeyer flasks
50-mL graduated cylinder
ice bath
melting-point apparatus
125-mL separatory funnel
spatula
vacuum filtration assembly

CHEMICALS:

Part A
crude benzaldehyde, 50 mL (about 52 g)

Part B
diethyl ether, 50 mL
conc. hydrochloric acid, 15 mL
10% sodium hydroxide solution, 50 mL

TIME REQUIRED: about three hours for part A; one-half hour for part B plus time to dry the crystals and determine the melting point

››› ***SAFETY NOTE 1*** Whenever vacuum is applied to a closed system, there is the danger of an implosion. *Safety glasses must be* *worn at all times*! Before setting up the vacuum distillation apparatus, check all glassware for stars or cracks. (Only round-bottom flasks, pressure flasks, or Erlenmeyer flasks of 50-mL size or smaller should ever be used in a vacuum distillation.)

>>>> ***SAFETY NOTE 2*** Benzaldehyde is a skin irritant and also has a penetrating, irritating odor. Weigh this compound in a well-ventilated area and clean any spills carefully. If some benzaldehyde spills on your skin, wash immediately with soap and water.

PROCEDURE

A. Vacuum Distillation of Benzaldehyde

Assemble a simple vacuum distillation apparatus as shown in Figure 22.1 and described in the discussion, using a tared receiving flask. With an empty distillation flask, test the seals by applying vacuum with the needle valve closed. You should be able to obtain a pressure of approximately 20 mm Hg with an aspirator. If you cannot obtain this low pressure, use the pinch clamp to isolate various parts of the system to find the leak(s).

After checking the empty system, break the vacuum *slowly* (see Experimental Note 1) and place 50 mL of crude benzaldehyde in the tared distillation flask, along with a few carbon boiling chips (see Safety Note 2). Apply full vacuum with the needle valve completely closed. You should be able to attain the same low pressure (or near to it) as when the distillation flask was empty. If not, a new leak has developed, and it will be necessary to seal it before proceeding.

Adjust the needle valve on the system so that the pressure is 40–50 mm Hg, then let the system stand for a few minutes until the pressure is no longer changing. Heat the mixture to a boil with a heating mantle, and distil the benzaldehyde. There should be only one fraction. (If the distillation is stopped and then resumed, fresh boiling chips should be added to the mixture. How to stop a vacuum distillation is discussed at the end of this procedure and in Experimental Note 1.)

A graph of the boiling part of the benzaldehyde versus pressure is shown in Figure 22.5. The boiling point of your benzaldehyde should fall near the curve on this graph. A boiling point determined at reduced pressure must include the pressure, for example, bp 90.0° (35.0 mm Hg).

The amount of benzaldehyde that distils will vary, depending upon the amount of benzoic acid present in the crude material. As the distillation proceeds, the distillation residue will become darker yellow. At the end of the distillation, drop the heating mantle and allow the system to approach room temperature before breaking the vacuum (see Experimental Notes 1 and 2). Do not turn off the aspirator until the vacuum has been broken.

Weigh the distilled benzaldehyde and determine its refractive index (see Experimental Note 3). Calculate the per cent recovery.

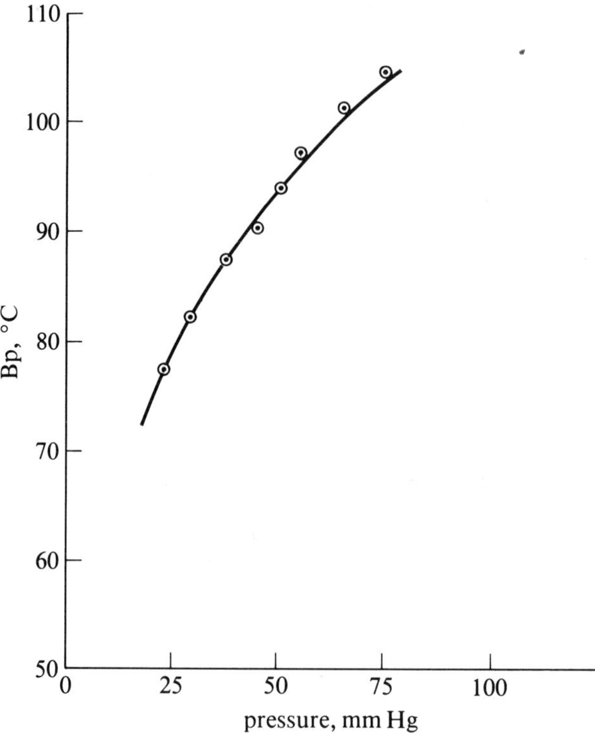

Figure 22.5 *Graph of boiling point versus pressure for benzaldehyde.*

B. Isolation of Benzoic Acid (Supplemental)

The distillation residue should be a solid at room temperature. To verify that this residue is principally benzoic acid, add 50 mL of 10% sodium hydroxide solution (*caustic*) to the residue. Break up the solid with a spatula and swirl until the bulk dissolves. The mixture will turn a milky yellow color. Chill the solution in an ice bath and transfer it to a separatory funnel. Extract the solution with two 25-mL portions of diethyl ether (*volatile and flammable*). Discard the ether extracts in the waste jug. Transfer the aqueous alkaline solution to an Erlenmeyer flask. In the hood, acidify the solution with concentrated hydrochloric acid (*strong acid with irritating vapors*). Chill the mixture in an ice bath, then isolate the solid by vacuum filtration. Air-dry the crude material, calculate the per cent of benzoic acid in the original crude benzaldehyde, and determine the melting point. If a more pure product is desired, crystallize the benzoic acid from water.

EXPERIMENTAL NOTES

1) Break the vacuum by opening the needle valve *slowly* while watching the mercury in the manometer rise. If the vacuum is broken very rapidly, the rising mercury can have sufficient momentum to break the glass in the manometer.

2) Always cool an apparatus after a vacuum distillation before allowing atmospheric oxygen to enter the distillation flask. Toward the end of the distillation, the distillation residue becomes superheated and may be above its flash point. It is possible that in-rushing oxygen could ignite this residue. At best, extensive charring will occur, making the flask very difficult to clean.

3) Crude benzaldehyde usually has a yellow cast, while pure benzaldehyde is colorless. The refractive index is not suitable for estimating the purity of this compound. Freshly distilled benzaldehyde and a mixture of freshly distilled benzaldehyde–10% benzoic acid have the same refractive index. (Can you suggest a way to check the purity of the benzaldehyde?)

»»» EXPERIMENT 22.2 «««

Column Chromatography: Separation of o- and p-Nitroaniline

Discussion

Column, or elution, chromatography is discussed in Section 6.1. In this experiment, column chromatography is used to separate a mixture of *o*-nitroaniline and *p*-nitroaniline. A silicic acid adsorbent and an eluting solvent of heptane, followed by a mixture of heptane–ethyl acetate and finally by pure ethyl acetate, are used for the separation.

The two anilines were selected for this experiment partly because *o*-nitroaniline is orange and *p*-nitroaniline is yellow. Therefore, the separation of these two compounds can be followed visually and any column imperfections, such as is evidenced by uneven movement of the bands (called "tailing"), can be readily detected.

The two anilines are easily separated on a silicic acid column because *p* nitroaniline forms hydrogen bonds with the solid silicic acid and its movement down the column is retarded. *o*-Nitroaniline undergoes *internal* hydrogen bonding; consequently, it can be eluted from the column using a relatively nonpolar solvent (3% ethyl acetate in heptane). Under these conditions, the *p*-nitroaniline remains near the top of the column. When the solvent system is changed to a polar solvent (pure ethyl acetate), the *p*-nitroaniline is eluted from the column and collected in a separate receiver.

The experiment can be conducted in one of two ways, depending upon the time available: (1) the column can be developed to the point where it is apparent that a separation has taken place and that the components could be isolated in their pure forms if the process were continued; or (2) a complete isolation of the components can be carried out. In either case, the experiment requires careful and correct preparation of the column, loading onto the column of the compounds to be separated, and elution of the components.

Successful separation of compounds by elution chromatography requires considerable time. The rate of elution must be kept low (about 1–2 mL per minute); otherwise, the solvent will flow unevenly through the solid packing and equilibration will not take place. Once the actual separation has begun, it must be continued until complete. If the eluting process is stopped before the components are eluted from the column, they will diffuse and possibly become remixed.

The characteristics of a specific elution chromatographic separation depend upon the exact procedure used; therefore, the experimental procedure is described in detail. If it is not possible to use the exact conditions described (size of chromatography column, type and particle size of the packing, etc.), then considerable differences from the reported times, optimal flow rates, mixed solvent composition and volumes needed, etc., can be expected. If other conditions must be used, then several trial runs should be made to determine the proper conditions.

Experimental

EQUIPMENT:

100-mL beaker
chromatography column, 3 cm × 38 cm with a capacity of 150 mL
disposable plastic gloves
three droppers
10-mL, 125-mL, and two 250-mL Erlenmeyer flasks
glass rod (able to reach to the bottom of the chromatography column)
10-mL and 100-mL graduated cylinders
spatula

CHEMICALS:

95% ethanol, 11 mL
ethyl acetate, about 10 mL (or 100 mL if complete separation is desired)
glass wool
n-heptane, about 300 mL (or 500 mL if complete separation is desired)
o-nitroaniline, 0.5 g
p-nitroaniline, 0.5 g
sand, 0.5 g
set of sieving screens
silicic acid, chromatography grade, 20 g (after sieving)

TIME REQUIRED: (a) $2\frac{1}{2}$ hours to the stage where the two components are separated on the column; (b) an additional 3 hours for the complete elution and isolation of both compounds

>>>> ***SAFETY NOTE 1*** Both *o*- and *p*-nitroaniline are toxic and can be absorbed through the skin. Disposable plastic gloves should be worn when working with these compounds.

>>>> ***SAFETY NOTE 2*** Large volumes of flammable solvents are used in this experiment, constituting a greater fire hazard than usual.

PROCEDURE

Preliminary Preparations

1) Prepare a stock solution containing 0.5 g of *o*-nitroaniline and 0.5 g of *p*-nitroaniline in 10 mL of 95% ethanol. (This solution can be used by several students.) Record the color of each solid before adding the ethanol.

2) Prepare an eluting solvent mixture containing 3% ethyl acetate in *n*-heptane (see Experimental Note 1).

3) Sieve the silicic acid (see Experimental Note 2); the final particle size should pass an 80-mesh screen but be retained on a 115-mesh screen (the Tyler Standard Screen Scale, or 180- to 125-μ screens using the U.S. Series Designation Standard).

Column Preparation. Mount a clean chromatography column (see Experimental Note 3) as shown in Figure 6.1, page 105. Check the column from two directions to make sure it is vertical. With the stopcock (or clamp) closed, pour 50 mL of *n*-heptane into the column, then push a small plug of glass wool to the bottom of the column, using a glass rod or tube. (Use the rod to push out any air entrapped in the glass wool.) Then add enough sand to the column to cover the glass wool. (See Experimental Note 4.) Using solvent from a dropper, wash down any sand clinging to the sides of the column. Finally, drain about 20 mL of the heptane from the column to ensure that all the air bubbles have been displaced.

Weigh 20 g of sieved silicic acid into a 100-mL beaker and add 50 mL of heptane. Mix the slurry with a spatula. When completely mixed, the slurry should appear almost translucent and should contain no white lumps of silicic acid or air bubbles. Adjust the stopcock on the column to allow about 5 mL/min to pass. Remix the slurry and pour it in a continuous stream into the chromatography column. Wash the beaker immediately with about 20 mL of heptane and add this wash to the top of the column.

While the heptane is draining through the column, wash any silicic acid on the sides of the column onto the packing, using heptane from a dropper. Allow the heptane to drain from the column until only about 1 cm remains on top of the packing, then close the stopcock. As the heptane drains from the base of the column, the silicic acid settles and forms the solid support for the chromatographic separation. The packing is delicate and will be useless if it is allowed to drain dry of solvent. The column preparation should take no more than 30 minutes and should result in a solid packing of silicic acid about 17 cm in length above the sand level (see Experimental Note 5).

Loading the Column. Allow the heptane to drain from the column until the top of the packing is just free of liquid. Measure exactly 1 mL of the stock solution of the two anilines and, using a dropper, transfer it directly onto the top of the packing. (Do not let the solution run down the sides of the glass tubing onto the column.) Open the stopcock to allow this solution to flow into the packing. Stop when the top of the packing is just free of liquid.

Measure out no more than 1 mL of 95% ethanol and, using a dropper, wash the solution, along with any splatters on the sides of the column, into the packing, again stopping when the top of the packing is just free of the liquid.

Using a clean dropper, add about 3 mL of heptane (see Experimental Note 6) and drain it into the packing as before, being careful not to disturb the top of the packing. Repeat the process with a second 3-mL portion of heptane. Finally, gently add 10 mL of heptane to the top of the column, using a dropper, then very carefully pour 90 mL of heptane down the side of the column onto the packing. The loading process should take about 15 minutes.

Developing the Column. Adjust the flow rate of the heptane from the column to 1–1.5 mL per minute (see Experimental Note 7). Continue the development until about 1 cm of heptane remains on top of the packing. The colored anilines should have moved about 7 cm down the column. The top of the colored band will be yellow, while the bottom will be orange.

Carefully add 100 mL of the 3% ethyl acetate–heptane eluting solvent to the top of the column and continue the eluting process. As the 3% ethyl acetate solution passes through the column, the orange band will move down the column while the yellow band will remain near the top. Eventually, the two colors will become completely separated and the pure white of the column packing will be observed between the two colors. It should be apparent that if sufficient eluting solvent were passed through the column, the orange band could be eluted into a collection flask and the yellow band would be retained on the column. If the chromatographic separation is being run only for demonstration purposes, it can be stopped at this point.

Complete Separation and Isolation of the Components. Add another 100 mL of 3% ethyl acetate–heptane to the top of the column and continue the eluting process. When the orange component starts to elute from the bottom of the column, collect it in a tared 250-mL Erlenmeyer flask. About 200 mL of 3% ethyl acetate–heptane will be required. After the orange compound has been completely eluted from the column, wash the compound from the tip of the stem into the collection flask with pure ethyl acetate.

Place a tared 125-mL Erlenmeyer flask under the column. Add pure ethyl acetate to the top of the column (see Experimental Note 8). The yellow band will move down the column with the solvent front and drain into the collection flask. This compound will be completely eluted with about 25–50 mL of solvent.

Evaporate the solvent from both flasks. Weigh the flasks to determine the amount of material collected. Calculate the per cent recovery. Compare the color of each component with those used to form the stock solution. Determine the melting points.

EXPERIMENTAL NOTES

1) Pure *n*-heptane will separate the two anilines; however, a week of eluting is required. An eluting solvent system containing more than 5% ethyl acetate in heptane is too polar to effect a separation using the conditions described.

The amount of eluting solvent needed depends upon the goal of the experiment. If only a demonstration of elution chromatography is carried out, then 100 mL is adequate. If complete elution and isolation is the goal, the about 350 mL will be required.

2) Baker Analyzed Reagent silicic acid (#0324) was used. Other packings, including alumina, are suitable; however, both the column flow rate and the percentage of ethyl acetate in heptane must be varied from the procedure described.

It is not imperative that the silicic acid be sieved. However, most chromatographic grade silicic acid contains a large range of particle sizes. By using a rather uniform particle size, the quality and ease of the separation is enhanced. Also, by using the larger-mesh-sized particles, the flow rate through the column can be increased, thus decreasing the time required for the separation.

3) A column of about 3 cm outside diameter and 38 cm long should be used. (These dimensions are not essential; however, all times and measurements described in the procedure are based upon a column of this size.) A Jones Reductor tube (Fischer #13-941; Kimble #39000) is a satisfactory chromatography column for this experiment.

4) Use a minimum amount of glass wool. Its function is to prevent the sand from passing into the stopcock. If too much glass wool is used, there will be a solvent holdup in this area, which can remix the components being separated.

Use sand that has been washed and dried and that will pass a 20-mesh sieve screen but be retained on a 100-mesh screen (Tyler Standard Screen Scale).

5) If necessary, the column can stand overnight at this point. However, if the stopcock (or clamped tubing) should spring a leak (a likely occurrence because of the combined effects of pressure and solvent action on the stopcock grease) and the column drain dry, the packing has been ruined. If you do stop here, make sure the top of the column is corked and the stopcock is very tight.

6) Do not start with the eluting solvent, which contains 3% ethyl acetate. The excess ethanol must first be washed away from the compounds to be separated; otherwise the initial solvent front will be too polar and both compounds will be eluted to the bottom of the column without separation.

7) Do not elute the solvent faster than the recommended rate. At this rate, equilibration can take place. If the flow rate is greater, the column will channel and a considerable amount of tailing will take place. Tailing prevents clean separation of the compounds. On the other hand, a slower flow rate than that recommended does not enhance the separation.

8) Do not add pure ethyl acetate to the top of the column before all the orange component has been eluted from the column. A premature addition of ethyl acetate will cause the yellow band to overrun the orange band and result in a ruined separation.

»»» EXPERIMENT 22.3 «««

Sublimation of Camphor

Discussion

The Principle of Sublimation. Sublimation is a purification process whereby a solid substance is converted to a vapor (without going through an intermediate liquid state) and then is condensed as a purified solid on a cool surface.

Solid compounds that evaporate (that is, pass directly from the solid phase to the gaseous phase) are rather rare; solid CO_2 (dry ice) is a familiar example of such a compound. Even though both solids and liquids have vapor pressures at any given temperature, most solids have very low vapor pressures. In order for a solid to evaporate, it must have an unusually high vapor pressure compared to other solids. For a solid compound to exhibit such a high vapor pressure, it must have relatively weak intermolecular attractions. One factor that contributes to weak intermolecular attractions is the shape of the molecules. Many compounds that evaporate readily contain molecules that are roughly spherical or cylindrical—shapes that do not lend themselves to strong intermolecular attractions.

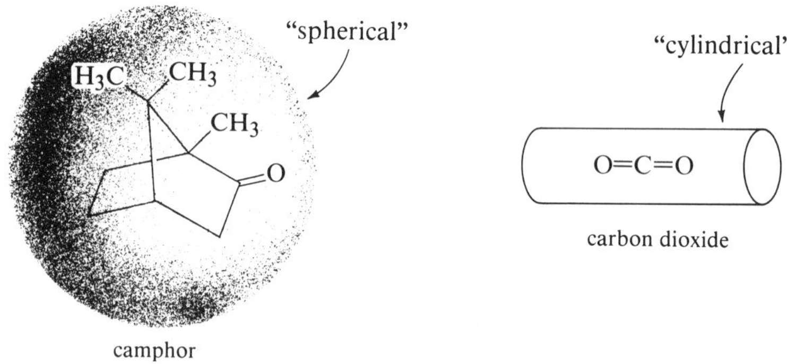

The vapor pressure of a solid increases as the temperature is increased, just as it does for liquids. Therefore, evaporation can be facilitated by heating the solid (but not to its melting point). The rate of evaporation can also be increased by conducting the sublimation under a vacuum; however, a very efficient cooling surface must be used so that the solid's vapor is not lost into the vacuum system.

Some Practical Aspects of Sublimation. The technique of sublimation can be used to purify appropriate solids in much the same way that distillation can be used to purify a liquid. In sublimation, nonvolatile solid impurities are left behind when the sample evaporates. Condensation of the vapor yields the pure solid compound. Sublimation has the advantages of being fast and clean (no solvent used). Unfortunately, most solid compounds have too low vapor pressures to be purified in this fashion. Also, sublimation is successful only if the impurities have much lower vapor pressures than that of the substance being purified. It would be practical to purify technical-grade iodine (which is contaminated with inorganic salts) by sublimation. It would *not* be practical to separate camphor from isoborneol (see Figure 22.6), however, because *both* compounds sublime readily.

A clue as to whether or not a compound can be purified by sublimation can often be obtained during a melting-point determination. A compound that can be purified by sublimation evaporates during a melting-point determination, often leaving nothing in the capillary! (To determine the melting point of such a solid, it is necessary to use a sealed capillary.)

Camphor. Camphor is used in the manufacture of plastics, in lacquers and varnishes, in explosives, as a moth repellent, as a preservative, and in embalming fluids. Medically, it finds limited use as a counterirritant, an anesthetic, and a mild antiseptic. It is used in some liniments.

Camphor is a natural product that can be isolated from the camphor tree, which grows in Japan and Taiwan. However, commercial needs exceed natural supply. For this reason, camphor is also synthesized commercially from α-pinene, a component of turpentine. A flow diagram of this synthetic route is shown in Figure 22.6.

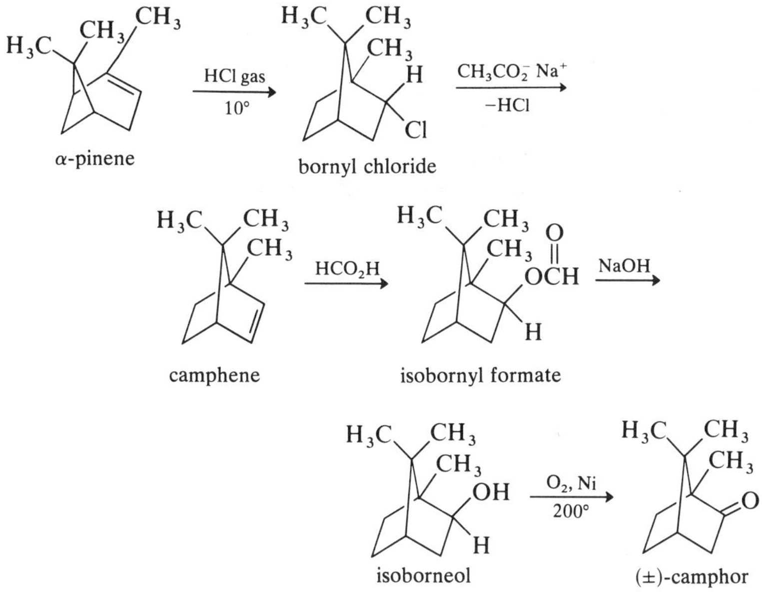

α-pinene $\xrightarrow[10^\circ]{\text{HCl gas}}$ bornyl chloride $\xrightarrow[-\text{HCl}]{CH_3CO_2^- Na^+}$

camphene $\xrightarrow{HCO_2H}$ isobornyl formate $\xrightarrow{\text{NaOH}}$

isoborneol $\xrightarrow[200^\circ]{O_2,\ \text{Ni}}$ (±)-camphor

Figure 22.6 The commercial synthesis of camphor. Note the rearrangement of the carbon skeleton in the first step, the conversion of α-pinene to bornyl chloride.

Experimental

EQUIPMENT:

cork or neoprene adapter
150-mL filter flask
ice
hot plate
150-mm test tube
thermometer

CHEMICALS:

camphor, 0.5 g

TIME REQUIRED: about three hours, plus time for a melting-point determination

STOPPING POINTS: any time during the experiment

PROCEDURE

Place 0.50 g of camphor in a 150-mL filter flask. Using either a cork or a neoprene adapter, fit the flask with an ice-filled test tube as shown in Figure 22.7. Place the flask on a hot plate at a temperature of approximately 80° (see Experimental Note 1). The camphor will sublime from the hot surface of the filter flask to the cold test tube. A white coating on the surface of the test tube, followed by crystal growth, will be observed. The camphor will also condense on the cooler surfaces of the filter flask.

The complete sublimation requires about $2\frac{1}{2}$–3 hours, depending upon the actual temperature of the hot plate. Every 15–20 minutes, remove the entire apparatus from the hot plate and allow it to cool. Carefully remove the test tube and scrape the crystals onto a tared watch glass with a spatula. Then, remount the test tube and continue the sublimation. Discontinue the sublimation when solid no longer remains on the bottom of the filter flask.

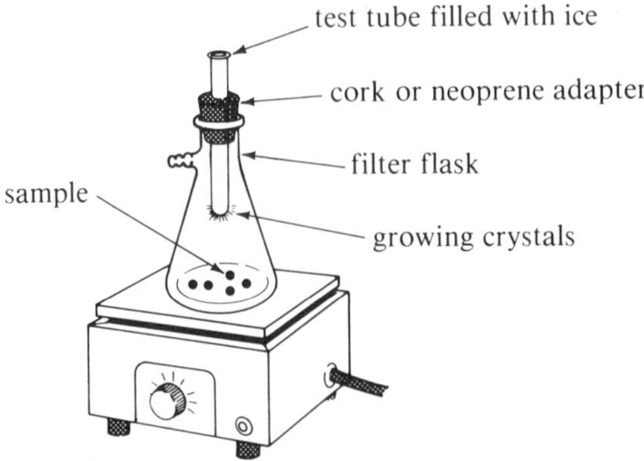

Figure 22.7 A sublimation apparatus using a filter flask, ice-filled test tube, and hot plate.

Weigh the watch glass and calculate the per cent recovery. With this type of apparatus, you can expect a 60% or higher recovery, depending upon the care you take in scraping the test tube. Determine the melting point of the sublimed camphor in a sealed tube (Experimental Note 2) and record your results.

EXPERIMENTAL NOTES

1) A simple (but rather inaccurate) method to obtain the surface temperature of a hot plate is to lay a thermometer down on the hot plate with the bulb resting on the hot surface.

2) To seal a melting-point capillary, first load the capillary in the usual way. Then, holding the capillary tube horizontal, twirl about a millimeter of the open end in the base of a flame. The edges of the open end will melt and collapse. The twirling of the tube back and forth ensures that the glass will collapse inward and seal the tube. After the tube has been sealed, inspect the closed end with a magnifying glass to be sure that it does not contain a pinhole.

APPENDIX I

Commonly Used Calculations

This appendix is a brief summary of calculations commonly performed in the laboratory. If you do not understand any one of the concepts presented here, you should review that concept in any good general chemistry text.

In any mathematical calculation, carry along the units of the numbers. In this way, you can determine which units cancel (and which do not cancel) as a check on how you have set up your equation. Also, before mixing any reagents or reporting any per cent yield, always look at your calculations and ask yourself, "Is this reasonable?"

Molarity

Molarity is defined as *the number of moles of solute in 1.00 liter of solution.* The following equations are used to determine molarity.

$$\text{molarity } (M) = \frac{\text{moles of solute}}{\text{liters of solution}}$$

$$= \frac{\text{g of solute/MW of solute}}{\text{liters of solution}}$$

$$= \frac{\text{g of solute}}{(\text{MW})\,(\text{liters})}$$

Example. What is the molarity of an aqueous solution of 5.0 g of NaOH in 100 mL of solution?

$$M = \frac{\text{g}}{(\text{MW})\,(\text{liters})} = \frac{5.0\text{ g}}{(40.0\text{ g/mol})\,(0.100\text{ L})} = 1.25\text{ mol/L}$$

Example. What weight of solid NaOH would you need to prepare 50 mL of a 2.0M aqueous solution?

$$M = \frac{\text{g}}{(\text{MW})\,(\text{liters})}$$

$$\text{g} = (M)\,(\text{MW})\,(\text{liters})$$

$$= (2.0\text{ mol/L})\,(40.0\text{ g/mol})\,(0.050\text{ L})$$

$$= 4.0$$

Normality

The **normality** of a solution is the *number of equivalents of solute in 1.00 liter of solution*. In the organic laboratory, normality is generally encountered only with acids and bases (for example, 6N HCl, 6N H_2SO_4, or 6N NH_4OH). One **equivalent** of acid or base is the weight of the substance that contains 1.00 mole of H^+ or OH^-.

for a monoprotic acid (such as HCl) or a base containing one ^-OH per molecule (such as NaOH):

equivalent weight = molecular weight

Therefore, normality = molarity

Examples.

6N HCl = 6M HCl

6N NaOH = 6M NaOH

for a diprotic acid, such as H_2SO_4, or a base containing two OH^- per molecule, such as $Ca(OH)_2$:

equivalent weight = $\frac{1}{2}$ molecular weight

Because each mole contains two equivalents,

normality = 2 × molarity

Example. What is the molarity of 6N H_2SO_4?

$$N = 2M$$

$$M = \frac{N}{2} = \frac{6N}{2} = 3$$

Example. Calculate the molarity of a 2.5N aqueous solution of H_3PO_4.

In this case, one mole of H_3PO_4 can theoretically supply three moles of H^+. The normality is three times the molarity.

$$N = 3M$$

$$M = \frac{N}{3} = \frac{2.5N}{3} = 0.83$$

Example. What weight of $Ca(OH)_2$ would be necessary to prepare 100 mL of a 0.25*N* solution?

$$N = \frac{\text{no. of equivalents}}{\text{liters}}$$

$$\begin{aligned}\text{no. of equivalents} &= N \times \text{liters} \\ &= (0.25 \text{ equivalent/L})\ (0.100 \text{ L}) \\ &= 0.025\end{aligned}$$

$$\text{equivalent weight} = \tfrac{1}{2}\,\text{MW}$$

Therefore, $0.025 \text{ equivalent} = \tfrac{1}{2}(0.025 \text{ mol})$

$$\begin{aligned}&= \tfrac{1}{2}(0.025 \times 74.0 \text{ g}) \\ &= 0.925 \text{ g}\end{aligned}$$

As a check on *N*–*M* conversions, remember that, for a given solution, ***N* is always equal to or larger than *M*.**

Dilutions

In practice, we are often required to dilute a more concentrated acid (or base) to a less concentrated solution. Because the number of moles or equivalents is not changed by dilution, the following simple equations allow us to calculate the amount of more concentrated solution needed.

$$M_1V_1 = M_2V_2 \quad \text{or} \quad N_1V_1 = N_2V_2$$

Example. What volume of 12*M* HCl is needed to prepare 100 mL of 1.5*M* HCl?

$$\begin{aligned}M_1V_1 &= M_2V_2 \\ V_1 &= \frac{M_2V_2}{M_1} \\ &= \frac{(1.5M)(0.100 \text{ L})}{12M} \\ &= 0.0125 \text{ L, or } 12.5 \text{ mL}\end{aligned}$$

Per Cent Concentrations

Many common laboratory manipulations require solutions with concentrations reported in per cents. These percentages generally refer to **weight–volume per cents.** For example, a 5% $NaHCO_3$ solution is an aqueous solution of 5 g of $NaHCO_3$ dissolved, then diluted *to* 100 mL (not *with* 100 mL).

$$\text{per cent (weight/volume)} = \frac{\text{g of solute}}{100 \text{ mL of solution}}$$

Example. What weight of NaOH is required to prepare 30 mL of a 15% aqueous solution?

This type of problem may be solved quickly by a simple proportion.

$$\frac{15\text{ g}}{100\text{ mL}} = \frac{x\text{ g}}{30\text{ mL}}$$

$$x = \frac{15\text{ g}}{100\text{ mL}} \times 30\text{ mL} = 4.5\text{ g}$$

In certain instances, it may be desirable to convert a concentrated solution to a more dilute solution.

Example. What volume of 5.0% $NaHCO_3$ is needed to prepare 7.0 mL of 2.0% $NaHCO_3$?

$$C_1V_1 = C_2V_2$$

$$V_1 = \frac{C_2V_2}{C_1}$$

$$= \frac{2.0\% \times 7.0\text{ mL}}{5.0\%}$$

$$= 2.8\text{ mL}$$

Per Cent Yields and Theoretical Yields

A **per cent yield** is simply the per cent of the theoretical amount of product actually obtained in a reaction.

$$\text{per cent yield} = \frac{\text{actual yield}}{\text{theoretical yield}} \times 100$$

Example. What is the per cent yield when 5.2 g of product is obtained out of a theoretical 7.5 g?

$$\text{per cent yield} = \frac{5.2}{7.5} \times 100 = 69\%$$

To calculate the theoretical yield, balance the equation for the reaction, calculate the moles of reactants, then calculate the theoretical yield based on the limiting reagent (the reagent present in the shortest supply). For example, in the oxidation of cyclohexanol (Experiment 10.4B), 20.0 g of the alcohol is treated with 23.8 g of $Na_2Cr_2O_7 \cdot 2H_2O$ and 26.5 g of H_2SO_4. The balancing of the equation is discussed in Section 10.3.

1) Calculate the molecular weights.
2) Calculate the numbers of moles.

3 (cyclohexyl)–OH + $Na_2Cr_2O_7 \cdot 2H_2O$ + $4H_2SO_4$ →

MW:	100.16	298.00	98.08
weight:	20.0 g	23.8 g	26.5 g
moles:	0.20	0.080	0.27

$$3\ \text{(cyclohexanone)}{=}O + Cr_2(SO_4)_3 + Na_2SO_4 + 9H_2O$$

MW:	98.15	—	—	—
weight:	?	—	—	—
moles:	?	—	—	—

3) Determine the limiting reagent. The required amounts of $Na_2Cr_2O_7 \cdot 2H_2O$ and H_2SO_4 to oxidize 0.20 mol of cyclohexanol are 0.20/3, or 0.067, mol of the chromate and (4/3) (0.20), or 0.27, mol of H_2SO_4. In this particular reaction, the alcohol and H_2SO_4 are limiting reagents, while the chromate is present in excess.

4) Determine the theoretical number of moles of product possible. In this case, 0.20 mol of cyclohexanone is the maximum, or theoretical, yield (the same as the number of moles of cyclohexanol, one of the limiting reagents).

5) Convert the theoretical yield of product to grams.

$$\begin{aligned}0.20 \text{ mol of cyclohexanone} &= 0.20 \text{ mol} \times 98.15 \text{ g/mol} \\ &= 19.6 \text{ g (theoretical yield)}\end{aligned}$$

APPENDIX II

Elemental Analyses

The **weight per cents of carbon and hydrogen** in an organic compound are determined by burning a weighed sample of the compound in a special apparatus. The resultant water vapor is collected in a tared chamber containing a drying agent, while the carbon dioxide is trapped in a chamber containing a strong base (which converts the carbon dioxide to the carbonate ion). The chambers are reweighed and the weights of water and carbon dioxide determined by difference. The %C and %H in the original compound can then be calculated.

In practice, the problems of ensuring complete combustion and collecting 100% of the gases (uncontaminated by outside moisture or carbon dioxide) are difficult to overcome without the proper equipment. Therefore, most organic chemists do not perform these analyses, but send samples to analytical chemists who specialize in this type of analysis. It is, of course, the responsibility of the organic chemist to submit a sample that is as pure as possible. Impurities amounting to 5% of the sample may not affect an infrared or nmr spectrum, but they will invalidate the results of a C and H analysis.

Most organic compounds contain only C, H, and O. A typical analysis report shows only the %C and %H. The %O is usually determined by difference.

for a compound containing only C, H, and O:

$$\%C + \%H + \%O = 100\%$$

therefore, $\%O = 100\% - \%C - \%H$

Example. A compound that contains only C, H, and O is subjected to C and H analysis with the following results: 38.72% C, 9.72% H. What is the %O?

$$\%O = 100 - 38.72 - 9.72$$
$$= 51.56$$

If requested to do so, an analytical laboratory can perform analyses for oxygen, the halogens, phosphorus, nitrogen, and other elements. Here, we will discuss compounds containing only C, H, and O.

Determining the Empirical Formula

From the weight per cents of the elements in a compound, the **molar ratio** of the elements can be calculated. This is accomplished by dividing each *weight per cent* by the *atomic mass* of that element. The following example describes this procedure for a compound containing 38.72% C, 9.72% H, and 51.56% O.

Step 1, divide per cent by atomic mass to determine the molar ratio:

$$\frac{38.72\%\ C}{12.01} = 3.22 \qquad \frac{9.72\%\ H}{1.008} = 9.64 \qquad \frac{51.56\%\ O}{16.00} = 3.22$$

We see that the molar ratio of C, H, and O is 3.22:9.64:3.22. These numbers must be converted to small whole numbers by dividing all three values by the smallest value.

Step 2, divide the values in the ratio by the smallest value:

$$\text{for C: } \frac{3.22}{3.22} = 1.00 \qquad \text{for H: } \frac{9.64}{3.22} = 2.99 \qquad \text{for O: } \frac{3.22}{3.22} = 1.00$$

When the molar ratio has been converted to small whole numbers, the ratio tells us the relative numbers of the atoms in a molecule. In our example, the molar ratio is very close to 1:3:1. From these numbers, we can write an **empirical formula.**

Step 3, write the empirical formula:

CH_3O

Note that this formula is an *empirical* formula, which shows only the *ratios* of the atoms, not their actual numbers in the molecule. The true **molecular formula** might be CH_3O, $C_2H_6O_2$, $C_3H_9O_3$, or any other formula with C, H, and O in a ratio of 1:3:1.

Example. A compound contains 65.50% C, 9.46% H, and 25.02% O. What is its empirical formula?

Step 1, divide per cent by atomic mass:

$$\frac{65.50\%\ C}{12.01} = 5.45 \qquad \frac{9.46\%\ H}{1.008} = 9.38 \qquad \frac{25.02\%\ O}{16.00} = 1.56$$

Step 2, divide by the smallest value:

$$\text{C: } \frac{5.45}{1.56} = 3.49 \qquad \text{H: } \frac{9.38}{1.56} = 6.01 \qquad \text{O: } \frac{1.56}{1.56} = 1.00$$

In this example, the values for the molar ratio at this point are *not* all close to small whole numbers, but can be rounded to 3.5 : 6 : 1. Multiplying this ratio by 2 does result in the necessary small whole numbers that are needed for an empirical formula—in this case, 7 : 12 : 2. The empirical formula is therefore $C_7H_{12}O_2$.

Determining the Molecular Formula

To convert an empirical formula to a molecular formula, we need to know the **molecular weight** of the compound. An analytical laboratory can determine the molecular weight along with the elemental analysis. Freezing-point depression is one way this is accomplished. In many cases, an organic chemist can determine the molecular weight of a compound from its mass spectrum.

In a molecular-formula determination, the molecular weight of the empirical formula is simply compared with the experimentally determined molecular weight. If the two weights are the same, the empirical formula represents the molecular formula. If the values are *not* the same, the experimental molecular weight should be a simple multiple of the molecular weight of the empirical formula. The molecular formula is determined by multiplying the empirical formula by this number.

Example. The theoretical molecular weight of CH_3O is 31.03. The actual molecular weight of the compound in question is found to be 62.11, which is very close to 2 × 31.03. Therefore, the molecular formula for the compound is $(CH_3O)_2$, or $C_2H_6O_2$.

Interpreting the Results

In research, an elemental analysis is often used as evidence to substantiate the identity of a proposed structure. In a chemical journal, an analysis might be reported as: "Anal. Calcd for $C_2H_6O_2$: C, 38.70; H, 9.74. Found: C, 38.79; H, 9.67."

Generally, an analysis that results in values within 0.3% of the calculated value for each element is an acceptable piece of evidence for structure proof. An extremely pure sample will usually analyze closer to the calculated values. An analysis that does not fall within 0.3% suggests either an impure sample or possibly a different compound from the one expected.

If a compound contains a large number of carbon and hydrogen atoms (often the case with molecules of biological or medical importance), it is difficult to determine the exact molecular formula from analytical data with any accuracy. For example, *cholesterol*, $C_{27}H_{46}O$, contains 83.87% C and 11.99% H. Cholesterol contains one double bond. When this double bond is hydrogenated, *cholestanol*, $C_{27}H_{48}O$, is obtained. Cholestanol contains 83.43% C and 12.45% H. The per cent compositions of these two compounds are fairly close, as you can see—much closer than the per cent compositions of comparable smaller molecules would be. It would be difficult to differentiate cholesterol and cholestanol by elemental analysis alone.

APPENDIX III

Toxicology of Organic Compounds

Virtually every chemical has the potential for toxicity. Even dietary necessities can be poisonous when consumed to excess. A few years ago, a Florida woman died from drinking too much water! Most compounds encountered in the organic laboratory are far more toxic than water. Yet, it has been only in the last two or three decades that toxicities of various types of organic compounds have been studied at all. It was not until the 1970's that comprehensive surveys were undertaken. Because of this, and because it is difficult to test toxicities with human subjects, much of the available data is tentative or is applicable directly only to rats, mice, dogs, or monkeys. The allowable, "harmless" levels of organic compounds in the immediate environments of workers and consumers are still being scrutinized by both government agencies and private corporations.

The techniques used in different studies are not always the same. One study may be concerned with the medical symptoms in a group of factory workers. Another study may involve feeding potentially toxic compounds to laboratory animals. Yet another piece of data may come from an isolated case reported in a medical journal.

Other problems that arise in toxicology are the differences between short-term and long-term exposures, individual allergic reactions, variable individual tolerances, and differences in individual habits (for example, does the subject smoke, or is he on any medication?). In time, comprehensive and directly comparable lists of toxicities may become available. Today, we must still extrapolate from one study to another, often with inadequate information.

There are several types of toxicity. A compound may be caustic and irritating, leading to burns and rashes. A compound may be relatively harmless in short-term exposure, but may cause cancer if the exposure is repeated or prolonged. A short-term exposure to a compound may be harmless to an adult, but cause serious defects in an unborn baby. (For this reason, pregnant women should pay strict attention to safety procedures.) Some compounds are deposited in the fatty tissue of the body instead of being eliminated and so may build up in concentration and lead to chronic toxicity. (This is frequently the case with organohalogen compounds.) The route of administration of a compound also affects its toxicity. *Inhalation* of vapors, aerosols, or dusts usually causes more-severe symptoms than other types of exposure, such as ingestion or dermal absorption.

Keeping all these variables in mind, let us consider some of the ways that toxicological data are reported. The mode of testing is usually self-evident, or should be; the route of administration and the species tested, as well as the toxicity levels, are a necessary part of any reliable report. For example, the term **LD_{50}** (lethal dose, 50% kill) means that half of a statistical sample of test animals died at a particular dose level. The route of administration will be designated as *inhalation*; *oral*; *dermal*; *iv* (intravenous); *im* (intramuscular); *ip* (intraperitoneal, or injected into the abdominal cavity); or *subcutaneous* (injected just under the surface of the skin). In Table III.1, you can see that the LD_{50} of diethyl ether is 2200 mg/kg when given orally to rats. This means that when diethyl ether was administered through a tube into the stomachs of a statistical sample of rats at the rate of 2200 mg of ether per kg of rat body weight, half the rats died. If only one rat from the statistical sample had died, the toxicity would be reported as the *lowest lethal dose*, or **LD_{Lo}** (oral, rats). If only one rat showed any ill effects at all, then the toxicity would be reported as the *lowest toxic dose*, or **TD_{Lo}** (oral, rats).

Because inhalation toxicity is of great importance to industrial workers and chemists exposed to fumes and dust, data are often reported as *toxic concentration*, or **TC**. The term **TC_{Lo}** means the lowest concentration known to cause any toxic effects in one or more members of the group of subjects. The term **TD_{50}** means, of course, that 50% of the subjects suffered from one or more toxic effects. The *threshold limit value* **TLV** is the maximum amount of vapor in the air that is reported safe for continuous exposure. Concentrations are generally reported as **parts per million (ppm)** or occasionally as **parts per billion (ppb)**. An inhalation concentration of 1.0 ppm is 1.0 μL of sample per liter of air, where 1.0 μL = 0.001 mL or 1.0×10^{-6} liter. (As a liquid concentration, the term 1.0 ppm usually means 1.0 mg of solute in 1.0 liter of solution.)

Table III.1 lists the reported toxicities of some popular organic solvents. Until a few years ago, *all* these solvents were used regularly in industrial, research, and student laboratories. In the past four to five years, methanol, benzene, chloroform, and carbon tetrachloride have been used less widely because their toxicities have been highly publicized. Today, chemists are more likely to choose ethanol over methanol, methylene chloride over chloroform or carbon tetrachloride, and toluene over benzene.

Table III.2 is a list of some toxic organic compounds. This list is not

Table III.1 Toxicities of some common solvents[a,b]

Name	*Formula*	*Reported human toxicity*[c]	*Maximum allowed occupational level (time weighted average)*[d]
methanol	CH_3OH	inhalation TC_{Lo}, 300 ppm	200 ppm
ethanol	CH_3CH_2OH	oral TD_{Lo}, 50 mg/kg oral LD_{Lo}, 1400 mg/kg	1000 ppm
acetone (propanone)	$(CH_3)_2C{=}O$	inhalation TC_{Lo}, 500 ppm	1000 ppm
diethyl ether	$(CH_3CH_2)_2O$	oral (rat) LD_{50}, 2200 mg/kg	400 ppm
benzene[e]	C_6H_6	inhalation TC_{Lo}, 210 ppm	10 ppm
carbon tetrachloride[e] (tetrachloromethane)	CCl_4	inhalation TC_{Lo}, 20 ppm	10 ppm
chloroform (trichloromethane)	$CHCl_3$	inhalation TC_{Lo}, 10 ppm	50 ppm
methylene chloride (dichloromethane)	CH_2Cl_2	inhalation TC_{Lo}, 500 ppm	500 ppm

[a] *Toxic Substances List*, H. E. Christensen and T. Luginbyhl, Eds. Rockville, Md.: U.S. Dept. of Health, Education, and Welfare, 1974.

[b] Ligroin and petroleum ether (both mixtures of alkanes) are not listed, but they are only mildly toxic.

[c] TC_{Lo} is the *lowest published toxic concentration*; TD_{Lo} is the *lowest published toxic dose*; LD_{Lo} is the *lowest published lethal dose*; and LD_{50} is the *lethal dose, 50% kill.*

[d] For comparison, the maximum allowed level of hydrogen cyanide (HCN) is 10 ppm.

[e] Suspected or known carcinogen, can be absorbed through the skin.

complete. Many compounds that are *suspected carcinogens* are not included. The list of known or suspected *teratogens* (compounds that can cause fetal damage) is too long to include here; therefore, we have shown only some of the more common ones. The list of *allergens* includes only a few of the compounds that frequently cause allergic reactions; unfortunately, a particular individual may be allergic to a compound that has no effect on most other people.

The fact that many toxic compounds can be *absorbed through the skin* was largely ignored until recent years. For example, a lethal dose of phenol ("carbolic acid"), once used as a surgical antiseptic, can be absorbed dermally. A laboratory worker would be wise to assume that *all* compounds can be dermally absorbed. In many cases, a relatively harmless solvent can carry other, nonabsorbable compounds through the skin. Dimethyl sulfoxide (DMSO), which can actually be tasted by a person after it has been applied to his or her hand, has been used to administer nonabsorbable drugs dermally. This solvent-carrying effect is a good reason not to use a solvent to cleanse your hands of other organic compounds unless absolutely necessary. If a solvent must be used, choose the least objectionable solvent appropriate: ethanol or an alkane solvent like petroleum ether (or, better, mineral oil). Then, scrub immediately with soap and water.

Table III.2 Some toxic organic compounds[a]

Known carcinogens[b]

acrylonitrile
4-nitrobiphenyl
α- and β-naphthylamine
methyl chloromethyl ether
benzidine
4–aminodiphenyl
1,2-dibromo-3-chloropropane
ethyleneimine
β-propiolactone
2-acetylaminofluorene
4-dimethylaminoazobenzene
N-nitrosodimethylamine
vinyl chloride
benzene
4,4-methylenebis(2-chloroaniline)

Known or suspected teratogens[c]

benzene
toluene
xylene
aniline
nitrobenzene
phenol
vinyl chloride
formaldehyde
dimethylformamide
dimethyl sulfoxide
N,N-dimethylacetamide
carbon disulfide

Allergens

pyrethrums[d]
diazomethane
p-phenylenediamine
some glues, gums, and resins

Compounds that can be absorbed dermally

methanol
1-propanol
1-pentanol
allyl alcohol
2-chloroethanol
dimethyl sulfoxide
acrylonitrile
benzene
nitrobenzene (and some other aromatic nitro compounds, including 2,4-dinitrophenylhydrazine)
bromobenzene
phenol (and some substituted phenols)
aniline (and some other aryl amines)

[a] This list is not intended to be complete. The reader is referred to *Carcinogens*, OSHA 2204, U.S. Dept. of Labor, Jan., 1975; *Safety in Academic Chemistry Laboratories*, 3rd ed., Washington, D.C.; American Chemical Society, 1979; *Hazards in the Chemical Laboratory*, 2nd ed., G. D. Muir, Ed., London: The Chemical Society, 1977; N. I. Sax, *Cancer Causing Chemicals*, New York: Van Nostrand Reinhold Co., 1981.

[b] A compound that can cause cancer in laboratory animals or humans.

[c] A compound that can cause a physical or functional defect in the embryo or fetus.

[d] A commonly used, naturally occurring insecticide that is isolated from one variety of chrysanthemum.

APPENDIX IV

Derivative Tables[a]

Table IV.1 Derivatives of liquid alcohols

Name	Bp, °C	Mp of derivative, °C: 3,5-Dinitrobenzoate	Mp of derivative, °C: α-Naphthylurethane
Alcohols with refractive indexes less than 1.40			
methanol	65	108	124
ethanol	78	93	79
2-propanol	82	123	106
1-propanol	97	74	80; 76
2-butanol (*sec*-butyl alcohol)	99	76	97
2-methyl-1-propanol (isobutyl alcohol)	108	87	104
3-methyl-2-butanol	114	—	109
1-butanol	117	64; 62	71
Alcohols with refractive indexes between 1.40 and 1.41			
2-methyl-2-butanol (*t*-amyl alcohol)	102	116; 118	72
3-pentanol	116	101; 97	95
2-pentanol (*sec*-amyl alcohol)	120	62	75
2-methoxyethanol (Methyl Cellosolve)	124	—	113
3-methyl-1-butanol	132	61	68
2-ethoxyethanol (ethylene glycol, monoethyl ether)	135	75	67

Table IV.1 (Cont)—Derivatives of liquid alcohols

Name	Bp, °C	Mp of derivative, °C	
		3,5-Dinitro-benzoate	α-Naphthyl-urethane
Alcohols with refractive indexes between 1.41 and 1.42			
2-propen-1-ol (allyl alcohol)	97	50	108
2,3-dimethyl-2-butanol	120	111	101
3,3-dimethyl-2-butanol	120	107	—
2-methyl-2-pentanol	123	72	—
3-methyl-3-pentanol	123	96	83
2-methyl-3-pentanol	127	85	—
2-methyl-1-butanol	129	70	82
1-pentanol (*n*-amyl alcohol)	138	46	68
2-hexanol	139	38	60
2-methyl-1-pentanol	148	50	76
3-methyl-1-pentanol	152	38	58
1-hexanol	157	58; 61	59; 62
Alcohols with refractive indexes between 1.42 and 1.44			
4-heptanol	156	64	80
2-heptanol	158	49	54
1-heptanol	177	46	62
2-octanol	179	32	63
2-ethyl-1-hexanol	184	—	60
1-octanol	195	62	67
1,2-ethanediol (ethylene glycol)	198	di-, 169	di-, 176
2-nonanol	198	43	55
1-nonanol	213	52	65
1,3-propanediol (trimethylene glycol)	215	di-, 178	di-, 164
1-decanol	231	58	73
Alcohols with refractive indexes between 1.44 and 1.47			
cyclopentanol	141	—	118
cis-2-methylcyclohexanol	165	99	—
trans-2-methylcyclohexanol	167	114	—
cis- and *trans*-2-methylcyclohexanol	—	85–90	—
cis-4-methylcyclohexanol	173	134	—
trans-4-methylcyclohexanol	173	140	—
cis- and *trans*-4-methylcyclohexanol	—	125–130	—
cis-3-methylcyclohexanol	173	92	129
trans-3-methylcyclohexanol	174–175	98	122
cis- and *trans*-3-methylcyclohexanol	—	80–85	—
tetrahydrofurfuryl alcohol	177	84	92
1,3-butanediol	207	—	184
1,4-butanediol	232	—	di-, 199
1,5-pentanediol	238	—	di-, 147
Alcohols with refractive indexes greater than 1.47			
furfuryl alcohol	172	81	130
benzyl alcohol	205	113	134
2-phenylethanol	219	108	119
α-terpineol	221	79	152; 147
geraniol	230	62	48
3-phenylpropanol (hydrocinnamyl alcohol)	236	92	91
2-phenoxyethanol (ethylene glycol, monophenyl ether)	237	—	120

[a] More than one melting point or boiling point indicates different reported values in the literature.

Table IV.2 Derivatives of solid alcohols

		Mp of derivative, °C	
Name	**Mp (Bp), °C**	**3,5-Dinitro-benzoate**	**α-Naphthyl-urethane**
1-phenylethanol	20 (202)	94	106
1-dodecanol	21; 24 (259)	60	80
cyclohexanol	25 (161)	113	129
4-methoxybenzyl alcohol	25 (259)	—	128
t-butyl alcohol	25 (82)	142	101
3-phenyl-2-propen-1-ol (cinnamyl alcohol)	33 (257)	121	114
(±)-fenchyl alcohol	39 (201)	104	149
1-tetradecanol (myristyl alcohol)	38	67	82
(−)-menthol	44	153	119; 126
1-hexadecanol	50	66	82
2,2-dimethyl-1-propanol (neopentyl alcohol)	52	—	100
1-heptadecanol	54	121	88
4-methylbenzyl alcohol	60	118	—
diphenylmethanol (benzhydrol)	68	141	136
2-hydroxy-1,2-diphenylethanone (benzoin)	137	—	140

Table IV.3 Derivatives of liquid phenols

		Mp of derivative, °C		
Name	**Bp, °C**	**3,5-Dinitro-benzoate**	**α-Naphthyl-urethane**	**Aryloxyacetic acid**
All refractive indexes greater than 1.4800				
2-chlorophenol	175	143	120	145
2-bromophenol	195	—	129	—
salicylaldehyde	197	—	—	132
3-methylphenol (*m*-cresol)	203	165	128	103
2-ethylphenol	207	108	—	141
3-ethylphenol	217	—	—	77
2-*n*-propylphenol	224; 220	96	—	100
2-*n*-butylphenol	237	97	—	105
5-isopropyl-2-methylphenol (carvacrol)	237	83; 77	116	151
3-methoxyphenol	243	—	129	118; 113
4-allyl-2-methoxyphenol (eugenol)	254	130	122	81
2-methoxy-4-propenylphenol (isoeugenol)	267	158	150	116

Table IV.4 Derivatives of solid phenols

		Mp of derivative, °C		
Name	*Mp (Bp), °C*	*3,5-Dinitro-benzoate*	*α-Naphthyl-urethane*	*Aryloxyacetic acid*
4-*n*-propylphenol	22 (232)	123	—	—
4-*n*-butylphenol	22 (248)	92	—	81
4-*n*-pentylphenol	23 (250)	—	—	90
3-*n*-propylphenol	26 (228)	75; 118	—	97
2,4-dimethylphenol	27 (211)	165	135	141
2-methylphenol (*o*-cresol)	31 (192)	138	142	152
2-methoxyphenol (guaiacol)	32 (205)	141	118; 132	116
3-bromophenol	33 (236)	—	108	108
3-chlorophenol	33 (214)	156	158	110
4-methylphenol (*p*-cresol)	36 (202)	187	146	135
2,4-dibromophenol	36; 40	—	—	153
4-chlorophenol	37; 43	186	166	156
3-iodophenol	40	183	—	115
phenol	42	146	133	99
2,4-dichlorophenol	45	143	—	141
4-ethylphenol	47	133	128	97
2,6-dimethylphenol	49	—	174	134
2-isopropyl-5-methyl-phenol (thymol)	50	103	160	149
4-methoxyphenol	55	—	135	112
3-hydroxy-6-nitrobiphenyl	58	171	—	—
3,4-dimethylphenol	62	182	142	162
4-bromophenol	64	191	169	157
3,5-dimethylphenol	68	195	171	111
2,4,5-trichlorophenol	68	—	158	157
2,4,6-trichlorophenol	68	—	172	186
2,4,6-trimethylphenol	70	—	—	142
2,4,5-trimethylphenol	71	—	—	132
2,5-dimethylphenol	75	137	173	118
2,3-dimethylphenol	75	—	—	187
4-hydroxy-3-methoxybenz-aldehyde (vanillin)	81	—	—	187
2-hydroxybenzyl alcohol (saligenin)	87	—	—	120
1-naphthol	94	217	152	193
4-iodophenol	94	—	—	156
2,4,6-tribromophenol	95	174	153; 170	200
n-propyl *p*-hydroxybenzoate	96	—	127	—
4-*t*-butylphenol	100	—	110; 149	86
1,2-dihydroxybenzene (catechol)	105	di-, 152	175	138
3-hydroxybenzaldehyde	104; 108	—	—	148

(continued)

Table IV.4 (Cont)—Derivatives of solid phenols

Name	Mp (Bp), °C	Mp of derivative, °C		
		3,5-Dinitro-benzoate	α-Naphthyl-urethane	Aryloxyacetic acid
3,5-dihydroxytoluene (5-methylresorcinol)	108	190	160	217
1,2-dihydroxybenzene (resorcinol)	110	di-, 201	—	175; 195
4-hydroxybenzaldehyde	117	—	—	198
3-aminophenol	122	179	—	—
2-naphthol	123	210	157	95
1,2,3-trihydroxybenzene (pyrogallol)	133	tri-, 205	—	198
2–hydroxy-2′-nitrobiphenyl	140	180	—	—
2-isopropyl-5-methyl-hydroquinone	143	—	148	—
salicylic acid	158	—	—	191
hydroquinone	171	di-, 317	—	250
p-aminophenol	184	178	—	—

Table IV.5 Derivatives of liquid ketones

Name	Bp, °C	Mp of derivative, °C	
		Semicarbazone	2,4-Dinitro-phenylhydrazone
Ketones with refractive index less than 1.40			
2-propanone (acetone)	56	190	126; 128
2-butanone (methyl ethyl ketone)	80	146	116
2,3-butanedione (biacetyl)	88	mono-, 235; di-, 278	di-, 314
2-methyl-3-butanone	94	113	120; 117
2-pentanone	102	112; 106	144
3-pentanone	102	139	156
3,3-dimethyl-2-butanone (pinacolone)	106	158	125; 131
4-methyl-2-pentanone	117	132; 135	95
Ketones with refractive indexes between 1.40 and 1.41			
3-buten-2-one (methyl vinyl ketone)	81	141	—
2,4-dimethyl-3-pentanone	124	160; 149	88; 86; 98
3-hexanone	125	113	130
2-hexanone	128	125; 121	106; 110
4-methyl-3-hexanone	136	137	78
4-heptanone	144	132	75
3-heptanone	148	97; 101; 152	80
2-heptanone	151	123; 127	74; 89

Table IV.5 (Cont)—Derivatives of liquid ketones

Name	Bp, °C	Mp of derivative, °C: Semicarbazone	Mp of derivative, °C: 2,4-Dinitrophenylhydrazone
Ketones with refractive indexes between 1.41 and 1.44			
cyclopentanone	131	210; 203; 217	146; 142
3-hydroxy-2-butanone (acetoin)	145	185; 202	di-, 318
1-hydroxy-2-propanone (acetol)	146	196	128
2,6-dimethyl-4-heptanone	168	122; 126	92; 66
methyl acetoacetate	170	152	—
2-octanone	173	123	58
acetoxyacetone	175	145	—
ethyl acetoacetate	181	133; 129d	93
2-undecanone (2-hendecanone)	228	122	63
Ketones with refractive indexes between 1.44 and 1.50			
4-methyl-3-penten-2-one (mesityl oxide)	130	164; 134	200; 203
2.4-pentanedione	139	mono-, 122; di-, 209	di-, 209
cyclohexanone	156	167	160; 162
2-methylcyclohexanone	165	191; 197d	137
3-methylcyclohexanone	168	179; 191d	155
4-methylcyclohexanone	171	199; 203d	132
2,5-hexanedione (acetonylacetone)	194	mono-, 185d; di-, 224	di-, 257
(+)-fenchone	196	184; 172	140
(−)-menthone	208	189; 187; 184	146
pulegone	224	174; 176	142
(+)-carvone	230	163; 142	191
Ketones with refractive indexes greater than 1.50			
acetophenone	205	199; 203	240; 250
2-methylacetophenone	214	205; 210	159
propiophenone	218	174; 182	190; 189
3-methylacetophenone	220	198; 203	207
isobutyrophenone	222	181	163
2,2-dimethyl-1-phenylpropanone (pivalophenone)	224	150	195
n-butyrophenone	230	188; 191	190
2,5-dimethylacetophenone	230	170	171
2,4-dimethylacetophenone	234	187	170
n-valerophenone	242	160	166
5-isopropyl-2-methyl-acetophenone (2-acetyl-*p*-cymene)	245	147	142
2-methoxyacetophenone (2-acetylanisole)	245	183	188
3,4-dimethylacetophenone	246	234; 224d	247

Table IV.6 Derivatives of solid ketones

Name	Mp (Bp), °C	Mp of derivative, °C	
		Semicarbazone	2,4-Dinitro-phenylhydrazone
4-methoxypropiophenone	27 (266)	170	194
n-pentyl phenyl ketone	24 (265)	132	168
phenylacetone	27 (216)	199	156
4-methylacetophenone	28 (266)	205; 197	260
2-hydroxyacetophenone	28 (215)	210	212
diisopropylideneacetone (phorone)	28 (198)	221; 186	118
2-acetylfuran	32 (170)	150	220
1-acetylnaphthalene	34 (302)	224	—
1,3-diphenyl-2-propanone	34	146; 125	100
4-methoxyacetophenone	38 (258)	198	220; 231
1-indanone (α-hydrindone)	42; 38	233; 239	258
benzalacetone	42	187; 198; 142	227; 223
benzophenone	48	165	229; 244
4-bromoacetophenone	51	208	230; 237
3,4-dimethoxyacetophenone	51	218	207
2-acetylnaphthalene	54	235	262d; 147
benzalacetophenone (chalcone)	58	169; 180	244d
2-indanone (β-hydrindone)	58	216; 212d	—
deoxybenzoin (desoxybenzoin)	60	148	204
4-methylbenzophenone	60; 55	122	200
4-methoxybenzophenone	64	176	180
2-phenylcyclohexanone	63; 60	190	139
cinnamalacetone	68	186	221
4-methoxybenzalacetone	73	—	229
1-naphthyl phenyl ketone	76	385	247; 220
4-chlorobenzophenone	78	—	185
4-methoxybenzalacetophenone	77	190; 168	—
dibenzoylmethane	81; 78	205	—
fluorenone	83	—	284
acetoacetanilide	85	165	—
4,4′-dimethylbenzophenone	95	144	219
benzil	95	mono-, 175d; di-, 244d	di-, 189
2-acetyl-1-hydroxynaphthalene	102; 98	250	—
p-aminoacetophenone	106	250	267; 263
4-hydroxyacetophenone	109	199	210
dibenzalacetone	112	190	180
4,4′-dimethoxybenzoin	113	185	—
α-hydroxyacetophenone	118	146	—

Table IV.6 (Cont)—Derivatives of solid ketones

Name	Mp (Bp), °C	Mp of derivative, °C	
		Semicarbazone	2,4-Dinitro-phenylhydrazone
4,4′-dimethoxybenzil	133	di-, 255	—
(±)-benzoin	133	206; 195d	245; 234
2,4-dihydroxyacetophenone	147	218d	208
4-hydroxypropiophenone	148	—	229

Table IV.7 Derivatives of liquid aldehydes

Name	Bp, °C	Mp of derivative, °C	
		Semicarbazone	2,4-Dinitro-phenylhydrazone
Aldehydes with refractive index less than 1.50			
propanal (propionaldehyde)	49	89; 154	150; 148; 154
ethanedial (glyoxal)	50	270	328
3-methylpropanal (isobutyraldehyde)	64	126	187; 182
butanal (*n*-butyraldehyde)	75	95; 106	123
2,2-dimethylpropanal (pivaldehyde)	75	190	210
3-methylbutanal (isovaleraldehyde)	92	107	123
methoxyethanal (methoxyacetaldehyde)	92	—	125
2-butenal (crotonaldehyde)	104	199	190
tetrahydrofurfural	142	166	134
Aldehydes with refractive index greater than 1.50			
furfural	162	202	214; 230
benzaldehyde	179	222; 235	237
2-hydroxybenzaldehyde (salicylaldehyde)	197	231	248; 252
3-methylbenzaldehyde (3-tolualdehyde)	199	204; 223	212; 194
2-methylbenzaldehyde (2-tolualdehyde)	200	209; 212; 218	194
4-methylbenzaldehyde (4-tolualdehyde)	204	234; 215	234
3-chlorobenzaldehyde	214	228	248; 256
2-chlorobenzaldehyde	214	146; 225; 229	213; 209
3-phenylpropanal (hydrocinnamaldehyde)	224	127	149
3-methoxybenzaldehyde (3-anisaldehyde)	230	233d	—
4-methoxybenzaldehyde (4-anisaldehyde)	248	210; 203	250; 254d
3-phenylpropenal (cinnamaldehyde)	252d	216	255d

Table IV.8 Derivatives of solid aldehydes

Name	*Mp, °C*	*Mp of derivative, °C* Semicarbazone	2,4-Dinitrophenylhydrazone
1-naphthaldehyde	34	221	—
piperonal	37	230; 234; 237	265d
2-methoxybenzaldehyde	39	215d	253
2-nitrobenzaldehyde	44	256	265
veratraldehyde (3,4-dimethoxybenzaldehyde)	44; 58	177	262
4-chlorobenzaldehyde	48	230; 233	254
2,3-dimethoxybenzaldehyde	54	231d	—
4-bromobenzaldehyde	57	228	128; 257
3-nitrobenzaldehyde	58	246	293d
2-naphthaldehyde	60	245	270
4-(dimethylamino)benzaldehyde	74	222	325
3,4,5-trimethoxybenzaldehyde	78; 75	220	—
vanillin	81	230; 240d	271d
2-hydroxy-1-naphthaldehyde	82	240	—
3-hydroxybenzaldehyde	104; 108	198	259
4-nitrobenzaldehyde	106	221; 211	320
4-hydroxybenzaldehyde	117	224; 280d	280d; 260
3,4-dihydroxybenzaldehyde	154	230d	275d

INDEX

Typical chemical shifts in nmr spectra

Structure	δ value (ppm)
proton on sp^3 carbon:	
$RC\underline{H}_3$	0.8–1.2
$R_2C\underline{H}_2$	1.1–1.5
$R_3C\underline{H}$	~1.5
$ArC\underline{H}_3$	2.2–2.5
$R_2NC\underline{H}_3$	2.2–2.6
$R_2C\underline{H}OR$	3.2–4.3
$R_2C\underline{H}Cl$	3.5–3.7
$RC(=O)C\underline{H}R_2$	2.0–2.7
$R_2C\underline{H}CR{=}CR_2$	~1.7
proton on sp^2 or sp carbon:	
$R_2C{=}C\underline{H}R$	4.9–5.9
$Ar\underline{H}$	6.0–8.0
$RC\underline{H}O$	9.4–10.4
$RC{\equiv}C\underline{H}$	2.3–2.9
proton on N or O:	
$R_2N\underline{H}$	2–4
$RO\underline{H}$	1–6
$ArO\underline{H}$	6–8
$RCO_2\underline{H}$	10–12